Springer Proceedings in Mathematics & Statistics

Volume 51

For further volumes:
http://www.springer.com/series/10533

Springer Proceedings in Mathematics & Statistics

This book series features volumes composed of select contributions from workshops and conferences in all areas of current research in mathematics and statistics, including OR and optimization. In addition to an overall evaluation of the interest, scientific quality, and timeliness of each proposal at the hands of the publisher, individual contributions are all refereed to the high quality standards of leading journals in the field. Thus, this series provides the research community with well-edited, authoritative reports on developments in the most exciting areas of mathematical and statistical research today.

Alexey Sorokin • Panos M. Pardalos
Editors

Dynamics of Information Systems: Algorithmic Approaches

Springer

Editors
Alexey Sorokin
Innovative Scheduling Inc.
Gainesville, FL, USA

Panos M. Pardalos
Department of Industrial and Systems Engineering
University of Florida
Gainesville, FL, USA

ISSN 2194-1009 ISSN 2194-1017 (electronic)
ISBN 978-1-4614-7581-1 ISBN 978-1-4614-7582-8 (eBook)
DOI 10.1007/978-1-4614-7582-8
Springer New York Heidelberg Dordrecht London

Library of Congress Control Number: 2013944685

Mathematics Subject Classification (2010): 49, 68, 90(90-06, 90Bxx: 90B10, 90B15, 90B18, 90B50), 92, 93

Printed on acid-free paper

Springer is part of Springer Science+Business Media (www.springer.com)

Preface

Information systems have been developed in parallel with computer science, although information systems have roots in different disciplines including mathematics, engineering, and cybernetics. Research in information systems is by nature very interdisciplinary. As it is evidenced by the chapters in this book, dynamics of information systems has several diverse applications.

The book presents the state-of-the-art work on theory and practice relevant to the dynamics of information systems. First, the book covers algorithmic approaches to numerical computations with infinite and infinitesimal numbers. Also the book presents important problems arising in service-oriented systems, such as dynamic composition, analysis of modern service-oriented information systems, and estimation of customer service times on a rail network from GPS data. After that, the book addresses the complexity of the problems arising in stochastic and distributed systems. In addition, the book discusses modulating communication for improving multi-agent learning convergence. Network issues, in particular minimum risk maximum clique problems, vulnerability of sensor networks, influence diffusion, community detection, and link prediction in social network analysis, as well as a comparative analysis of algorithms for transmission network expansion planning are described in subsequent chapters.

We thank all the authors and anonymous referees for their advice and expertise in providing valuable contributions, which improved the quality of this book. Furthermore, we want to thank Springer for helping us to produce this book.

Gainesville, FL, USA — Alexey Sorokin
Gainesville, FL, USA — Panos M. Pardalos

Contents

Numerical Computations with Infinite and Infinitesimal Numbers: Theory and Applications

Yaroslav D. Sergeyev

Abstract A new computational methodology for executing calculations with infinite and infinitesimal quantities is described in this chapter. It is based on the principle "The part is less than the whole" introduced by Ancient Greeks and applied to all numbers (finite, infinite, and infinitesimal) and to all sets and processes (finite and infinite). It is shown that it becomes possible to write down finite, infinite, and infinitesimal numbers by a finite number of symbols as particular cases of a unique framework that is not related to non-standard analysis theories. The Infinity Computer working with numbers of a new kind is described (its simulator has already been realized). The concept of accuracy of mathematical languages and its importance for a number of theoretical and practical issues regarding computations is discussed. Numerous examples dealing with divergent series, infinite sets, probability, limits, fractals, etc. are given.

Keywords Numerical infinities and infinitesimals • Numbers and numerals • Infinity computer • Numerical analysis • Infinite sets • Divergent series • Fractals

1 Introduction

In different periods of human history, mathematicians and physicists in order to solve theoretical and applied problems existing in their times developed mathematical languages that use different approaches to the ideas of infinity and infinitesimals

---

Y.D. Sergeyev (✉)
University of Calabria, Via P. Bucci, Cubo 42-C, 87030 Rende, Italy
N.I. Lobatchevsky State University, Nizhni Novgorod, Russia

Institute of High Performance Computing and Networking of the National Research Council of Italy, Rende, Italy
e-mail: yaro@si.deis.unical.it

A. Sorokin and P.M. Pardalos (eds.), *Dynamics of Information Systems: Algorithmic Approaches*, Springer Proceedings in Mathematics & Statistics 51,
DOI 10.1007/978-1-4614-7582-8_1,

(see [1,2,5,12,14,16,19,20,25,28,51] and references given therein). To emphasize the importance of the subject it is sufficient to mention that the Continuum Hypothesis related to infinity has been included by David Hilbert as the Problem Number One in his famous list of 23 unsolved mathematical problems (see [16]) that have influenced strongly development of Mathematics in the twentieth century. However, arithmetics developed for working with infinities are quite different with respect to the finite arithmetic we are used to deal with. Moreover, very often they leave undetermined many operations where infinite numbers take part (for example, $\infty - \infty$, $\frac{\infty}{\infty}$, sum of infinitely many items, etc.) or use representation of infinite numbers based on infinite sequences of finite numbers.

Many approaches describing manipulations with infinities and infinitesimals are rather old: ancient Greeks following Aristotle distinguished the potential infinity from the actual infinity; John Wallis (see [51]) credited as the person who has introduced the infinity symbol, ∞, has published his work *Arithmetica infinitorum* in 1655; the foundations of analysis we use nowadays have been developed more than 200 years ago with the goal to develop mathematical tools allowing one to solve problems that were emerging in the world at that remote time; Georg Cantor (see [2]) has introduced his cardinals and ordinals more than 100 years ago, as well. As a result, mathematical languages that we use now while work with infinities and infinitesimals do not reflect numerous achievements made by Physics of the twentieth century.[1] Let us illustrate this observation by a couple of examples.

We know from the modern Physics that the same object can be viewed as either discrete or continuous in dependence on the instrument used for the observation [we see a table continuous when we look at it by eye and we see it discrete (consisting of molecules, atoms, etc.) when we observe it under a microscope. In addition, physicists do not give some absolute results of their observations in sense that together with the result of the observation they always supply *the accuracy of the instrument* used for this observation.

In Mathematics, both facts are absent: each mathematical object (e.g., function) is either discrete or continuous and nothing is said about the accuracy of the observation of the mathematical objects and about tools used for these observations. The mathematical notion of continuity itself is from nineteenth century. Many of the mathematical notions have an absolute character and the ideas of relativity are almost not present in them. The ideas of the influence of the instrument of an observation on the object of the observation are almost absent in Mathematics, as well.

[1] Even the brilliant efforts of the creator of the nonstandard analysis Robinson that were made in the middle of the twentieth century have been also directed to a reformulation of the classical analysis (i.e., analysis created 200 years before Robinson) in terms of infinitesimals and not to the creation of a new kind of analysis that would incorporate new achievements of Physics. In fact, he wrote in Sect. 1.1 of his famous book [28]: "It is shown in this book that Leibniz's ideas can be fully vindicated and that they lead to a novel and fruitful approach to *classical analysis* and to many other branches of mathematics" (the words *classical analysis* have been emphasized by the author of this chapter).

In some sense, there exists a gap between the physical achievements made in the last 200 years (especially during the twentieth century) and their mathematical models that continue to be written using the mathematical language developed two centuries ago on the basis of (among other things) physical ideas of that remote time that now are absolutely outdated.

As was already mentioned, in relation to the concepts of infinite and infinitesimal we have an analogous situation. In fact, the point of view on infinity accepted nowadays takes its origins from the famous ideas of Cantor (see [2]) who has shown that there exist infinite sets having different number of elements. This has been done during the second half of the nineteenth century. Infinitesimals have been developed even earlier when, in the early history of calculus, arguments involving infinitesimals played a pivotal role in the differential calculus developed by Leibniz and Newton (see [19, 25]). At that time the notion of an infinitesimal, however, lacked a precise mathematical definition and in order to provide a more rigorous foundation for the calculus infinitesimals were gradually replaced by the d'Alembert–Cauchy concept of a limit (see [4, 6]).

The creation of a rigorous mathematical theory of infinitesimals on which it would be possible to construct Calculus remained an open problem until the end of the 1950s when Robinson (see [28]) has introduced his famous nonstandard analysis approach. He has shown that non-archimedean ordered field extensions of the reals contained numbers that could serve the role of infinitesimals and their reciprocals could serve as infinitely large numbers. Robinson then has derived the theory of limits, and more generally of calculus, and has found a number of important applications of his ideas in many other fields of Mathematics (see [28]).

It is important to emphasize that in his approach Robinson used Cantor's mathematical tools and terminology (cardinal numbers, countable sets, continuum, one-to-one correspondence, etc.) incorporating so advantages and disadvantages of Cantor's approach into nonstandard analysis. In particular, we are reminded that it is well known that Cantor's approach leads to some situations that often are called by non mathematicians "paradoxes". The most famous and simple of them is, probably, Hilbert's paradox of the Grand Hotel. In a normal hotel having a finite number of rooms no more new guests can be accommodated if it is full. Hilbert's Grand Hotel has an infinite number of rooms (of course, the number of rooms is countable, because the rooms in the Hotel are numbered). Due to Cantor, if a new guest arrives at the Hotel where every room is occupied, it is, nevertheless, possible to find a room for him. To do so, it is necessary to move the guest occupying room 1 to room 2, the guest occupying room 2 to room 3, etc. In such a way room 1 will be ready for the newcomer and, in spite of our assumption that there are no available rooms in the Hotel, we have found one.

This result is very difficult to be fully realized by anyone who is not a mathematician since in our every day experience in the world around us the part is always less than the whole and if a hotel is complete, there are no places in it. In order to understand how it is possible to tackle the problem of infinity in such a way that Hilbert's Grand Hotel would be in accordance with the principle "the part is less than the whole" let us consider a study published in *Science* by Peter Gordon (see [13]) where he describes a primitive tribe living in Amazonia—Pirahã—that uses a very simple numeral system[2] for counting: one, two, many.

For Pirahã, all quantities larger than two are just "many" and such operations as $2+2$ and $2+1$ give the same result, i.e., "many". Using their weak numeral system Pirahã are not able to see, for instance, numbers 3, 4, 5, and 6, to execute arithmetical operations with them, and, in general, to say anything about these numbers because in their language there are neither words nor concepts for that. Moreover, the weakness of their numeral system leads to such results as

$$\text{“many”} + 1 = \text{“many”}, \qquad \text{“many”} + 2 = \text{“many”},$$

which are very familiar to us in the context of views on infinity used in the traditional calculus

$$\infty + 1 = \infty, \qquad \infty + 2 = \infty$$

and in the context of Cantor's infinite cardinals[3] we also have

$$\aleph_0 + 1 = \aleph_0, \quad \aleph_1 + 1 = \aleph_1. \tag{1}$$

These observations lead us to the following idea: *Probably our difficulty in working with infinity is not connected to the nature of infinity but is a result of inadequate numeral systems used to express numbers.*

In this chapter, we describe a new methodology for treating infinite and infinitesimal quantities (examples of its usage can be found in [31–37, 39, 41]). It has a strong numerical character and is closer to the point of view on the world accepted by modern Physics.[4] In particular, it incorporates the following two ideas borrowed from the modern Physics: relativity and interrelations holding between the object of an observation and the tool used for this observation. The latter is directly related

---

[2] We remind that *numeral* is a symbol or group of symbols that represents a *number*. The difference between numerals and numbers is the same as the difference between words and the things they refer to. A *number* is a concept that a *numeral* expresses. The same number can be represented by different numerals. For example, the symbols "3," "three," and "III" are different numerals, but they all represent the same number.

[3] In connection with Cantor's $\aleph_0$ and $\aleph_1$ it makes sense to remind another Amazonian tribe—Mundurukú (see [27]) who fail in exact arithmetic with numbers larger than 5 but are able to compare and add large approximate numbers that are far beyond their naming range. Particularly, they use the words "some, not many" and "many, really many" to distinguish two types of large numbers. Their arithmetic with "some, not many" and "many, really many" reminds strongly the rules Cantor uses to work with $\aleph_0$ and $\aleph_1$, respectively. For instance, compare "some, not many" + "many, really many" = "many, really many" with $\aleph_0 + \aleph_1 = \aleph_1$.

[4] As it was already mentioned, in 1900, at the second Mathematical Congress in Paris, David Hilbert has presented his 23 problems for the twentieth century promoting the abstract philosophy in Mathematics that was close to Kant. However, before this event, at the first Congress 3 years earlier Henri Poincaré has given a general talk emphasizing the connection of Mathematics with Physics sharing this point of view with Fourier, Laplace, and many others. Clearly, in this dispute between Poincaré and Hilbert the present chapter is closer to the position of Poincaré.

to connections between numeral systems used to describe mathematical objects and the objects themselves. Numerals that we use to write down numbers, functions, etc. are among our tools of investigation and, as a result, they strongly influence our capabilities to study mathematical objects.

Since new numeral systems appear very rarely, in each concrete historical period people tend to think that *any* number can be expressed by the current numeral system and the importance of numeral systems for Mathematics is very often underestimated (especially by pure mathematicians). However, if we observe the situation in the historical prospective we can immediately see limitations that various numeral systems induce. In order to illustrate this assertion, it is sufficient to think about Pirahã. We can also remind the Roman numeral system that does not allow one to express zero and negative numbers. In this system, the expression III–X is an indeterminate form. As a result, before appearing the positional numeral system and inventing zero (by the way, the second event was several hundred years later with respect to the first one) mathematicians were not able to create theorems involving zero and negative numbers and to execute computations with them. Thus, developing new (more powerful than existing ones) numeral systems can help a lot both in theory and practice of computations.

If we compare the usage of numeral systems in Mathematics when one works, on the one hand, with finite quantities and, on the other hand, with infinities and infinitesimals, then we can see immediately an important difference. In our everyday activities with finite numbers the *same* finite numerals are used for *different* purposes (e.g., the same numeral 6 can be used to express the number of elements of a set, to indicate the position of an element in a finite sequence, and to execute practical computations). In contrast, when we face the necessity to work with infinities or infinitesimals, the situation changes drastically. In fact, in this case *different* numerals are used to work with infinities and infinitesimals in *different* situations:

- ∞ in standard analysis
- ω for working with ordinals
- $\aleph_0, \aleph_1, \ldots$ for dealing with cardinalities
- Nonstandard numbers using a generic infinitesimal h in nonstandard analysis, etc.

In particular, since the mainstream of the traditional Mathematics very often does not pay a great attention to the distinction between numbers and numerals (in this occasion it is necessary to recall constructivists who studied this issue), many theories dealing with infinite and infinitesimal quantities have a symbolic (not numerical) character. For instance, many versions of nonstandard analysis are symbolic, since they have no numeral systems to express their numbers by a finite number of symbols (the finiteness of the number of symbols is necessary for organizing numerical computations). Namely, if we consider a finite n, then it can be taken $n = 7$, or $n = 108$ or any other numeral used to express finite quantities and consisting of a finite number of symbols. In contrast, if we consider a nonstandard infinite m, then it is not clear which numerals can be used to assign a concrete value to m.

Analogously, in nonstandard analysis, if we consider an infinitesimal h, then it is not clear which numerals consisting of a finite number of symbols can be used to assign a value to h and to write $h = \ldots$ In fact, very often in nonstandard analysis texts, a *generic* infinitesimal h is used and it is considered as a symbol, i.e., only symbolic computations can be done with it. Approaches of this kind leave unclear such issues, e.g., whether the infinite $1/h$ is integer or not or whether $1/h$ is the number of elements of an infinite set. Another problem is related to comparison of values. When we work with finite quantities then we can compare x and y if they assume numerical values, e.g., $x = 4$ and $y = 6$ then, by using rules of the numeral system the symbols 4 and 6 belong to, we can compute that $y > x$. If one wishes to consider two infinitesimals h_1 and h_2, then it is not clear how to compare them because numeral systems that can express infinitesimals are not provided by nonstandard analysis techniques.

The approach developed in [31, 37, 43] proposes a numeral system that uses *the same numerals* for several different purposes for dealing with infinities and infinitesimals: in analysis for working with functions that can assume different infinite, finite, and infinitesimal values (functions can also have derivatives assuming different infinite or infinitesimal values); for measuring infinite sets; for indicating positions of elements in ordered infinite sequences; in probability theory, etc. It is important to emphasize that the new numeral system avoids situations like that of Pirahã and (1) providing results ensuring that if a is a numeral written in this system then for any a (i.e., a can be finite, infinite, or infinitesimal) it follows $a+1 > a$. The new methodology has allowed the author to introduce the Infinity Computer (see the patent [41]) working numerically with infinite and infinitesimal numbers.

In order to see the place of the new approach in the historical panorama of ideas dealing with infinite and infinitesimal, see [21,22,40,42,47]. The new methodology has been successfully applied for studying percolation (see [17, 50]), Euclidean and hyperbolic geometry (see [23, 29]), fractals (see [36, 38, 46, 50]), numerical differentiation and optimization (see [8,39,44,53]), infinite series (see [40,45,52]), the first Hilbert problem, Riemann zeta function, and Turing machines (see [42,45, 47]), cellular automata (see [7]), etc.

The rest of the chapter is structured as follows. An introduction to the new methodology is given in Sect. 2. It allows us to introduce in Sect. 3 a new infinite unit of measure that is then used as the radix of a new positional numeral system. Section 4 shows that this system gives a possibility to express finite, infinite, and infinitesimal numbers in a unique framework and to execute arithmetical operations with all of them. Section 5 discusses first applications of the new methodology. Section 6 establishes relations of the new methodology to some of the results of Cantor. New computational possibilities for mathematical modeling supplied by the new approach are discussed in Sect. 7. A quantitative analysis of fractals executed by using infinite and infinitesimal numbers is given in Sect. 8. Concepts of continuity in Physics and Mathematics from the point of view of the new methodology are discussed in Sect. 9. Finally, Sect. 10 concludes the chapter.

We close this Introduction by emphasizing that the new approach is not a contraposition to the ideas of Cantor, Levi–Civita, and Robinson. In contrast, it is introduced as an applied evolution of their ideas. The problem of infinity is considered from positions of applied Mathematics and theory and practice of computations—fields being among the main scientific interests (see, e.g., monographs [48, 49]) of the author. The new computational methodology introduces the notion of the accuracy of mathematical languages and shows that different tools (numeral systems) can express different sets of numbers (and other mathematical objects) with different accuracies. It can be shown that Cantor's alephs and new numerals have different accuracies and cases where the new tools are more accurate can be provided. Thus, the traditional approaches and the new one do not contradict one another, they are just different instruments having different accuracies for observations of mathematical objects.

2 A New Computational Methodology and Accuracy of Numeral Systems

The aim of this section is to introduce a new methodology that would allow one to work with infinite and infinitesimal quantities *in the same way* as one works with finite numbers. Evidently, it becomes necessary to define what does it mean *in the same way*. Usually, in modern Mathematics, when it is necessary to define a concept or an object, logicians try to introduce a number of axioms describing the object. However, this way is fraught with danger because of the following reasons. First of all, when we describe a mathematical object or concept we are limited by the expressive capacity of the language we use to make this description. A more rich language allows us to say more about the object and a weaker language—less (remind Pirahã that are not able to say a word about number 4). Thus, development of the mathematical (and not only mathematical) languages leads to a continuous necessity of a transcription and specification of axiomatic systems. Second, there is no any guarantee that the chosen axiomatic system defines "sufficiently well" the required concept and a continuous comparison with practice is required in order to check the goodness of the accepted set of axioms. However, there cannot be again any guarantee that the new version will be the last and definitive one. Finally, the third limitation latent in axiomatic systems has been discovered by Gödel in his two famous incompleteness theorems (see [11]).

In this chapter, we introduce a different, significantly more applied and less ambitious view on axiomatic systems related only to utilitarian necessities to make calculations. We start by introducing three postulates that will fix our methodological positions with respect to infinite and infinitesimal quantities and Mathematics, in general. In contrast to the modern mathematical fashion that tries to make all axiomatic systems more and more precise (decreasing so degrees of freedom of the studied part of Mathematics), we just define a set of general rules describing how practical computations should be executed leaving so as much space as possible for further, dictated by practice, changes and developments of the introduced mathematical language. Speaking metaphorically, we prefer to make a hammer and to use it instead of describing what is a hammer and how it works.

Usually, when mathematicians deal with infinite objects (sets or processes) it is supposed [even by constructivists (see, for example, [24])] that human beings are able to execute certain operations infinitely many times. For example, in a fixed numeral system it is possible to write down a numeral with *any* number of digits. However, this supposition is an abstraction (courageously declared by constructivists in [24]) because we live in a finite world and all human beings and/or computers finish operations they have started. In this chapter, this abstraction is not used and the following postulate is adopted.

Postulate 1. *We postulate existence of infinite and infinitesimal objects but accept that human beings and machines are able to execute only a finite number of operations.*

Thus, we accept that we shall never be able to give a complete description of infinite processes and sets due to our finite capabilities. Particularly, this means that we accept that we are able to write down only a finite number of symbols to express numbers. However, we do not agree with finitists who deny infinite mathematical objects. We accept their existence and shall try to study them using our finite capabilities.

The second postulate is adopted following the way of reasoning used in natural sciences where researchers use tools to describe the object of their study and the used instrument influences the results of the observations. When a physicist uses a weak lens A and sees two black dots in his/her microscope he/she does not say: the object of the observation *is* two black dots. The physicist is obliged to say: the lens used in the microscope allows us to see two black dots and it is not possible to say anything more about the nature of the object of the observation until we change the instrument—the lens or the microscope itself—by a more precise one. Suppose that he/she changes the lens and uses a stronger lens B and is able to observe that the object of the observation is viewed as ten (smaller) black dots. Thus, we have two different answers: (a) the object is viewed as two dots if the lens A is used; (b) the object is viewed as ten dots by applying the lens B. Which of the answers is correct? Both. Both answers are correct but with the *different accuracies* that depend on the lens used for the observation.

The same happens in Mathematics studying natural phenomena, numbers, and objects that can be constructed by using numbers. Numeral systems used to express numbers are among the instruments of observations used by mathematicians. The usage of powerful numeral systems gives the possibility to obtain more precise results in Mathematics in the same way as usage of a good microscope gives the possibility of obtaining more precise results in Physics. However, even for the best existing tool the capabilities of this tool will be always limited due to Postulate 1 (we are able to write down only a finite number of symbols when we wish to describe a mathematical object) and due to Postulate 2 we shall never tell, **what is**, for example, a number but shall just observe it through numerals expressible in a chosen numeral system.

Postulate 2. *We shall not tell* ***what are*** *the mathematical objects we deal with; we just shall construct more powerful tools that will allow us to improve our capacities to observe and to describe properties of mathematical objects.*

This Postulate means that we emphasize that mathematical results are not absolute, they depend on mathematical languages used to formulate them, i.e., there always exists an accuracy of the description of a mathematical result, fact, object, etc. imposed by the mathematical language used to formulate this result. For instance, the result of Pirahã $2+2=$ "many" is not wrong, it is just *inaccurate*. The introduction of a stronger tool (in this case, a numeral system that contains a numeral for a representation of the number four) allows us to have a more precise answer.

The concept of the accuracy allows us to look at paradoxes in a new way: *paradox* is a situation where the accuracy of the used language is not sufficient to describe the phenomenon we are interested in. For instance, the answers of Pirahã $2+1=$ "many" and $2+2=$ "many" can be viewed as a paradox because from these two records one could conclude that $2+1=2+2$. This paradox shows us the borderline that separates the zone where the language has the high precision from the region where the language cannot be applied because it does not allow one to distinguish different objects within "many". Analogously, the records "many" + 1= "many", $\infty+1=\infty$, $1+\omega=\omega\neq\omega+1$, (1), etc. can also be viewed as situations where the accuracy of the used numeral systems is not sufficient.

It is necessary to comment upon another important aspect of the distinction between a mathematical object and a mathematical tool used to observe this object. Postulates 1 and 2 impose us to think always about *the possibility to execute* a mathematical operation by applying a numeral system. They tell us that there always exist situations where we are not able to express the result of an operation. Let us consider, for example, the operation of construction of the successive element widely used in number and set theories. In the traditional Mathematics, the aspect whether this operation can be executed is not taken into consideration, it is supposed that it is always possible to execute the operation $k=n+1$ starting from any integer n. Thus, there is no any distinction between the existence of the number k and the possibility to execute the operation $n+1$ and to express its result, i.e. to have a numeral that can express k.

Postulates 1 and 2 emphasize this distinction and tell us that: (a) in order to execute the operation it is necessary to have a numeral system allowing one to express both numbers, n and k; (b) for any numeral system there always exists a number k that cannot be expressed in it. For instance, for Pirahã $k=3$, for

Mundurukú $k = 6$. Even for modern powerful numeral systems there exist such a number k (for instance, nobody is able to write down a numeral in the decimal positional system having 10^{100} digits). Hereinafter we shall always emphasize the triad—researcher, object of the investigation, and tools used to observe the object—in various mathematical and computational contexts paying a special attention to the accuracy of the obtained results.

Another important issue related to Postulate 2 consists of the fact that, from our point of view, axiomatic systems *do not define* mathematical objects but just determine formal rules for operating with certain numerals reflecting some (not all) properties of the studied mathematical objects using a certain mathematical language L. We are aware that the chosen language L has its accuracy and there always can exist a richer language $\tilde{L}$ that would allow us to describe the studied object better. As has already been discussed above, any language has a limited expressibility, in particular, there always exist situations where the accuracy of the answers expressible in this language is not sufficient.

Numerals that we use to write down numbers, functions, etc. are among our tools of the investigation and, as a result, they strongly influence our capabilities to study mathematical objects. This separation (having an evident physical spirit) of mathematical objects from the tools used for their description is crucial for our study, but it is used rarely in contemporary Mathematics. In fact, the idea of finding an adequate (absolutely the best) set of axioms for one or another field of Mathematics continues to be among the most attractive goals for contemporary mathematicians. Usually, when it is necessary to define a concept or an object, logicians try to introduce a number of axioms *defining* the object. However, this way is fraught with danger because of the following reasons.

First, when one describes a mathematical object or concept he or she is limited by the expressive capacity of the language that is used to make this description. A richer language allows one to say more about the object and a weaker language—less. Thus, development of the mathematical (and not only mathematical) languages leads to a continuous necessity of a transcription and specification of axiomatic systems. Second, there is no guarantee that the chosen axiomatic system defines "sufficiently well" the required concept and a continuous comparison with practice is required in order to check the goodness of the accepted set of axioms. However, there cannot be again any guarantee that the new version will be the last and definitive one. Finally, the third limitation has been discovered by Gödel in his two famous incompleteness theorems (see [11]).

It should be emphasized that in both Philosophy and Linguistics, the relativity of the language (the instrument) with respect to the world around (the object of study) is a well-known thing. It is sufficient to mention Wittgenstein: "The limits of my language are the limits of my mind. All I know is what I have words for." In Linguistics, it is sufficient to remind the Sapir–Whorf thesis (see [3,30]), also known as the "linguistic relativity thesis". As becomes clear from its name, the thesis does not accept the idea of the universality of language and postulates that the nature of a particular language influences the thought of its speakers. The thesis challenges the possibility of perfectly representing the world with language, because it implies that the mechanisms of any language condition the thoughts of its speakers.

Thus, due to Postulate 2, our point of view on axiomatic systems is significantly more applied with respect to the modern mathematical fashion that tries to make all axiomatic systems more and more precise (decreasing so degrees of freedom of the studied part of Mathematics). We just define a set of general rules describing how practical computations should be executed leaving so as much space as possible for further, dictated by practice, changes and developments of the introduced mathematical language. Speaking metaphorically, we prefer to make a hammer and to use it instead of trying to define what the hammer *is* and how it works.

For example, from this applied point of view, axioms for real numbers are considered together with a particular numeral system $\mathcal{S}$ used to write down numerals and are viewed as practical rules (associative and commutative properties of multiplication and addition, distributive property of multiplication over addition, etc.) describing operations with the numerals. The completeness property is interpreted as a possibility to extend $\mathcal{S}$ with additional symbols (e.g., e, π, $\sqrt{2}$, etc.) taking care of the fact that the results of computations with these symbols agree with the facts observed in practice. As a rule, the assertions regarding numbers that cannot be expressed in a numeral system are avoided (e.g., it is not supposed that real numbers form a field).

Finally, before we switch our attention to Postulate 3, it should be noticed the key difference distinguishing our approach from the constructivism. Constructivists assert that it is necessary to construct (in some sense) a mathematical object to prove that it exists. Following Physics, we do not discuss the questions of *existence* of mathematical objects at all. We discuss just what can be observed through our tools (languages, numeral systems, etc.).

Let us now start to introduce the last Postulate. We want to treat infinite and infinitesimal numbers in the same manner as we are used to deal with finite ones, i.e., by applying the philosophical principle of Ancient Greeks "The part is less than the whole." This principle, in our opinion, very well reflects organization of the world around us but is not incorporated in many traditional infinity theories where it is true only for finite numbers. The reason of this traditional discrepancy (as the example with Pirahã advices) is related to the accuracy of numeral systems used to work with infinity.

Postulate 3. *We adopt the principle "The part is less than the whole" to all numbers (finite, infinite, and infinitesimal) and to all sets and processes (finite and infinite).*

Due to this Postulate, the traditional point of view on infinity accepting such results as $\infty - 1 = \infty$ should be substituted in a way that $\infty - 1 < \infty$. One of the motivations *pro* this substitution has already been discussed in detail in connection with the numerals of Pirahã. We can introduce another simple argument.

Suppose that we are at a point A and at another point, B, being infinitely far from A there is an object. Let us see what will happen if we shall change our position and will move, let us say, 1 m forward in the direction of the point B. The traditional numeral system using the symbol ∞ will not be able to register this movement in a quantitative way because $\infty - 1 = \infty$. This numeral system allows us to say only that the object was infinitely far before the movement and remains to be infinitely far after the movement, i.e., the accuracy of the answer is very low. In practice, due to this traditional way of doing, we are forced to negate the finite movement that we have executed. Hereinafter, our goal will be to avoid similar situations by the introduction of a new numeral system that instead of the traditional numerals $\infty, \aleph_0, \omega, \aleph_1$, etc. would use a new kind of numerals satisfying Postulates 1–3 introduced above.

Due to Postulates 1–3, such concepts as bijection, numerable and continuum sets, cardinal and ordinal numbers cannot be used in this chapter because they belong to theories working with different assumptions. It can seem at first glance that Postulate 3 contradicts Cantor's one-to-one correspondence principle. However, as it will be shown hereinafter, this is not the case. Instead, the situation is similar to the example from Physics described above where we have considered two lenses having different accuracies. We have here just two different instruments (numeral systems) having different accuracies: Cantor's approach and the new one based on Postulates 1–3. Analogously, in the finite case, when we observe a garden with 123 trees, then our answer, i.e., 123 trees, and the answer of Pirahã, i.e., many trees, are both correct, but the accuracy of our answer is higher.

It is important to notice that the adopted Postulates impose also the style of exposition of results in the chapter: we first introduce new mathematical instruments, then show how to use them in several areas of Mathematics, introducing each item as soon as it becomes indispensable for the problem under consideration.

Let us introduce now the new way of counting by studying a situation arising in practice and related to the necessity to operate with extremely large quantities (see [31] for a detailed discussion). Imagine that we are in a granary and the owner asks us to count how much grain he has inside it. In this occasion, nobody counts the grain seed by seed because the number of seeds is enormous.

To overcome this difficulty, people take sacks, fill them in with seeds, and count the number of sacks. In this situation, we suppose that: (a) the number of seeds in each sack is the same but it is so huge that we are not able to count seed by seed how many they are and (b) in any case the resulting number would not be expressible by available numerals.

Then, if the granary is huge and it becomes difficult to count the sacks, then trucks or even big train waggons are used. In this model, we suppose that all sacks contain the same number of seeds, all trucks—the same number of sacks, and all waggons—the same number of trucks, however, these numbers are so huge that it becomes impossible to determine them. At the end of the counting of this type we obtain a result in the following form: the granary contains 14 waggons, 54 trucks, 18 sacks, and 47 seeds of grain. Note, that if we add, for example, one seed to the granary, we can count it and see that the granary has more grain. If we take out one waggon, we again are able to say how much grain has been subtracted.

Thus, in our example it is necessary to count large quantities. They are finite but it is impossible to count them directly by using an elementary unit of measure, u_0, (seeds in our example) because the quantities expressed in these units would be too large. Therefore, people are forced to behave as if the quantities were infinite.

To solve the problem of "infinite" quantities, new units of measure, u_1, u_2, and u_3, are introduced (units u_1—sacks, u_2—trucks, and u_3—waggons). The new units have the following important peculiarity: all the units u_{i+1} contain a certain number K_i of units u_i but this number, K_i, is unknown. Naturally, it is supposed that K_i is the same for all instances of the units u_{i+1}. Thus, numbers that were impossible to express using only the initial unit of measure are perfectly expressible in the new units we have introduced in spite of the fact that the numbers K_i are unknown.

This key idea of counting by introduction of new units of measure will be used in the chapter to deal with infinite quantities together with the idea of separate count of units with different exponents used in traditional positional numeral systems.

3 A New Way of Counting and the Infinite Unit of Measure

The infinite unit of measure is expressed by the numeral ① called *grossone* and is introduced as the number of elements of the set, $\mathbb{N}$, of natural numbers. Remind that the usage of a numeral indicating totality of the elements we deal with is not new in mathematics. It is sufficient to mention the theory of probability (axioms of Kolmogorov) where events can be defined in two ways. First, as union of elementary events; second, as a sample space, Ω, of all possible elementary events (or its parts $\Omega/2, \Omega/3$, etc.) from which some elementary events have been excluded (or added in case of parts of Ω). Naturally, the latter way to define events becomes particularly useful when the sample space consists of infinitely many elementary events.

Grossone is introduced by describing its properties (similarly, in order to pass from natural to integer numbers a new element—zero—is introduced by describing its properties) postulated by the *Infinite Unit Axiom* (IUA) consisting of three parts: Infinity, Identity, and Divisibility. This axiom is added to axioms for real numbers (remind that we consider axioms in sense of Postulate 2). Thus, it is postulated that associative and commutative properties of multiplication and addition, distributive property of multiplication over addition, existence of inverse elements with respect to addition, and multiplication hold for grossone as for finite numbers.[5] Let us introduce the axiom and then give comments on it.

Infinity. Any finite natural number n is less than grossone, i.e., $n <$ ①.

[5] It is important to emphasize that we speak about axioms of real numbers in sense of Postulate 2, i.e., axioms define formal rules of operations with numerals in a given numeral system. Therefore, if we want to have a numeral system including grossone, we should fix also a numeral system to express finite numbers. In order to concentrate our attention on properties of grossone, this point will be investigated later.

Identity. The following relations link ① to identity elements 0 and 1

$$0 \cdot ① = ① \cdot 0 = 0, \quad ① - ① = 0, \quad \frac{①}{①} = 1, \quad ①^0 = 1, \quad 1^{①} = 1, \quad 0^{①} = 0. \tag{2}$$

Divisibility. For any finite natural number n sets $\mathbb{N}_{k,n}, 1 \leq k \leq n$, being the nth parts of the set, $\mathbb{N}$, of natural numbers have the same number of elements indicated by the numeral $\frac{①}{n}$ where

$$\mathbb{N}_{k,n} = \{k, k+n, k+2n, k+3n, \ldots\}, \qquad 1 \leq k \leq n, \qquad \bigcup_{k=1}^{n} \mathbb{N}_{k,n} = \mathbb{N}. \tag{3}$$

The first part of the introduced axiom—Infinity—is quite clear. In fact, we want to describe an infinite number, thus, it should be larger than any finite number. The second part of the axiom—Identity—tells us that ① behaves itself with identity elements 0 and 1 as all other numbers. In reality, we could even omit this part of the axiom because, due to Postulate 3, all numbers should be treated in the same way and, therefore, at the moment we have told that grossone is a number, we have fixed usual properties of numbers, i.e., the properties described in Identity, associative and commutative properties of multiplication and addition, distributive property of multiplication over addition, existence of inverse elements with respect to addition and multiplication. The third part of the axiom—Divisibility—is the most interesting, it is based on Postulate 3. Let us first illustrate it by an example.

Example 1. If we take $n = 1$, then $\mathbb{N}_{1,1} = \mathbb{N}$ and Divisibility tells that the set, $\mathbb{N}$, of natural numbers has ① elements. If $n = 2$, we have two sets $\mathbb{N}_{1,2}$ and $\mathbb{N}_{2,2}$

$$\begin{array}{l} \mathbb{N}_{1,2} = \{1, \quad 3, \quad 5, \quad 7, \ldots\}, \\ \mathbb{N}_{2,2} = \{\quad 2, \quad 4, \quad 6, \quad \ldots\} \end{array} \tag{4}$$

and they have $\frac{①}{2}$ elements each. Pay attention that we are not able to count the number of elements of the sets $\mathbb{N}$, $\mathbb{N}_{1,2}$, and $\mathbb{N}_{2,2}$ one by one because due to Postulate 1 we are able to execute only a finite number of operations and these sets are infinite. To define their number of elements we apply Postulate 3 and determine the number of the elements of the parts using the whole.

Then, if $n = 3$, we have three sets

$$\begin{array}{l} \mathbb{N}_{1,3} = \{1, \quad 4, \quad 7, \ldots\}, \\ \mathbb{N}_{2,3} = \{\quad 2, \quad 5, \quad \ldots\}, \\ \mathbb{N}_{3,3} = \{\quad 3, \quad 6, \quad \ldots\} \end{array} \tag{5}$$

and they have $\frac{①}{3}$ elements each. Note that in formulae (4), (5) we have added extra spaces writing down the elements of the sets $\mathbb{N}_{1,1}, \mathbb{N}_{1,2}, \mathbb{N}_{1,3}, \mathbb{N}_{2,3}, \mathbb{N}_{3,3}$ just to emphasize Postulate 3 and to show visually that $\mathbb{N}_{1,1} \cup \mathbb{N}_{1,2} = \mathbb{N}$ and $\mathbb{N}_{1,3} \cup \mathbb{N}_{2,3} \cup \mathbb{N}_{3,3} = \mathbb{N}$. □

We emphasize again that to introduce $\frac{①}{n}$ we do not try to count elements $k, k+n, k+2n, k+3n, \ldots$ one by one in (3). In fact, we cannot do this due to Postulate 1. By using Postulate 3, we construct the sets $\mathbb{N}_{k,n}, 1 \leq k \leq n$, by separating the whole, i.e., the set $\mathbb{N}$, in n parts [this separation is highlighted visually in formulae (4) and (5)]. Again due to Postulate 3, we affirm that the number of elements of the nth part of the set, i.e., $\frac{①}{n}$, is n times less than the number of elements of the whole set, i.e., than ①.

In terms of our granary example ① can be interpreted as the number of seeds in the sack. In that example, the number K_0 of seeds in each sack was fixed and finite but impossible to be expressed in units u_0, i.e., seeds, by counting seed by seed because we have supposed that sacks were very big and the corresponding number would not be expressible by available numerals. In spite of the fact that K_0 and $K_1, K_2, \ldots$ were inexpressible and unknown, by using new units of measure (sacks, trucks, etc.) it was possible to count easier and to express the required quantities. Now our sack has the infinite but again *fixed* number of seeds. It is fixed because it has a strong link to a concrete set—it is the number of elements of this set, precisely, of the set of natural numbers. This number is inexpressible by existing numeral systems with the same high accuracy as we do it with finite small sets[6] and we introduce a new number—grossone—expressible by a new numeral—①. Then, we apply Postulate 3 and say that if the sack contains ① seeds, its nth part contains n times less quantity, i.e., $\frac{①}{n}$ seeds. Note that, since the numbers $\frac{①}{n}$ have been introduced as numbers of elements of sets $\mathbb{N}_{k,n}$, they are integer.

The new unit of measure allows us to calculate easily the number of elements of sets being union, intersection, difference, or product of other sets of the type $\mathbb{N}_{k,n}$. Due to our accepted methodology, we do it in the same way as these measurements are executed for finite sets. Let us consider two simple examples (a general rule for determining the number of elements of infinite sets having a more complex structure will be given in Sect. 5) showing how grossone can be used for this purpose.

Example 2. Let us determine the number of elements of the set $A_{k,n} = \mathbb{N}_{k,n} \backslash \{a\}$, $a \in \mathbb{N}_{k,n}, n \geq 1$. Due to the IUA, the set $\mathbb{N}_{k,n}$ has $\frac{①}{n}$ elements. The set $A_{k,n}$ has

[6]First, this quantity is inexpressible by numerals used to count the number of elements of finite sets because $\mathbb{N}$ is infinite. Second, traditional numerals existing to express infinite numbers do not have the required high accuracy (remind that we would like to be able to register the alteration of the number of elements of infinite sets even when one element has been excluded). For example, by using Cantor's alephs we say that cardinality of the sets $\mathbb{N}$ and $\mathbb{N} \setminus \{1\}$ is the same—$\aleph_0$. This answer is correct but its accuracy is low—we are not able to register the fact that one element was excluded from the set $\mathbb{N}$. Analogously, we can say that both of the sets have "many" elements. Again, this answer is correct but its accuracy is low.

been constructed by excluding one element from $N_{k,n}$. Thus, the set $A_{k,n}$ has $\frac{①}{n}-1$ elements. The granary interpretation can be also given for the number $\frac{①}{n}-1$: the number of seeds in the nth part of the sack minus one seed. For $n=1$ we have $①-1$ interpreted as the number of seeds in the sack minus one seed. □

Divisibility and Example 2 show us that in addition to the usual way of counting, i.e., by adding units, that has been well formalized in Mathematics, there exist also the way to count by taking parts of the whole and by subtracting units or parts of the whole. The following example shows a little bit more complex situation (other more sophisticated examples will be given later after the reader will got accustomed with the concept of grossone).

Example 3. Let us consider the following two sets

$$B_1 = \{4,9,14,19,24,29,34,39,44,49,54,59,64,69,74,79,\ldots\},$$
$$B_2 = \{3,14,25,36,47,58,69,80,91,102,113,124,135,\ldots\}$$

and determine the number of elements in the set $B = (B_1 \cap B_2) \cup \{3,4,5,69\}$. It follows immediately from the IUA that $B_1 = \mathbb{N}_{4,5}, B_2 = \mathbb{N}_{3,11}$. Their intersection

$$B_1 \cap B_2 = \mathbb{N}_{4,5} \cap \mathbb{N}_{3,11} = \{14,69,124,\ldots\} = \mathbb{N}_{14,55}$$

and, therefore, due to the IUA, it has $\frac{①}{55}$ elements. Finally, since 69 belongs to the set $\mathbb{N}_{14,55}$ and 3, 4, and 5 do not belong to it, the set B has $\frac{①}{55}+3$ elements. The granary interpretation: this is the number of seeds in the 55th part of the sack plus three seeds. □

One of the important differences of the new approach with respect to the nonstandard analysis consists of the fact that $① \in \mathbb{N}$ because grossone has been introduced as the quantity of natural numbers. Similarly, the number 5 being the number of elements of the set

$$A = \{1,2,3,4,5\} \qquad (6)$$

is the largest element in this set. The new numeral ① allows one to write down the set, $\mathbb{N}$, of natural numbers in the form

$$\mathbb{N} = \left\{ 1,2, \ \ldots \ \frac{①}{2}-2, \frac{①}{2}-1, \frac{①}{2}, \frac{①}{2}+1, \frac{①}{2}+2, \ \ldots \ ①-2, \ ①-1, \ ① \right\}. \qquad (7)$$

Note that traditional numeral systems did not allow us to see infinite natural numbers

$$\ldots \ \frac{①}{2}-2, \frac{①}{2}-1, \frac{①}{2}, \frac{①}{2}+1, \frac{①}{2}+2, \ \ldots \ ①-2, ①-1, ①. \qquad (8)$$

It is important to emphasize that in the new approach the set (7) is the same set of natural numbers

$$\mathbb{N} = \{1, 2, 3, \ \dots \ \} \tag{9}$$

we are used to deal with and infinite numbers (8) also take part of $\mathbb{N}$. Both records, (7) and (9), are correct and do not contradict each other. They just use two different numeral systems to express $\mathbb{N}$. Traditional numeral systems do not allow us to see infinite natural numbers that we can observe now thanks to ①. Similarly, Pirahã are not able to see finite natural numbers greater than 2. In spite of this fact, these numbers (e.g., 3 and 4) belong to $\mathbb{N}$ and are visible if one uses a more powerful numeral system. Thus, we have the same object of observation—the set $\mathbb{N}$—that can be observed by different instruments—numeral systems—with different accuracies (see Postulate 2).

This example illustrates also the fact that when we speak about sets (finite or infinite) it is necessary to take care about tools used to describe a set (remind Postulate 2). In order to introduce a set, it is necessary to have a language (e.g., a numeral system) allowing us to describe its elements and the number of the elements in the set. For instance, the set A from (6) cannot be defined using the mathematical language of Pirahã.

Analogously, the words "the set of all finite numbers" do not define a set completely from our point of view, as well. It is always necessary to specify which instruments are used to describe (and to observe) the required set and, as a consequence, to speak about "the set of all finite numbers expressible in a fixed numeral system." For instance, for Pirahã "the set of all finite numbers" is the set $\{1,2\}$ and for Mundurukú "the set of all finite numbers" is the set A from (6). As it happens in Physics, the instrument used for an observation bounds the possibility of the observation. It is not possible to say how we shall see the object of our observation if we have not clarified which instruments will be used to execute the observation.

Now the following obvious question arises: which natural numbers can we express by using the new numeral ①? Suppose that we have a numeral system, $\mathcal{S}$, for expressing finite natural numbers and it allows us to express $K_{\mathcal{S}}$ numbers (not necessary consecutive) belonging to a set $\mathcal{N}_{\mathcal{S}} \subset \mathbb{N}$. Note that due to Postulate 1, $K_{\mathcal{S}}$ is finite. Then, addition of ① to this numeral system will allow us to express also infinite natural numbers $\frac{i①}{n} \pm k \leq ①$ where $1 \leq i \leq n$, $k \in \mathcal{N}_{\mathcal{S}}$, $n \in \mathcal{N}_{\mathcal{S}}$ (note that since $\frac{①}{n}$ are integers, $\frac{i①}{n}$ are integers too). Thus, the more powerful system $\mathcal{S}$ is used to express finite numbers, the more infinite numbers can be expressed but their quantity is always finite, again due to Postulate 1. The new numeral system using grossone allows us to express more numbers than traditional numeral systems thanks to the introduced new numerals but, as it happens for all numeral systems, its abilities to express numbers are limited.

Example 4. Let us consider the numeral system, $\mathcal{P}$, of Pirahã able to express only numbers 1 and 2 (the only difference will be in the usage of numerals "1" and "2" instead of original numerals *I* and *II* used by Pirahã). If we add to $\mathcal{P}$ the new numeral ①, we obtain a new numeral system (we call it $\hat{\mathcal{P}}$) allowing us to express only ten numbers represented by the following numerals

$$\underbrace{1,2}_{\text{finite}}, \quad \ldots \quad \underbrace{\frac{①}{2}-2, \frac{①}{2}-1, \frac{①}{2}, \frac{①}{2}+1, \frac{①}{2}+2}_{\text{infinite}}, \quad \ldots \quad \underbrace{①-2, ①-1, ①}_{\text{infinite}}. \quad (10)$$

The first two numbers in (10) are finite, the remaining eight are infinite, and dots show natural numbers that are not expressible in $\hat{\mathcal{P}}$. As a consequence, $\hat{\mathcal{P}}$ does not allow us to execute such operation as $2+2$ or to add 2 to $\frac{①}{2}+2$ because their results cannot be expressed in it. Of course, we do not say that results of these operations are equal (as Pirahã do for operations $2+2$ and $2+1$). We just say that the results are not expressible in $\hat{\mathcal{P}}$ and it is necessary to take another, more powerful numeral system if we want to execute these operations. □

Note that crucial limitations discussed in Example 4 hold for sets, too. As a consequence, the numeral system $\mathcal{P}$ allows us to define only the sets $\mathbb{N}_{1,2}$ and $\mathbb{N}_{2,2}$ among all possible sets of the form $\mathbb{N}_{k,n}$ from (3) because we have only two finite numerals, "1" and "2", in $\mathcal{P}$. This numeral system is too weak to define other sets of this type, for instance, $\mathbb{N}_{4,5}$, because numbers greater than 2 required for these definition are not expressible in $\mathcal{P}$. These limitations have a general character and are related to all questions requiring a numerical answer (i.e., an answer expressed only in numerals, without variables). In order to obtain such an answer, it is necessary to know at least one numeral system able to express numerals required to write down this answer.

We are ready now to formulate the following important result being a direct consequence of the accepted methodological postulates.

Theorem 1. *The set* $\mathbb{N}$ *is not a monoid under addition.*

Proof. Due to Postulate 3, the operation $①+1$ gives us the result a number greater than ①. Thus, by definition of grossone, $①+1$ does not belong to $\mathbb{N}$ and, therefore, $\mathbb{N}$ is not closed under addition and is not a monoid. □

This result also means that adding the IUA to the axioms of natural numbers defines the set of *extended natural numbers* indicated as $\hat{\mathbb{N}}$ and including $\mathbb{N}$ as a proper subset

$$\hat{\mathbb{N}} = \{1,2,\ldots,①-1,①,①+1,\ldots,①^2-1,①^2,①^2+1,\ldots\}. \quad (11)$$

The extended natural numbers greater than grossone are also linked to sets of numbers and can be interpreted in the terms of grain.

Example 5. Let us determine the number of elements of the set

$$C_m = \{(a_1,a_2,\ldots,a_{m-1},a_m) : a_i \in \mathbb{N}, 1 \le i \le m\}, \quad 2 \le m \le ①.$$

The elements of C_m are m-tuples of natural numbers. It is known from combinatorial calculus that if we have m positions and each of them can be filled in by one of l symbols, the number of the obtained m-tuples is equal to l^m. In our case, since $\mathbb{N}$ has grossone elements, $l = ①$. Thus, the set C_m has $①^m$ elements. In the particular case, $m = 2$, we obtain that the set

$$C_2 = \{(a_1, a_2) : a_i \in \mathbb{N}, i \in \{1, 2\}\},$$

being the set of couples of natural numbers, has $①^2$ elements. These couples are shown below

$$\begin{array}{ccccc}
(1,1), & (1,2), & \dots & (1,①-1), & (1,①), \\
(2,1), & (2,2), & \dots & (2,①-1), & (2,①), \\
\dots & \dots & \dots & \dots & \dots \\
(①-1,1), & (①-1,2), & \dots & (①-1,①-1), & (①-1,①), \\
(①,1), & (①,2), & \dots & (①,①-1), & (①,①).
\end{array}$$

Another interesting particular case is the set

$$C_① = \{(a_1, a_2, \dots, a_{①-1}, a_①) : a_i \in \mathbb{N}, 1 \leq i \leq ①\}$$

having $①^①$ elements.

Note that we can also give the granary interpretation for the numbers of the type $①^m$: if we accept that the numbers K_i from page 13 are such that $K_i = ①, 1 \leq i \leq m-1$, then $①^2$ can be viewed as the number of seeds in the truck, $①^3$ as the number of seeds in the train waggon, etc. □

The set, $\hat{\mathbb{Z}}$, of *extended integer numbers* can be constructed from the set, $\mathbb{Z}$, of integer numbers by a complete analogy and inverse elements with respect to addition are introduced naturally. For example, $7①$ has its inverse with respect to addition equal to $-7①$.

It is important to notice that, due to Postulates 1 and 2, the new system of counting cannot give answers to *all* questions regarding infinite sets. What can we say, for instance, about the number of elements of the sets $\hat{\mathbb{N}}$ and $\hat{\mathbb{Z}}$? The introduced numeral system based on ① is too weak to give answers to these questions. It is necessary to introduce in a way a more powerful numeral system by defining new numerals (for instance, ②, ③, etc).

We conclude this section by the following remark. The IUA introduces a new number—the quantity of elements in the set of natural numbers—expressed by the new numeral ①. However, other numerals and sets can be used to state the idea of the axiom. For example, the numeral ❶ can be introduced as the number of elements of the set, $\mathbb{E}$, of even numbers and can be taken as the base of a numeral system. In this case, the IUA can be reformulated using the numeral ❶ and numerals using it will be used to express infinite numbers. For example, the number of elements of the set, $\mathbb{O}$, of odd numbers will be expressed as $|\mathbb{O}| = |\mathbb{E}| = \text{❶}$ and $|\mathbb{N}| = 2\cdot \text{❶}$. We emphasize through this note that infinite numbers (similarly to the finite ones) can be expressed by various numerals and in different numeral systems.

4 Arithmetical Operations in the New Numeral System

We have already started to write down simple infinite numbers and to execute arithmetical operations with them without concentrating our attention upon this question. Let us consider it systematically.

4.1 Positional Numeral System with Infinite Radix

Different numeral systems have been developed to describe finite numbers. In positional numeral systems, fractional numbers are expressed by the record

$$(a_n a_{n-1} \dots a_1 a_0 \cdot a_{-1} a_{-2} \dots a_{-(q-1)} a_{-q})_b, \qquad (12)$$

where numerals a_i, $-q \leq i \leq n$, are called *digits*, belong to the alphabet $\{0, 1, \dots, b-1\}$, and the dot is used to separate the fractional part from the integer one. Thus, the numeral (12) is equal to the sum

$$a_n b^n + a_{n-1} b^{n-1} + \dots + a_1 b^1 + a_0 b^0 + a_{-1} b^{-1} + \dots + a_{-(q-1)} b^{-(q-1)} + a_{-q} b^{-q}. \qquad (13)$$

Record (12) uses numerals consisting of one symbol each, i.e., digits $a_i \in \{0, 1, \dots, b-1\}$, to express how many finite units of the type b^i belong to the number (13). Quantities of finite units b^i are counted separately for each exponent i and all symbols in the alphabet $\{0, 1, \dots, b-1\}$ express finite numbers.

To express infinite and infinitesimal numbers we shall use records that are similar to (12) and (13) but have some peculiarities. In order to construct a number C in the new numeral positional system with base ①, we subdivide C into groups corresponding to powers of ①:

$$C = c_{p_m} ①^{p_m} + \dots + c_{p_1} ①^{p_1} + c_{p_0} ①^{p_0} + c_{p_{-1}} ①^{p_{-1}} + \dots + c_{p_{-k}} ①^{p_{-k}}. \qquad (14)$$

Then, the record

$$C = c_{p_m} ①^{p_m} \dots c_{p_1} ①^{p_1} c_{p_0} ①^{p_0} c_{p_{-1}} ①^{p_{-1}} \dots c_{p_{-k}} ①^{p_{-k}} \qquad (15)$$

represents the number C, where all numerals $c_i \neq 0$, they belong to a traditional numeral system and are called *grossdigits*. They express finite positive or negative numbers and show how many corresponding units ①^{p_i} should be added or subtracted in order to form the number C. Grossdigits can be expressed by several symbols using positional systems, the form $\frac{Q}{q}$ where Q and q are integer numbers, or in any other finite numeral system.

Numbers p_i in (15) called *grosspowers* can be finite, infinite, and infinitesimal (the introduction of infinitesimal numbers will be given soon), they are sorted in the decreasing order

$$p_m > p_{m-1} > \cdots > p_1 > p_0 > p_{-1} > \cdots p_{-(k-1)} > p_{-k}$$

with $p_0 = 0$.

In the traditional record (12), there exists a convention that a digit a_i shows how many powers b^i are present in the number and the radix b is not written explicitly. In the record (15), we write ①^{p_i} explicitly because in the new numeral positional system the number i in general is not equal to the grosspower p_i. This gives possibility to write, for example, such a number as $7.6$①$^{244.5}\,34$①^{32} having grosspowers $p_2 = 244.5, p_1 = 32$ and grossdigits $c_{244.5} = 7.6, c_{32} = 34$ without indicating grossdigits equal to zero corresponding to grosspowers less than 244.5 and greater than 32. Note also that if a grossdigit $c_{p_i} = 1$, then we often write ①^{p_i} instead of 1①^{p_i}.

The term having $p_0 = 0$ represents the finite part of C because, due to (2), we have $c_0$①$^0 = c_0$. The terms having finite positive grosspowers represent the simplest infinite parts of C. Analogously, terms having negative finite grosspowers represent the simplest infinitesimal parts of C. For instance, the number ①$^{-1} = \frac{1}{①}$ is infinitesimal. It is the inverse element with respect to multiplication for ①:

$$①^{-1} \cdot ① = ① \cdot ①^{-1} = 1. \tag{16}$$

Note that all infinitesimals are not equal to zero. Particularly, $\frac{1}{①} > 0$ because it is a result of division of two positive numbers. It also has a clear granary interpretation. Namely, if we have a sack containing ① seeds, then one sack divided by the number of seeds in it is equal to one seed. Vice versa, one seed, i.e., $\frac{1}{①}$, multiplied by the number of seeds in the sack, ①, gives one sack of seeds.

All of the numbers introduced above can be grosspowers, as well, giving so a possibility to have various combinations of quantities and to construct terms having a more complex structure.[7]

[7] At the first glance the record (14) [and, therefore, the numerals (15)] can remind numbers from the Levi–Civita field (see [20]) that is a very interesting and important precedent of algebraic manipulations with infinities and infinitesimals. However, the two mathematical objects have several crucial differences. They have been introduced for different purposes by using two mathematical languages having different accuracies and on the basis of different methodological foundations. In fact, Levi–Civita does not discuss the distinction between numbers and numerals and works with generic numbers while each numeral (15) represents a concrete number. His numbers have neither cardinal nor ordinal properties; they are built using a generic infinitesimal

Example 6. The left-hand expression below shows how to write down numbers in the new numeral system and the right-hand shows how the value of the number is calculated:

$$15①^{1.4①}(-17.2045)①^{3}7①^{0}52.1①^{-6} = 15①^{1.4①} - 17.2045①^{3} + 7①^{0} + 52.1①^{-6}.$$

The number above has one infinite part having the infinite grosspower, one infinite part having the finite grosspower, a finite part, and an infinitesimal part. □

Finally, numbers having a finite and infinitesimal parts can be also expressed in the new numeral system, for instance, the number $-3.5①^{0}(-37)①^{-2}11①^{-15①+2.3}$ has a finite and two infinitesimal parts, the second of them has the infinite negative grosspower equal to $-15①+2.3$.

4.2 Arithmetical Operations

We start the description of arithmetical operations for the new positional numeral system by the operation of *addition* (*subtraction* is a direct consequence of addition and is thus omitted) of two given infinite numbers A and B, where

$$A = \sum_{i=1}^{K} a_{k_i}①^{k_i}, \qquad B = \sum_{j=1}^{M} b_{m_j}①^{m_j}, \qquad C = \sum_{i=1}^{L} c_{l_i}①^{l_i}, \tag{17}$$

and the result $C = A+B$ is constructed by including in it all items $a_{k_i}①^{k_i}$ from A such that $k_i \neq m_j, 1 \le j \le M$, and all items $b_{m_j}①^{m_j}$ from B such that $m_j \neq k_i, 1 \le i \le K$. If in A and B there are items such that $k_i = m_j$, for some i and j, then this grosspower k_i is included in C with the grossdigit $b_{k_i} + a_{k_i}$, i.e., as $(b_{k_i} + a_{k_i})①^{k_i}$.

Example 7. We consider two infinite numbers A and B, where

$$A = 16.5①^{44.2}(-12)①^{12}17①^{0}, \qquad B = 6.23①^{3}10.1①^{0}15①^{-4.1}.$$

Their sum C is calculated as follows:

$$\begin{aligned} C = A+B &= 16.5①^{44.2} + (-12)①^{12} + 17①^{0} + 6.23①^{3} + 10.1①^{0} + 15①^{-4.1} \\ &= 16.5①^{44.2} - 12①^{12} + 6.23①^{3} + 27.1①^{0} + 15①^{-4.1} \\ &= 16.5①^{44.2}(-12)①^{12}6.23①^{3}27.1①^{0}15①^{-4.1}. \end{aligned}$$

□

---

and only its rational powers are allowed; he uses symbol ∞ in his construction; there is no numeral system that would allow one to assign numerical values to his numbers; it is not explained how it would be possible to pass from d a generic infinitesimal h to a concrete one (see also the discussion above on the distinction between numbers and numerals).

In no way the said above should be considered as a criticism with respect to results of Levi–Civita. The above discussion has been introduced in this text just to underline that we are in front of two different mathematical tools that should be used in different mathematical contexts.

The operation of *multiplication* of two numbers A and B in the form (17) returns, as the result, the infinite number C constructed as follows:

$$C = \sum_{j=1}^{M} C_j, \qquad C_j = b_{m_j} ①^{m_j} \cdot A = \sum_{i=1}^{K} a_{k_i} b_{m_j} ①^{k_i+m_j}, \qquad 1 \le j \le M. \tag{18}$$

Example 8. We consider two infinite numbers

$$A = 1①^{18}(-5)①^{2.4}(-3)①^{1}, \qquad B = -1①^{1}0.7①^{-3}$$

and calculate the product $C = B \cdot A$. The first partial product C_1 is equal to

$$\begin{aligned} C_1 &= 0.7①^{-3} \cdot A = 0.7①^{-3}(①^{18} - 5①^{2.4} - 3①^{1}) \\ &= 0.7①^{15} - 3.5①^{-0.6} - 2.1①^{-2} = 0.7①^{15}(-3.5)①^{-0.6}(-2.1)①^{-2}. \end{aligned}$$

The second partial product, C_2, is computed analogously

$$C_2 = -①^{1} \cdot A = -①^{1}(①^{18} - 5①^{2.4} - 3①^{1}) = -①^{19}5①^{3.4}3①^{2}.$$

Finally, the product C is equal to

$$C = C_1 + C_2 = -1①^{19}0.7①^{15}5①^{3.4}3①^{2}(-3.5)①^{-0.6}(-2.1)①^{-2}. \qquad \square$$

In the operation of *division* of a number C by a number B from (17), we obtain a result A and a reminder R (that can be also equal to zero), i.e., $C = A \cdot B + R$. The number A is constructed as follows. The first grossdigit a_{k_K} and the corresponding maximal exponent k_K are established from the equalities

$$a_{k_K} = c_{l_L} / b_{m_M}, \qquad k_K = l_L - m_M. \tag{19}$$

Then the first partial reminder R_1 is calculated as

$$R_1 = C - a_{k_K} ①^{k_K} \cdot B. \tag{20}$$

If $R_1 \neq 0$, then the number C is substituted by R_1 and the process is repeated with a complete analogy. The grossdigit $a_{k_{K-i}}$, the corresponding grosspower k_{K-i} and the partial reminder R_{i+1} are computed by formulae (21) and (22) obtained from (19) and (20) as follows: l_L and c_{l_L} are substituted by the highest grosspower n_i and the corresponding grossdigit r_{n_i} of the partial reminder R_i that, in turn, substitutes C:

$$a_{k_{K-i}} = r_{n_i} / b_{m_M}, \qquad k_{K-i} = n_i - m_M, \tag{21}$$

$$R_{i+1} = R_i - a_{k_{K-i}} ①^{k_{K-i}} \cdot B, \qquad i \ge 1. \tag{22}$$

The process stops when a partial reminder equal to zero is found (this means that the final reminder $R = 0$) or when a required accuracy of the result is reached.

Example 9. Let us divide the number $C = -10①^3 16①^0 42①^{-3}$ by the number $B = 5①^3 7$. For these numbers we have

$$l_L = 3, \quad m_M = 3, \quad c_{l_L} = -10, \quad b_{m_M} = 5.$$

It follows immediately from (19) that $a_{k_K}①^{k_K} = -2①^0$. The first partial reminder R_1 is calculated as

$$R_1 = -10①^3 16①^0 42①^{-3} - (-2①^0) \cdot 5①^3 7$$
$$= -10①^3 16①^0 42①^{-3} + 10①^3 14①^0 = 30①^0 42①^{-3}.$$

By a complete analogy we should construct $a_{k_{K-1}}①^{k_{K-1}}$ by rewriting (19) for R_1. By doing so we obtain equalities

$$30 = a_{k_{K-1}} \cdot 5, \quad 0 = k_{K-1} + 3$$

and, as the result, $a_{k_{K-1}}①^{k_{K-1}} = 6①^{-3}$. The second partial reminder is

$$R_2 = R_1 - 6①^{-3} \cdot 5①^3 7 = 30①^0 42①^{-3} - 30①^0 42①^{-3} = 0.$$

Thus, we can conclude that the reminder $R = R_2 = 0$ and the final result of division is $A = -2①^0 6①^{-3}$.

Let us now substitute the grossdigit 42 by 40 in C and divide this new number $\tilde{C} = -10①^3 16①^0 40①^{-3}$ by the same number $B = 5①^3 7$. This operation gives us the same result $\tilde{A}_2 = A = -2①^0 6①^{-3}$ (where subscript 2 indicates that two partial reminders have been obtained) but with the reminder $\tilde{R} = \tilde{R}_2 = -2①^{-3}$. Thus, we obtain $\tilde{C} = B \cdot \tilde{A}_2 + \tilde{R}_2$. If we want to continue the procedure of division, we obtain $\tilde{A}_3 = -2①^0 6①^{-3}(-0.4)①^{-6}$ with the reminder $\tilde{R}_3 = 0.28①^{-6}$. Naturally, it follows $\tilde{C} = B \cdot \tilde{A}_3 + \tilde{R}_3$. The process continues until a partial reminder $\tilde{R}_i = 0$ is found or when a required accuracy of the result will be reached. □

A working software simulator of the Infinity Computer has been implemented and the first application—the Infinity Calculator—has been realized. Figure 1 shows operation of multiplication executed at the Infinity Calculator that works using the Infinity Computer technology. The left operand has two infinitesimal parts and the right operand has an infinite part and a finite one.

We conclude this section by emphasizing the following important issue: the Infinity Computer works with infinite, finite, and infinitesimal numbers *numerically*, not symbolically (see [41]).

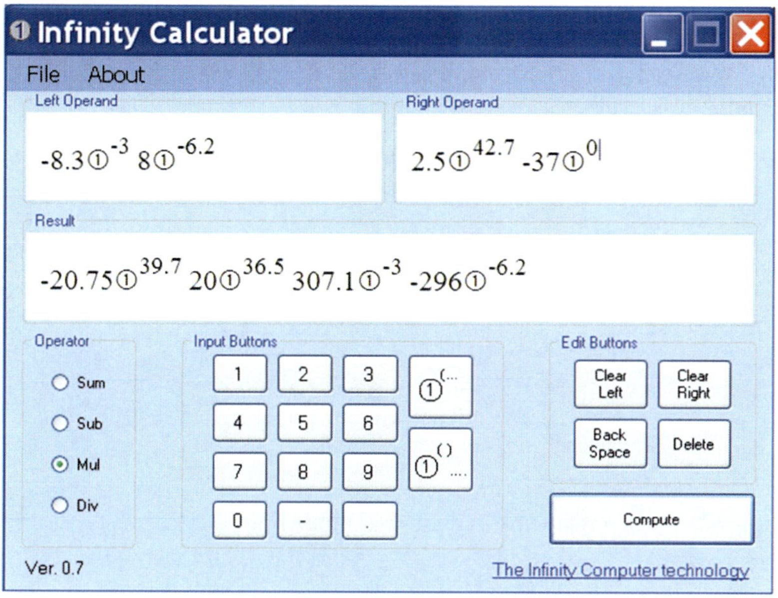

Fig. 1 Operation of multiplication executed at the Infinity Calculator

5 Examples of Problems Where Computations with New Numerals Can Be Useful

5.1 The Work with Infinite Sequences

We start by reminding traditional definitions of the infinite sequences and subsequences. An *infinite sequence* $\{a_n\}, a_n \in A, n \in \mathbb{N}$, is a function having as the domain the set of natural numbers, $\mathbb{N}$, and as the codomain a set A. A *subsequence* is a sequence from which some of its elements have been removed. In a sequence $a_1, a_2, \ldots, a_n$ the number n is the number of elements of the sequence. Then, the IUA allows us to consider sequences having n that can assume different finite or infinite values and to prove the following result.

Theorem 2. *The number of elements of any infinite sequence is less or equal to ①.*

Proof. The IUA states that the set $\mathbb{N}$ has ① elements. Thus, due to the sequence definition given above, any sequence having $\mathbb{N}$ as the domain has ① elements.

The notion of subsequence is introduced as a sequence from which some of its elements have been removed. Thus, this definition gives infinite sequences having the number of members less than grossone. □

One of the immediate consequences of the understanding of this result is that any sequential process can have at maximum ① elements. Due to Postulate 1, it depends on the chosen numeral system which numbers among ① members of the process we can observe.

Example 10. For example, if we consider the set, $\hat{\mathbb{N}}$, of extended natural numbers, then starting from the number 1, it is possible to arrive at maximum to ①

$$\underbrace{1,2,3,4,\ \ldots\ ①-2,\ ①-1,①}_{①},①+1,①+2,①+3,\ldots \tag{23}$$

Starting from 2 it is possible to arrive at maximum to $①+1$

$$1,\underbrace{2,3,4,\ \ldots\ ①-2,\ ①-1,①,①+1}_{①},①+2,①+3,\ldots \tag{24}$$

Starting from 3 it is possible to arrive at maximum to $①+2$

$$1,2,\underbrace{3,4,\ \ldots\ ①-2,\ ①-1,①,①+1,①+2}_{①},①+3,\ldots \tag{25}$$

Of course, since we have postulated that our possibilities to express numerals are finite, it depends on the chosen numeral system which numbers among ① members of these processes we can observe. □

It is also very important to notice a deep relation of this observation to the Axiom of Choice. The IUA postulates that any process can have at maximum ① elements, thus the process of choice too and, as a consequence, it is not possible to choose more than ① elements from a set. This observation also emphasizes the fact that the parallel computational paradigm is significantly different with respect to the sequential one because p parallel processes can choose p① elements from a set. Note also that the new more precise definition of sequences allows us to obtain a new vision of Turing machines (see [47]).

It becomes appropriate now to define the *complete sequence* as an infinite sequence containing ① elements. For example, the sequence of natural numbers is complete, the sequences of even and odd natural numbers are not complete. Thus, the IUA imposes a more precise description of infinite sequences. To define a sequence $\{a_n\}$ it is not sufficient just to give a formula for a_n, we should determine (as it happens for sequences having a finite number of elements) the first and the last elements of the sequence. If the number of the first element is equal to one, we can use the record $\{a_n : k\}$ where a_n is, as usual, the general element of the sequence and k is the number (that can be finite or infinite) of members of the sequence.

Example 11. Let us consider the following two sequences, $\{a_n\}$ and $\{c_n\}$:

$$\{a_n\} = \{5, \quad 10, \quad \ldots \quad 5(①-1), \quad 5①\},$$

$$\{b_n\} = \left\{5, \quad 10, \quad \ldots \quad 5\left(\frac{2①}{5}-1\right), \quad 5\cdot\frac{2①}{5}\right\}, \tag{26}$$

$$\{c_n\} = \left\{5, \quad 10, \quad \ldots \quad 5\left(\frac{4①}{5}-1\right), \quad 5\cdot\frac{4①}{5}\right\}. \tag{27}$$

They have the same general element $a_n = b_n = c_n = 5n$ but they are different because they have different numbers of members. The first sequence has ① elements and is thus complete, the other two sequences are not complete: $\{b_n\}$ has $\frac{2①}{5}$ elements and $\{c_n\}$ has $\frac{4①}{5}$ members. □

In connection with this definition the following natural question arises inevitably. Suppose that we have two sequences, for example, $\{b_n : \frac{2①}{5}\}$ and $\{c_n : \frac{4①}{5}\}$ from (26) and (27). Can we create a new sequence, $\{d_n : k\}$, composed from both of them, for instance, as it is shown below

$$b_1, \; b_2, \; \ldots \; b_{\frac{2①}{5}-2}, \; b_{\frac{2①}{5}-1}, \; b_{\frac{2①}{5}}, \; c_1, \; c_2, \; \ldots \; c_{\frac{4①}{5}-2}, \; c_{\frac{4①}{5}-1}, \; c_{\frac{4①}{5}}$$

and which will be the value of the number of its elements k?

The answer is "no" because due to the definition of the infinite sequence, a sequence can be at maximum complete, i.e., it cannot have more than ① elements. Starting from the element b_1 we can arrive at maximum to the element $c_{\frac{3①}{5}}$ being the element number ① in the sequence $\{d_n : k\}$ which we try to construct. Therefore, $k = ①$ and

$$\underbrace{b_1, \; \ldots \; b_{\frac{2①}{5}}, \; c_1, \; \ldots c_{\frac{3①}{5}}}_{①\text{ elements}}, \; \underbrace{c_{\frac{3①}{5}+1}, \ldots \; c_{\frac{4①}{5}}}_{\frac{①}{5}\text{ elements}}.$$

The remaining members of the sequence $\{c_n : \frac{4①}{5}\}$ will form the second sequence, $\{g_n : l\}$ having $l = \frac{4①}{5} - \frac{3①}{5} = \frac{①}{5}$ elements. Thus, we have formed two sequences, the first of them is complete and the second is not.

To conclude this subsection, let us return to Hilbert's paradox of the Grand Hotel presented in Sect. 1. In the paradox, the number of the rooms in the Hotel is countable. In our terminology this means that it has ① rooms. When a new guest arrives, it is proposed to move the guest occupying room 1 to room 2, the guest occupying room 2 to room 3, etc. Under the IUA this procedure does not help because the guest from room ① should be moved to room $①+1$ and the Hotel has only ① rooms. Thus, when the Hotel is full, no more new guests can be accommodated—the result corresponding perfectly to Postulate 3 and the situation taking place in normal hotels with a finite number of rooms.

5.2 *From Divergent Series to Expressions Evaluated at Different Points in Infinity*

Let us show how the new approach can be applied in such an important area as theory of divergent series. We consider two infinite series $S_1 = 7+7+7+\cdots$ and $S_2 = 3+3+3+\cdots$ The traditional analysis gives us a very poor answer that both of them diverge to infinity. Such operations as, $\frac{S_2}{S_1}$ and $S_2 - S_1$ are not defined.

Now, when we are able to express not only different finite numbers but also different infinite numbers such records as $S_1 = a_1 + a_2 + \cdots$ or $\sum_{i=1}^{\infty} a_i$ become unprecise (by continuation the analogy with Pirahã the record $\sum_{i=1}^{\infty} a_i$ becomes a kind of $\sum_{i=1}^{\text{many}} a_i$). It is therefore necessary to indicate explicitly the number of items in the sums S_1 and S_2 and it is not important whether k is finite or infinite.

We emphasize again that in order to be able to calculate a sum it is necessary that the number of items and the result are expressible in the numeral system used for calculations. It is important to notice that even though a sequence cannot have more than ① elements, the number of items in a series can be greater than grossone because the process of summing up is not necessary executed by a sequential adding items.

Example 12. Let us consider the infinite series S_1 and S_2 mentioned above. In order to use our approach, it is necessary to indicate explicitly the number of their items.

Suppose that the sum S_1 has k items and S_2 has n items:

$$S_1(k) = \underbrace{7+7+7+\cdots+7}_{k}, \qquad S_2(n) = \underbrace{3+3+3+\cdots+3}_{n}.$$

Then $S_1(k) = 7k$ and $S_2(n) = 3n$ and by giving different numerical values (finite or infinite) to k and n we obtain different numerical values for the sums. For chosen k and n it becomes possible to calculate $S_2(n) - S_1(k)$ (analogously, the expression $\frac{S_1(k)}{S_2(n)}$ can be calculated). If, for instance, $k = 5①$ and $n = ①$, we obtain $S_1(5①) = 35①$, $S_2(①) = 3①$ and it follows

$$S_2(①) - S_1(5①) = 3① - 35① = -32① < 0.$$

If $k = 3①$ and $n = 7①+2$, we obtain $S_1(3①) = 21①$, $S_2(①) = 21①+6$ and it follows

$$S_2(7①+2) - S_1(3①) = 21①+6-21① = 6.$$

It is also possible to sum up sums having an infinite number of infinite or infinitesimal items

$$S_3(l) = \underbrace{2①+2①+\cdots+2①}_{l}, \qquad S_4(m) = \underbrace{4①^{-1}+4①^{-1}+\cdots+4①^{-1}}_{m}.$$

For $l = m = 0.5①$ it follows $S_3(0.5①) = ①^2$ and $S_4(0.5①) = 2$ [remind that $① \cdot ①^{-1} = ①^0 = 1$ (see (16))]. It can be seen from this example that it is possible to obtain finite numbers as the result of summing up infinitesimals. This is a direct consequence of Postulate 3. □

The infinite and infinitesimal numbers allow us to calculate also arithmetic and geometric sums with an infinite number of items. Traditional approaches tell us that if $a_n = a_1 + (n-1)d$ then for a finite n it is possible to use the formula

$$\sum_{i=1}^{n} a_i = \frac{n}{2}(a_1 + a_n).$$

Due to Postulate 3, we can use it also for infinite n.

Example 13. The sum of all natural numbers from 1 to ① can be calculated as follows

$$1+2+3+\cdots+(①-1)+① = \sum_{i=1}^{①} i = \frac{①}{2}(1+①) = 0.5①^2 0.5①. \qquad (28)$$

Let us calculate now the following sum of infinitesimals where each item is ① times less than the corresponding item of (28)

$$①^{-1}+2①^{-1}+\cdots+(①-1)\cdot①^{-1}+①\cdot①^{-1} = \sum_{i=1}^{①} i①^{-1} = \frac{①}{2}(①^{-1}+1) = 0.5①^1 0.5.$$

Obviously, the obtained number, $0.5①^1 0.5$ is ① times less than the sum in (28). This example shows, particularly, that infinite numbers can also be obtained as the result of summing up infinitesimals. □

Let us consider now the geometric series $\sum_{i=0}^{\infty} q^i$. Traditional analysis proves that it converges to $\frac{1}{1-q}$ for q such that $-1 < q < 1$. We are able to give a more precise answer for *all* values of q. To do this we should fix the number of items in the sum. If we suppose that it contains n items, then

$$Q_n = \sum_{i=0}^{n} q^i = 1 + q + q^2 + \cdots + q^n. \qquad (29)$$

By multiplying the left-hand and the right-hand parts of this equality by q and by subtracting the result from (29) we obtain

$$Q_n - qQ_n = 1 - q^{n+1}$$

and, as a consequence, for all $q \neq 1$ the formula

$$Q_n = (1-q^{n+1})(1-q)^{-1} \qquad (30)$$

holds for finite and infinite n. Thus, the possibility to express infinite and infinitesimal numbers allows us to take into account infinite n and the value q^{n+1} being infinitesimal for a finite q. Moreover, we can calculate Q_n for infinite and finite values of n and $q = 1$, because in this case we have just

$$Q_n = \underbrace{1+1+1+\cdots+1}_{n+1} = n+1.$$

Example 14. As the first example we consider the divergent series

$$1+3+9+\cdots = \sum_{i=0}^{\infty} 3^i.$$

To fix it, we should decide the number of items, n, at the sum and, for example, for $n = ①^2$ we obtain

$$\sum_{i=0}^{①^2} 3^i = 1+3+9+\cdots+3^{①^2} = \frac{1-3^{①^2+1}}{1-3} = 0.5(3^{①^2+1}-1).$$

Analogously, for $n = ①^2+1$ we obtain

$$1+3+9+\cdots+3^{①^2}+3^{①^2+1} = 0.5(3^{①^2+2}-1).$$

If we now find the difference between the two sums

$$0.5(3^{①^2+2}-1)-(0.5(3^{①^2+1}-1)) = 3^{①^2+1}(0.5\cdot 3-0.5) = 3^{①^2+1},$$

we obtain the newly added item $3^{①^2+1}$. □

Example 15. In this example, we consider the series $\sum_{i=1}^{\infty}\frac{1}{2^i}$. It is well known that it converges to one. However, we are able to give a more precise answer. In fact, due to Postulate 3, the formula

$$\sum_{i=1}^{n}\frac{1}{2^i} = \frac{1}{2}\left(1+\frac{1}{2}+\frac{1}{2^2}+\cdots+\frac{1}{2^{n-1}}\right) = \frac{1}{2}\cdot\frac{1-\frac{1}{2}^n}{1-\frac{1}{2}} = 1-\frac{1}{2^n}$$

can be used directly for infinite n, too. For example, if $n = ①$, then

$$\sum_{i=1}^{①}\frac{1}{2^i} = 1-\frac{1}{2^{①}},$$

where $\frac{1}{2^{①}}$ is infinitesimal. Thus, the traditional answer $\sum_{i=1}^{\infty} \frac{1}{2^i} = 1$ was just a finite approximation to our more precise result using infinitesimals. □

Example 16. In this example, we consider divergent series with alternate signs. Let us start from the famous series

$$S_5 = 1 - 1 + 1 - 1 + 1 - 1 + \cdots$$

In literature there exist many approaches giving different answers regarding the value of this series (see [18]). All of them use various notions of average. However, the notions of sum and average are different. In our approach we do not appeal to average and calculate the required sum directly. To do this we should indicate explicitly the number of items, k, in the sum. Then

$$S_5(k) = \underbrace{1 - 1 + 1 - 1 + 1 - 1 + 1 - \ldots}_{k} = \begin{cases} 0, & \text{if } k = 2n, \\ 1, & \text{if } k = 2n + 1, \end{cases}$$

and it is not important whether k is finite or infinite. For example, $S_5(①) = 0$ because the number $\frac{①}{2}$ being the result of division of ① by 2 has been introduced as the number of elements of a set and, therefore, it is integer. As a consequence, ① is even number. Analogously, $S_5(① - 1) = 1$ because $① - 1$ is odd. □

It is important to emphasize that, as it happens in the case of the finite number of items in a sum, the obtained answers do not depend on the way the items in the entire sum are rearranged. In fact, if we know the exact infinite number of items in the sum and the order of alternating the signs is clearly defined, we know also the exact number of positive and negative items in the sum.

Let us illustrate this point by supposing, for instance, that we want to re-arrange the items in the sum $S_1(2①)$ in the following way

$$S_1(2①) = 1 + 1 - 1 + 1 + 1 - 1 + 1 + 1 - 1 + \cdots$$

However, we know that the sum has 2① items and the number 2① is even. This means that in the sum there are ① positive and ① negative items. As a result, the re-arrangement considered above can continue only until the positive items present in the sum will not finish and then it will be necessary to continue to add only negative numbers. More precisely, we have

$$S_1(2①) = \underbrace{1 + 1 - 1 + 1 + 1 - 1 + \cdots + 1 + 1 - 1}_{① \text{ positive and } \frac{①}{2} \text{ negative items}} \underbrace{-1 - 1 - \cdots - 1 - 1 - 1}_{\frac{①}{2} \text{ negative items}} = 0,$$

where the result of the first part in this rearrangement is calculated as $(1 + 1 - 1) \cdot \frac{①}{2} = \frac{①}{2}$ and the result of the second part is equal to $-\frac{①}{2}$.

Example 17. Let us consider now the following divergent series

$$S_6 = 1 - 2 + 3 - 4 + \cdots$$

It can be easily considered as the difference of two arithmetic progressions after we have fixed the number of items, k, in the sum $S_6(k)$. Suppose that it contains grossone items. Then it follows

$$\begin{aligned} S_6(①) &= 1 - 2 + 3 - 4 + \cdots - (① - 2) + (① - 1) - ① \\ &= (1 + 3 + 5 + \cdots + (① - 3) + (① - 1)) - (2 + 4 + 6 + \cdots + (① - 2) + ①) \\ &= \frac{(1 + ① - 1)①}{4} - \frac{(2 + ①)①}{4} = \frac{①^2 - 2① - ①^2}{4} = -\frac{①}{2}. \end{aligned}$$ □

5.3 *Calculating Limits and Expressing Irrational Numbers*

Let us now discuss the problem of calculation of limits from the point of view of our approach. In traditional analysis, if a limit $\lim_{x \to a} f(x)$ exists, then it gives us a very poor—just one value—information about the behavior of $f(x)$ when x tends to a. Now we can obtain significantly richer information because we are able to calculate $f(x)$ directly at any finite, infinite, or infinitesimal point that can be expressed by the new positional system even if the limit does not exist.

Thus, limits equal to infinity can be substituted by precise infinite numerals and limits equal to zero can be substituted by precise infinitesimal numerals.[8] This is very important for practical computations because these substitutions eliminate indeterminate forms.

Example 18. Let us consider the following two limits

$$\lim_{x \to +\infty} (5x^3 - x^2 + 10^{61}) = +\infty, \qquad \lim_{x \to +\infty} (5x^3 - x^2) = +\infty.$$

Both give us the same result, $+\infty$, and it is not possible to execute the operation

$$\lim_{x \to +\infty} (5x^3 - x^2 + 10^{61}) - \lim_{x \to +\infty} (5x^3 - x^2).$$

that is an indeterminate form of the type $\infty - \infty$ in spite of the fact that for any finite x it follows

$$5x^3 - x^2 + 10^{61} - (5x^3 - x^2) = 10^{61}. \tag{31}$$

[8] Naturally, if we speak about limits of sequences, $\lim_{n \to \infty} a(n)$, then $n \in \mathbb{N}$ and, as a consequence, it follows that n should be less than or equal to grossone.

The new approach allows us to calculate exact values of both expressions, $5x^3 - x^2 + 10^{61}$ and $5x^3 - x^2 + 10$, at any infinite (and infinitesimal) x expressible in the chosen numeral system. For instance, the choice $x = 3①^2$ gives the value

$$5(3①^2)^3 - (3①^2)^2 + 10^{61} = 135①^6\text{-}9①^4 10^{61}$$

for the first expression and $135①^6$-$9①^4$ for the second one. We can easily calculate the difference of these two infinite numbers, thus obtaining the same result as we had for finite values of x in (31):

$$135①^6 - 9①^4 10^{61} - (135①^6 - 9①^4) = 10^{61}. \qquad \square$$

An additional advantage of the usage of the Infinity Computer for calculating limits arises in the following situations. Suppose that we have a computer procedure calculating $f(x)$, we do not know the corresponding analytic formulae for $f(x)$, for a certain argument a the value $f(a)$ is not defined (or a traditional computer produces an overflow or underflow message), and it is necessary to calculate the $\lim_{x \to a} f(x)$. Traditionally, this situation requires a human intervention and an additional theoretical investigation whereas the Infinity Computer is able to process it automatically working numerically with the expressions involved in the procedure. It is sufficient to calculate $f(x)$, for example, at a point $x = a + ①^{-1}$ in cases of finite a or $a = 0$ and $x = ①$ in the case when we are interested in the behavior of $f(x)$ at infinity. Obviously, if the limit does not exist but there exist limits from the right and from the left, it is sufficient to calculate $x = a + ①^{-1}$ and $x = a - ①^{-1}$, respectively.

Example 19. Suppose that we have two procedures evaluating $f(x) = \frac{x^2+2x}{x}$ and $g(x) = \frac{34}{x}$. Obviously, $f(0)$ and $g(0)$ are not defined and it is not possible to calculate $\lim_{x \to 0} f(x)$, $\lim_{x \to \infty} f(x)$ and $\lim_{x \to 0} g(x)$, $\lim_{x \to \infty} g(x)$ using traditional computers. Then, suppose that we are interested in evaluating the expression

$$h(x) = (f(x) - 2) \cdot g(x).$$

It is easy to see that $h(x) = 34$ for any finite value of x. On the other hand, the following limits

$$\lim_{x \to 0} h(x) = (\lim_{x \to 0} f(x) - 2) \cdot \lim_{x \to 0} g(x),$$

$$\lim_{x \to \infty} h(x) = (\lim_{x \to \infty} f(x) - 2) \cdot \lim_{x \to \infty} g(x)$$

cannot be evaluated. The Infinity Computer can calculate $h(x)$ numerically for different infinitesimal and infinite values of x obtaining the same result that takes place for finite x. For example, it follows

$$h(①^{-1}) = \left(\frac{(①^{-1})^2 + 2①^{-1}}{①^{-1}} - 2 \right) \cdot \frac{34}{①^{-1}} = (①^{-1} + 2 - 2) \cdot 34① = 34,$$

$$h(①) = \left(\frac{①^2 + 2①}{①} - 2 \right) \cdot \frac{34}{①} = (① + 2 - 2) \cdot 34①^{-1} = 34.$$ □

It is necessary to emphasize the fact that expressions can be calculated even when their limits do not exist. Thus, we obtain a very powerful tool for studying divergent processes.

Example 20. The limit $\lim_{n\to+\infty} f(n)$, $f(n) = (-1)^n n^3$, does not exist. However, we can easily calculate expression $(-1)^n n^3$ at different infinite points n. For instance, for $n = ①$ it follows $f(①) = ①^3$ because grossone is even and for the odd $n = 0.5① - 1$ it follows

$$f(0.5① - 1) = -(0.5① - 1)^3 = -0.125①^3 0.75①^2 - 1.5①^1 1.$$ □

Limits with the argument tending to zero can be considered analogously. In this case, we can calculate the corresponding expression at any infinitesimal point using the new positional system and obtain a significantly more reach information.

Example 21. If x is a fixed finite number, then

$$\lim_{h\to 0} \frac{(x+h)^2 - x^2}{h} = 2x. \tag{32}$$

In the new positional system we obtain

$$\frac{(x+h)^2 - x^2}{h} = 2x + h. \tag{33}$$

If, for instance, $h = ①^{-1}$, the answer is $2x①^0①^{-1}$, if $h = 4.2①^{-2}$ we obtain the value $2x①^0 4.2①^{-2}$, etc. Thus, the value of the limit (32), for a finite x, is just the finite approximation of the number (33) having finite and infinitesimal parts. □

Let us make a remark regarding irrational numbers. Among their properties, they are characterized by the fact that we do not know any numeral system that would allow us to express them by a finite number of symbols used to express other numbers. Thus, special numerals ($e, \pi, \sqrt{2}, \sqrt{3}$, etc.) are introduced by describing their properties in a way (similarly, all other numerals, e.g., symbols "0" or "1," are introduced also by describing their properties). These special symbols are then used in analytical transformations together with ordinary numerals.

For example, it is possible to work directly with the symbol e in analytical transformations by applying suitable rules defining this number together with numerals taking part in a chosen numeral system $\mathcal{S}$. At the end of transformations, the obtained result will be expressed in numerals from $\mathcal{S}$ and, probably, in terms of e. If it is then required to execute some *numerical* computations, this means that it is necessary to substitute e by a numeral (or numerals) from $\mathcal{S}$ that will allow us to approximate e in some way.

The same situation takes place when one uses the new numeral system, i.e., while we work analytically we use just the symbol e in our expressions and then, if we wish to work numerically we should pass to approximations. The new numeral system opens a new perspective on the problem of the expression of irrational numbers. Let us consider one of the possible ways to obtain an approximation of e, i.e., by using the limit

$$e = \lim_{n \to +\infty} \left(1 + \frac{1}{n}\right)^n = 2.71828182845904\ldots \tag{34}$$

In our numeral system the expression $(1 + \frac{1}{n})^n$ can be written directly for finite and/or infinite values of n. For $n = ①$ we obtain the number e_0 designated so in order to distinguish it from the record (34)

$$e_0 = \left(1 + \frac{1}{①}\right)^{①} = (①^0①^{-1})^{①}. \tag{35}$$

It becomes clear from this record why the number e cannot be expressed in a positional numeral system with a finite base. Due to the definition of a sequence under the IUA, such a system can have at maximum ① numerals—digits—to express fractional part of a number (see Sect. 5.5 for details) and, as it can be seen from (35), this quantity is not sufficient for e because the item $\frac{1}{①^{①}}$ is present in it.

Naturally, it is also possible to construct more exotic e-type numbers by substituting ① in (35) by any infinite number written in the new positional system with infinite base. For example, if we substitute ① in (35) by $①^2$, we obtain the number

$$e_1 = \left(1 + \frac{1}{①^2}\right)^{①^2} = (①^0①^{-2})^{①^2}.$$

The numbers considered above take their origins in the limit (34). Similarly, other formulae leading to approximations of e expressed in traditional numeral systems give us other new numbers that can be expressed in the new numeral system. The same way of reasoning can be used with respect to other irrational numbers, too.

5.4 Measuring Infinite Sets with Elements Defined by Formulae

We have already discussed in Sect. 3 how we calculate the number of elements for sets being results of the usual operations (intersection, union, etc.) with finite sets and infinite sets of the type $\mathbb{N}_{k,n}$. In order to have a possibility to work with infinite sets having a more general structure than the sets $\mathbb{N}_{k,n}$, we need to develop more powerful instruments. Suppose that we have an integer function $g(i) > 0$ strictly increasing on indexes $i = 1, 2, 3, \dots$ and we wish to know how many elements are there in the set

$$G = \{g(1), g(2), g(3), \dots\}.$$

In our terminology this question has no any sense because of the following reason.

In the finite case, to define a set it is not sufficient to say that it is finite. It is necessary to indicate its number of elements explicitly as, e.g., in this example

$$G_1 = \{g(i) : 1 \leq i \leq 5\},$$

or implicitly, as it is made here:

$$G_2 = \{g(i) : i \geq 1,\ \ 0 < f(i) \leq b\}, \tag{36}$$

where b is finite.

Now we have mathematical tools to indicate the number of elements for infinite sets, too. Thus, analogously to the finite case and due to Postulate 3, it is not sufficient to say that a set has infinitely many elements. It is necessary to indicate its number of elements explicitly or implicitly. For instance, the number of elements of the set

$$G_3 = \{g(i) : 1 \leq i \leq ①^{10}\}$$

is indicated explicitly: the set G_3 has $①^{10}$ elements.

If a set is given in the form (36) where b is infinite, then its number of elements, J, can be determined as

$$J = \max\{i : g(i) \leq b\} \tag{37}$$

if we are able to determine the inverse function $g^{-1}(x)$ for $g(x)$. Then, $J = [g^{-1}(b)]$, where $[u]$ is integer part of u. Note that if $b = ①$, then the set $G_2 \subseteq \mathbb{N}$ since all its elements are integer, positive, and $g(i) \leq ①$ due to (37).

Example 22. Let us consider the following set, $A_1(k,n)$, having $g(i) = k + n(i-1)$,

$$A_1(k,n) = \{g(i) : i \geq 1,\ \ g(i) \leq ①\}, \qquad 1 \leq k \leq n, \quad n \in \mathbb{N}.$$

It follows from the IUA that $A_1(k,n) = \mathbb{N}_{k,n}$ from (3). By applying (37) we find for $A_1(k,n)$ its number of elements

$$J_1(k,n) = [\tfrac{①-k}{n} + 1] = [\tfrac{①-k}{n}] + 1 = \tfrac{①}{n} - 1 + 1 = \tfrac{①}{n}. \qquad \square$$

Example 23. Analogously, the set

$$A_2(k,n,j) = \{k+ni^j : i \geq 0,\ 0 < k+ni^j \leq ①\}, \qquad 0 \leq k < n,\ \ n \in \mathbb{N},\ \ j \in \mathbb{N},$$

has $J_2(k,n,j) = \left[\sqrt[j]{\frac{①-k}{n}}\right]$ elements. □

5.5 *Measuring Infinite Sets of Numerals and Their Comparison*

Let us calculate the number of elements in some well-known infinite sets of numerals using the designation $|A|$ to indicate the number of elements of a set A.

Theorem 3. *The number of elements of the set,* $\mathbb{Z}$*, of integers is* $|\mathbb{Z}| = 2①1$.

Proof. The set $\mathbb{Z}$ contains ① positive numbers, ① negative numbers, and zero. Thus,

$$|\mathbb{Z}| = ① + ① + 1 = 2①1. \qquad □$$

Traditionally, rational numbers are defined as ratio of two integer numbers. The new approach allows us to calculate the number of numerals in a fixed numeral system. Let us consider a numeral system $\mathbb{Q}_1$ containing numerals of the form

$$\frac{p}{q}, \qquad p \in \mathbb{Z},\quad q \in \mathbb{Z},\ q \neq 0. \tag{38}$$

Theorem 4. *The number of elements of the set,* $\mathbb{Q}_1$*, of rational numerals of the type (38) is* $|\mathbb{Q}_1| = 4①^2 2①^1$.

Proof. It follows from Theorem 3 that the numerator of (38) can be filled in by 2①1 and the denominator by 2① numbers. Thus, the number of all possible combinations is

$$|\mathbb{Q}_1| = 2①1 \cdot 2① = 4①^2 2①^1. \qquad □$$

It is necessary to notice that in Theorem 4 we have calculated different numerals and not different numbers. For example, in the numeral system $\mathbb{Q}_1$ the number 0 can be expressed by 2① different numerals

$$\frac{0}{-①}, \frac{0}{-①+1}, \frac{0}{-①+2}, \ldots \frac{0}{-2}, \frac{0}{-1}, \frac{0}{1}, \frac{0}{2}, \ldots \frac{0}{①\text{-}2}, \frac{0}{①\text{-}1}, \frac{0}{①}$$

and numerals such as $\frac{-1}{-2}$ and $\frac{1}{2}$ have been calculated as two different numerals. The following theorem determines the number of elements of the set $\mathbb{Q}_2$ containing numerals of the form

$$-\frac{p}{q}, \frac{p}{q}, \qquad p \in \mathbb{N}, \quad q \in \mathbb{N}, \tag{39}$$

and zero is represented by one symbol 0.

Theorem 5. *The number of elements of the set, $\mathbb{Q}_2$, of rational numerals of the type (39) is $|\mathbb{Q}_2| = 2①^2 1$.*

Proof. Let us consider positive rational numerals. The form of the rational numeral $\frac{p}{q}$, the fact that $p, q \in \mathbb{N}$, and the IUA impose that both p and q can assume values from 1 to ①. Thus, the number of all possible combinations is $①^2$. The same number of combinations we obtain for negative rational numbers and one is added because we count zero as well. □

Let us now calculate the number of elements of the set, $\mathbb{R}_b$, of real numbers expressed by numerals in the positional system by the record

$$(a_{n-1}a_{n-2}\ldots a_1 a_0 \cdot a_{-1}a_{-2}\ldots a_{-(q-1)}a_{-q})_b \tag{40}$$

where the symbol b indicates the radix of the record and $n, q \in \mathbb{N}$.

Theorem 6. *The number of elements of the set, $\mathbb{R}_b$, of numerals (40) is $|\mathbb{R}_b| = b^{2①}$.*

Proof. In formula (40) defining the type of numerals we deal with there are two sequences of digits: the first one, $a_{n-1}a_{n-2}\ldots a_1 a_0$, is used to express the integer part of the number and the second, $a_{-1}a_{-2}\ldots a_{-(q-1)}a_{-q}$, for its fractional part. Due to definition of sequence and the IUA, each of them can have at maximum ① elements. Thus, it can be at maximum ① positions on the left of the dot and, analogously, ① positions on the right of the dot. Every position can be filled in by one of the b digits from the alphabet $\{0, 1, \ldots, b-1\}$. Thus, we have $b^{①}$ combinations to express the integer part of the number and the same quantity to express its fractional part. As a result, the positional numeral system using the numerals of the form (40) can express $b^{2①}$ numbers. □

Note that the result of Theorem 6 does not consider the practical situation of writing down concrete numerals. Obviously, the number of numerals of the type (40) that can be written in practice is finite and depends on the chosen numeral system for writing digits.

It is worthwhile to notice also that all the numerals of the type (40) represent different numbers. In addition, minimal and maximal numbers expressible in $\mathbb{R}_b$ can be explicitly indicated.

Example 24. For instance, in the decimal positional system $\mathbb{R}_{10}$ the numerals

$$1.\underbrace{999\ldots99}_{①\text{ digits}}, \quad 2.\underbrace{000\ldots00}_{①\text{ digits}}$$

represent different numbers and their difference is equal to

$$2.\underbrace{000\ldots00}_{\text{① digits}}-1.\underbrace{999\ldots9}_{\text{① digits}}=0.\underbrace{000\ldots01}_{\text{① digits}}.$$

Analogously the smallest and the largest numbers expressible in $\mathbb{R}_{10}$ can be easily indicated. They are, respectively,

$$-\underbrace{999\ldots9}_{\text{① digits}}.\underbrace{999\ldots9}_{\text{① digits}}, \qquad \underbrace{999\ldots9}_{\text{① digits}}.\underbrace{999\ldots9}_{\text{① digits}}. \qquad \square$$

On the other hand, the traditional point of view on real numbers tells that there exist real numbers that can be represented in positional systems by two different infinite sequences of digits, for instance, in the decimal positional system the records 2.000000... and 1.99999... represent the same number. Note that there is no contradiction between the traditional and the new points of view. They just use different lens in their mathematical microscopes to observe numbers. The instruments used in the traditional point of view for this purpose was just too weak to distinguish two different numbers in the records 2.000000... and 1.99999....

Note that traditionally it was accepted that *any* positional numeral system is able to represent *all* real numbers ("the whole real line"). In this section, we have shown that any numeral system is just an instrument that can be used to observe *certain* real numbers. This instrument can be more or less powerful, e.g., the positional system (40) with the radix 10 is more powerful than the positional system (40) with the radix 2 but neither of the two is able to represent irrational numbers. Two numeral systems can allow us to observe either the same sets of numbers, or sets of numbers having an intersection, or two disjoint sets of numbers. Due to Postulate 2, we are not able to answer the question "What is the whole real line?" because this is the question asking "What is the object of the observation?", we are able just to invent more and more powerful numeral systems that will allow us to improve our observations of numbers by using newly introduced numerals.

Theorem 7. *The sets $\mathbb{Z}, \mathbb{Q}_1, \mathbb{Q}_2$, and $\mathbb{R}_b$ are not monoids under addition.*

Proof. The proof is obvious and is so omitted. □

6 Relations to Results of Georg Cantor

We start this subsection by calculating the number of points at the interval $[0,1)$. To do this we need a definition of the term "point" and mathematical tools to indicate a point. Since this concept is one of the most fundamental, it is very difficult to find an adequate definition for it. If we accept (as is usually done in modern mathematics) that a *point* in $[0,1)$ is determined by a numeral x called the *coordinate of the point* where $x \in \mathcal{S}$ and $\mathcal{S}$ is a set of numerals, then we can indicate the point by its coordinate x and are able to execute required calculations.

It is important to emphasize that we have not postulated that x belongs to the set, $\mathbb{R}$, of real numbers as it is usually done. Since we can express coordinates only by numerals, then different choices of numeral systems lead to various sets of numerals and, as a consequence, to different sets of points we can refer to. The choice of a numeral system will define what is the *point* for us and we shall not be able to work with those points which coordinates are not expressible in the chosen numeral system (remind Postulate 2). Thus, we are able to calculate the number of points if we have already decided which numerals will be used to express the coordinates of points.

Different numeral systems can be chosen to express coordinates of the points in dependence on the precision level we want to obtain. For example, Pirahã are not able to express any point. If the numbers $0 \leq x < 1$ are expressed in the form $\frac{p-1}{①}, p \in \mathbb{N}$, then the smallest positive number we can distinguish is $\frac{1}{①}$ and the interval $[0,1)$ contains the following points

$$0, \; \frac{1}{①}, \; \frac{2}{①}, \; \ldots \; \frac{①-2}{①}, \; \frac{①-1}{①}. \tag{41}$$

It is easy to see that they are ①. If we want to count the number of intervals of the form $[a-1,a), a \in \mathbb{N}$, on the ray $x \geq 0$, then, due to Postulate 3, the definition of sequence, and Theorem 2, not more than ① intervals of this type can be distinguished on the ray $x \geq 0$. They are

$$[0,1), [1,2), [2,3), \ldots [①-3, ①-2), [①-2, ①-1), [①-1, ①).$$

Within each of them we are able to distinguish ① points and, therefore, at the entire ray $①^2$ points can be observed. Analogously, the ray $x < 0$ is represented by the intervals

$$[-①, -①+1), [-①+1, -①+2), \ldots [-2,-1), [-1,0).$$

Hence, this ray also contains $①^2$ such points and on the whole line $2①^2$ points of this type can be represented and observed.

Note that the point $-①$ is included in this representation and the point ① is excluded from it. Let us slightly modify our numeral system in order to have ① representable. For this purpose, intervals of the type $(a-1,a], a \in \mathbb{N}$, should be considered to represent the ray $x > 0$ and the separate symbol, 0, should be used to represent zero. Then, on the ray $x > 0$ we are able to observe $①^2$ points and, analogously, on the ray $x < 0$ we also are able to observe $①^2$ points. Finally, by adding the symbol used to represent zero we obtain that on the entire line $2①^2+1$ points can be observed.

It is important to stress that the situation with counting points is a direct consequence of Postulate 2 and is typical for natural sciences where it is well known that instruments influence results of observations. It is similar to the work with microscope or fractals (see [26]): we decide the level of the precision we need and obtain a result dependent on the chosen level of accuracy. If we need a more precise or a more rough answer, we change the lens of our microscope.

In our terms this means to change one numeral system with another. For instance, instead of the numerals considered above, let us choose a positional numeral system with the radix b

$$(a_1a_2\ldots a_{q-1}a_q)_b, \qquad q \in \mathbb{N}, \tag{42}$$

to calculate the number of points within the interval $[0,1)$.

Theorem 8. *The number of elements of the set of numerals of the type (42) is equal to* $b^{①}$.

Proof. Formula (42) defining the type of numerals we deal with contains a sequence of digits $a_1a_2\ldots a_{q-1}a_q$. Due to the definition of the sequence and Theorem 2, this sequence can have at maximum ① elements, i.e., $q \leq ①$. Thus, it can be at maximum ① positions on the the right of the dot. Every position can be filled in by one of the b digits from the alphabet $\{0,1,\ldots,b-1\}$. Thus, we have $b^{①}$ combinations. As a result, the positional numeral system using the numerals of the form (42) can express $b^{①}$ numbers. □

Corollary 1. *The entire line contains* $2①b^{①}$ *points of the type (42).*

Proof. We have already seen above that it is possible to distinguish 2① unit intervals within the line. Thus, the whole number of points of the type (42) on the line is equal to $2①b^{①}$. □

In this example of counting, we have changed the tool to calculate the number of points within each unit interval from (41) to (42), but used the old way to calculate the number of intervals, i.e., by natural numbers. If we are not interested in subdividing the line at intervals and want to obtain the number of the points on the line directly by using positional numerals of the type (40), then, as it has already has been established in Theorem 6, the number of points expressible by the numerals (40) is $|\mathbb{R}_b| = b^{2①}$.

It is obligatory to say in this occasion that the results presented above should be considered as a more precise analysis of the situation discovered by the genius of Cantor. He has proved, by using his famous diagonal argument, that the number of elements of the set $\mathbb{N}$ is less than the number of real numbers at the interval $[0,1)$ *without calculating the latter*. To do this he expressed real numbers in a positional numeral system. We have shown that this number will be different depending on the radix b used in the positional system (42) to express real numbers. However, all of the obtained numbers, $b^{①}$, are more than the number of elements of the set of natural numbers, ①, and, therefore, the diagonal argument maintains its force.

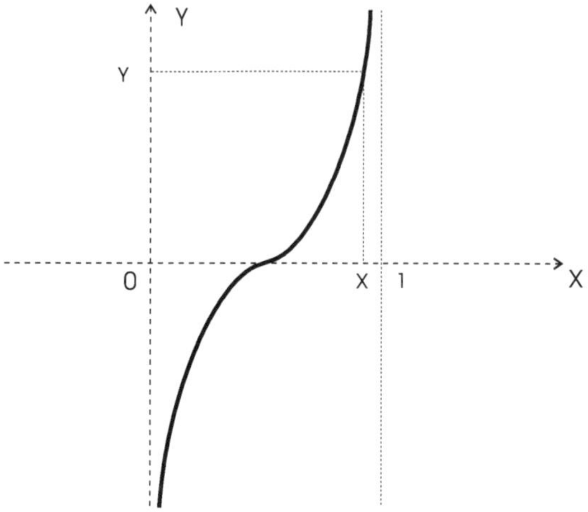

Fig. 2 Due to Cantor, the interval $(0,1)$ and the entire real number line have the same number of points

Let us now return to the problem of comparison of infinite sets and consider Cantor's famous result showing that the number of points over the interval $(0,1)$ is equal to the number of points over the whole real line, i.e.,

$$|\mathbb{R}| = |(0,1)|. \tag{43}$$

The proof of this counterintuitive fact is given by establishing a one-to-one correspondence between the elements of the two sets. Such a mapping can be done by using, for example, the function

$$y = \tan(0.5\pi(2x-1)), \qquad x \in (0,1), \tag{44}$$

illustrated in Fig. 2. Cantor shows by using Fig. 2 that to any point $x \in (0,1)$ a point $y \in (-\infty, \infty)$ can be associated and vice versa. Thus, he concludes that the requested one-to-one correspondence between the sets $\mathbb{R}$ and $(0,1)$ has been established and, therefore, this proves (43).

Our point of view is different: the number of elements is an intrinsic characteristic of each set (for both finite and infinite cases) that does not depend on any object outside the set. Thus, in Cantor's example from Fig. 2 we have (see Fig. 3) three mathematical objects: (a) a set, $X_{\mathcal{S}_1}$, of points over the interval $(0,1)$ which we are able to distinguish using a numeral system $\mathcal{S}_1$; (b) a set, $Y_{\mathcal{S}_2}$, of points over the vertical real line which we are able to distinguish using a numeral system $\mathcal{S}_2$; (c) the function (44) described using a numeral system $\mathcal{S}_3$. All these three mathematical objects are independent each other. The sets $X_{\mathcal{S}_1}$ and $Y_{\mathcal{S}_2}$ can have the same or different number of elements.

Thus, we are not able to evaluate $f(x)$ at *any* point x. We are able to do this only at points from $X_{\mathcal{S}_1}$. Of course, in order to be able to execute these evaluations it is necessary to conciliate the numeral systems $\mathcal{S}_1, \mathcal{S}_2$, and $\mathcal{S}_3$. The fact that we

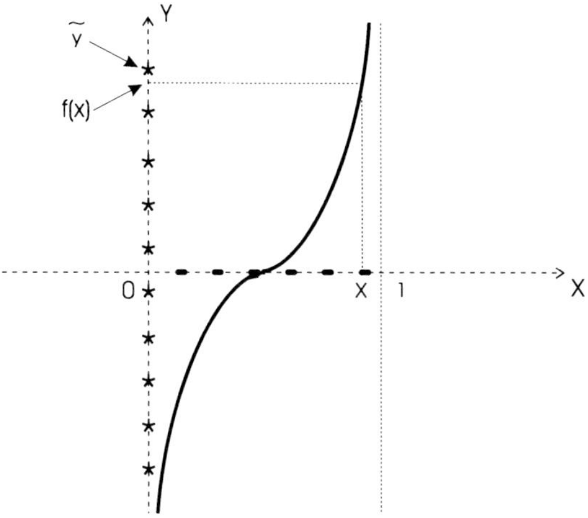

Fig. 3 Three independent mathematical objects: the set $X_{\mathcal{S}_1}$ represented by *dots*, the set $Y_{\mathcal{S}_2}$ represented by *stars*, and function (44)

have made evaluations of $f(x)$ and have obtained the corresponding values does not influence minimally the numbers of elements of the sets $X_{\mathcal{S}_1}$ and $Y_{\mathcal{S}_2}$. Moreover, it can happen that the number $y = f(x)$ cannot be expressed in the numeral system $\mathcal{S}_2$ and it is necessary to approximate it by a number $\tilde{y} \in \mathcal{S}_2$. This situation, very well known to computer scientists, is represented in Fig. 3.

Let us remind one more famous example related to the one-to-one correspondence and taking its origins in studies of Galileo Galilei: even numbers can be put in a one-to-one correspondence with all natural numbers in spite of the fact that they are a part of them:

$$\begin{array}{lcccccccc} \text{even numbers:} & 2, & 4, & 6, & 8, & 10, & 12, & \dots \\ & \updownarrow & \updownarrow & \updownarrow & \updownarrow & \updownarrow & \updownarrow & \\ \text{natural numbers:} & 1, & 2, & 3, & 4 & 5, & 6, & \dots \end{array} \qquad (45)$$

Again, our view on this situation is different since we cannot establish a one-to-one correspondence between the sets because they are infinite and we, due to Postulate 1, are able to execute only a finite number of operations. We cannot use the one-to-one correspondence as an executable operation when it is necessary to work with infinite sets.

However, we already know that the number of elements of the set of natural numbers is equal to ① and ① is even. Since the number of elements of the set of even numbers is equal to $\frac{\text{①}}{2}$, we can write down not only initial (as it is usually done traditionally) but also the final part of (45)

$$\begin{array}{cccccccccc} 2, & 4, & 6, & 8, & 10, & 12, & \dots & \text{①}-4, & \text{①}-2, & \text{①} \\ \updownarrow & \updownarrow & \updownarrow & \updownarrow & \updownarrow & \updownarrow & & \updownarrow & \updownarrow & \updownarrow \\ 1, & 2, & 3, & 4 & 5, & 6, & \dots & \frac{\text{①}}{2}-2, & \frac{\text{①}}{2}-1, & \frac{\text{①}}{2} \end{array} \qquad (46)$$

concluding so (45) in a complete accordance with Postulate 3. Note that record (46) does not affirm that we have established the one-to-one correspondence among *all* even numbers and a half of natural ones. We cannot do this due to Postulate 1. The symbols "..." indicate an infinite number of numbers and we can execute only a finite number of operations. However, record (46) affirms that for any even number expressible in the chosen numeral system it is possible to indicate the corresponding natural number in the lower row of (46).

We conclude this section by the following remark. With respect to our methodology, the mathematical results obtained by Pirahã, Cantor, and those presented in this chapter do not contradict to each other. *They all are correct with respect to mathematical languages used to express them.* This relativity is very important and it has been emphasized in Postulate 2. For instance, the result of Pirahã $1 + 2 =$ "many" is correct in their language in the same way as the result $1 + 2 = 3$ is correct in the modern mathematical languages. Analogously, the result (45) is correct in Cantor's language and the more powerful language developed in this chapter allows us to obtain a more precise result (46) that is correct in the new language.

The choice of the mathematical language depends on the practical problem that are to be solved and on the accuracy required for such a solution. Again, the result of Pirahã "many"+1="many" is correct. If one is satisfied with its accuracy, the answer "many" can be used (and is used by Pirahã) in practice. However, if one needs a more precise result, it is necessary to introduce a more powerful mathematical language (a numeral system in this case) allowing one to express the required answer in a more accurate way.

7 New Computational Possibilities for Mathematical Modelling

The computational capabilities of the Infinity Computer allow one to construct new and more powerful mathematical models able to take into account infinite and infinitesimal changes of parameters. In this section, the main attention is given to infinitesimals that can increase the accuracy of models and computations, in general. It is shown that the introduced infinitesimal numerals and the formalization of the concept "point" given in the previous sections can be successfully used in practical calculations. Examples related to computations of probabilities and areas (and volumes) of objects having several parts of different dimensions are given.

It becomes also possible in several occasions to automatize the process of the solving of computational problems avoiding an interruption of the work of computer procedures and the necessity of a human intervention required when one works with traditional computers. It is necessary to emphasize that the examples described in this section are related to *numerical* computations at the Infinity Computer. No symbolic computations are required to work with infinite and infinitesimal numbers when one uses the Infinity Computer.

7.1 *Usage of Infinitesimals for Solving Systems of Linear Equations*

Very often in computations, an algorithm performing calculations encounters a situation where the problem to divide by zero occurs. Then, obviously, this operation cannot be executed. If it is known that the problem under consideration has a solution, then a number of additional computational steps trying to avoid this division is performed. A typical example of this kind is the operation of pivoting used when one solves systems of linear equations by an algorithm such as Gauss–Jordan elimination. Pivoting is the interchanging of rows (or both rows and columns) in order to avoid division by zero and to place a particularly "good" element in the diagonal position prior to a particular operation.

The following two simple examples give just an idea of a numerical usage of infinitesimals and show that the usage of infinitesimals can help to avoid pivoting in cases when the pivotal element is equal to zero. We emphasize again that the Infinity Computer (see [41]) works with infinite and infinitesimal numbers expressed in the positional numeral system (14), (15) numerically, not symbolically.

Example 25. Solution to the system

$$\begin{bmatrix} 0 & 1 \\ 2 & 2 \end{bmatrix} \begin{bmatrix} x_1 \\ x_2 \end{bmatrix} = \begin{bmatrix} 2 \\ 2 \end{bmatrix}$$

is obviously given by $x_1^* = -1$, $x_2^* = 2$. It cannot be found by the method of Gauss without pivoting since the first pivotal element $a_{11} = 0$.

Since all the elements of the matrix are finite numbers, let us substitute the element $a_{11} = 0$ by ①$^{-1}$ and perform exact Gauss transformations without pivoting:

$$\left[\begin{array}{cc|c} ①^{-1} & 1 & 2 \\ 2 & 2 & 2 \end{array}\right] \to \left[\begin{array}{cc|c} 1 & ① & 2① \\ 0 & -2①+2 & -4①+2 \end{array}\right] \to \left[\begin{array}{cc|c} 1 & ① & 2① \\ 0 & 1 & \frac{-4①+2}{-2①+2} \end{array}\right]$$

$$\left[\begin{array}{cc|c} 1 & 0 & 2① - ① \cdot \frac{-4①+2}{-2①+2} \\ 0 & 1 & \frac{-4①+2}{-2①+2} \end{array}\right] \to \left[\begin{array}{cc|c} 1 & 0 & \frac{2①}{-2①+2} \\ 0 & 1 & \frac{-4①+2}{-2①+2} \end{array}\right] \to \left[\begin{array}{cc|c} 1 & 0 & -1 + \frac{1}{1-①} \\ 0 & 1 & 2 - \frac{1}{1-①} \end{array}\right].$$

It follows immediately that the solution to the initial system is given by the finite parts of numbers $-1 + \frac{1}{1-①}$ and $2 - \frac{1}{1-①}$.

We have introduced the number ①$^{-1}$ once and, as a result, we have obtained expressions where the maximal power of grossone is one and there are rational expressions depending on grossone, as well. It is possible to manage these rational expressions in two ways: (i) to execute division in order to obtain its result in the form (14), (15); (ii) without executing division. In the latter case, we just continue to work with rational expressions. In the case (i), since we need finite numbers as final results, in the result of division it is not necessary to store the parts c_p①^p

with $p < -1$. These parts can be forgotten because in any way the result of their successive multiplication with the numbers of the type $c_1\text{①}^1$ (remind that 1 is the maximal exponent present in the matrix under consideration) will give exponents less than zero, i.e., numbers with these exponents will be infinitesimals that are not interesting for us in this computational context.

Thus, by using the positional numeral system (14), (15) with the radix grossone we obtain

$$\left[\begin{array}{cc|c} 1 & \text{①} & 2\text{①} \\ 0 & 1 & \frac{-4\text{①}+2}{-2\text{①}+2} \end{array}\right] \rightarrow \left[\begin{array}{cc|c} 1 & \text{①} & 2\text{①} \\ 0 & 1 & 2\text{①}^0{+}1\text{①}^{-1} \end{array}\right]$$

$$\left[\begin{array}{cc|c} 1 & 0 & 2\text{①} - \text{①}\cdot(2\text{①}^0 1\text{①}^{-1}) \\ 0 & 1 & 2\text{①}^0 1\text{①}^{-1} \end{array}\right] \rightarrow \left[\begin{array}{cc|c} 1 & 0 & -1\text{①}^0 \\ 0 & 1 & 2\text{①}^0 1\text{①}^{-1} \end{array}\right].$$

The finite parts of numbers -1①^0 and $2\text{①}^0 1\text{①}^{-1}$, i.e., -1 and 2, respectively, then provide the required solution. □

Example 26. Solution to the system

$$\begin{bmatrix} 0 & 0 & 1 \\ 2 & 0 & -1 \\ 1 & 2 & 3 \end{bmatrix} \begin{bmatrix} x_1 \\ x_2 \\ x_3 \end{bmatrix} = \begin{bmatrix} 1 \\ 3 \\ 1 \end{bmatrix}$$

is the following: $x_1^* = 2$, $x_2^* = -2$, and $x_3^* = 1$. The coefficient matrix of this system has the first two leading principal minors equal to zero. Consequently, the first two pivots, in the Gauss transformations, are zero. We solve the system without pivoting by substituting the zero pivot by ①^{-1}, when necessary.

Let us show how the exact computations are executed:

$$\left[\begin{array}{ccc|c} 0 & 0 & 1 & 1 \\ 2 & 0 & -1 & 3 \\ 1 & 2 & 3 & 1 \end{array}\right] \rightarrow \left[\begin{array}{ccc|c} 1 & 0 & \text{①} & \text{①} \\ 0 & 0 & -2\text{①}-1 & -2\text{①}+3 \\ 0 & 2 & -\text{①}+3 & -\text{①}+1 \end{array}\right]$$

$$\left[\begin{array}{ccc|c} 1 & 0 & \text{①} & \text{①} \\ 0 & 1 & -2\text{①}^2-\text{①} & -2\text{①}^2+3\text{①} \\ 0 & 2 & -\text{①}+3 & -\text{①}+1 \end{array}\right] \rightarrow \left[\begin{array}{ccc|c} 1 & 0 & \text{①} & \text{①} \\ 0 & 1 & -2\text{①}^2-\text{①} & -2\text{①}^2+3\text{①} \\ 0 & 0 & 4\text{①}^2+\text{①}+3 & 4\text{①}^2-7\text{①}+1 \end{array}\right]$$

$$\left[\begin{array}{ccc|c} 1 & 0 & \text{①} & \text{①} \\ 0 & 1 & -2\text{①}^2-\text{①} & -2\text{①}^2+3\text{①} \\ 0 & 0 & 1 & \frac{4\text{①}^2-7\text{①}+1}{4\text{①}^2+\text{①}+3} \end{array}\right] \rightarrow \left[\begin{array}{ccc|c} 1 & 0 & 0 & \frac{8\text{①}^2+2\text{①}}{4\text{①}^2+\text{①}+3} \\ 0 & 1 & 0 & \frac{-8\text{①}^2+10\text{①}}{4\text{①}^2+\text{①}+3} \\ 0 & 0 & 1 & \frac{4\text{①}^2-7\text{①}+1}{4\text{①}^2+\text{①}+3} \end{array}\right].$$

It is easy to see that the finite parts of the numbers

$$\tilde{x}_1^* = \frac{8①^2+2①}{4①^2+①+3} = 2 - \frac{6}{4①^2+①+3},$$

$$\tilde{x}_2^* = \frac{-8①^2+10①}{4①^2+①+3} = -2 + \frac{12①+6}{4①^2+①+3},$$

$$\tilde{x}_3^* = \frac{4①^2-7①+1}{4①^2+①+3} = 1 - \frac{8①+2}{4①^2+①+3},$$

coincide with the corresponding solution $x_1^* = 2$, $x_2^* = -2$, and $x_3^* = 1$.

In this procedure we have introduced the number $①^{-1}$ two times. As a result, we have obtained expressions where the maximal power of grossone is equal to 2 and there are rational expressions depending on grossone, as well. By reasoning analogously to Example 26, when we execute divisions, in the obtained results it is not necessary to store the parts of the type $c_p①^p, p < -2$, because in any way the result of their successive multiplication with the numbers of the type $c_2①^2$ will give finite exponents less than zero. That is, numbers with these exponents will be infinitesimals that are not interesting for us in this computational context. Thus, by using the positional numeral system (14), (15), we obtain

$$\left[\begin{array}{ccc|c} 1 & 0 & ① & ① \\ 0 & 1 & -2①^2-① & -2①^2+3① \\ 0 & 0 & 1 & \frac{4①^2-7①+1}{4①^2+①+3} \end{array}\right] \rightarrow \left[\begin{array}{ccc|c} 1 & 0 & ① & ① \\ 0 & 1 & -2①^2-① & -2①^2+3① \\ 0 & 0 & 1 & 1①^0-2①^{-1} \end{array}\right].$$

Note that the number $1①^0$-$2①^{-1}$ does not contain the part of the type $c_{-2}①^{-2}$ because the coefficient c_{-2} obtained after the executed division is such that $c_{-2} = 0$. Then we proceed as follows

$$\left[\begin{array}{ccc|c} 1 & 0 & ① & ① \\ 0 & 1 & 0 & -2 \\ 0 & 0 & 1 & 1①^0-2①^{-1} \end{array}\right] \rightarrow \left[\begin{array}{ccc|c} 1 & 0 & 0 & 2 \\ 0 & 1 & 0 & -2 \\ 0 & 0 & 1 & 1①^0-2①^{-1} \end{array}\right].$$

The obtained solutions $x_1^* = 2$ and $x_2^* = -2$ have been obtained exactly without infinitesimal parts and $x_3^* = 1$ is derived from the finite part of $1①^0 - 2①^{-1}$. □

We conclude this section by emphasizing that zero pivots in the matrix are substituted dynamically by $①^{-1}$. Thus, the number of the introduced infinitesimals $①^{-1}$ depends on the number of zero pivots.

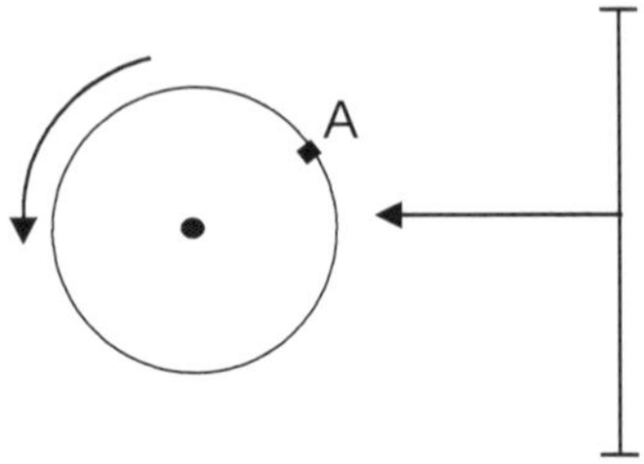

Fig. 4 What is the probability that the rotating disk stops in such a way that the point A will be exactly in front of the arrow?

7.2 Applications in Probability Theory and Calculating Volumes

A formalization of the concept "point" introduced above allows us to execute more accurately computations having relations with this concept. Very often in scientific computing and engineering it is required to construct mathematical models for multi-dimensional objects. Usually this is done by partitioning the modelled object in several parts having the same dimension and each of the parts is modelled separately. Then additional efforts are made in order to provide a correct functioning of a model unifying the obtained sub-models and describing the entire object.

Another interesting applied area is linked to stochastic models dealing with events having probability equal to zero. In this subsection, we first show that the new approach allows us to distinguish the impossible event having the probability equal to zero (i.e., $P(\varnothing) = 0$) and events having an infinitesimal probability. Then we show how infinitesimals can be used in calculating volumes of objects consisting of parts having different dimensions.

Let us consider the problem presented in Fig. 4 from the traditional point of view of probability theory. We start to rotate a disk having radius r with the point A marked at its border and we would like to know the probability $P(E)$ of the following event E: the disk stops in such a way that the point A will be exactly in front of the arrow fixed at the wall. Since the point A is an entity that has no extent it is calculated by considering the following limit

$$P(E) = \lim_{h \to 0} \frac{h}{2\pi r} = 0.$$

where h is an arc of the circumference containing A and $2\pi r$ is its length.

However, the point A can stop in front of the arrow, i.e., this event is not impossible and its probability should be strictly greater than zero, i.e., $P(E) > 0$. The new approach allows us to calculate this probability numerically.

First of all, in order to state the experiment more rigorously, it is necessary to choose a numeral system to express the points on the circumference. This choice will fix the number of points, K, that we are able to distinguish on the circumference. Definition of the notion *point* allows us to define elementary events in our experiment as follows: the disk has stopped and the arrow indicates a point. As a consequence, we obtain that the number, $N(\Omega)$, of all possible elementary events, e_i, in our experiment is equal to K where $\Omega = \cup_{i=1}^{N(\Omega)} e_i$ is the sample space of our experiment. If our disk is well balanced, all elementary events are equiprobable and, therefore, have the same probability equal to $\frac{1}{N(\Omega)}$. Thus, we can calculate $P(E)$ directly by subdividing the number, $N(E)$, of favorable elementary events by the number, $K = N(\Omega)$, of all possible events.

For example, if we use numerals of the type $\frac{i}{①}, i \in \mathbb{N}$, then $K = ①$. The number $N(E)$ depends on our decision about how many numerals we want to use to represent the point A. If we decide that the point A on the circumference is represented by m numerals, we obtain

$$P(E) = \frac{N(E)}{N(\Omega)} = \frac{m}{K} = \frac{m}{①} > 0,$$

where the number $\frac{m}{①}$ is infinitesimal if m is finite. Note that this representation is very interesting also from the point of view of distinguishing the notions "point" and "arc". When m is finite then we deal with a point, when m is infinite we deal with an arc.

In the case we need a higher accuracy, we can choose, for instance, numerals of the type $i①^{-2}, 1 \leq i \leq ①^2$, for expressing points at the disk. Then it follows $K = ①^2$ and, as a result, we obtain $P(E) = m①^{-2} > 0$.

This example with the rotating disk, of course, is a particular instance of the general situation taking place in the traditional probability theory where the probability that a continuous random variable X attains a given value a is zero, i.e., $P(X = a) = 0$. While for a discrete random variable one could say that an event with probability zero is impossible, this cannot be said in the case of a continuous random variable. As we have shown by the example above, in our approach this situation does not take place because this probability can be expressed by infinitesimals. As a consequence, probabilities of such events can be computed and used in numerical models describing the real world (see [36] for a detailed discussion on the modelling continuity by infinitesimals in the framework of the approach using grossone).

Moreover, the obtained probabilities are not absolute, *they depend on the accuracy chosen for the mathematical model describing the experiment.* There is again a straight analogy with physics where it is not possible to obtain results that have a precision higher than the accuracy of the measurement of the data. We also cannot obtain a precision that is higher than the precision of numerals used in the mathematical model.

Let us now consider two examples showing that the new approach allows us to calculate areas and volumes of a more general class of objects than the traditional one. In Fig. 5 two figures are shown. The traditional approach tells us that both of them have area equal to one. In the new approach, if we use numerals of the type $i①^{-1}, i \in \mathbb{N}$, to express points within a unit interval, then the unit interval consists of ① points and in the plane each point has the infinitesimal area $①^{-1} \cdot ①^{-1} = ①^{-2}$.

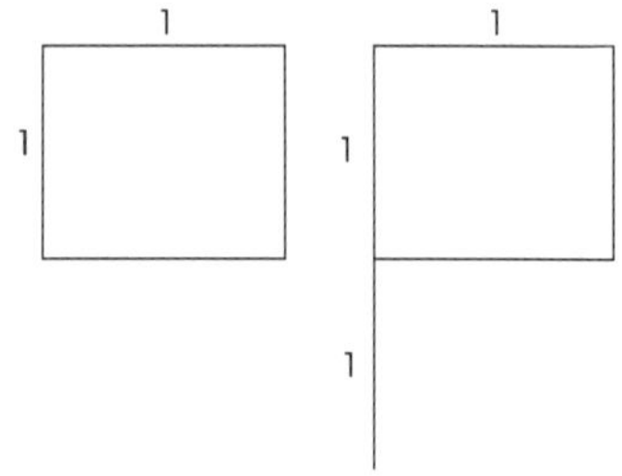

Fig. 5 It is possible to calculate and to distinguish areas of these two objects

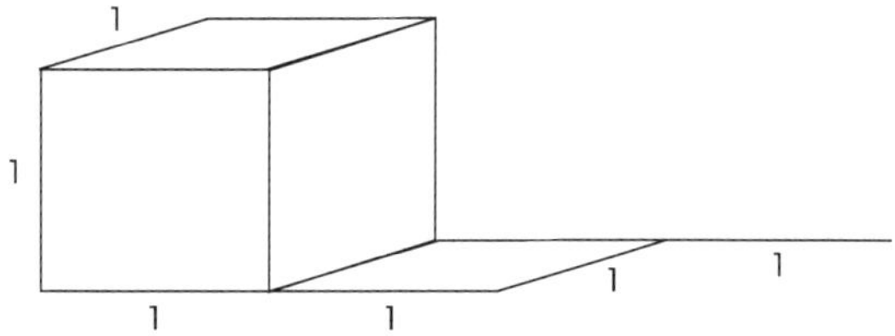

Fig. 6 New possibilities for calculating volumes of objects

As a consequence, this value will be our accuracy in calculating areas in this example. Suppose now that the vertical line added to the square at the right figure in Fig. 5 has the width equal to one point. Then we are able to calculate the area, S_2, of the right figure and it will be possible to distinguish it from the area, S_1, of the square on the left

$$S_1 = 1 \cdot 1 = 1, \qquad S_2 = 1 \cdot 1 + 1 \cdot ①^{-1} = 1①^0 1①^{-1}.$$

If the added vertical line has the width equal to three points, then it follows

$$S_2 = 1 \cdot 1 + 3 \cdot ①^{-1} = 1①^0 3①^{-1}.$$

The volume of the figure shown in Fig. 6 is calculated analogously:

$$V = 1 \cdot 1 \cdot 1 + 1 \cdot 1 \cdot ①^{-1} + 1 \cdot ①^{-1} \cdot ①^{-1} = 1①^0 1①^{-1} 1①^{-2}.$$

If the accuracy of the considered numerals of the type $i①^{-1}, i \in \mathbb{N}$, is not sufficient, we can increase it by using, for instance, numerals of the type $i①^{-2}, 1 \leq i \leq ①^2$. Then the unit interval consists of $①^2$ points and at the plane each point has the infinitesimal area $①^{-2} \cdot ①^{-2} = ①^{-4}$. As a result, by a complete analogy with the previous case we obtain for lines having the width, for instance, equal to five points in all three dimensions that

$$S_2 = 1 \cdot 1 + 5 \cdot ①^{-2} = 1①^0 5①^{-2},$$

$$V = 1 \cdot 1 \cdot 1 + 1 \cdot 1 \cdot 5 \cdot ①^{-2} + 1 \cdot 5 \cdot ①^{-2} \cdot 5 \cdot ①^{-2} = 1①^0 5①^{-2} 25①^{-4}.$$

8 Traditional and Blinking Fractals and Their Quantitative Analysis Using Infinite and Infinitesimal Numbers

Fractal objects have been very well studied during the last few decades (see, e.g., [10, 26] and references given therein) and have been applied in various fields (see numerous applications given in [9, 10, 15, 26, 49]). However, mathematical analysis of fractals (except, of course, a very well-developed theory of fractal dimensions) very often continues to have mainly a qualitative character and tools for a quantitative analysis of fractals at infinity are not very rich yet.

In this section, we propose to apply the methodology developed above for a quantitative analysis of traditional and newly introduced blinking fractals.

8.1 Blinking Fractals

Let us start by introducing the new class of objects—blinking fractals—that are not covered by traditional theories studying self-similarity processes. Traditional fractals are constructed using the principle of self-similarity that infinitely many times repeats a basic object (some times slightly modified in time). However, there exist processes and objects that evidently are very similar to classical fractals but cannot be covered by the traditional approaches because several self-similarity mechanisms participate in the process of their construction. Before going to a general definition of blinking fractals let us give just three examples of them.

The first example is derived from one of the famous fractal constructions—the coast of Britain—as follows. Suppose that we have made a picture of the coast two times using the same scale of the map: at the moment of the early sunrise and at the moment of late sunset. Then, due to the long shadows present at these moments and directed to the opposite sides we shall have two different pictures. If we suppose, for example, that sunset corresponds to shadows on the left and sunrise to shadows on the right, then we can indicate them as L and R, correspondingly. If now we start to make pictures (starting from sunrise) alternating moments of the photographing from sunrise to sunset and decreasing the scale each time, we shall obtain a series of pictures being very similar to traditional fractals but different because left shadows will alternate right shadows at this sequence as follows: $R, L, R, L, R, L, \dots$ Thus, there are two fractal mechanisms working in our process. Each of them can be represented by one of subsequences $R, R, R, \dots$ and $L, L, L, \dots$ and the traditional analysis does not allow us to say what will be the limit fractal object and will it have L or R type of shadow.

The second example is constructed as follows. Let us take a prism (see Fig. 7) that is rotating around its vertical axis and observe it at two different moments. The first is the moment when we see its face being the blue rectangular with sides 1 and $\sqrt{2}$. Since we look exactly at the front of the prism we see the rectangular as the square with the length one on side. The second moment is when we look at the face

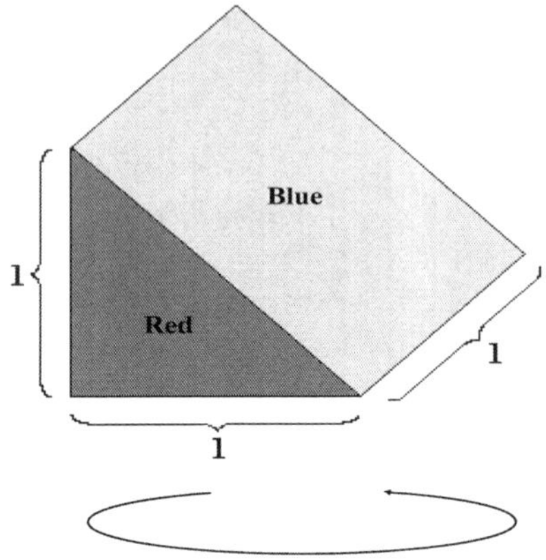

Fig. 7 The rotating prism having the *triangular face red* and the *rectangular face blue*

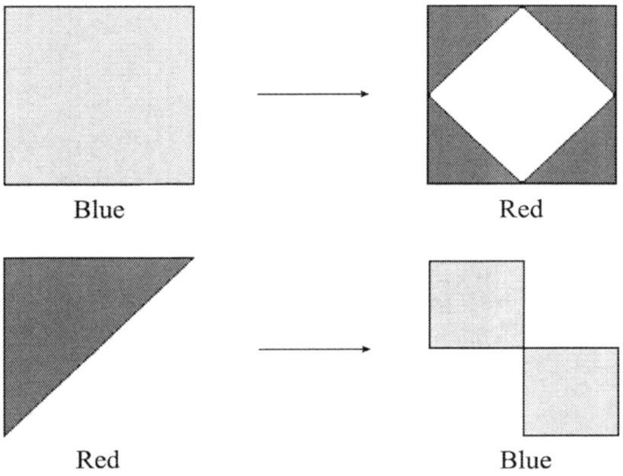

Fig. 8 We observe that each *blue square* is transformed in four *red triangles* and each *red triangle* is transformed in two *blue squares*

being the red right isosceles triangle with the legs equal to one. Then we apply to this three-dimensional object the two following self-similarity rules: we substitute each prism by four smaller prisms during the time passing between each even and odd observation and by two smaller prisms during the time passing between each odd and even observation. Thus, at the odd iterations we observe application of the first mechanism shown in the top of Fig. 8. The second mechanism shown in the bottom of this figure is applied during the even iterations. As a result, starting from the blue square one on side at iteration 0 we observe the pictures (see Fig. 9) with alternating blue squares and red triangles. Again, as it was with the above example related to the coast of Britain, we can extract two fractal subsequences being traditional fractals. The mechanism of the first one dealing with blue squares is shown in Fig. 10. The second mechanism dealing with red triangles is presented in Fig. 11. Traditional approaches are not able to say anything about the behavior of this process at infinity. Does there exist a limit object of this process? If it exists, what can we say about its structure? Does it consist of red triangles or blue squares? What is the area of this (again, if it exists) limit object? All these questions remain without answers.

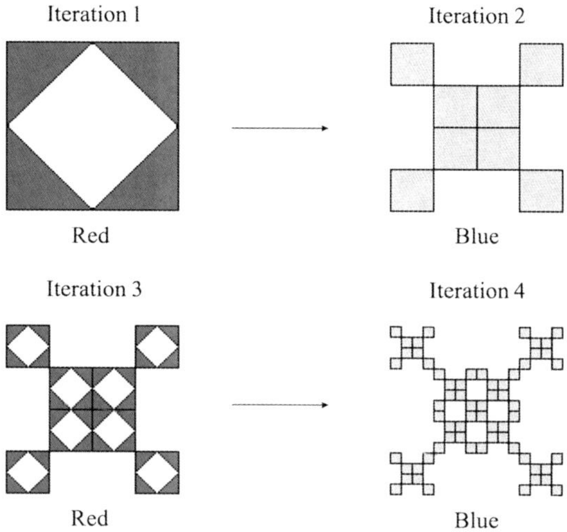

Fig. 9 The first four iterations of the process that has started from one *blue square* and uses two self-similarity mechanisms

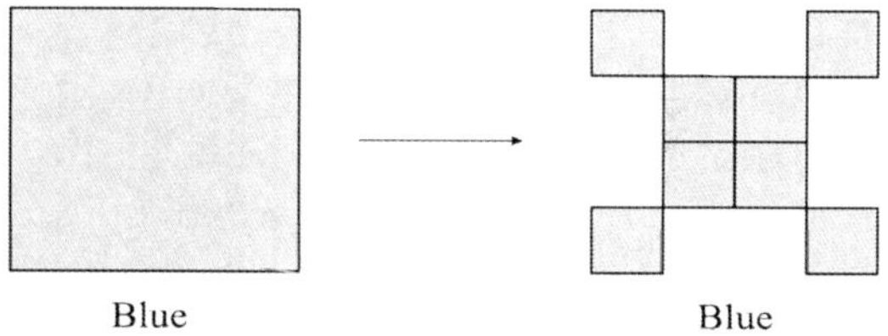

Fig. 10 The first traditional fractal mechanism regarding *blue squares* that can be separated from the process shown in Fig. 9

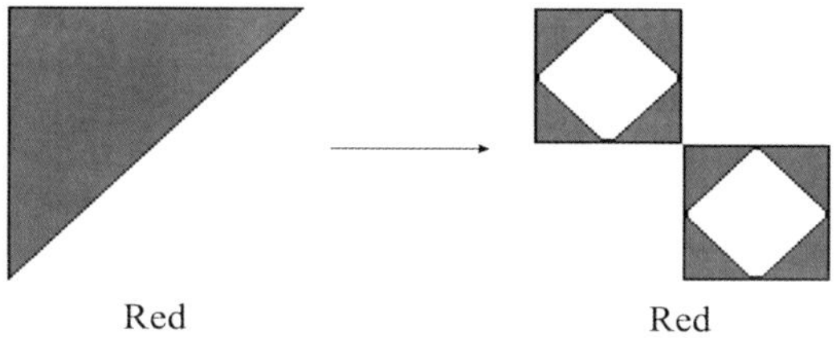

Fig. 11 The second traditional fractal mechanism regarding red triangles that can be separated from the process shown in Fig. 9

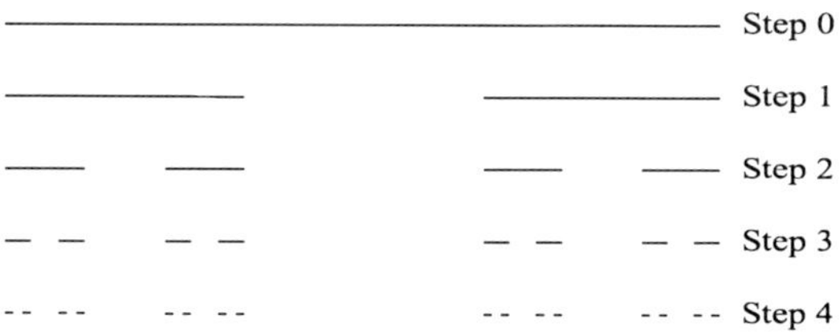

Fig. 12 Cantor's set

Before we discuss the last example linked, as it was with our first example, to another famous fractal construction—Cantor's set (see Fig. 12)—let us make a few comments reminding that very often we can give certain numerical answers to questions regarding fractals only if a finite number of steps in the procedure

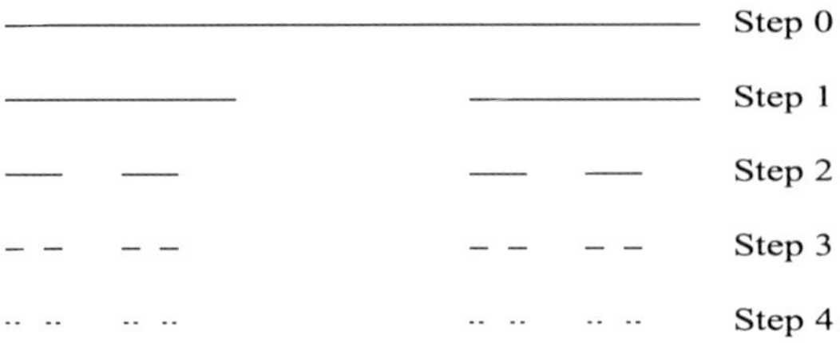

Fig. 13 At each odd iteration we remove the open interval being the middle third part from each of the intervals present in the construction and at each even iteration from each interval present in the set we remove open intervals being the second and the last fourth parts

of their construction has been considered. The same questions very often remain without any answer if we consider an infinite number of steps. If a finite number of steps, n, has been done in construction of Cantor's set, then we are able to describe numerically the set being the result of this operation. It will have 2^n intervals having the length $\frac{1}{3^n}$ each. Obviously, the set obtained after $n+1$ iterations will be different and we also are able to measure the lengths of the intervals forming the second set. It will have 2^{n+1} intervals having the length $\frac{1}{3^{n+1}}$ each. The situation changes drastically in the limit because we are not able to distinguish results of n and $n+1$ steps of the construction if n is infinite.

We also are not able to distinguish at infinity the results of the following two processes that both use Cantor's construction but start from different positions. The first one is the usual Cantor's set and it starts from the interval $[0,1]$, the second starts from the couple of intervals $[0,\frac{1}{3}]$ and $[\frac{2}{3},1]$. In spite of the fact that for any given finite number of steps, n, the results of the constructions will be different for these two processes we have no tools to distinguish them at infinity.

Let us now slightly change the process of construction used in Cantor's set to create a new example of a blinking fractal. At each odd iteration we shall maintain Cantor's rule, i.e., we remove the open interval being the middle third part from each of 2^n intervals present in the construction at the n-th iteration, where $n = 2k-1$. In contrast, if $n = 2k$ from each interval present in the set corresponding to the n-th iteration, we remove open intervals being the second and the last fourth parts (see Fig. 13). Again, as it was in the two previous examples, we have two different mechanisms working in this process and we are not able to say anything with respect to the structure of the resulting object at infinity. All the examples considered above have two different fractal mechanisms participating in their construction. Naturally, examples with more than two such mechanisms can be easily given.

To conclude this subsection we give the following general definition of objects that will be studied in this chapter together with traditional fractals. Objects constructed using the principle of self-similarity with an infinite cyclic application of *several fractal rules* are called *blinking fractals*.

8.2 *Quantitative Analysis of Traditional and Blinking Fractals*

Starting from Cantor's set we show how lengths of traditional fractals can be calculated at infinity.

We remind that if a finite number of steps, n, has been executed in Cantor's construction starting from the interval $[0,1]$, then we are able to describe numerically the set being the result of this operation. It will have 2^n intervals having the length $\frac{1}{3^n}$ each. Obviously, the set obtained after $n+1$ iterations will be different and we also are able to measure the lengths of the intervals forming the second set. It will have 2^{n+1} intervals having the length $\frac{1}{3^{n+1}}$ each. The situation changes drastically in the limit because traditional approaches are not able to distinguish results of n and $n+1$ steps of the construction if n is infinite. Now, we can do it using the introduced infinite and infinitesimal numbers.

Since the construction of Cantor's set is a process, it cannot contain more than ① steps [see discussion related to the example (23)–(25)]. Thus, if we start the process from the interval $[0,1]$, after ① steps Cantor's set consists of $2^{①}$ intervals and their total length, $L(n)$, is expressed in infinitesimals: $L(①) = (\frac{2}{3})^{①}$, i.e., the set has a well-defined infinite number of intervals and each of them has the infinitesimal length equal to $3^{-①}$. Analogously, after $① - 1$ steps Cantor's set consists of $2^{①-1}$ intervals and their total length is expressed in infinitesimals: $L(①) = (\frac{2}{3})^{①-1}$. Thus, the length $L(n)$ for any (finite or infinite) number of steps, n, where $1 \leq n \leq ①$ and is expressible in the chosen numeral system can be calculated.

It is important to notice here that [again due to the limitation illustrated by the example (23)–(25)] it is not possible to count one by one all the intervals at Cantor's set if their number is superior to ①. For instance, after ① steps it has $2^{①}$ intervals and they cannot be counted one by one because $2^{①} > ①$ and any process (including that of the sequential counting) cannot have more than ① steps.

It becomes possible to study by a complete analogy other classical fractals. For instance, we immediately obtain that the length of the Koch Curve starting from the interval $[0,1]$ after ① steps has the infinite length equal to $(\frac{4}{3})^{①}$ because it consists of $4^{①}$ segments having the length $(\frac{1}{3})^{①}$ each. In the same way we can calculate the area of the Sierpinski Carpet. If its construction starts from the unit square, then after ① steps we obtain the set of squares having the total infinitesimal area equal to $(\frac{8}{9})^{①}$ because it consists of $8^{①}$ squares and each of them has area equal to $(\frac{1}{9})^{①}$.

Consider now two processes that both use Cantor's construction but start from different initial conditions. Traditional approaches do not allow us to distinguish them at infinity in spite of the fact that for any given finite number of steps, n, the results of the constructions are different and can be calculated. Using the new approach we are able to study the processes numerically also at infinity. For example, if the first process is the usual Cantor's set and it starts from the interval $[0,1]$ and the second one starts from the couple of intervals $[0,\frac{1}{3}]$ and $[\frac{2}{3},1]$, then after $\frac{①}{2}$ steps the result of the first process will be the set consisting of $2^{\frac{①}{2}}$ intervals and its length $L(\frac{①}{2}) = (\frac{2}{3})^{\frac{①}{2}}$. The second set after $\frac{①}{2}$ steps will consist of $2^{\frac{①}{2}+1}$ intervals and its length $L(\frac{①}{2}+1) = (\frac{2}{3})^{\frac{①}{2}+1}$.

Let us answer now to the following traditional problem: How many points are there at Cantor's set? From our new point of view this formulation is not sufficiently precise. Now, when it becomes possible to distinguish different sets at different iterations we should say: How many points there are at Cantor's set being the result of n steps of Cantor's procedure started from the initial set consisting of k intervals? In the following without loss of generality we consider the case $k = 1$ and calculate the number of the points in the set C_n being the result of n steps of Cantor's procedure starting from the interval $[0,1]$.

Then, as it has been shown in Sect. 6, we are able to calculate the number of points if we have decided which numerals will be used to express the coordinates of the points within the interval $[0,1]$. Moreover, we shall be able to do such calculations only if the numeral system chosen to express coordinates of the points will be powerful enough to distinguish the points within the intervals generated during this process. Obviously, we shall be able to distinguish within C_n no more points than our chosen numeral system will allow us. For instance, if we give to a person from our primitive Pirahã tribe the set C_2 consisting of four intervals, this person operating with his poor numeral system consisting of the numerals I, II, and "many" will not be able to say us how many intervals there are in this set and which are coordinates of, for example, their end points. This happens because his system is too poor both for counting the intervals and for expressing coordinates of their end points. His answer will be just "many" for the number of intervals and he will be able to indicate the coordinate of only one point—1. However, if we give him the set C_0 his answer will be correct for the intervals—there is one interval—and he will be able to indicate the coordinate of the same point—1.

Thus, the situation with counting points in Cantor's set again is similar to the work with a microscope: we decide the level of the precision we need and obtain a result dependent on the chosen level. If we need a more precise or a more rough answer, we change the level of accuracy of our microscope. If we need a high precision and need to distinguish many points, we should take a powerful numeral system to express the coordinates. In the case when we need a low precision, a weak numeral system can be taken.

The introduced mathematical tools allow us to give answers to similar questions not only for traditional but for blinking fractals, too. We start by considering the blinking fractal described in Figs. 7–11. Since the answers depend on the initial conditions, we suppose without loss of generality that the process starts from the blue square one unit of length on side. This means that during any (finite or infinite) even iteration we observe blue squares and during any odd iteration we see red triangles. We shall indicate the set obtained after n iterations by P_n. The area A_n of the set P_n is calculated as follows. For any (finite or infinite) $n = 2k, k \geq 0$, it consists of 2^{3k} squares with the side equal to 2^{-2k}. Thus, the area of P_n is

$$A_{2k} = (2^{-2k})^2 \cdot 2^{3k} = 2^{-k}.$$

For $n = 2k-1, k \geq 1$, the set P_n consists of 2^{3k-1} right isosceles triangles with the legs equal to 2^{-2k+1}. In this case the area of P_n is calculated as follows

$$A_{2k-1} = 0.5(2^{-2k+1})^2 \cdot 2^{3k-1} = 2^{-k}. \tag{47}$$

For example, for the infinite $n = 0.5①$ the set $P_{0.5①}$ consists of $2^{0.75①}$ blue squares (because the number $0.5①$ is even), their total area is infinitesimal and is equal to $A_{0.5①} = 2^{-0.25①}$. Analogously, if the number of iterations is $n = 0.5① + 1$, then the set $P_{0.5①+1}$ consists of red triangles and k from (47) is equal to $-0.25① + 1$. The number of triangles is $2^{0.75①+2}$ and their total area is infinitesimal and is equal to $A_{0.5①+1} = 2^{-0.25①+1}$.

Finally, let us consider the blinking fractal from Fig. 13. We shall indicate the set obtained after n iterations by F_n. The length L_n of the set F_n is calculated as follows. For any (finite or infinite) $n = 2k, k \geq 0$, it consists of 2^{2k} intervals and each of them has the length $3^{-k} \cdot 4^{-k}$. Thus,

$$L_{2k} = 2^{2k} \cdot 3^{-k} \cdot 4^{-k} = 3^{-k}.$$

Analogously, for $n = 2k-1, k \geq 1$, we obtain that F_n consists of 2^{2k-1} intervals and each of them has the length $3^{-k} \cdot 4^{-k+1}$. Thus,

$$L_{2k-1} = 2^{2k-1} \cdot 3^{-k} \cdot 4^{-k+1} = 2 \cdot 3^{-k}.$$

For example, for the infinite odd $n = 0.5① - 1$ the set $F_{0.5①-1}$ consists of $2^{0.5①-1}$ intervals and their total length is infinitesimal and is equal to $L_{0.5①-1} = 2 \cdot 3^{-0.25①}$.

9 Concepts of Continuity in Physics and Mathematics

The goal of this section is to discuss mathematical and physical definitions of continuity and to develop a new, more physical point of view on this notion using the infinite and infinitesimal numbers introduced above. The new point of view is illustrated by a detailed consideration of one of the most fundamental mathematical definitions—function.

In physics, the "continuity" of an object is relative. For example, if we observe a table by eyes, then we see it continuous. If we use a microscope for our observation, we see that the table is discrete. This means that we decide how to see the object, as a continuous or as a discrete, by the choice of the instrument for observation. A weak instrument—our eyes—is not able to distinguish its internal small separate parts (e.g., molecules) and we see the table as a continuous object. A sufficiently strong microscope allows us to see the separate parts and the table becomes discrete but each small part now is viewed as continuous.

In this connection, fractals become a very useful tool for describing physical objects. Let us return to Figs. 12 and 13 and suppose that we observe two beams consisting of two different materials at Step 0 by eye and we see both of them continuous. Then we take a microscope with a weak lens number 1, look at the microscope and see the pictures corresponding to Step 1 in Figs. 12 and 13, i.e., that the beams are not continuous but consist of two smaller parts that, in their turn, now seem to us to be continuous. Then we proceed by taking a stronger lens number 2, look again at the microscope and see the pictures corresponding to Step 2 in Figs. 12 and 13. First, we see now that the beams consist of four smaller parts and each of them seems to be continuous. Second, we see that their locations are different (remind that we have supposed that the beams have been made using different materials). By increasing the force of lenses we can observe pictures viewed at Steps 3, 4, etc. obtaining higher levels of discretization. Thus, continuity in physics is resolution dependent and fractal ideas can serve as a good tool for modeling the physical relative continuity.

In contrast, in the traditional mathematics any mathematical object is either continuous or discrete. For example, the same function cannot be both continuous and discrete. Thus, this contraposition of discrete and continuous in the traditional mathematics does not reflect properly the physical situation that we observe in practice. For fortune, the infinite and infinitesimal numbers introduced in the previous sections give us a possibility to develop a new theory of continuity that is closer to the physical world and better reflects the new discoveries made by physicists (remind that the foundations of the mathematical analysis have been established centuries ago and, therefore, do not take into account the subsequent revolutionary results in physics, e.g., appearance of quantum physics). We start by introducing a definition of the one-dimensional continuous set of points based on the above consideration and Postulate 2 and establish relations to such a fundamental notion as function using the infinite and infinitesimal numbers.

We remind that traditionally a function $f(x)$ is defined as a binary relation among two sets X and Y (called the *domain* and the *codomain* of the relation) with the additional property that to each element $x \in X$ corresponds exactly one element $f(x) \in Y$. We consider now a function $f(x)$ defined over a one-dimensional interval $[a,b]$. It follows immediately from the previous sections that to define a function $f(x)$ over an interval $[a,b]$ it is not sufficient to give a rule for evaluating $f(x)$ and the values a and b because we are not able to evaluate $f(x)$ at *any* point $x \in [a,b]$ (for example, traditional numeral systems do not allow us to express any irrational number ζ and, therefore, we are not able to evaluate $f(\zeta)$). However, the traditional definition of a function includes in its domain points at which $f(x)$ cannot be evaluated, thus introducing ambiguity.

In order to be precise in the definition of a function, it is necessary to indicate explicitly a numeral system, $\mathcal{S}$, we intend to use to express points from the interval $[a,b]$. Thus, a function $f(x)$ is defined when we know a rule allowing us to obtain $f(x)$ given x and its domain, i.e., the set $[a,b]_{\mathcal{S}}$ of points $x \in [a,b]$ expressible in the chosen numeral system $\mathcal{S}$. We suppose hereinafter that the system $\mathcal{S}$ is used to write down $f(x)$ (of course, the choice of $\mathcal{S}$ determines a class of formulae and/or procedures we are able to express using $\mathcal{S}$) and it allows us to express any number

$$y = f(x), \qquad x \in [a,b]_{\mathcal{S}}.$$

The number of points of the domain $[a,b]_{\mathcal{S}}$ can be finite or infinite but the set $[a,b]_{\mathcal{S}}$ is always discrete. This means that for any point $x \in [a,b]_{\mathcal{S}}$ it is possible to determine its closest right and left neighbors, x^+ and x^-, respectively, as follows

$$x^+ = \min\{z : z \in [a,b]_{\mathcal{S}}, \quad z > x\}, \qquad x^- = \max\{z : z \in [a,b]_{\mathcal{S}}, \quad z < x\}. \tag{48}$$

Apparently, the obtained discrete construction leads us to the necessity to abandon the nice idea of continuity, which is a very useful notion used in different fields of mathematics. But this is not the case. In contrast, the new approach allows us to introduce a new definition of continuity very well reflecting the physical world.

Let us consider $n+1$ points at a line

$$a = x_0 < x_1 < x_2 < \cdots < x_{n-1} < x_n = b \tag{49}$$

and suppose that we have a numeral system $\mathcal{S}$ allowing us to calculate their coordinates using a unit of measure μ (for example, meter, inch, etc.) and to construct so the set $X = [a,b]_{\mathcal{S}}$ expressing these points.

The set X is called *continuous in the unit of measure* μ if for any $x \in (a,b)_{\mathcal{S}}$ it follows that the differences $x^+ - x$ and $x - x^-$ from (48) expressed in units μ are equal to infinitesimal numbers. In our numeral system with radix grossone this means that all the differences $x^+ - x$ and $x - x^-$ contain only negative grosspowers. Note that it becomes possible to differentiate types of continuity by taking into account values of grosspowers of infinitesimal numbers (continuity of order ①$^{-1}$, continuity of order ①$^{-2}$, etc.).

This definition emphasizes the physical principle that there does not exist an absolute continuity: it is relative (see discussion in page 57) with respect to the chosen instrument of observation which in our case is represented by the unit of measure μ. Thus, the same set can be viewed as a continuous or not in dependence of the chosen unit of measure.

Example 27. The set of six equidistant points

$$X_1 = \{a, x_1, x_2, x_3, x_4, x_5\} \tag{50}$$

from Fig. 14 can have the distance d between the points equal to ①$^{-1}$ in a unit of measure μ and to be, therefore, continuous in μ. Usage of a new unit of measure $\nu =$ ①$^{-3}\mu$ implies that $d =$ ①^2 in ν and the set X_1 is not continuous in ν. □

Note that the introduced definition does not require that all the points from X are equidistant. For instance, if in Fig. 14 for a unit measure μ the largest over the set $[a,b]_{\mathcal{S}}$ distance $x_6 - x_5$ is infinitesimal, then the whole set is continuous in μ.

The set X is called *discrete in the unit of measure* μ if for all points $x \in (a,b)_{\mathcal{S}}$ it follows that the differences $x^+ - x$ and $x - x^-$ from (48) expressed in units μ are not infinitesimal numbers. In our numeral system with radix grossone this means that in all the differences $x^+ - x$ and $x - x^-$ negative grosspowers cannot be the largest ones. For instance, the set X_1 from (50) is discrete in the unit of measure ν from

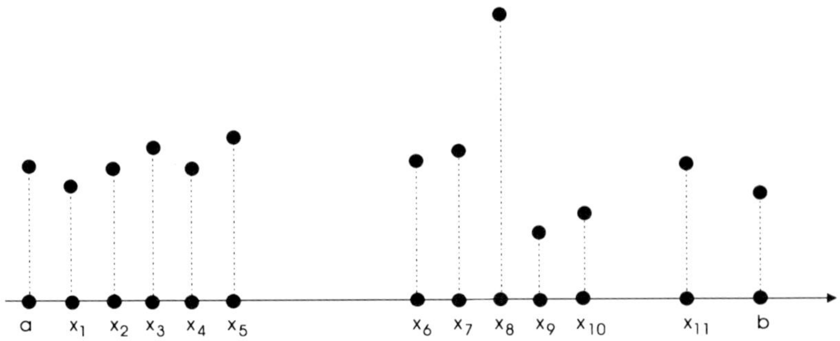

Fig. 14 It is not possible to say whether this function is continuous or discrete until we have not introduced a unit of measure and a numeral system to express distances between the points

Example 27. Of course, it is also possible to consider intermediate cases where sets have continuous and discrete parts (see again discussion in page 57 related to beams from Figs. 12 and 13).

The introduced notions allow us to give the following very simple definition of a function continuous at a point. A function $f(x)$ defined over a set $[a,b]_{\mathcal{S}}$ continuous in a unit of measure μ is called *continuous in the unit of measure* μ *at a point* $x \in (a,b)_{\mathcal{S}}$ if both differences $f(x) - f(x^+)$ and $f(x) - f(x^-)$ are infinitesimal numbers in μ, where x^+ and x^- are from (48). For the continuity at points a, b it is sufficient that one of these differences is infinitesimal. The notions of continuity from the left and from the right in a unit of measure μ at a point are introduced naturally. Similarly, the notions of a function discrete, discrete from the right, and discrete from the left can be defined.

The function $f(x)$ is *continuous in the unit of measure* μ *over the set* $[a,b]_{\mathcal{S}}$ if it is continuous in μ at all points of $[a,b]_{\mathcal{S}}$. Again, it becomes possible to differentiate types of continuity by taking into account values of grosspowers of infinitesimal numbers (continuity of order ①$^{-1}$, continuity of order ①$^{-2}$, etc.) and to consider functions in such units of measure that they become continuous or discrete over certain subintervals of $[a,b]$. In the further consideration we shall often fix the unit of measure μ and write just "continuous function" instead of "continuous function in the unit of measure μ." Let us give three simple examples illustrating the introduced definitions.

Example 28. We start by showing that the function $f(x) = x^2$ is continuous over the set X_2 defined as the interval $[0,1]$ where numerals $\frac{i}{①}, 0 \leq i \leq ①$, are used to express its points in units μ. First of all, note that the set X_2 is continuous in μ because its points are equidistant with the distance $d = ①^{-1}$. Since this function is strictly increasing, to show its continuity it is sufficient to check the difference $f(x) - f(x^-)$ at the point $x = 1$. In this case, $x^- = 1 - ①^{-1}$ and we have

$$f(1) - f(1 - ①^{-1}) = 1 - (1 - ①^{-1})^2 = 2①^{-1}(-1)①^{-2}.$$

This number is infinitesimal, thus $f(x) = x^2$ is continuous over the set X_2. □

Example 29. Consider the same function $f(x) = x^2$ over the set X_3 defined as the interval $[①-1, ①]$ where numerals $① - 1 + \frac{i}{①}, 0 \leq i \leq ①$, are used to express its points in units μ. Analogously, the set X_3 is continuous and it is sufficient to check the difference $f(x) - f(x^-)$ at the point $x = ①$ to show continuity of $f(x)$ over this set. In this case,

$$x^- = ① - 1 + \frac{① - 1}{①} = ① - ①^{-1},$$

$$f(x) - f(x^-) = f(①) - f(① - ①^{-1}) = ①^2 - (① - ①^{-1})^2 = 2①^0(-1)①^{-2}.$$

This number is not infinitesimal because it contains the finite part $2①^0$ and, as a consequence, $f(x) = x^2$ is not continuous over the set X_3. □

Example 30. Consider $f(x) = x^2$ defined over the set X_4 being the interval $[① - 1, ①]$ where numerals $① - 1 + \frac{i}{①^2}, 0 \leq i \leq ①^2$, are used to express its points in units μ. The set X_4 is continuous and we check the difference $f(x) - f(x^-)$ at the point $x = ①$. We have

$$x^- = ① - 1 + \frac{①^2 - 1}{①^2} = ① - ①^{-2},$$

$$f(x) - f(x^-) = f(①) - f(① - ①^{-2}) = ①^2 - (① - ①^{-2})^2 = 2①^{-1}(-1)①^{-4}.$$

Since the obtained result is infinitesimal, $f(x) = x^2$ is continuous over X_4. □

Let us consider now a function $f(x)$ defined by formulae over a set $X = [a, b]_{\mathcal{S}}$ so that different expressions can be used over different subintervals of $[a, b]$. The term "formula" hereinafter indicates a single expression used to evaluate $f(x)$.

Example 31. The function $g(x) = 2x^2 - 1, x \in [a, b]_{\mathcal{S}}$, is defined by one formula and function

$$f(x) = \begin{cases} \max\{-10x, 5x^{-1}\}, & x \in [c, 0)_{\mathcal{S}} \cup (0, d]_{\mathcal{S}}, \\ 4x, & x = 0, \end{cases} \qquad c < 0, \qquad d > 0, \tag{51}$$

is defined by three formulae, $f_1(x), f_2(x)$, and $f_3(x)$ where

$$\begin{aligned} f_1(x) &= -10x, \quad x \in [c, 0)_{\mathcal{S}}, \\ f_2(x) &= 4x, \qquad x = 0, \\ f_3(x) &= 5x^{-1}, \quad x \in (0, d]_{\mathcal{S}}. \end{aligned} \tag{52}$$

□

Consider now a function $f(x)$ defined in a neighborhood of a point x as follows

$$f(\xi) = \begin{cases} f_1(\xi), & x-l \le \xi < x, \\ f_2(\xi), & \xi = x, \\ f_3(\xi), & x < \xi \le x+r, \end{cases} \tag{53}$$

where the number l is any number such that the same formula $f_1(\xi)$ is used to define $f(\xi)$ at all points ξ such that $x-l \le \xi < x$. Analogously, the number r is any number such that the same formula $f_3(\xi)$ is used to define $f(\xi)$ at all points ξ such that $x < \xi \le x+r$. Of course, as a particular case it is possible that the same formula is used to define $f(\xi)$ over the interval $[x-l, x+r]$, i.e.,

$$f(\xi) = f_1(\xi) = f_2(\xi) = f_3(\xi), \qquad \xi \in [x-l, x+r]. \tag{54}$$

It is also possible that (54) does not hold but formulae $f_1(\xi)$ and $f_3(\xi)$ are defined at the point x and are such that at this point they return the same value, i.e.,

$$f_1(x) = f_2(x) = f_3(x). \tag{55}$$

If condition (55) holds, we say that function $f(x)$ has *continuous formulae* at the point x. Of course, in the general case, formulae $f_1(\xi), f_2(\xi)$, and $f_3(\xi)$ can be or cannot be defined out of the respective intervals from (53). In cases where condition (55) is not satisfied we say that function $f(x)$ has *discontinuous formulae* at the point x. Definitions of functions having formulae which are continuous or discontinuous from the left and from the right are introduced naturally.

Example 32. Let us study the following function

$$f(x) = \begin{cases} ①^2 + \frac{x^2-1}{x-1}, & x \neq 1, \\ a, & x = 1, \end{cases} \tag{56}$$

at the point $x = 1$. By using designations (53) and the fact that for $x \neq 1$ it follows $\frac{x^2-1}{x-1} = x+1$ we have

$$f(\xi) = \begin{cases} f_1(\xi) = ①^2 + \xi + 1, & \xi < 1, \\ f_2(\xi) = a, & \xi = 1, \\ f_3(\xi) = ①^2 + \xi + 1, & \xi > 1, \end{cases}$$

Since

$$f_1(1) = f_3(1) = ①^2 + 2, \qquad f_2(1) = a,$$

we obtain that if $a = ①^2 + 2$, then the function (56) has continuous formulae[9] at the point $x = 1$. Analogously, the function (51) has continuous formulae at the point $x = 0$ from the left and discontinuous from the right. □

Thus, functions having continuous formulae at a point can be continuous or discrete at this point in dependence of the chosen unit of measure. Analogously, functions having discontinuous formulae at a point can be continuous or discrete at this point again in dependence of the chosen unit of measure. The notion of continuity of a function depends on the chosen unit of measure and numeral system $\mathcal{S}$ and it can be used for functions defined by formulae, computer procedures, tables, etc. In contrast, the notion of a function having continuous formulae works only for functions defined by formulae and does not depend on units of measure or numeral systems chosen to express its domain. It is related only to properties of formulae.

We conclude this section by the note that the expressed numerical point of view on the definition of continuity has been then extended in [40] to the differential calculus for one-dimensional functions assuming finite, infinite, and infinitesimal values over finite, infinite, and infinitesimal domains.

10 A Brief Conclusion

In this chapter, a new computational methodology has been introduced. It allows us to express, by a finite number of symbols, not only finite numbers but infinite and infinitesimals, as well, and to execute numerical computations with all of them. A number of theoretical and applied problems where the new way of counting helps a lot has been discussed.

It has been emphasized that the philosophical triad—researcher, object of investigation, and tools used to observe the object—existing in such natural sciences as physics and chemistry, exists in mathematics, too. In natural sciences, the instrument used to observe the object influences the results of observations. The same happens in mathematics where numeral systems used to express numbers are among the instruments of observations used by mathematicians. The usage of powerful numeral systems gives the possibility to obtain more precise results in mathematics, in the same way as the usage of a good microscope gives the possibility to obtain more precise results in physics.

When a mathematician chooses a mathematical language (an instrument), in this moment he/she chooses both a set of numbers that can be observed through the numerals available in the chosen numeral system and the accuracy of results that can be obtained during computations. In the cases where two languages having different accuracies can be applied, it does not usually make sense to mix the languages,

[9] Note that even if $a = ①^2 + 2 + \varepsilon$, where ε is an infinitesimal number (remind that all infinitesimals are not equal to zero), we are able to establish that the function has discontinuous formulae.

i.e., to compose mathematical expressions using symbols from both languages, because the result of such a mixing either has no sense or has the lower of the two accuracies.

The analysis done in the chapter shows that the traditional mathematical language using for computations the symbol ∞ very often does not possess a sufficiently high accuracy when one deals with problems having their interesting properties at infinity. However, the new numeral system and the new way of counting described in this chapter do not contradict the traditional approaches. They just describe objects with different accuracies. It has been discovered that situations that can be illustrated by the following metaphor can take place. Suppose that we have measured two distances A and B with the accuracy equal to 1 m and we have found that both of them are equal to 25 m. Suppose now that we want to measure them with the accuracy equal to 1 cm. Then, very probably, we shall obtain something like $A = 2,487$ cm and $B = 2,538$ cm, i.e., $A \neq B$. Both answers, $A = B$ and $A \neq B$, are correct but with different accuracies and both of them can be used successfully in different situations. For instance, if one just wants to go for a walk, then the accuracy of the answer $A = B$ expressed in meters is sufficient. However, if one needs to connect some devices with a cable, then a higher accuracy is required and the answer expressed in centimeters should be used.

Acknowledgements This research was partially supported by the project "High accuracy supercomputations and solving global optimization problems using the information approach" of the Russian Federal Program "Scientists and Educators in Russia of Innovations," project 14.B37.21.0878.

References

1. V. Benci and M. Di Nasso. Numerosities of labeled sets: a new way of counting. *Advances in Mathematics*, 173:50–67, 2003.
2. G. Cantor. *Contributions to the founding of the theory of transfinite numbers*. Dover Publications, New York, 1955.
3. J.B. Carroll, editor. *Language, Thought, and Reality: Selected Writings of Benjamin Lee Whorf*. MIT Press, 1956.
4. A.L. Cauchy. *Le Calcul infinitésimal*. Paris, 1823.
5. J.H. Conway and R.K. Guy. *The Book of Numbers*. Springer-Verlag, New York, 1996.
6. J. d'Alembert. Différentiel. *Encyclopédie, ou dictionnaire raisonné des sciences, des arts et des métiers*, 4, 1754.
7. L. D'Alotto. Cellular automata using infinite computations. *Applied Mathematics and Computation*, 218(16):8077–8082, 2012.
8. S. De Cosmis and R. De Leone. The use of grossone in mathematical programming and operations research. *Applied Mathematics and Computation*, 218(16):8029–8038, 2012.
9. R.L. Devaney. *An Introduction to Chaotic Dynamical Systems*. Westview Press Inc., New York, 2003.
10. K. Falconer. *Fractal Geometry: Mathematical foundations and applications*. John Wiley & Sons, Chichester, 1995.
11. K. Gödel. Über formal unentscheidbare Sätze der Principia Mathematica und verwandter Systeme. *Monatshefte für Mathematik und Physik*, 38:173–198, 1931.

12. K. Gödel. *The Consistency of the Continuum-Hypothesis*. Princeton University Press, Princeton, 1940.
13. P. Gordon. Numerical cognition without words: Evidence from Amazonia. *Science*, 306(15 October):496–499, 2004.
14. G.H. Hardy. *Orders of infinity*. Cambridge University Press, Cambridge, 1910.
15. H.M. Hastings and G. Sugihara. *Fractals: A user's guide for the natural sciences*. Oxford University Press, Oxford, 1994.
16. D. Hilbert. Mathematical problems: Lecture delivered before the International Congress of Mathematicians at Paris in 1900. *Bulletin of the American Mathematical Society*, 8:437–479, 1902.
17. D.I. Iudin, Ya.D. Sergeyev, and M. Hayakawa. Interpretation of percolation in terms of infinity computations. *Applied Mathematics and Computation*, 218(16):8099–8111, 2012.
18. K. Knopp. *Theory and Application of Infinite Series*. Dover Publications, New York, 1990.
19. G.W. Leibniz and J.M. Child. *The Early Mathematical Manuscripts of Leibniz*. Dover Publications, New York, 2005.
20. T. Levi-Civita. Sui numeri transfiniti. *Rend. Acc. Lincei, Series 5a*, 113:7–91, 1898.
21. G. Lolli. Infinitesimals and infinites in the history of mathematics: A brief survey. *Applied Mathematics and Computation*, 218(16):7979–7988, 2012.
22. M. Margenstern. Using grossone to count the number of elements of infinite sets and the connection with bijections. *p-Adic Numbers, Ultrametric Analysis and Applications*, 3(3):196–204, 2011.
23. M. Margenstern. An application of grossone to the study of a family of tilings of the hyperbolic plane. *Applied Mathematics and Computation*, 218(16):8005–8018, 2012.
24. A.A. Markov Jr. and N.M. Nagorny. *Theory of Algorithms*. FAZIS, Moscow, second edition, 1996.
25. I. Newton. *Method of Fluxions*. 1671.
26. H.-O. Peitgen, H. Jürgens, and D. Saupe. *Chaos and Fractals*. Springer-Verlag, New York, 1992.
27. P. Pica, C. Lemer, V. Izard, and S. Dehaene. Exact and approximate arithmetic in an amazonian indigene group. *Science*, 306(15 October):499–503, 2004.
28. A. Robinson. *Non-standard Analysis*. Princeton Univ. Press, Princeton, 1996.
29. E.E. Rosinger. Microscopes and telescopes for theoretical physics: How rich locally and large globally is the geometric straight line? *Prespacetime Journal*, 2(4):601–624, 2011.
30. E. Sapir. *Selected Writings of Edward Sapir in Language, Culture and Personality*. University of California Press, Princeton, 1958.
31. Ya.D. Sergeyev. *Arithmetic of Infinity*. Edizioni Orizzonti Meridionali, CS, 2003.
32. Ya.D. Sergeyev. http://www.theinfinitycomputer.com. 2004.
33. Ya.D. Sergeyev. A few remarks on philosophical foundations of a new applied approach to Infinity. *Scheria*, 26–27:63–72, 2005.
34. Ya.D. Sergeyev. Mathematical foundations of the Infinity Computer. *Annales UMCS Informatica AI*, 4:20–33, 2006.
35. Ya.D. Sergeyev. Misuriamo l'infinito. *Periodico di Matematiche*, 6(1):11–26, 2006.
36. Ya.D. Sergeyev. Blinking fractals and their quantitative analysis using infinite and infinitesimal numbers. *Chaos, Solitons & Fractals*, 33(1):50–75, 2007.
37. Ya.D. Sergeyev. A new applied approach for executing computations with infinite and infinitesimal quantities. *Informatica*, 19(4):567–596, 2008.
38. Ya.D. Sergeyev. Evaluating the exact infinitesimal values of area of Sierpinski's carpet and volume of Menger's sponge. *Chaos, Solitons & Fractals*, 42(5):3042–3046, 2009.
39. Ya.D. Sergeyev. Numerical computations and mathematical modelling with infinite and infinitesimal numbers. *Journal of Applied Mathematics and Computing*, 29:177–195, 2009.
40. Ya.D. Sergeyev. Numerical point of view on Calculus for functions assuming finite, infinite, and infinitesimal values over finite, infinite, and infinitesimal domains. *Nonlinear Analysis Series A: Theory, Methods & Applications*, 71(12):e1688–e1707, 2009.

41. Ya.D. Sergeyev. *Computer system for storing infinite, infinitesimal, and finite quantities and executing arithmetical operations with them.* USA patent 7,860,914, 2010.
42. Ya.D. Sergeyev. Counting systems and the First Hilbert problem. *Nonlinear Analysis Series A: Theory, Methods & Applications*, 72(3–4):1701–1708, 2010.
43. Ya.D. Sergeyev. Lagrange Lecture: Methodology of numerical computations with infinities and infinitesimals. *Rendiconti del Seminario Matematico dell'Università e del Politecnico di Torino*, 68(2):95–113, 2010.
44. Ya.D. Sergeyev. Higher order numerical differentiation on the infinity computer. *Optimization Letters*, 5(4):575–585, 2011.
45. Ya.D. Sergeyev. On accuracy of mathematical languages used to deal with the Riemann zeta function and the Dirichlet eta function. *p-Adic Numbers, Ultrametric Analysis and Applications*, 3(2):129–148, 2011.
46. Ya.D. Sergeyev. Using blinking fractals for mathematical modelling of processes of growth in biological systems. *Informatica*, 22(4):559–576, 2011.
47. Ya.D. Sergeyev and A. Garro. Observability of Turing machines: A refinement of the theory of computation. *Informatica*, 21(3):425–454, 2010.
48. Ya.D. Sergeyev and D. E. Kvasov. *Diagonal Global Optimization Methods*. FizMatLit, Moscow, 2008. In Russian.
49. R.G. Strongin and Ya.D. Sergeyev. *Global Optimization and Non-Convex Constraints: Sequential and Parallel Algorithms*. Kluwer Academic Publishers, Dordrecht, 2000.
50. M.C. Vita, S. De Bartolo, C. Fallico, and M. Veltri. Usage of infinitesimals in the Menger's Sponge model of porosity. *Applied Mathematics and Computation*, 218(16):8187–8196, 2012.
51. J. Wallis. *Arithmetica infinitorum*. 1656.
52. A.A. Zhigljavsky. Computing sums of conditionally convergent and divergent series using the concept of grossone. *Applied Mathematics and Computation*, 218(16):8064–8076, 2012.
53. A. Žilinskas. On strong homogeneity of two global optimization algorithms based on statistical models of multimodal objective functions. *Applied Mathematics and Computation*, 218(16):8131–8136, 2012.

Dynamic Composition and Analysis of Modern Service-Oriented Information Systems

Habib Abdulrab, Eduard Babkin, and Jeremie Doucy

Abstract Despite all the advantages brought by service-oriented architecture (SOA), experts argue that SOA introduces more complexity into information systems rather than resolving it. The problem of service integration challenges modern companies taking the risk of implementing SOA. One of important aspects of this problem relates to dynamic service composition, which has to take into account many types of information and restrictions existing in each enterprise. Moreover, all the changes in business logic should also be promptly reflected. This chapter proposes the approach to solution of the stated problem based on such concepts as model-driven architecture (MDA), ontology modelling and logical analysis. The approach consists of several steps of modelling and finite scope logical analysis for automated translation of business processes into the sequence of service invocations. Formal language of relational logic is proposed as a key element of the proposed approach which is responsible for logical analysis and service workflow generation. We present a logical theory to automatically specialize generic orchestration templates which are close to semantic specification of abstract services in OWL-S. The developed logical theory is described formally in terms of Relational Logic. Our approach is implemented and tested using MIT Alloy Analyzer software.

H. Abdulrab
INSA de Rouen, LITIS Laboratory, BP08 Avenue de l'Université, 76801 Saint-Étienne-deRouvray, France
e-mail: abdulrab@insa-rouen.fr

E. Babkin (✉)
National Research University Higher School of Economics, B. Pechorskaya 25/12, 603155 Nizhny Novgorod, Russia
e-mail: eababkin@hse.ru

J. Doucy
EADS Defence & Security, Information Processing Control and Cognition, Parc d'Affaire des Portes BP 613, 27106 Val De Reuil, France
e-mail: jdoucy@gmail.com

A. Sorokin and P.M. Pardalos (eds.), *Dynamics of Information Systems: Algorithmic Approaches*, Springer Proceedings in Mathematics & Statistics 51,
DOI 10.1007/978-1-4614-7582-8_2,

Keywords Information systems • Web-service • Composition • Formal analysis • Relational logic

1 Introduction

Almost all modern companies use distributed and heterogeneous information systems (IS). It implies a variety of various applications based on different information technologies (IT) which interact with each other using diverse interfaces. In addition, modern companies are much interested in IS integration with their business partners in order to speed up and improve operational business processes. All mentioned above set new requirements to IT integration which is currently one of the top priorities for modern software engineering.

However in practice traditional approaches to software engineering can hardly meet the requirements of dynamic business environment, such as flexibility and simplicity of IT operations. A significant shortcoming of traditional integration approaches is software redundancy and difficulty of software reuse. A concept of service-oriented architecture (SOA) offers new solutions to the mentioned problems.

SOA provides the opportunity of abstraction from software and hardware implementation [8]. This makes IS solutions much more flexible and capable of quick adaptation to business process changes. According to IBM, SOA can be defined as an application architecture in which functions are represented by independent and coarse-grained services with triggered interfaces. SOA enables business process automation in the form of workflow—a predefined sequence of business process activities [3]. Several machine languages were proposed for definition and enactment. Among them BPEL [4] and OWL-S [19] are mostly used in industry and research.

In SOA terminology a coordinated aggregate of services refers to service composition [9]. The following types of service composition are distinguished: choreography and orchestration. Orchestration is the management of services within a single run of a business process. Choreography is defined as the management of services during the asynchronous run of the business process working simultaneously with several data flows [8]. Our work deals with automation of service orchestration solely. Of course such a limitation does not allow modeling the complete life cycle of information systems; however, we believe that the proposed methodology can be further developed to automate service choreography as well.

However, in spite of SOA advantages and its popularity within business and IT communities, practical benefits of SOA are still intensively discussed. Many enterprises argue that SOA introduced more complexity into their information systems rather than resolving it. Heterogeneity of individual services and the need for multi-aspect modeling of complete distributed systems contribute to extreme complexity and high costs of SOA solutions. The solution for that problem has been looked for by means of enhancing semantic description of web services and wide application of formal analysis and reasoning methods for semantic integration of web services. Among the recent projects and standards we need to mention SUPER project [17], OWL-S [20], and IRS-III framework [5]. Such works as [21, 22, 28, 34, 35] mainly focus on using planning algorithms to address the automatic services composition challenge. The work of Duan et al. [7] was one of the first to introduce a formal logic-based model for specification and refinement of abstract business processes using the concepts of BPEL and program logic. Recent researches use Semantic Web Service Ontology (OWL-S) [19] for description of static and dynamics properties of abstract and concrete services, as well as different formal methods [10, 13].

Our analysis shows that most of the existing reasoning algorithms are based on theorem proving which influences the scope of solvable tasks. They do not enable analysis based on counterexamples generation, and this significantly restricts the capabilities of analysis and practical impact. The first significant problem is the presence of a multi-layered hierarchy of services. Their intrinsic interdependences and complex connections of pre- and post-conditions require application of more sophisticated formalisms and logic analysis. The second problem arises from the practice of software engineering. Software engineers frequently require support for the iterative process of service development and orchestration of the group of the services in the customized end-user application. Connection of software engineering methods of program verification and formal principles of semantic service composition give us a hint to study opportunities of traditional model checking tools for application in the context of service-oriented software systems. Several research works provide us with the background. For example, the works [23,32] investigate opportunities of Object-Z and Rewrite Logic to specify formally semantics of OWL-S. Such works as [31,33] show great potential of relational logic and Alloy Analyzer for formal specification and refinement of business processes.

The objective of this research is to develop a new formal approach which uses benefits of finite model checking and automates service orchestration based on business-process logic taking into account existing constraints in the context of an enterprise. Such an approach makes it possible to skip the manual analytical step and to automate service orchestration process based on business requirements and changes of business processes. Our work proposes a solution based on such concepts as model-driven approach, ontology modeling, business process modeling, and relational logic. These concepts, belonging to different areas, allow the creation of an advantageous integrated approach.

The chapter has the following structure. In Sect. 2 we present the main foundations of the proposed approach in the realm of IT architecture and formal logic analysis. Section 3 offers an overview of used formal tools and the proposed approach. In Sect. 4 we give detailed explanation of our approach using two specific case studies. In Sect. 5 we discuss the achieved results, compare them with the related research works, and determine further research directions.

2 Foundations

Our research fuses the concepts of IT architecture and of particular methods of formal analysis. For better comprehension we provide for the key foundational elements of these two realms.

2.1 IT Architecture

Large-scale enterprises entail a great number of business processes and procedures linked with each other on different levels and evolving constantly in the turbulent business environment. The complexity of business processes manifests itself in the number of participants, documents, interactions, and different scenarios that may happen. In the result the business process analysis and modeling become very difficult. Additionally, there are quite a lot of formally fixed policies and rules, informal restrictions, and constraints that should be taken into account as well in order to automate the processes so that they comply with the business reality. Different factors are spread across various conceptual layers of organizational structure and cannot be analyzed within the single modeling practice. That is why design of multi-layered models of description and composition of services is the central task for development of service-oriented information systems. One widely accepted software engineering approach is model-driven architecture (MDA) introduced by Object Management Group (OMG) [24]. MDA combines the advantages of modeling and SOA. According to MDA there are several abstraction levels of modeling. Three layers of models are distinguished in MDA [25]:

- The computation independent model (CIM) describes a system from the computation-independent viewpoint, addressing structural aspects of the system. A CIM is often called a domain model.
- The platform independent model (PIM) can be seen as defining a system in terms of a technology-neutral virtual machine or a computational abstraction.
- The platform specific model (PSM) usually consists of a platform model that captures the technical concepts and services that make up the platform and an implementation-specific model geared towards the concrete implementation technique.

Large number of various enterprise models are used to capture different facets of knowledge about processes, their automation, and software implementations. The variation of notations leads to the great complexity of service composition process. Unified Modeling Language (UML) was proposed to offer the single common modeling notation. However, currently UML cannot fully support domain models; therefore, means for high level modeling are required.

The approach proposed in this work utilizes the ontology modeling concept [12]. It has proved itself as an efficient knowledge management approach intended to simplify and structure the composition process. Ontologies were applied for web services description and integration by introducing the concept of Semantic Web Service (SWS). It stands for Web Service enhanced with semantic description. According to Bhiri [1] there are four main SWS initiatives, namely WSMO/L/X Framework [26], OWL-S, [20], IRS-III framework [5], and METEOR-S system [30].

Also ontology dialects for widely spread modeling methodologies EPC, BPMN, BPEL were created. sEPC, sBPMN, and sBPEL ontologies have been developed and proposed within the framework of SUPER project as part of Semantic Business Process Modeling methodology [17]. Such achievements provided a practical opportunity to link business process models to the domain ontology concept, to set users objectives and services capabilities in terms of the domain, and to automate service orchestration based on formal reasoning tools.

Ontology and MDA approaches are tightly coupled. As for description of SWSs, ontologies are mostly used for creation of PSMs (WSMO, OWL-S, etc.). However, ontologies can be successfully used for model creation on all MDA abstraction levels.

Application of ontologies and other conceptual models requires a well-developed language for ontology and models descriptions supported with formal reasoning. Syntax of such language should be intuitively clear for non-experts and compatible with existing SOA standards. Semantics of such language should be formally defined as it should provide consistent interpretability. In other words, no varied interpretations should exist. Expressive power of the ontology language should allow for fine-grained detailed description without overcomplications preventing a formal reasoning process [2].

Formal reasoning tools are required to control and support ontology quality. For example, such tools can be used during the ontology developing phase for testing model consistency and adequacy.

Formal semantics and meta-language are the key differences between formal logic languages and other languages. Main advantages of formal logic application for knowledge representation are the following:

- Consistency, lack of expression interpretation ambiguity
- Ability of distinguishing of logic expressions from the conclusions of their validity

Leading notations and methodologies for ontology development (OWL, WSMO) are based on Descriptive Logics (DLs) which are the formal extensions of first-order logic [2]. However, despite many important results achieved in using DLs in knowledge representation, there are still a number of principal scientific and engineering issues to be solved. For example, in reality many attractable DLs are ExpTime-complete (e.g., SHIQ) and only a few polynomial-complexity DL dialects for very strict domains have been developed so far. Additionally, in order to support effective practical engineering activities during design and evolution of service-oriented systems reasoning mechanisms should also provide model finding and counterexample generation capabilities, while DL-based reasoning does not allow them.

2.2 Relational Logic and Alloy Analyzer

In our research we use Alloy Analyzer [16] which gives an opportunity to enhance the logic-based method with the features of constraint-based approach. Theoretical foundations of Alloy Analyzer include a mathematical theory of relational logic and finite scope logical analysis. The first Alloy prototype came out in 1997 from MIT Software Design Group and to the moment it has evolved to the matured simulation and verification system. The formalism of Relational Logic is based on the first-order logical theory; it facilitates rigorous definition of the structure and constraints of data structures in the generic form of relations. Alloy Analyzer consists of the structural modeling language based on first-order logic and of a Java-based constraint solver for models analysis and verification. That modeling language is rooted in well-known formal language Z for program specifications, but it uses different modeling capabilities like inheritance and reuse of formulas, which facilitate declarative object-oriented description of the problem. The users of Alloy Analyzer can select the most appropriate formal approach to define the structure and behavior of the system among the following alternatives: predicate calculus, relational calculus, navigation expression style. Simple heuristic of incremental grow of the instances bound was used to find the minimal number of the instances which satisfy the specified theory.

In Alloy language all universe of discourse is modeled in terms of atoms and relations. Atoms model indivisible, constant entities, while relations with multiple arities represent meaningful relationships and dynamical aspects. Expressive means of logic include:

- Set constants (empty set, universal set, identity set)
- Commonly accepted set-theoretical operators (union, intersection, subset inclusion, etc.)
- Relational operators (product, join, transpose, transitive closure, etc.)

Non-trivial constraints against the relations can be made from usual logical operators, quantifiers, specific multiplicity constraints restricting the basic relational operators, and cardinality constraints.

Practice-oriented modeling language of Alloy facilitates declaration of logic sentences in the text object-oriented form and provides convenient means to organize large models into tractable components and to manage simulation or verification. In the Alloy modeling language the basic building block is referred to as signature. In principle each signature represents a set of atoms. Specific constraints, defined in the modeled domain, are expressed in terms of facts, predicates, and functions. Assertions denote studied properties of the domain, which are verified by the means of the model simulation. Finally, testing of the model is performed with the help of a pair of control commands: run and check. Run command starts Alloy analyzer in order to find a correspondent model for a given number of instances.

In Alloy the flexible separation of concerns principle is used for implementation of simulation and checking. To verify a model against constraints Alloy analyzer translates the definition and constraints of the model into binary constraints and passes them to an external satisfiability solver in order to solve the classical finite domain constraint satisfaction problem via the search in the state space. Thus simulation of the model gives back such instances of states of executions that satisfy a given constraint (the model of the logical theory), and checking gives instances of the model, which violate the specified constraints. As an example of Alloy syntax, we consider the following definitions of signatures which describe basic constituents of service-oriented information systems:

```
sig Condition {}

sig WorkflowElem {}

sig ControlElem extends WorkflowElem  {}

sig Service extends WorkflowElem {
  preconditions:  set Condition,
  effects:   set Condition
}
```

In terms of relational logic these signatures define such mathematical structures as four sets (Condition, WorkflowElem, ControlElem and Service) and two binary relations between the set "Service" and the set "Condition" (the relations has name preconditions and effects correspondingly). An additional signature and more relations may be defined to model a generic linked structure of services in terms of relational logic.

```
sig WTemplate {
elems  : set WorkflowElem,
  first   : one  elems,
  last   : one  elems,
  transition: (elems - last)   ->    (elems - first)}
}
```

Relation "transition" demonstrates how ternary relations can be defined in Alloy language. In this case the relation is defined between WTemplate, and two subsets of WorkflowElem, thus determining the order of control flow transition from one service to another. The following fact sets a logical constraint on the transition tuples.

```
fact f1 {
  all p: WTemplate |  let t = p.transition |
  all s1,s2: Service | s1->s2 in t =>
s2.preconditions in s1.effects
}
```

According to that fact a pair of services may be linked in a workflow template only if the preconditions of the former service are in the set of effects of the former service. Later in the text of the chapter we describe in more detail how these and other similar definitions allow to facilitate service composition.

It is important to understand that Alloy logic is unresolvable [23]. Therefore, Alloy logical means for resolution and reasoning are based on automated generation of examples (for predicate validation) and counterexamples (for constraints assertion) within the finite scope. However, finite scope of Alloy logic does not lead to poor reliability of such analysis because Alloy allows modeling of an infinite number of objects and relations between them. In addition to the capability of simultaneous analysis of a great number of objects, Alloy provides opportunity to model and analyze such complex data structures as trees, which makes it convenient and applicable for real domain objects modeling. Moreover, the structure of the Alloy language allows integration of its models with the models described in other notations and languages such as UML.

3 Proposed Approach to Service Composition

In order to manage and orchestrate program services and ensure sufficient quality of SOA implementation business analysts should take the role of a handler and an integrator of all information gathered during the studies of the application domain and analysis of the business processes. Also for creation and further maintenance of information systems the majority of industrial developers perform system modeling in different styles, and finally execute manual or semi-automated composition of independent program services. Besides the costs associated with this kind of analytical work, such a huge amount of manual work is accompanied by high risk of losing or skipping relevant information that needed to be taken into account in the course of software implementation.

Changes in business processes require deep analytical work including interviews with involved employees, analysis of different models and restrictions. This does not allow the information system to be as flexible as business demands. Such a manual and loosely controlled process of service orchestration leads to inefficiency of information systems usage and high costs of SOA implementation.

3.1 New Basic Principles of Composition and Modeling

Most of the SOA solution vendors define at least two levels of SOA: level of business process and level of services [9]. The correspondence between services and business operations is the core of SOA concept. Despite all the results in researches of using semantics to service composition, existing approaches are more focused on technical description of service-related data and neglect a considerable part of business requirements.

However, SOA does not cover only one business process, but the whole range of enterprise business processes in a given application domain. All business processes in one business environment have a common set of notions and features that are defined by functional and industrial specificity of the application domain. In order

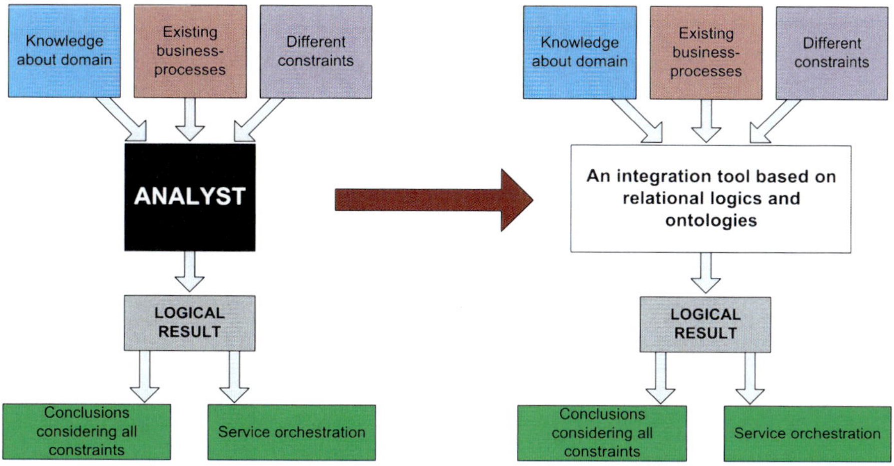

Fig. 1 Human factor elimination in the proposed approach to service composition

to achieve the common approach to modeling and consequently to automation of business processes in one domain, it is necessary to introduce concepts of a given application domain.

On the contrary, our approach to service composition is based not only on the service description but also on the description of the domain in the form of ontology, business process, and business requirements (see Fig. 1). Using ontology it is possible to define a thesaurus, which can function as a "translator" between the system and its users, between business representatives and software developers, and between the business process model and the model of technical realization. It makes possible to combine together different types of models created and analyzed during the implementation and change processes and to ensure consistency of all models.

The proposed approach unites the restrictions of all levels during the service orchestration. As a result, our approach features the modeling hierarchy with three layers: domain ontology, business processes, and program services (see Fig. 2). All layers and dependencies between them can be described in the logic language using Alloy Analyzer.

Domain ontology defines the basic notions of business environment and its restrictions. This layer corresponds to the CIM model in MDA. The analysis performed on Layer 1 results in the consistency checking of different restrictions.

Layer 2 models describe the dynamics and logic of business process based on the concepts defined on the Layer 1 model. This layer corresponds to the PIM model in MDA. The analysis performed on Layer 2 shows different scenarios of business process execution taking into account all the restrictions from the Layer 1 and Layer 2 models. Again during the analysis the consistence of the restrictions is checked.

Layer 3 model defines the core notions of the service environment and its restrictions. This layer corresponds to the PSM model in MDA. The analysis

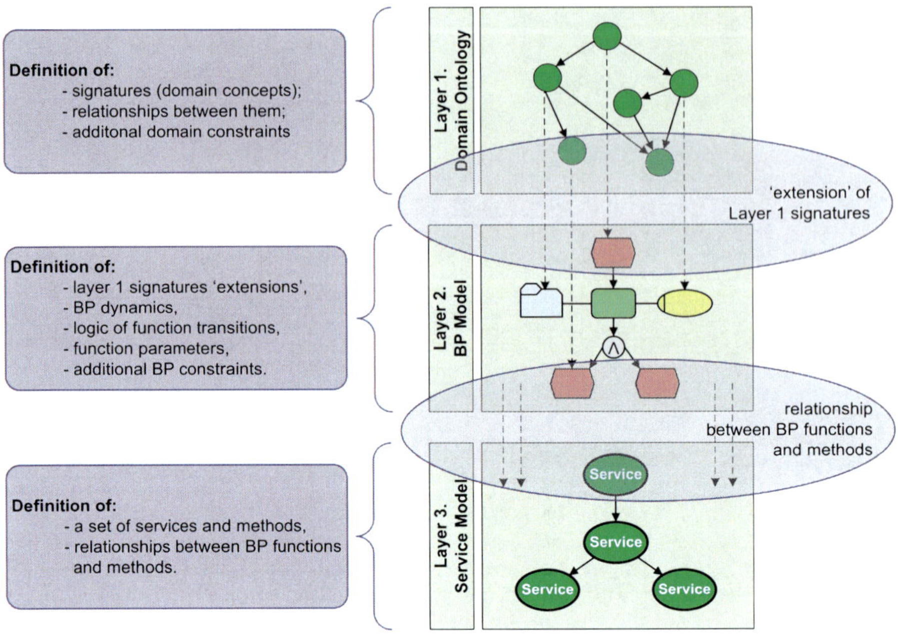

Fig. 2 Three layers of modeling hierarchy in the proposed approach

performed on Layer 3 shows multiple scenarios of service invocations taking into account all the restrictions from the Layer, Layer 2, Layer 3 models. Also the analysis checks the consistence of the restrictions. Alloy allows for scenarios to be generated in XML format. This enables different scenarios to be "glued" and translated into BPEL executed code according to architecture described above.

3.2 Mapping to Existing Approaches in the Realm of Business Process Modeling

To show how mapping to existing approaches may be implemented, we utilize widely used event-driven process chain (EPC) notation. The objective of the first task in our approach is to define the way of translation from a business process model to a particular logic language of Alloy Analyzer. According to EPC, business processes consist of a sequence of events and actions. Therefore, we introduce two concepts of ontology: Event and Action. For modeling the transitions between these concepts, an additional object has been introduced—Obligation Instance which was taken from the work [27]. The transition between these three concepts is defined by the relations between appropriate signatures (see Fig. 3).

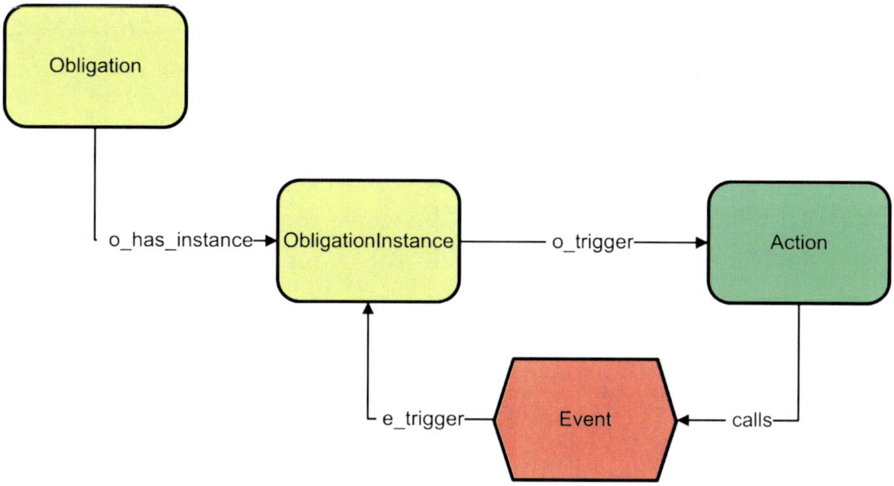

Fig. 3 Business process logic. Relation between objects Event, Action, ObligationInstance

Proposed approach for business process description gives a possibility to describe real business processes with all branches of the scenarios. First of all, the model signatures have fields with quantitative keywords ensuring a possibility to define complex conditions for triggering events. Secondly, using classical logical operators such as OR and AND it is possible to describe the logic of complex process. For example, one Action can trigger several Events. That is modeled by Alloy relational logic language as follows:

```
abstract sig Action  { // definition of signature Action
a_pre, a_post: one Time,
contains: Operation ->  DocumentInstance,
// one Action can trigger several Events
calls: set Event,
performed_by: one Role
}
```

An advantage of the proposed approach is modeling of the business process dynamics. For that an additional concept Time is introduced, which is linked with other business process notions according to the scheme in Fig. 4.

Using the following Alloy fact, the synchronization of all business process notions is achieved:

```
fact Times
{
all a: Action | all e: Event | e=a.calls =>
                        a.a_post=e.event_time

all o: ObligationInstance | all e: Event | o in e.e_trigger =>
                        e.event_tim = o.oblig_time
```

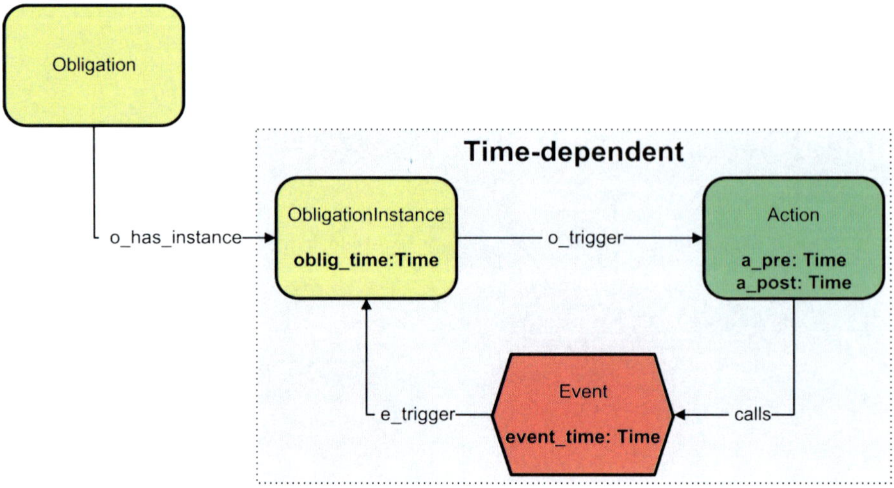

Fig. 4 Business process logic. Dynamics modeling

```
all o: ObligationInstance | all a: Action | o.o_trigger=a  =>
                              a.a_pre=o.oblig_time

all a:Action | all t:Time | t=a.a_pre=>a.a_post=t.TO/next
}
```

All described methods of business process modeling are defined on Layer 1 of our modeling hierarchy. This ensures that the common rules for business processes are set on a general level and they are valid for all introduced business processes within the business environment. Modeling a typical business process and its analysis are carried out on Layer 2 of the modeling hierarchy. The main point is that the concepts of Layer 1 are defined as Abstract, which means that there is no real object related to them. Using the extensions of such an abstract signatures specific business process elements are defined on Layer 2. For example, on Layer 1 an action is defined as an abstract concept. Particularly in Alloy it is expressed as: `abstract sig Action`. At the same time on Layer 2 a specific concept is defined as an extension of the signature Action introduced on Layer 1:`sig F1_OrderRegistration extends Action`. Usage of extensions ensures the consistency between concepts and models of the two layers (see Fig. 5).

3.3 *Program Services Modeling*

One of the main objectives in our approach is to ensure relations between models of different abstraction levels and to automate service sequence generation based on the business process logic. Therefore, Layer 3 of the modeling hierarchy in

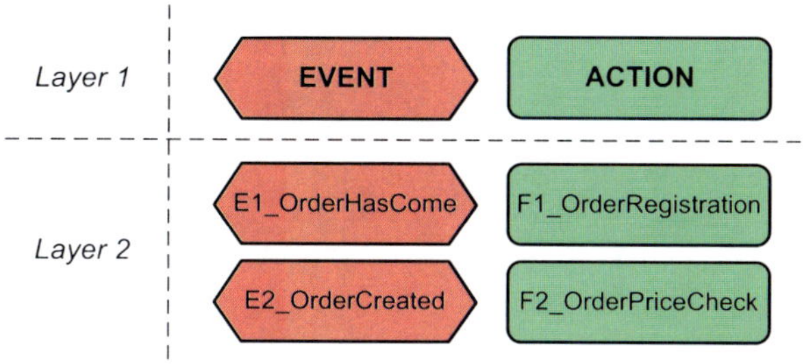

Fig. 5 Correspondence of Layer 1 and Layer 2 signatures

our approach, which is represented by services, is strictly linked with the business process layer and depends on the business process flow. As already mentioned above, a particular service represents the technical implementation of a particular business function. Service-oriented architecture implies existence of a number of services and technical methods which can be reused during the business function run. So, it is logical to introduce concepts Service and Method as an element of a given Service. In Alloy language, Methods are linked with Services using the field `consists_of` in Service signature, and Methods are linked with the corresponding business function from Layer 2 using field `corresponds` in Method signature. Based on such a simple approach a link between Layer 2 and Layer 3 of the modeling hierarchy is implemented (Fig. 6).

In addition to the introduced relation between the business function and the implementation method, there is also a fact which defines the logic of methods triggering in real time:

```
fact Synchronisation
{
all s:Service |all m:Method| m in s.consists_of =>
      m.(s.current_method_pre)= (m.corresponds).a_pre

all s:Service |all m: Method | m in s.consists_of =>
    m.(s.current_method_post) = (m.corresponds).a_post
}
```

Such an approach ensures the link between the technical and the business parts of SOA. This link enables automatic service triggering in the sequence defined by the business process flow.

3.4 Template-Based Service Orchestration

Proposed approach also allows for developing a specific method to automatically specialize generic web-service orchestration templates which are close to semantic specification of abstract services in OWL-S. In our case two main principles shape the proposed method for template-based service orchestration. First, abstract composite business processes take the role of a workflow template, thus reducing

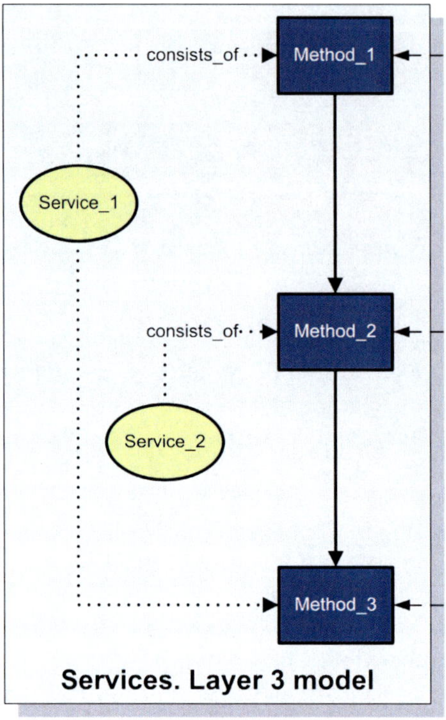

Fig. 6 The link between Layer 2 and Layer 3

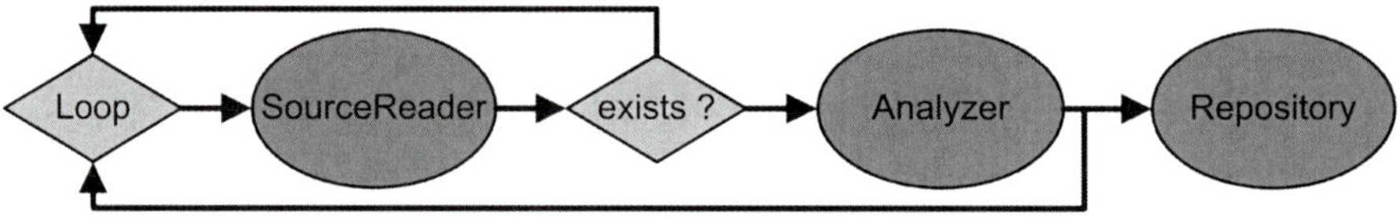

Fig. 7 One example of a generic workflow template

complexity and representing a repeating service workflow. Mainly inspired by the OWL-S definition of a composite process, for definition of workflow templates we use the limited set of such Control Constructs as: Sequence, Split, Split + Join, Choice, Any-Order, Condition, If-Then-Else, Iterate, Repeat-While, and Repeat-Until. Using these constructs business analysts and software architects may design general workflow templates on the basis of commonly used practices and existing processes of enterprise information systems. In most practical cases each workflow template consists of four to five building elements with a clear control structure. Figure 7 examplifies such a template. Diamonds define control flow elements and ovals represent abstract services.

Second, the logical theory and the principles of model finding facilitate specialization and validation of previously defined workflow templates which fit within the existing service infrastructure.

According to the stated principles we create a logical model which is capable of formal description of all needed abstract or specific service specifications. The model may conceptually be separated in four parts. The first, generic, part of our model describes such cornerstone concepts of our approach as composite processes (workflow templates), services, and control elements. This part of the model remains the same for all particular patterns and orchestration scenarios. The following Alloy statements represent key generic concepts: Condition, WorkflowElem, ControlElem, and Service.

```
abstract sig Condition { }
abstract sig WorkflowElem { }
abstract sig ControlElem extends WorkflowElem  { }
abstract sig Service extends WorkflowElem {
  preconditions:  set Condition,
  effects:   set Condition
}
```

Service and ControlElem are defined as extensions of WorkFlowElem. In other words, services and control elements are defined as workflow elements, so they could be included into the workflow template as follows:

```
abstract sig WTemplate {
elems : set WorkflowElem,
  first  : one  elems,
  last  : one  elems,
  transition: (elems - last)   ->   (elems - first) }
```

This formal description represents the workflow template mainly as a set of links between workflow elements, in other words, as a Directed Graph. Two other Alloy statements are needed for the purpose of services chaining in the workflow template.

```
fact f1 {
  all p: WTemplate |  let t = p.transition |
  all s1,s2: Service | s1->s2 in t =>
        s2.preconditions in s1.effects
}

fact f2  {
  all p: WTemplate |  all s1, s2: (Service-p.last) |
  all c: ControlElem | let t = p.transition |
  (s1->c  in t) and (c->s2 in t)  =>
         s2.preconditions in s1.effects
}
```

Fact f1 states that two services are linked if the preconditions of the former are included in the effects of the later. Fact f2 is quite similar, but it enables this chain through one control element. In other words, if we add a control element between two services, this control element becomes transparent in terms of service chaining.

The second part of the logical theory is domain-specific. It specifies existing services in accordance with the current configuration of the service infrastructure. One practical example of such specification is given in Sect. 4. The third part of the logical theory consists of formal specification of available abstract composite services, or workflow templates.

The following Alloy statements describe a frequently used workflow template with six nodes (including mandatory Start and Finish) and seven transitions. Further down, Start, Finish, SourceReader, and Analyser are defined as services, whereas Loop and IfCond are defined as control elements.

```
one sig P1 extends WTemplate { }  {
one Start
one Loop
one Finish
one SourceReader
one Analyzer
one IfCond

transition =
  Start->Loop +
  Loop->Finish +
  Loop->SourceReader +
  SourceReader -> IfCond +
  IfCond -> Analyser +
  IfCond -> Loop +
  Analyzer ->Loop }
```

The final fourth part of the logical theory defines specific requirements for particular refinement of the workflow template and the implementation details of the service orchestration. Users will have to specify some parameters to enable selection of the relevant concrete services. These parameters are translated directly to the formal language interpretable by Alloy Analyzer. The following constraints define how to express the condition ContentExtracted.

```
pred CyclicProcess {
one p:  WPattern | p.first = Start
one p:  WPattern | p.last  = Finish
Start.effects = none
Start.preconditions = none

Finish.effects = none
Finish.preconditions = ContentExtracted
one ContentExtracted
```

```
one p:  WPattern | some s:SourceReader |
 s->s in ^(p.transition) }
```

4 The Illustrative Examples

The practical application of the proposed approach is demonstrated by the creation of described models for two particular business cases. The first business case shows the practical mapping of EPC diagrams to the relational logic and further logical analyses in order to verify information security constraints. In the second case we show another application of our template-based method of service orchestration to a particular service-oriented information system in the domain of multi-media processing.

4.1 EPC Mapping and Security Constraints Checking

In this case we examine business process of sales order creation in a large sales company. Employees of sales department register a sales order in the system-based on an MS Excel request sent by email. Once an order is saved the system automatically checks if the price was lower than minimal acceptable one, taking into account information on discounts. Minimal acceptable prices entered in the system by Finance controller are used for order check. If price of at least one item from the order violates the norms, the order is blocked. In this case the information is sent to the employee of the Finance controlling department. The employee can approve the existing price or change it to the proper one. As the price is approved by the Finance controlling department, the order is sent to the Credit controlling department. The described business process can be seen in Fig. 8.

First, the domain ontology was created and described in the Alloy language. The main entities and relationships between them were defined. Second, the business process was modeled in terms defined at Layer 1 of the modeling hierarchy, in other words, the instance of Layer 2 Model was created. Then, the program services model (Layer 3 Model) was created and mapped with the business process model. The methods for automated orchestration of services were defined and the workflow order was automatically generated. Moreover, information security restrictions were checked both on the level of business process and on the level of program services.

According to the proposed approach, the Layer 1 Model was created. Its concepts and relations are shown in Fig. 9.

For each general concept from the domain ontology the successor was defined in Layer 2 Model representing the specific process function, event, principal, and document. For example, specific functional roles were defined. Order Desk specialist was represented by the role `R1_OrderDeskClerk`,

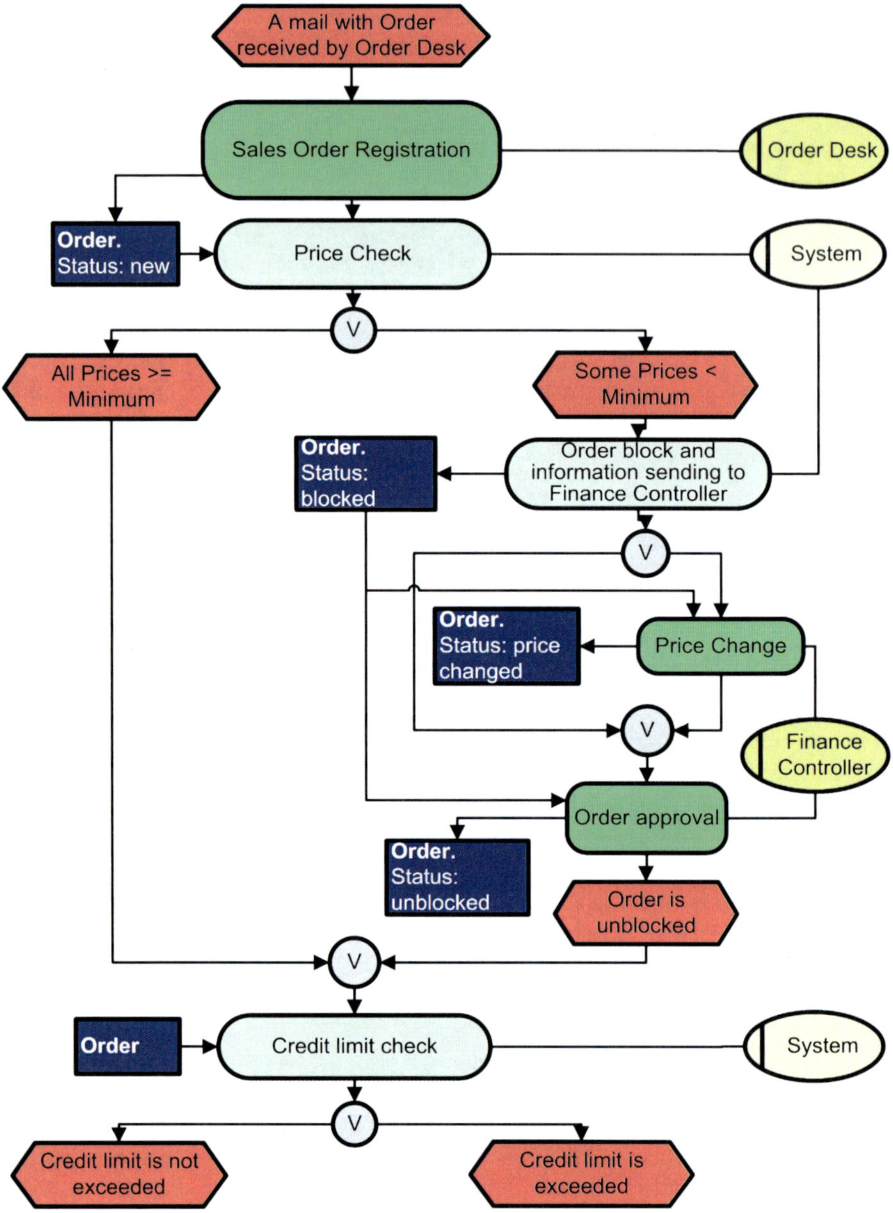

Fig. 8 The considered business process

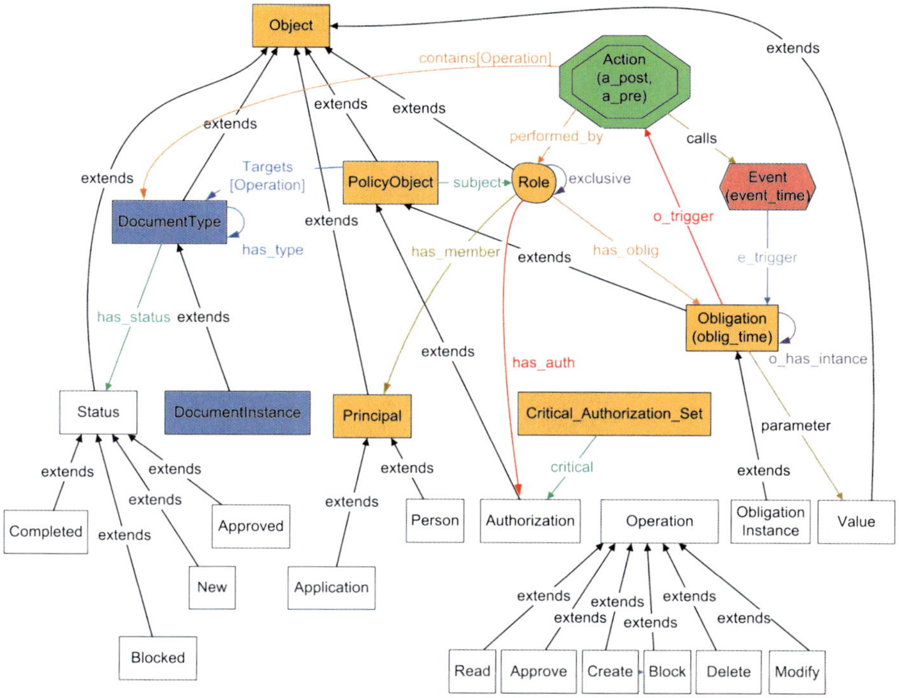

Fig. 9 Domain ontology

Finance controller—R2_FinanceController, Credit controller—R3_CreditController. Similarly, specific Authorizations, Obligations, and Documents were defined and relationships between them were set based on the ontology relations. For example, the role R1_OrderDeskClerk was set as exclusive to the roles R2_FinanceController and R3_CreditController using the relationship "exclusive" defined in Role signature:

```
one sig R1_OrderDeskClerk extends Role{}
{
exclusive = R2_FinanceController+R3_CreditController
has_oblig = O1_ToRegisterOrder
has_auth = A1_ToRegisterOrder
has_member = J.Johnson
}
```

According to the EPC model, the process starts with the reception of the client's order. The signature E1_OrderHasCome which is a successor of the Event concept was defined. The time of event was set as the very first moment TO/first represented by the successor of the Time signature. E1_OrderHasCome calls Obligation to register the order.

```
one sig E1_OrderHasCome extends Event {}
{
o_has_instance.(e_trigger) = O1_ToRegisterOrder
event_time=TO/first
 }
```

O1_ToRegisterOrder is a successor of the Obligation signature; however, according to the modeling methodology, the event calls not Obligation but ObligationInstance that refers to a more general Obligation. Therefore, to link the current Event with Obligation the relationship o_has_instance which connects signatures Obligation and ObligationInstance is used. A special fact is defined to link obligation with the ObligationInstance events.

```
fact ObligationInstanceFromO1
{
all oi: ObligationInstance |
        o_has_instance.oi = O1_ToRegisterOrder =>
        oi.o_trigger = F1_OrderRegistration
}
```

If the event triggers the obligation to register the order it means the execution of the function F1_OrderRegistration. Current model defines a strict logic of the business process, thus a possibility of accidental events and actions is not taken into account. This significantly restricts expressiveness of modeling; however, this aspect can be resolved while further model development.

Inside the Action of order registration several parameters are defined including Event that will follow Action execution, Role, which will execute Action and Document which will be the object of Action. The result of Action execution is a change of Document status.

```
one sig  F1_OrderRegistration extends Action {}
{
calls=E2_OrderCreated
performed_by=R1_OrderDeskClerk
contains.DocumentInstance =
Create ((Create.contains).has_status).a_post = New
}
```

The current description corresponds to the business process part from Fig. 10. The same business process with all the connections is presented in Fig. 11. Finally, the result generated by Alloy Analyzer can be seen in Fig. 12.

In addition to the static modeling, the modeling of dynamics using Alloy may be demonstrated. A particular model is developed to show changes of the business process flow in the different moments of time. The concept of modeling dynamics using Alloy was initially described by Daniel Jackson in [16]. His concept with some modifications was used for the developed dynamics modeling algorithm. This algorithm is defined on the ontology level; however, as all the entities of the second level are successors of the first level signatures, all rules and restrictions are applied to them and dynamics on the business process level is also demonstrated.

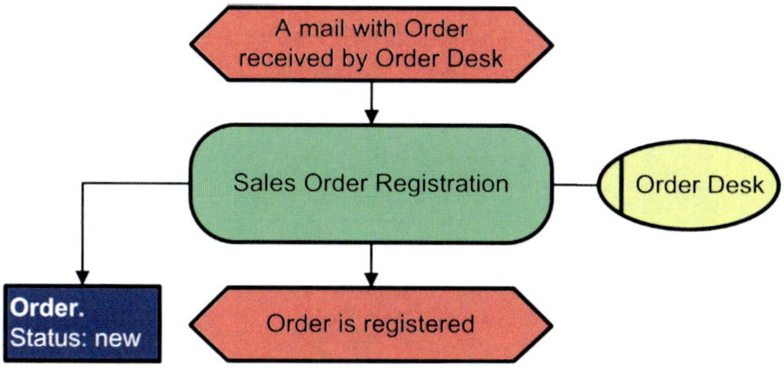

Fig. 10 The original EPC Diagram

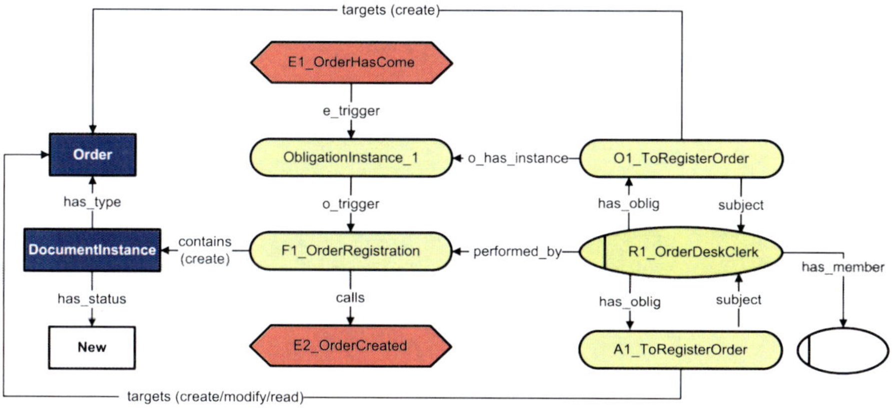

Fig. 11 Detailed information in EPC notation

MIT Alloy Analyzer generates the set of all possible scenarios of the business process flow that meet defined restrictions. The results of the particular scenario generation can be represented in several ways, including a graphical preview. This kind of representation shows graphical figures for all the instances entities described in the model. When the object (such as `Event`, `ObligationInstance`, and `Action`) occurs/starts or finishes executing, the corresponding time indicators appear in the figures. For example, at the moment `Time0` the model shows which instances of `Event`, `ObligationInstance`, and `Function` occur/start executing at the first moment of time. In Fig. 10 `E1_OrderHasCome` has an indicator "event_time", `ObligationInstance2` has an indicator "oblig_time", and `F1_OrderRegistration` has an indicator "a_pre". The indicators show instantiation of these objects at the current moment of time (`Time0`) (Fig. 13).

At the next moment `Time1` function `F1_OrderRegistration` finishes its work (it has an indicator "a_pre"). The function calls the next `Event`: `E2_OrderCreated`. The `Event` triggers `ObligationInstance`

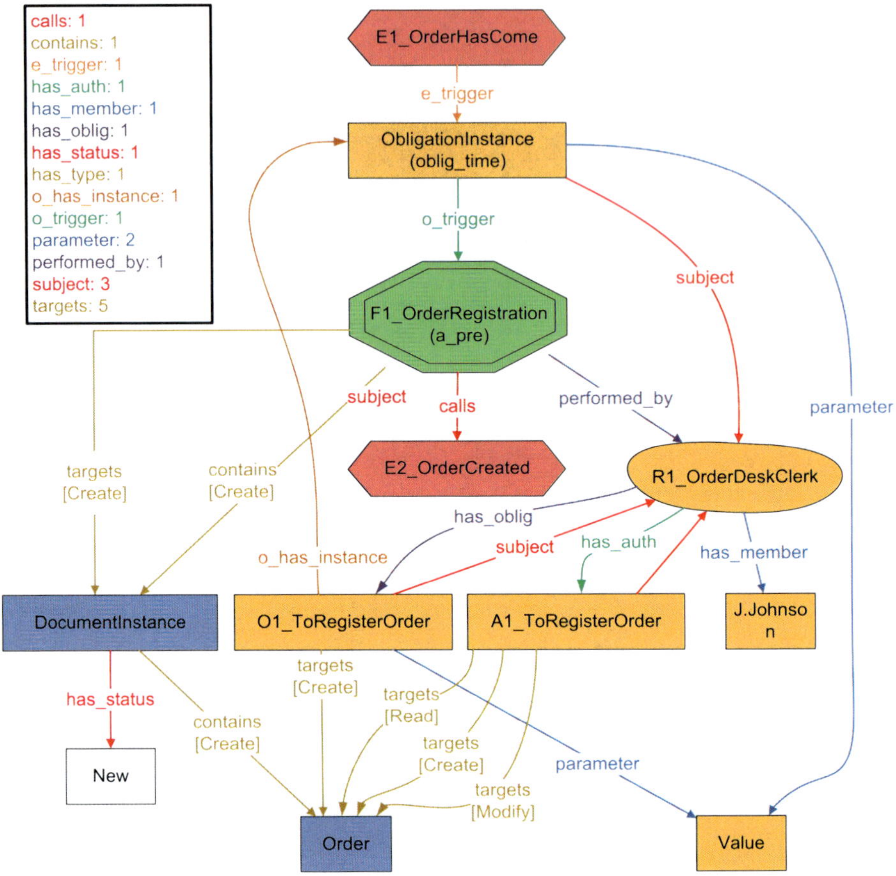

Fig. 12 The generated Alloy diagram

of `O2_ToCheckPrices`, which in turn initiates the start of function `F2_OrderPriceCheck`. (see Fig. 14).

The next step of the work is to model an algorithm which will automatically define the service orchestration scenarios using the approach to service modeling described above. Each function of the business process is associated with some program methods. Also the mechanism of synchronization of a service and a business function is defined enabling the automated service orchestration.

Figure 15 shows how at `Time0` moment the function `F1_Order Registration` and the corresponding method `M1_OrderRegistration (Service1)` start execution.

At `Time1` moment `M1_OrderRegistration` finishes its work and according to the business process the function `F2_OrderPriceCheck` should start execution. That, in turn, causes the method `M2_OrderPriceCheck (Service2)` to start execution (see Fig. 16).

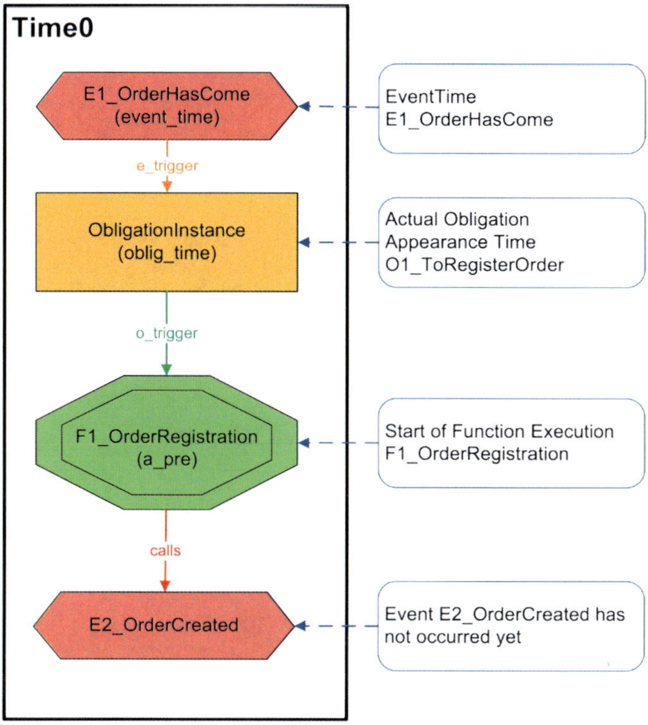

Fig. 13 The business process state at `Time0`

As a result of modeling at Layer 3, the sequence of services is defined for each scenario of the business process. In order to get the complete BPEL code those scenarios should be compiled together in one common XML code. The compilation of XML code and translation into BPEL is beyond the scope of the current case study.

Nevertheless, applicability of the developed method for automated orchestration of services was proved. Alloy assertions enable check of information security controls. These controls are thoroughly described in [27]. For example, the created model enables verification of the following assumptions:

- One principal does not have two exclusive roles;
- One principal does not have a set of critical authorizations;
- One role does not have a set of critical authorizations;
- No role can execute all the actions in the same document;
- Exclusive roles have different authorization sets;
- One service cannot execute all the actions in the document.

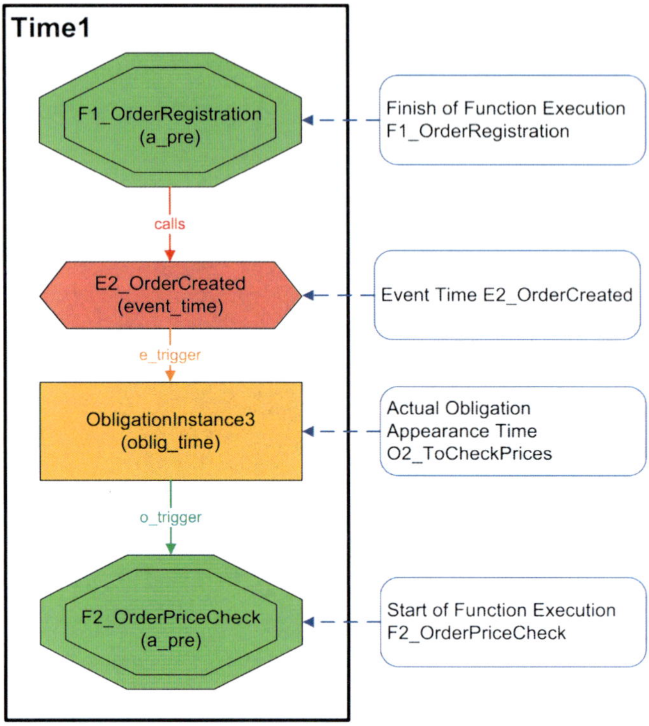

Fig. 14 The business process state at `Time1`

4.2 Service Template Specialization for WebLab Platform

In this sub-section we will focus on exploring our method of template-based orchestration (see Sect. 3.4) in the particular context of the WebLab platform providing a real-life set of abstract and concrete services. The WebLab platform [11] facilitates the development of multimedia projects leveraging a rich set of tools to create complex processing chains and dynamic graphical front-ends for domain users. The platform has different groups of users interested in design and using complex service chains of multimedia processing services. Each group of the users has own preferences and domains of discourse.

The WebLab platform has a service-oriented architecture and uses a common data model and generic interfaces. However the design of customized applications and sustainable management of individual services or composite processes in the WebLab platform still require considerable intellectual efforts and time. Firstly, the WebLab platform contains a large and frequently changing set of heterogeneous multimedia components which are represented as services and may be integrated in processing chains in different ways. Secondly, different groups of the WebLab users prefer different approaches to description of service functionality.

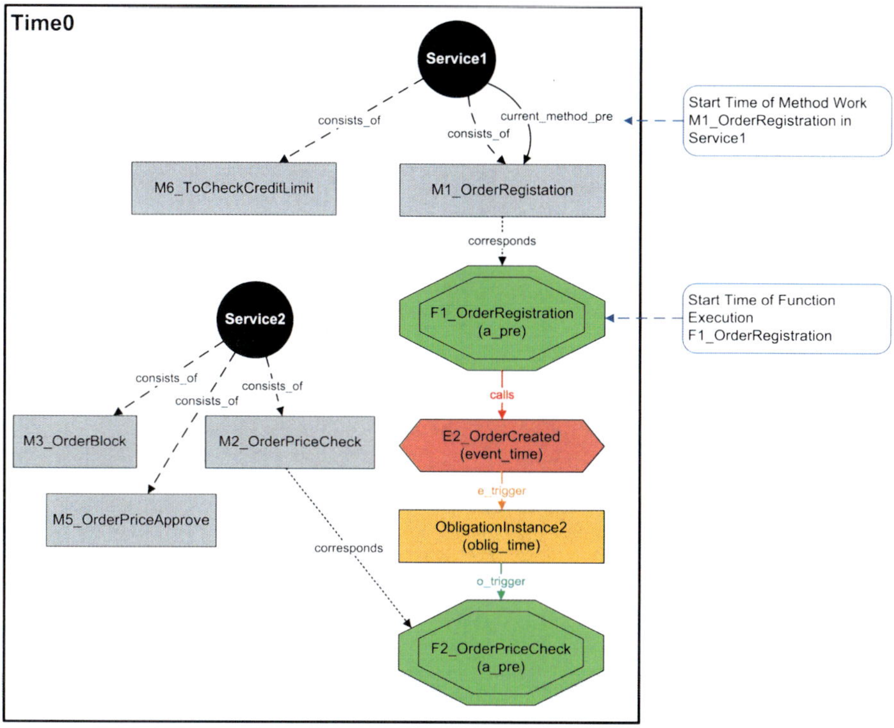

Fig. 15 Service launch at `Time0`

The semantically enriched service registry of the WebLab Platform [6] facilitates integration of the different viewpoints and maintains links between OWL-S and the domain-oriented WebLab vocabulary. An upper level ontology and specific inference rules help to seek a particular service and automatically infer the set of implementation-specific characteristics of that service (in particular, Input, Output, Preconditions, and Effects of the service—IOPE) on the basis of domain-oriented initial information. In turn, WebLab experts (like automated agents) are now able to search this registry for the services which fit with the needed IOPE (even indirectly).

However, currently the semantically enriched service registry of WebLab platform still does not have an opportunity to populate abstract OWL-S processes with particular instances of appropriate simple processes, thus the registry cannot implement service orchestration tasks. To do this we propose using formal methods of Relational Logic and specific tools of formal analysis such as Alloy Analyzer.

At first, we need to express the second part of the logical model in the particular terms of the existing WebLab services. For this purpose we use information from the semantically enriched WebLab service registry which stores information about service taxonomy and IOPE for each available service. An example given below shows the Alloy declaration of some known IOPEs of WebLab services.

```
sig AsNativeContent, ContentExtracted extends Condition{}
```

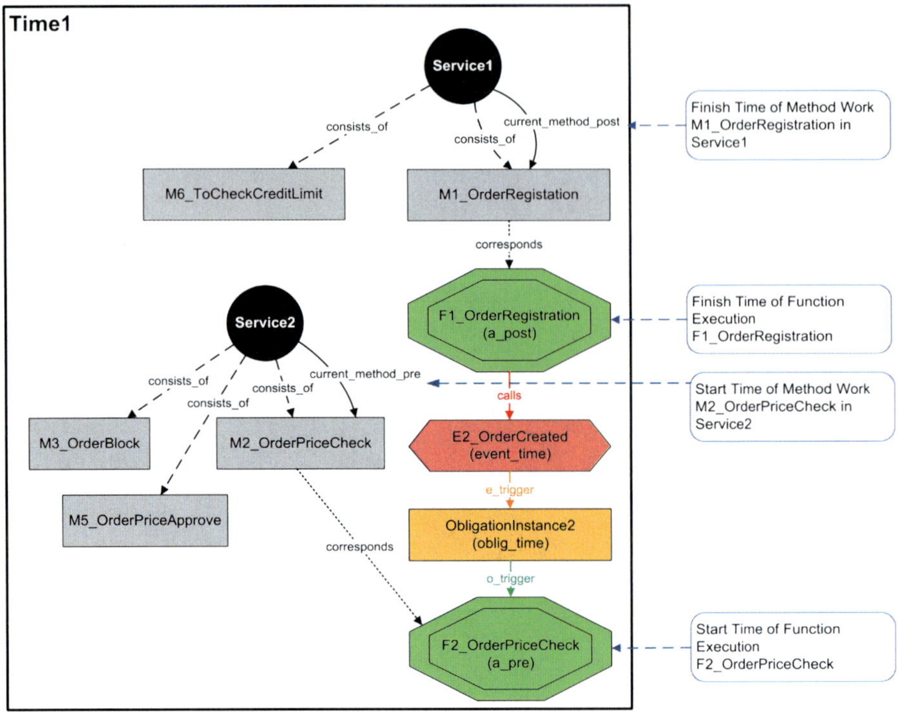

Fig. 16 Service launch at `Time1`

We also have to define service specification for each item in the registry. Such specifications are done in accordance with the following contents of the registry.

Service name	Preconditions	Effects
WebCrawler		asNativeContent
RSSCrawler		asNativeContent
Normaliser	asNativeContent	contentExtracted
EventExtractor	geoExtracted neExtracted	eventsExtracted
NEExtractor	contentExtracted	neExtracted
GeoExrtactor	contentExtracted	geoExtracted
FullTextIndexer	uniqueURI	searchable
FileRepository		uniqueURI
TextCleaner	contentExtracted	textCleaned

WebCrawler and RSSCrawler are two robots which are able to download content, respectively, from web sites and RSS feeds. Normaliser is able to extract content from an original file. In other words, it extracts a normalized text from web pages, pdf files, doc files, etc. EventExtractor, NEExtractor, and GeoExrtactor are text parsers which are able to extract information from this text. NEExtractor service focuses on the named entities part. The GeoExtractor service is able to extract a location (city, country, etc.) and to add its geographical position. EventExtractor processes geographical information and named entities in order to notify events in the text. FullTextIndexer is a classical text indexer which enables document retrieving from a text request. FileRepository manages the resources persistence, guarantees an unique URI on each resource it saves, and delivers the previously saved resource using its URI. TextCleaner is used to remove unused empty text (multiple new lines, etc.).

Each registry definition of the service including effects and preconditions is translated into corresponding logical statements. A simple extract from the complete formal specification of the services is given below.

```
sig Normalizer extends Service {} {
  preconditions = AsNativeContent
  effects = ContentExtracted
  one AsNativeContent
  one ContentExtracted
}
```

The specification of all services defines the second part of the logical theory in our approach. Specification of available workflow templates is the next step in our method. In the explored use case we limited available templates to only one, already presented in Sect. 3.4. That template represents a classic processing workflow which satisfies multiple user needs. It is composed of a loop which could be interpreted as "while there is some resource to compute, compute it." So the first treatment gets a resource from a reader, the second checks if this resource is already present in our platform. The third one analyzes this resource in order to extract certain information and the fourth one stores this analyzed resource.

Finally we need to specify and formalize requirements which will define the fourth part of the logical model. In our use case a user has needs in a service-oriented application which is able to make a continuous event extraction from different web sources about disasters on the world. The user also needs to search this collected data. Using this description, we can find high level preconditions and effects: "continuous" means cycle, "event extraction" means eventExtracted, and "need to search" means searchable. The specification of that requirement may be translated in terms of Relational Logic as it was shown in Sect. 3.4.

Having the defined logical theory we may execute logical analysis using Alloy Analyzer modeling tool: `run cyclicProcess for 10`. In the result we get several variants of instantiation of the signatures and relations defined in our logical theory. The collection of instantiated signatures and relations describes each possible transition between workflow elements (control elements and concrete services).

For visual overview the same result of logical analysis can be represented in the form of the graph. For example, in Fig. 17 one possible variant of instantiation is represented. Projection over instances of P1 signatures allows for intuitively

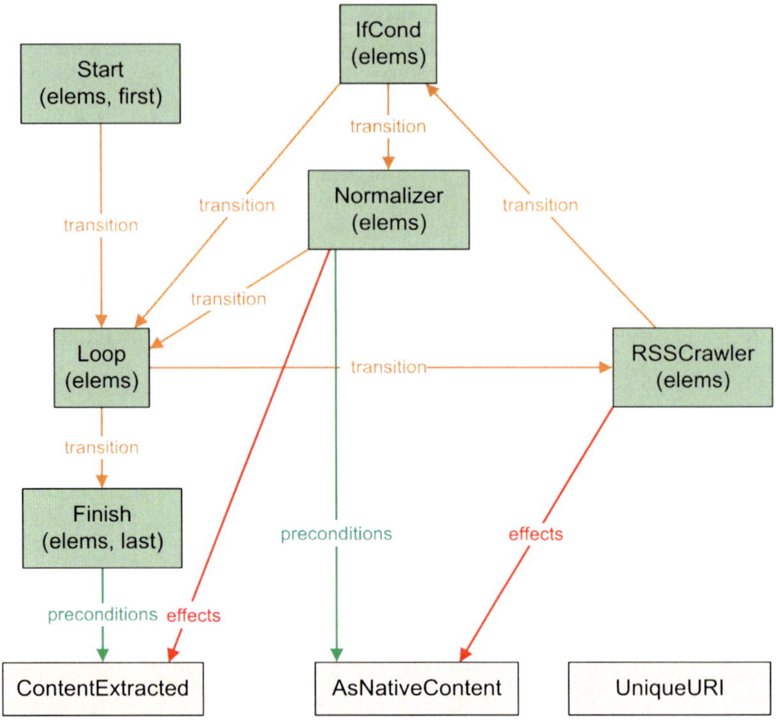

Fig. 17 The required specification of the condition for service orchestration

clear representing the results. There are two types of nodes in the graph: workflow elements and Condition elements. The start of workflow is indicated by including the signature instance to the relation "first" (in our case it is the instance of the Start signature). The end of workflow is indicated accordingly, by including the instance to the relation "last" (it is the instance of the Finish signature). Tracing over the relation "transition" determines the control flow of the obtained solution. In our case it corresponds to the control structure of the orchestration pattern. Relations "preconditions" and "effects" show which particular preconditions are required for service execution, and which effects are present after service execution.

5 Conclusion

Our work introduces new approach for service orchestration. The developed approach covers main conceptual organizational levels required for business process automation: the level of enterprise application domain, the business process level, and the program service level. A formal logic language is used for modeling structural properties and dynamics of business processes.

An algorithm based on finite scope logical analysis and relational logic is developed to ensure consistency and relations between different level models. To implement the algorithm MIT Alloy Analyzer is chosen.

Finally, the application of proposed approach for two concrete cases was demonstrated. Usage of proposed hierarchical modeling, domain ontologies, and Alloy relational logics as its key element were justified.

One of the most important tasks of IT and business societies is to establish reliable knowledge management processes including usage of gathered knowledge. The described approach proposes the practical and structural way of knowledge gathering, analysis, and further using business automation based on SOA. Proposed approach provides for an opportunity to integrate different level models (enterprise, BP, service levels) and to take into account all constraints that can affect the IT solution in an automated and semi-automated way. Moreover, counterexamples showing violation of constraints can be found and demonstrated. The main advantage of the proposed approach is increased reusability of expert knowledge of domain experts and IT experts expressed through the workflow templates and an ability to use information about complex interdependences within the existing hierarchy of services. Because the requirements to the service orchestration may be redefined in an iterative manner the proposed approach satisfies the needs of software engineers who design customized service-oriented applications.

The risk of human mistakes and poor analysis inefficiency can be significantly mitigated due to decreasing the number of operations performed by people at the modeling and analysis stage during IT solutions implementation or change. In addition, time for development and implementation of new processes can be reduced. All listed benefits give an opportunity to make IT solutions based on SOA much more flexible, easier, and cheaper to implement and maintain.

Our work is significantly influenced by such SWS initiatives as OWL-S and WSMO. MDA approach [24] served as a basis for multi-layer methodology development. SUPER project [17], which had similar objectives and tasks, also significantly contributed to our work. However, our approach combines the means of MIT Alloy Analyzer with the advantages brought by joining different layer models into single conceptual model.

A correspondence to some other researches may be found, mainly to [7, 13, 29] which describe how to generate executable processes from automatic semantic services composition. However our approach has several differences. At first, our method of template-based orchestration is based on combination of a taxonomy of concrete and abstract services, and reusable domain expert knowledge in the form of workflow templates. Other approaches use only a flat set of services which is named library. Second, we propose to express requirements for service orchestration and the structure of reusable workflow templates directly using an end-user vocabulary for translation to the statements of the Alloy Analyzer language, whereas previously cited research works choose more complex dialects of logic.

As for the modeling methodology, concepts of policy objects, such as authorization, obligation, and obligation instance are taken from [27]. Also the organizational controls to be checked were worked out by Kuhn [18]. However, while the emphasis in [27] was made on organizational control principles, our work is mainly aimed at developing architecture and a methodology for service composition. As for the modeling methods, including dynamics modeling, our work is contributed by the works of Daniel Jackson, creator of MIT Alloy Analyzer [14–16].

The SUPER project mentioned above also aims at developing a semantic-based and context-aware framework based on SWSs technology. It is intended to make companies more adaptive by management of business processes knowledge embedded in within IT systems and employees' heads [17]. However, our approach has several distinctive features. Means of MIT Alloy Analyzer are applied and different layer models are united in the single conceptual model. This enables:

- Ability of automated expansion of restrictions between all modeling levels
- Reduction of ontology transformation procedures
- Generation of examples and counterexamples for processes execution and for sequence of service invocation
- Simplification of reasoning

This work determines the key aspects of practical implementation of the proposed approach, which is the main task for further work in the current direction. In our further research we will work on developing original software architecture for a special middleware component which uses main principles of our approach for real-time composition of web services in complex distributed environments. We plan to evaluate the created component in several practical cases that will give us extra information about practical applicability of our approach to service composition.

References

1. Bhiri, S., Gaaloul, W., Rouached, M., Hauswirth, M.: Semantic Web Services for Satisfying SOA Requirements. In: Advances in Web Semantics I, LNCS **4891**, 374–395, Springer, Heidelberg (2008)
2. Borgida, A., Brachman, R.J.: Conceptual Modeling with Description Logics. In: Baader, F., Calvanese, D., McGuinness, D.L.,Nardi, D.,Patel-Schneider, P.F. (eds.) The description logic handbook: theory, implementation, and applications, pp. 349–372. Cambridge University Press (2003)
3. Channabasavaiah, K., Holley, K., Tuggle, E.M.: Migrating to a service-oriented architecture. IBM DeveloperWorks (2003)
4. Curbera, F., Goland Y., Klein, J., Leymann, F., Roller, D., Thatte, S., Weerawarana, S.: Business Process Execution Language for Web Services, Version 1.1. http://www-106.ibm.com/developerworks/library/ws-bpel/ Cited 01 May 2012
5. Domingue, J., Cabral, L., Galizia, S., Tanasescu, V., Gugliotta, A., Norton, B., Carlos, P.:IRS-III: A broker-based approach to semantic Web services. J. Web Sem. **6(2)**, 109–132, (2008)
6. Doucy, J., Abdulrab, H., Giroux, P., Kotowicz, J.: A new approach to populate a semantic service registry. LNCS, **6724**, pp. 112–125, Springer, Heidelberg (2011)
7. Duan, Z., Bernstein, A., Lewis, P., Lu, S.: A model for abstract process specification, verification and composition. In: ICSOC '04: Proceedings of the Second International Conference on Service Oriented Computing. pp. 232–241. (2004)
8. Emig, C., Langer, K.: The SOAs Layers, Cooperation and Management, Universitt Karlsruhe (TH) (2006)

9. Erl, T.: What is SOA – Service-Oriented Architecture http://www.whatissoa.com/p10.asp. Cited 01 May 2012
10. Eshuis, R., Grefen, P., Till,: Structured service composition. Vol. 4102 LNCS. (2006).
11. Giroux, P., Brunessaux, S., Brunessaux, S., Doucy, J., Dupont, G., Grilheres, B., Mombrun, Y., Saval, A.: Weblab – An integration infrastructure to ease the development of multimedia processing applications. In: International Conference on Software and System Engineering and their Applications (ICSSEA) (2008)
12. Gruber, T. R.: A Translation Approach to Portable Ontology Specifications. Knowledge Acquisition, **5(2)**, 199–220 (1993)
13. Halle, S., Villemaire, R., Cherkaoui, O., Ghandour, B.: Model-checking data-aware temporal workflow properties with CTL-FO+. In: Proceedings - IEEE International Enterprise Distributed Object Computing Workshop, EDOC. pp. 267–278. (2007)
14. Jackson, D.:Automating First-Order Relational Logic, ACM SIGSOFT Software Engineering Notes, Volume 25, Issue 6, 130–139, November (2000)
15. Jackson, D., Shlyakhter, I.,Sridharan,M.: A Micromodularity Mechanism, In Proceedings of the ACM SIGSOFT Symposium on the Foundations of Software Engineering, pp. 62–73. (2001)
16. Jackson, D.: Software Abstractions: Logic, Language, and Analysis. The MIT Press Cambridge, Massachusetts (2006)
17. Karastoyanova, D., Lessen, T. van, Leymann, F., Ma, Zh., Nitzsche, J., Wetzstein, B., Bhiri, S., Hauswirth, M., Zaremba, M.: A Reference Architecture for Semantic Business Process Management Systems. In: Multikonferenz Wirtschaftsinformatik (2008)
18. Kuhn, D.R.: Mutual exclusion of roles as a means of implementing separation of duty in role-based access control systems. In: Proceedings of the second ACM workshop on Role-based access control, pp. 23–30. (1997)
19. Martin, D., Paolucci, M., Mcilraith, S., Burstein, M., Mcdermott, D., Mcguinness, D., Parsia, B., Payne, T., Sabou, M., Solanki, M., Srinivasan, N., Sycara, K.: Bringing semantics to web services: The OWL-S approach. LNCS **3387**, pp. 26–42, Springer, Heidelberg (2005)
20. Martin, D., Burstein, M., McDermott, D., et al.: OWL-S 1.2, http://www.daml.org/services/owl-s/1.2/. Cited 4 May 2012
21. McIlraith, S., Son, S., Zeng, H.: Semantic Web Services. IEEE Intelligent Systems, 16(2):46–53, (2001)
22. McIlraith, S., Son, S.: Adapting Golog for composition of semantic web Services. In: Proc. 8th International Conference on Principles of Knowledge Representation and Reasoning (2002)
23. Ning, H., Yongyi, P., Camilo, R.: Extend OWL-S dynamic semantics with rewrite logic. In:Proc. International Conference on Computer Science and Software Engineering, CSSE 2008, Vol. 2, pp. 346–349. (2008)
24. Object Management Group. MDA Guide V1.0.1. OMG http://www.ultradark.com/01mda13userguide.htm. Cited 4 May 2012
25. Pahl, C.: Ontology Transformation and Reasoning for Model-Driven Architecture. In: Meersman, R., Tari, Z. (eds.) CoopIS/DOA/ODBASE 2005, LNCS, **3761**, pp. 1170–1187, Springer, Heidelberg (2005)
26. Roman, D., de Bruijn, J., Mocan, A., Lausen, H., Bussler, C., Fensel, D.: WWW: WSMO,WSML, and WSMX in a nutshell. In: 1st Asian Semantic Web Conference, pp. 516–522. Springer, Beijing (2006)
27. Schaad, A.: A Framework for Organizational Control Principles, PhD Thesis, Department of Computer Science. University of York (2003)
28. Sheshagiri, M., des Jardins, M., Finin, T.: A Planner for Composing Services Described in DAML-S. In: Proc. of Workshop on Web Services and Agent-based Engineering - AAMAS03, (2003)
29. Traverso, P., Pistore, M.: Automated composition of semantic web services into executable processes. In: McIlraith, S.A., Plexousakis, D., van Harmelen, F., eds.: International Semantic Web Conference. LNCS, **3298**, pp. 380–394, Springer, Heidelberg (2004)

30. Verma, K., Gomadam, K., Sheth, A.P., Miller, J.A., Wu, Z.: The METEOR-S Approach for Configuring and Executing Dynamic Web Processes. LSDSIS technical report http://lsdis.cs.uga.edu/projects/meteor-s/. Cited 10 May 2012 (2005)
31. Wallace, C.: Using Alloy in process modelling. Information and Software Technology, **45(15)**, pp. 1031–1043. (2003)
32. Wang, H. H., Saleh, A., Payne, T., Gibbins, N.: Formal specification of OWL-S with object-Z: The static aspect. In: IEEE/WIC/ACM International Conference on Web Intelligence, WI 2007, pp. 431–434. (2007)
33. Wegmann, A., Lł, L. -., De La Cruz, J. D., Rychkova, I., Regev, G.: An example of a hierarchical system model using SEAM and its formalization in Alloy.In: Proceedings - IEEE International Enterprise Distributed Object Computing Workshop, EDOC. (2007)
34. Wu, D., Parsia, B., Sirin, E. Hendler, J., Nau, D.: Automating DAML-S Web Services Composition using SHOP2. In: Proc. of the Second International Semantic Web Conference (ISWC2003), 2003
35. Yang, B., Qin, Z.: Composing semantic web services with PDDL. Information Technology Journal 9 (1), 48–54. (2010)

Estimating Customer Service Times on a Rail Network from GPS Data

Shantih M. Spanton and Joseph Geunes

Abstract A key input to managing and scheduling customer service on a rail network is an accurate characterization of the time a particular train takes to serve a customer, based on the quantity of work the customer requires. The algorithm presented here estimates the customer service times of a train on a given day utilizing global positioning system (GPS) data from train locomotives, the known customers to be served on the day and some geographic knowledge of the customer's tracks. The algorithm was built and evaluated on real data sets provided by CSX. This research was conducted as a joint effort between the University of Florida in Gainesville, FL and CSX Transportation, Inc. in Jacksonville, Florida.

Keywords Railroad • Vehicle tracking • GPS • Geofence • Experimentation • Algorithm • Polyline

1 Introduction

The majority of customer contact on the CSX Transportation, Inc. rail network occurs on what are known as local trains. These trains transport blocks of cars directly from yards to customer facilities, and vice versa. Unfortunately, local yard congestion and variability in work volume and crew availability create complexity in planning and allocating local work activities. A key input to efficiently managing local trains is an accurate characterization of the time it takes a train to serve a customer, based on the quantity of work the customer requires as well as the travel time between subsequent customers. For each train on a given day, the arrival and departure times at a served customer provide one point estimate of the service time

S.M. Spanton (✉) • J. Geunes
University of Florida, Gainesville, FL, USA
e-mail: sspanton@ufl.edu; geunes@ise.ufl.edu

A. Sorokin and P.M. Pardalos (eds.), *Dynamics of Information Systems: Algorithmic Approaches*, Springer Proceedings in Mathematics & Statistics 51,
DOI 10.1007/978-1-4614-7582-8_3,

for that customer. After applying this analysis to a sufficiently large number of days of historical data, a customer will have multiple point estimates of the time required for service. A statistical estimation technique may then be used to characterize the relationships between the dependent variable of service time and the many possible influencing parameters (type of work, number of cars handled, crew, day of week and train). Once the customer service time has been functionally estimated there are numerous potential industry uses such as optimizing customer to train scheduling, dynamic work load balancing, investigating service days that have deviated from expected service times, and real time announcements of expected delivery times to customers.

In the current system the train crew enters a single time stamp value to indicate when work was performed at a customer. This information is recorded on what is called a work order and contains information such as number of cars, movement type (place or pickup), customer identification number, train, and work assignment date. For example, the crew would report that today at 14:20 they placed 15 cars at the first customer to be served. This does not provide the time the work began or ended and thus does not provide the total time spent serving this customer. While it is possible to modify the data entry procedure to require the crew to enter both a start and stop time several factors make this undesirable. Most importantly, were the crew to enter both starting and stop times of work, the required time spent reporting information would double. The job of the train crew in a yard location is an extremely challenging one. Each day presents a new set of challenges which require spontaneous modifications and constant alertness. It may be burdensome to perform such secondary administrative duties at certain times resulting in post reporting of work which implies the times entered are determined at the discretion of the crew. Relying on an individual performing detailed complex job maneuvers to remember starting and ending service times may result in unintended inaccuracies. In addition to the service times, the work order must also record the serviced customer and work information. CSX serves thousands of customers on its network, several of whom have multiple facilities at proximate locations denoted by their own unique codes. Ensuring that the work order is completed for the correct customer ID, with the correct times, the correct number of cars as well as the quality of work (pick up or deliver or relocate) creates an added burden on a train crew. Other incentive-based reasons may cause under/over or missed reporting of customer service, such as break times, daily performance goals, or difficulties in overriding entered data. An ideal determination of customer service times will minimize crew data entry.

The widespread use of global positioning system (GPS) devices for vehicle tracking and routing implies several potential solution methods for estimating customer service time. GPS technology is widely available today and allows tracking of the position of any equipped vehicle. A known sequence of customers served in combination with GPS data captured from the train's locomotives can provide a measure of how long the train was physically near a customer's location. GPS devices are installed on 85–90 % of all CSX locomotives (operational or otherwise). CSX utilizes several devices including Trimble Crosschecks, the new Trimble TVG660, GE ATS, and Pinpoint I and II. When a locomotive is turned on, the GPS initiates and begins recording its location. Normal frequency of rccording is one measurement per minute. Numerous factors, such as purpose and proximity to industry defined areas of interest, determine how often the device transmits location information. In an area that requires real-time reporting, the position measurements may be transmitted at one measurement per minute. While for other transmissions, the device may record its position several times before transmitting the data. Depending on the type and age of the GPS unit and the function of the locomotive, 10 data recordings may be sent per 10 min, 20 per 20 min, or 60 per 60 min.

All locomotives have Wi-Fi, cellular and satellite communication devices on board. Depending on the availability of a signal, the GPS measurements will be transferred via one of these methods. While Wi-Fi transmission is preferred, it may not be available outside of yards or urban areas. The data is communicated directly to CSX with the exception of the Trimble devices which pass through a third party preprocessing at Trimble. The quantity of data received and stored on a single day is roughly three million records per day. The quantity of useful data in the analysis is very large and a substantial amount of time must be devoted to cleaning, interpreting, and learning how to correctly utilize the numerous required data sources. The physical nature of each customer and the way in which each is served varies widely making it difficult to apply traditional travel time/stop concepts. The creation of a data mining algorithm that can repeatedly extract service time information must be carefully constructed.

While the path the train follows can be known from the GPS measurements, several factors complicate the analysis and make determining customer service times from this information alone difficult. GPS spatial and temporal measurements are accurate only to within the error tolerance of the GPS device. Consider a single latitude and longitude position measurement in the set of all readings for a train on a given day. The error in the GPS measurements implies that if several tracks are present at the location of this reading, it may not be possible to determine on which track the train was truly located, even if the underlying track structure is known. Thus a train on a customer track is often indistinguishable from one on a nearby mainline track. For a discussion on the inherent performance issues of GPS which occur on the rail network, see [10].

This paper presents an algorithm that estimates customer service using only GPS data from the locomotives, the sequence of customers who were served on a given day, and geographic information about each customer. The algorithm requires no user input (other than GPS data, a sequential list of customers served and customer site information which is known a priori) and scales well for trains traveling routes of varied length. The algorithm is by no means unique to locomotives and can be modified to determine the customer service events of any GPS enabled vehicle. The paper then presents the method used to determine potential service time events followed by the methods used to classify these events as customer related or otherwise. A discussion on accuracy and testing of the method follows.

2 Solution Motivation

To an individual familiar with the problems of GPS travel time performance measures and classification, the determination of railroad customer service times may, at first glance, appear to be identical to the estimation of any other vehicle's activities/stops. However, important differences in the way in which trains service customers as opposed to cars or trucks render the same solution methods ineffective.

Presumably, when a train performs service for a customer two things will occur: the train will be near the customer and the train will also slow down or stop to complete the work. Much time has been spent investigating how to extract these two attributes from GPS data.

2.1 Fixed Geofence from Customer Geography

Commonly, to determine the time a vehicle spends near or in a particular location, a fixed spatial boundary or polygon (often called a "geofence") is drawn around the location of interest. When the vehicle enters the boundary (as observed from the GPS data), the vehicle is said to have arrived and its last GPS reading within the boundary denotes its departure. This approach has been used to track vehicles in various industries [11, 13, 17, 21, 23].

An obvious way to define a geofence to track customer service stops is to use the customer's facilities as a geofence. CSX maintains detailed GPS profiles of each customer facility on its network. Included in this information are outlines of each track in a customer facility. Since a train is constrained to customer track or mainline track outside the customer, approximating the customer by its tracks only (as opposed to considering the entire customer facility) is appropriate. However, the customer track information is insufficient to define a conventional geofence. Customers may not be serviced within their physical facility or along their tracks. Cars may be picked up or placed along the main line track near the customer to be moved later, and also multiple customers may have facilities in such close proximity that their tracks are shared or parallel.

2.2 Fixed Geofence from Historical Train Stop Locations

To define a more reliable geofence, a service design team at CSX considered locations near customers where GPS information showed that trains spent considerable time. In a pilot study, the team considered only a few customer locations. For a customer, 2 years of GPS data for a train whose route served that customer was examined. Dense geographic areas of GPS readings indicated locations where the train spent significantly more time. Those near to the customer facility could be interpreted as the service area for the customer. A fixed polygon was drawn around

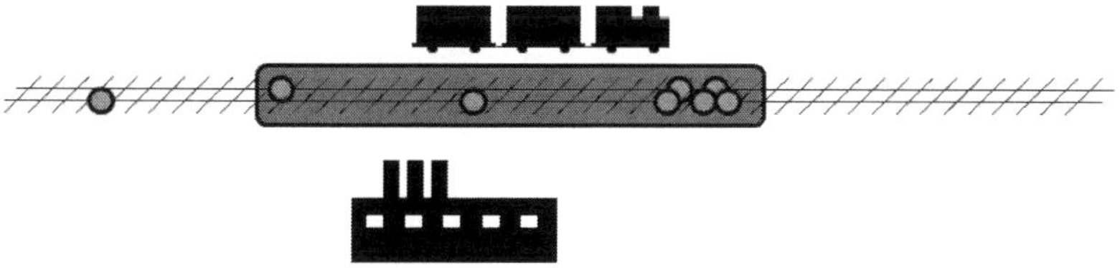

Fig. 1 A fixed geofence estimates service time by considering GPS readings which fall inside

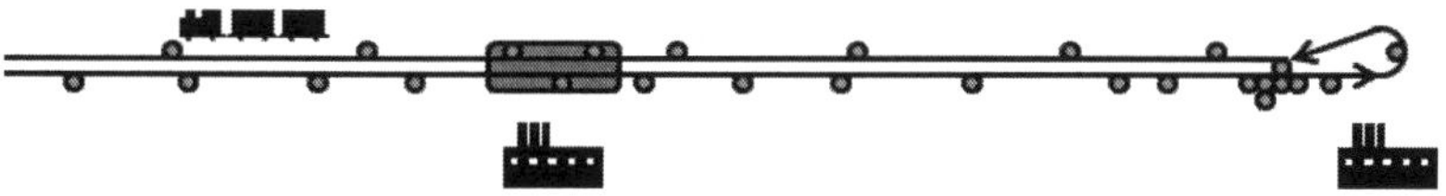

Fig. 2 A train passes through a customer's fixed geofence without servicing the customer

this estimated service area to represent the geofence for this customer's service events. On any given day, the time a train spent within the geofence would be considered its service time; see Fig. 1. The hope was that this method could be generalized to create an algorithmic way that would automatically determine the customer services times for each customer on the entire service network.

Several factors complicate this approach and make determining customer service times difficult. Firstly, this method relies on several years worth of data to define "dense" areas of GPS readings near customers. Since the data includes all days of service, GPS readings near a customer may not truly be from a service event at that customer. While the data can be filtered to remove days on which that customer was not served, this method is still unable to distinguish whether a cluster of readings represents a service event for the customer or one of several nearby customers. Also identifying "dense" areas of GPS readings attributed to customer service in frequently traveled areas such as yards or urban areas (where numerous measurements exist) is nearly impossible.

Secondly, the location where the train must perform work for a particular customer may vary from day to day, depending on the work the customer has requested. Certain customers have cars picked up in one location and placed at another location. Even if it were possible to correctly identify locations where service occurred, this could result in a very large area defined for the customer's geofence. For any day, if the service time is estimated by selecting the first and last GPS readings in the geofence, the value could be over estimated.

Thirdly, false positives and overestimation can occur. When attempting to determine the service time for a customer, the first and last GPS readings observed inside the geofence define the length of the service event. Due to the restrictive nature of rail tracks, a train may have to pass through a customer's geofence several times daily. This can result in a large overestimation of service time, or even a falsely reported service event if one did not occur; see Fig. 2. The issue of falsely recording a service time if a train did not even stop in a geofence can be resolved by only

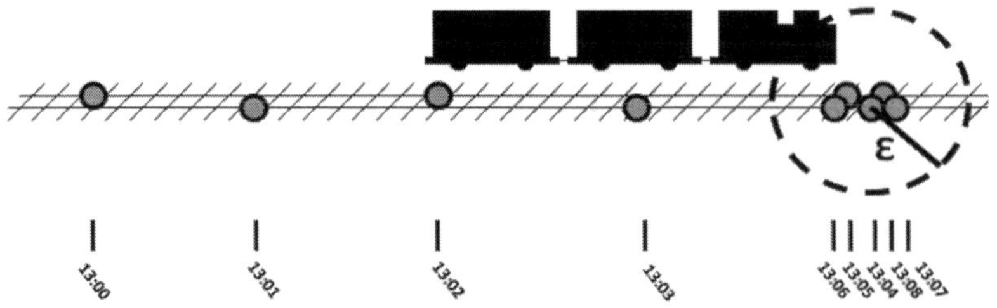

Fig. 3 A stop is measured such that subsequent GPS measurements are within a tolerance ε

considering those GPS readings which occur when the train is stopped. So to truly identify customer service events all times when the train slows or stops near the customer's facility must be identified.

2.3 *Determining Locomotive Stops and Slows*

Research that determines when a vehicle is stopped from GPS data is especially important in public bus transit [2, 5] and household travel activity research [1, 14, 19, 20, 24, 25]. On a day of service, a GPS device in a vehicle periodically records data containing the location and a time stamp for the time of day when the corresponding position data was recorded. By examining the change in position between each successive GPS measurement, the train's path can be traced as a function of time over a given day. If successive GPS recordings appear close to one another, the vehicle has moved very little. In the literature above if it must be determined if a vehicle is stopped, a tolerance level is used such that if subsequent GPS position measurements are less than this distance apart, the vehicle is considered to be stopped; see Fig. 3. If the GPS device of the vehicle being tracked records the velocity as well, this information can be used to determine stops by setting a minimum velocity value below which the vehicle is considered stopped. While the GPS devices used on CSX locomotives do record velocity, locomotives on the local trains considered here often travel at extremely low speeds for safety and logistical reasons. The velocity and directional measurements recorded by the GPS devices in the locomotives are inaccurate and thus unusable at these low speeds. We must rely on the positions of the train traced out by the locomotive GPS devices.

The discussion above leads us towards an obvious initial algorithmic attempt to determine customer service times for a particular day. For a given train on a given day and for each customer served by the train that day, find all the stops in the customer's geofence. If all these stops can be considered part of the service event, then the total service time should be the difference between the latest and earliest times in GPS measurements associated with the stops.

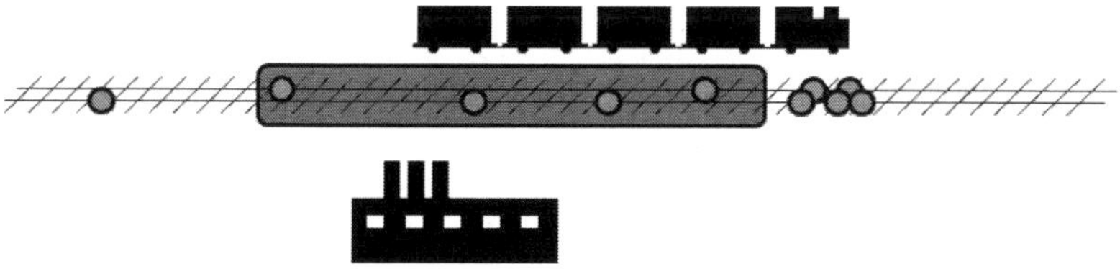

Fig. 4 A long train may indicate a stop further away from a customer than would a shorter train

The difficulties in determining stops on a locomotive arise from the inherent ways in which a train moves to serve its customers. Unlike a delivery truck, much of the time a train spends serving a customer the train is not stopped. When arriving at a customer, the train may have to wait near and then travel between a series of switches to arrive at the customer. Once at the customer, the train (which may already be several car lengths long) must pull clear of the track at which it needs to place or pick up cars and wait for additional switches to be aligned to its path. The train then backs onto the track with the waiting cars, attaches them, and pulls forward. This processes of pulling forward, aligning track segments with switches, and backing onto the track is repeated for one or more of several often parallel tracks containing cars for the train. The train often spends very little of its visit to a customer stationary.

This description of the train's movement brings to light another difficulty in locating service stops near customers: the length of the train is constantly changing. The GPS devices on the train are housed in the locomotives of each. Trains may reach over half a mile in length, and thus the head of the train (and GPS measured location of the train) may be very far away from the individual cars at the end of the train that are being picked up or placed at the customer. Once a stop or slow is found by inspecting the GPS measurements, the determination of its "close" proximity to the customer changes with the length of the train. And if the train is long enough on a day no stops will appear in the customer's geofence as the head of the train may lay outside this geofence for the entire service event; see Fig. 4. In this case, one obvious solution is to extend the geofence of the customer along the track. Extending the track far enough should ensure all service events will always occur within the enlarged geofence. However, this idea fails if several customers with overlapping service locations are visited on the same day. It may be impossible to distinguish between each customer's individual service using GPS data alone; see Fig. 5. Likewise, even if a customer service stop was found inside a geofence, the use of a fixed size geofence does not allow assignments of stops occurring just outside (or overlapping) the fixed boundary; see Fig. 6. While an observed stop inside the customer's geofence could perhaps be correctly understood to be part of the customer's service event, observed stops outside or on the edge of the geofence could be classified as either the service events of nearby customers, or stops near road crossings, as well as possibly being attributed to the customer. The initial pilot inspection of customers found that trains often enter and exit geofence boundaries

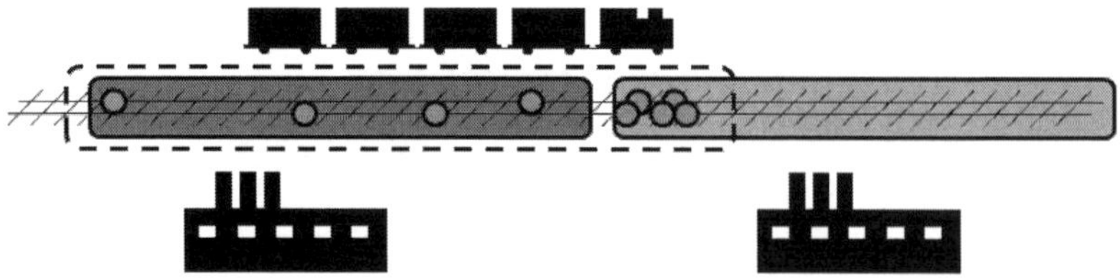

Fig. 5 Customer geofences may overlap causing ambiguity when assigning service stops

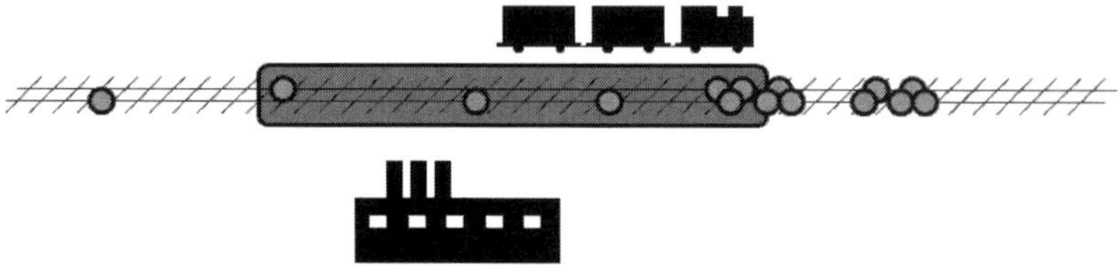

Fig. 6 Service event stops may appear on or immediately outside of a fixed geofence making classification difficult

several times during a visit as the train is repositioned during service. It is possible in singular cases to add logic to this preliminary algorithm to account for some of the numerous issues mentioned above. Unfortunately no uniform set of rules could be found which would accommodate all customers. Certain customer facilities are very large (possibly containing miles of track) while other customers are very small. Urban customers may be very densely populated while rural customers may be several miles apart. All this makes a consistent uniform interpretation of the GPS data impossible; the number of rules required to account for the varied customer geographies and sizes would be akin to processing each of thousands of customers on the network individually.

The desire for an automated means of determining customer service times from the locomotive's GPS data led us to the algorithm which will be described in the next section. It relies on a construction of a distance-versus-time graph from the GPS position and time measurements. Speed-versus-time and speed-versus-distance graphs have been constructed and analyzed for investigating travel time on automotive road networks [6,7,16]. However, as mentioned in [16], the speed varies drastically even across segments of the same road rendering such speed-versus-time and speed-versus-distance graphs difficult to interpret. Also, as mentioned above, the velocity information recorded in the GPS data of locomotives is inaccurate at low speeds. As the train nears a customer we expect the speed of the train to be slow, and it is these areas of low velocity that particularly interest us. The distance-versus-time graph we construct will help us identify times of low velocity where the train remains in the same location over time. These are potential service time intervals. Once these intervals have been identified an attempt at assigning them to customers listed on the day's work order is made. It should be noted that the distance-versus-time graph constructed in [16] is quite different and is applied to solve a different problem.

3 Stop Extraction

Here we summarize the creation of the distance-versus-time graph, which will allow us to identify areas where the train may have stopped or slowed. For any train on a given day the GPS measurements provide a time ordered sequence of latitude and longitude coordinates, see Fig. 7. We will refer to this sequence as S. The process starts by selecting a reference point somewhere outside the range of the latitude and longitude values of all the GPS measurements in S. For example in Fig. 7, we have arbitrarily selected a point P with latitude equal to the minimum latitude and longitude equal to the maximum longitude observed in all the GPS points. This will be a point to the lower right of all the points in S. The distance-versus-time graph is found by taking the Haversine distance between all points $x \in S$ and this point P and plotting these distances versus the time values of each point. Figure 8 shows the distance-versus-time graph for the GPS measurements of the imaginary train route shown in Fig. 7. Observe the two dense areas of points in Fig. 7 in the middle and end of the train's route. As the train slows down or stops near these geographic locations the distance relative to the reference point P changes very little. This is seen as a visually flat spot on the corresponding distance versus time graph Fig. 8.

To identify these flat regions of the graph, the graph was fit with a linear curve approximation (as shown in Fig. 9) known as split-and-merge or polyline fitting in computer visualization. In such an approximation, a curve can be approximated initially by a single line connecting the two endpoints. The approximation is refined by "splitting": identifying which of the other intermediate data points lies furthest from the current linear approximation. The initial line is replaced by two lines: each connecting the initial end points to the point identified as lying furthest from the approximation. Clearly, if this procedure were repeated infinitely, the approximation would be identical to a line traced between all of the data points. After splitting a number of times, areas of the graph which may have been overly fitted are corrected by applying various "merging" algorithms. We have used the Douglas–Peucker algorithm [4, 9] and a vertex reduction algorithm. Such techniques are easy to implement and can be found in any introductory text on computer visualization; for a survey of polygonal simplification algorithms see [8].

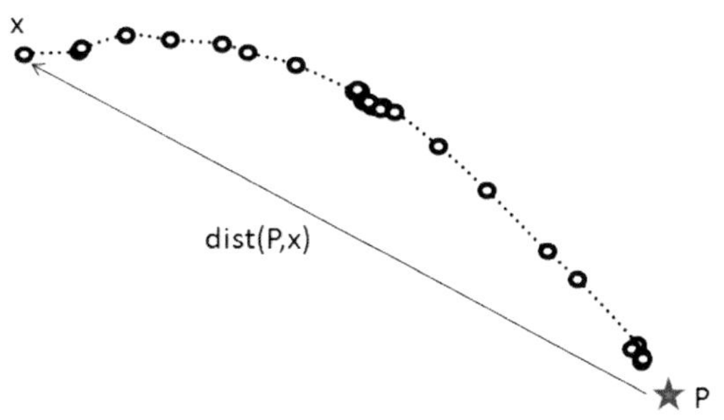

Fig. 7 GPS latitude and longitude measurements of a train and their distance from a reference point P

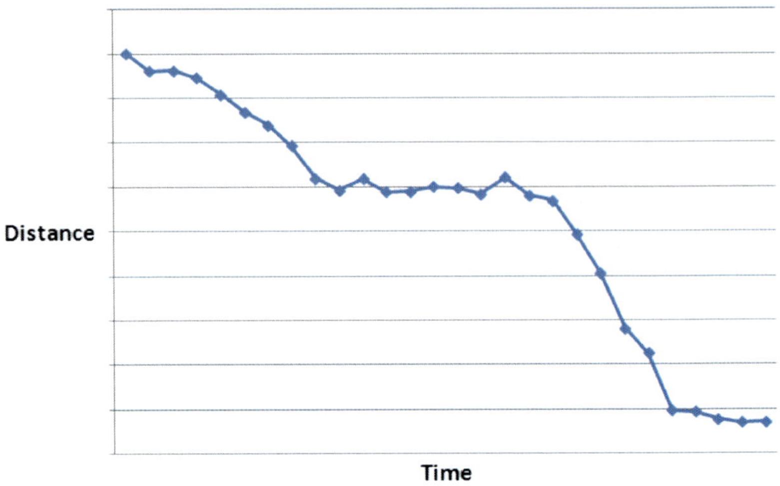

Fig. 8 Distance-versus-time graph (arbitrary units) relative to the reference point P

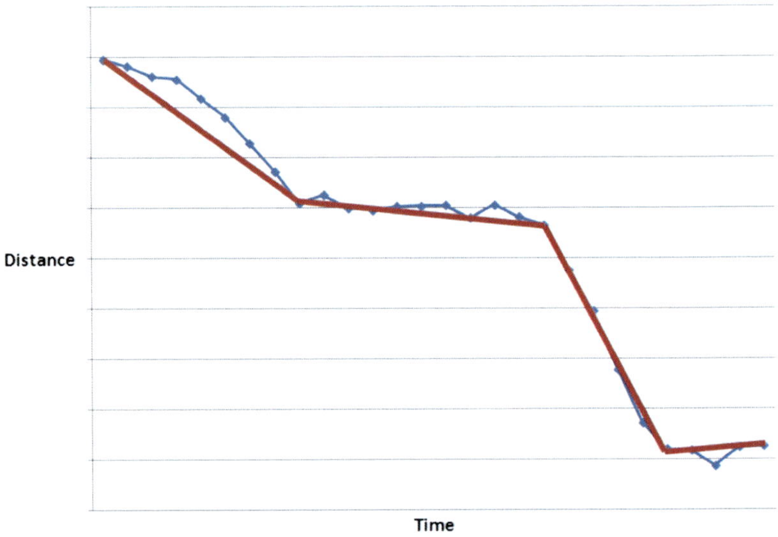

Fig. 9 The linear approximation of the distance-versus-time graph

The linear approximation of the distance-versus-time graph is a list of ordered line segments $L = \{l_1, l_2, \ldots, l_{|L|},\}$ whose contiguous end points are data points in the distance-versus-time graph. Once the linear approximation of the distance-versus-time graph has been created, the line segments of the approximation can be examined. We wish to identify those segments where the train has moved slowly or stopped, which visually appear flat on this graph.

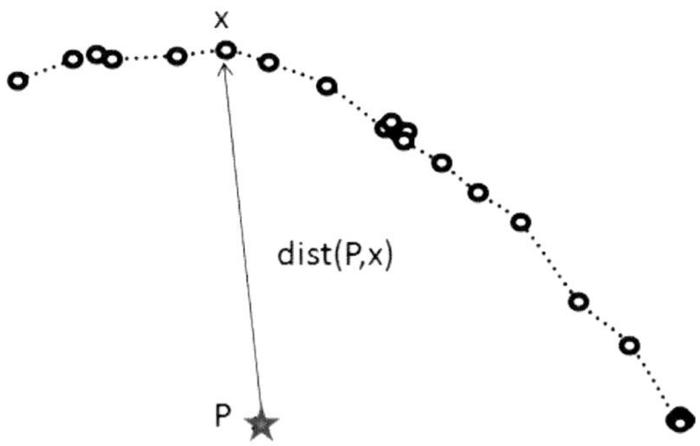

Fig. 10 GPS latitude and longitude measurements of a train lie at a constant distance from a reference point P

The local trains we are considering operate on very different geographies. Some trains traverse distances of nearly one hundred miles while others do not venture more than one mile from their originating yard. Because of this the distance-versus-time graphs can have largely different scales. To create a consistent definition of flatness on these graphs, all were scaled so that both the values of time and distance were between 0 and 10. This allowed a consistent interpretation of flatness across all trains. It was experimentally determined that a line segment should be considered flat if its slope was less than 0.8. However, if the line had a distance of longer than 0.5 it was considered flat if the slope was less than 0.2. These values are larger than required for most practical cases. The line segments on the distance-versus-time graph which correspond to the train moving have quite steep slopes in practice and it is unusual that the linear approximation of the curve would have fit in such a way that a flat line segment would be misidentified.

As the reader may have noted there is an exceptional case wherein a flat line segment on the distance-versus-time graph would correspond to a train which is not stopped or slowed. If the train is moving at any velocity yet always maintains a constant distance from the reference point P the distance-versus-time graph will appear flat, see Figs. 10 and 11. Thus, once line segments have been identified as flat they must be verified as corresponding to a truly stopped or slowed train to rule out such cases. In order to verify that the line segment is in fact flat we will calculate the distances of all points in the GPS data S relative to a second opposite reference point, essentially a triangulation. Define D_P as the set of distance values calculated from all the GPS measurements to the original reference point P (see Fig. 12). And define $D_{P'}$ as the set of distance values calculated from all the GPS measurements to a second reference point P' which lies opposite P. After the polyline fit has been completed on the graph of the distance values in D_P versus time, the graph will be approximated by the line segments L. Consider a line segment $l \in L$ whose end points correspond to data points i_s and i_t, respectively. Line segment l has a very low slope and has thus been identified as flat. If line segment l corresponds to a time when the train stopped or slowed, the line segment between i_s and i_t on the second distance-versus-time graph formed using the values $D_{P'}$ should also be flat. If this line is not also flat, then this line segment should not be considered to correspond to a time when the train slowed or stopped. If the line segments between i_s and i_t are

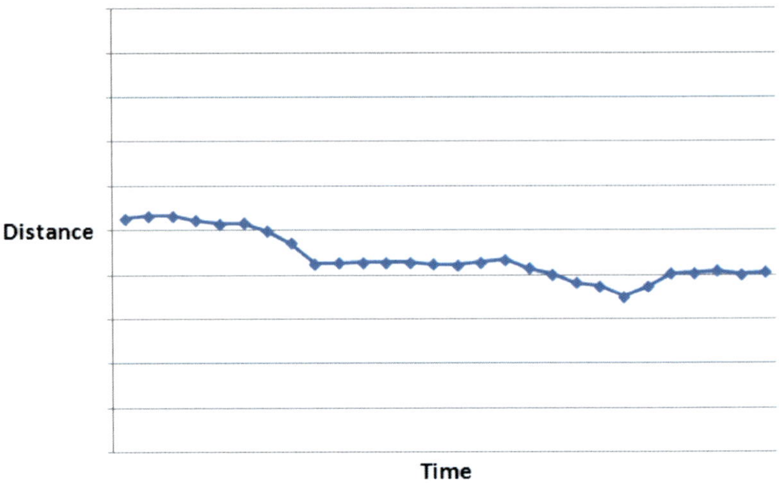

Fig. 11 Distance-versus-time graph (arbitrary units) relative to the reference point P appears very flat

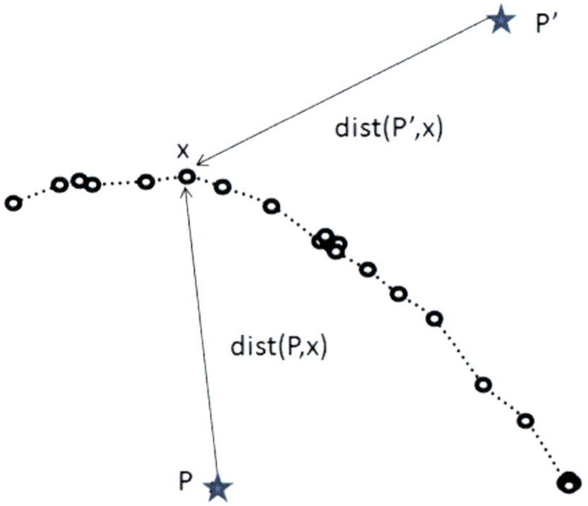

Fig. 12 The distances from all points in the data to the new reference point P' are not constant

flat on both graphs we would expect the covariance between both sets of distance values to be equal. The Pitman–Morgan [12, 15] test for equality of variance between two data sets X and Y works by testing the correlation between $X - Y$ and $X + Y$. To determine if both line segments are flat we have used a robust variant of this modified from the Box–Scheffe [3, 18] technique which is presented by Wilcox [22] on comparisons of methods for testing equivalence of variances for dependent data sets (for our purposes D_P and D'_P). If any line segment $l \in L$ which was identified as flat fails to pass this statistical test it will no longer be considered flat.

After this check, all the line segments that are considered flat are intervals which potentially correspond to a customer service event. To determine which (if any) interval corresponds to which customer, an assignment problem will be solved.

4 Assignment of Stop Intervals to Customers

Once the stop/slow intervals for a train route have been identified it must be determined if the interval corresponds to a scheduled customer service event. The customers served on a given day by a train are listed sequentially on the work order. Several issues complicate the assignment, the most important being that, due to data entry errors, a customer present on the work order may have been skipped or customers may have been serviced in an order which does not match the work order. Thus the assignment of customers to stop intervals must account for the fact that a customer may not be assigned. We address this by including a *null* assignment to represent when no time interval was found which would be a valid assignment for a customer. It is preferential to assign a customer a service time if at all possible, and thus a "penalty" term will be added to our assignment model to discourage such a null assignment. Also, since a train does not always remain stationary when serving a customer, numerous stop intervals may be assigned to a particular customer. Initially, our assignment associates only one stop interval with each customer and such additional stopped intervals are added in post processing.

Two assumptions have been made to address complications in the assignment procedure. Firstly, the procedure makes the assumption that the sequence ordering of the customers on the work order is correct. While a seemingly minor assumption, when a data entry error proves this false the customer service times may not be calculated correctly for a customer worked out of sequence with respect to the work order. There are certainly cases when customers are serviced out of order that are easy to identify and the correct assignment of stops made to the customers. One such case is a train which makes only one lengthy stop directly in front of each customer, no other stops are made by the train, and the train does not recross any section of track multiple times. In this case the number of stop intervals matches the number of customers and a clear assignment of the nearest stop to each customer can be made. However, this ideal scenario is unlikely in practice. The quantity of stop intervals which may appear to be associated with customers is large. Numerous sequential stop intervals may be assigned to a customer as the train adjusts position during service. Trains make frequent stops which are unaccounted for on the work order which may be near a customer location. This is especially true of trains in dense urban or suburban areas in which the train must observe proper right-of-way restrictions with regard to other rail and automobile traffic. Also, local trains frequently begin and end service at the same yard, often doubling back across track they had previously covered, passing (often stopping or slowing down) near customers already served. Thus, a reliable assignment of stop intervals to customers is considerably unlikely if the ordering of customers on the work order is not observed.

The second assumption made for the assignment procedure is that any customers served by the train were present on the work order. While unlikely, it is possible that a data entry error may result in a customer failing to be recorded on the work order or if a customer has numerous facilities, that the work order recorded another location. Due to the large number of customers on the CSX customer service network, it is impossible to ensure a valid assignment could be made to a customer not listed on the work order. However, since all stop intervals for each train have been identified for a given day, unaccounted for stop time is recorded allowing for further investigation.

Also, customers whose geofences overlap at any point and are served sequentially are grouped into a single customer. A total time will be assigned to both which can be apportioned separately as per business rules in post processing.

With these assumptions, the assignment of stop intervals to customers (allowing some customers to be unassigned) can be solved by finding the solution to a shortest path problem on a positively weighted acyclic directed network.

For a given train on a given day, consider the set of stop/slow intervals found as above $I = \{i_1, i_2, \ldots, i_{|I|}\} \in L$ and also the set of customers recorded on the work order for the day $C = \{c_1, c_2, \ldots, c_{|C|}\}$. To quantify how close a customer $c \in C$ is to the series of GPS measurements corresponding to an interval $i = [s,t] \in I$ we define a score w_{ci} for each $i \in I$ and each $c \in C$ which is the exponent of the average minimum Haversine distance of all pings on the interval i to the customer c's trace. Or,

$$w_{ci} = \exp\left\{ \frac{1}{t-s+1} \sum_{j=s}^{t} \min\{\text{dist}(j,c)\} \right\}, \tag{1}$$

where $\text{dist}(j,c)$ is the Haversine distance between the jth GPS measurement and the customer service area for customer c. Intervals of GPS measurements that lie nearer to customers will have smaller scores than those further from the customer. The exponential nature of this weight causes intervals nearest to customers to be highly favored. As intervals increase in distance from the customer, the weight becomes exponentially worse. Note that the correct assignment of customers to intervals is not necessarily to assign customer c the interval with the smallest w_{ci}. When customers are near to each other, it is possible that a single interval may be the minimal assignment for two customers, yet there is another nearby interval which is relatively proximate to one or both of these. With this in mind we formulate the assignment of customers to intervals. The weights w_{ci} will be used as the arc weights of our shortest path problem.

The graph on which we will solve the shortest path problem can be considered to be constructed in layers. The first layer of nodes consists of $|I|+1$ nodes corresponding to the possible assignment of the first customer c_1 to intervals $\{i_1, i_2, \ldots, i_{|I|}\}$ and the possibility of c_1 not being assigned to any interval. Label the first $|I|$ nodes for the potential assignment of the first customer to the intervals as $(1, j)$ for $j = 1, \ldots, |I|$ and the last node for the null assignment $(1, X)$. Each of these

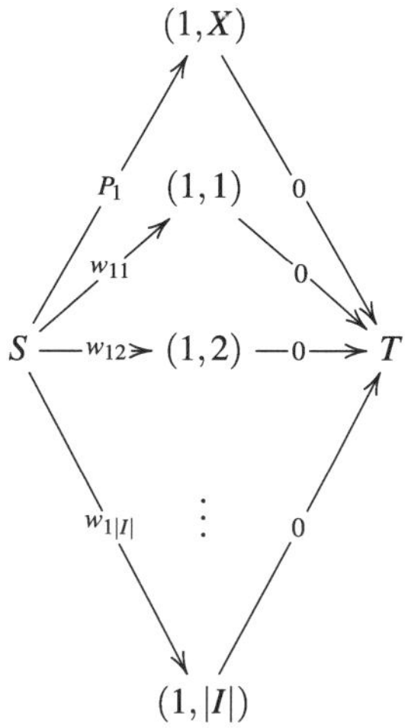

Fig. 13 Solving the assignment as a shortest path problem for a single customer

nodes is connected to a dummy start node S by directed arcs. The arc from node S to node $(1,j)$ will have weight w_{1j} for $j = 1,\ldots,|I|$. The weight of the arc from $(1,X)$ will be a penalty P_1 which is the penalty that will be incurred for failing to assign the first customer to a customer service event. If there is only a single customer on the work order, add a dummy end node T and connect nodes $(1,X)$ and $(1,j)$ for $j = 1,\ldots,|I|$ by directed arcs of weight zero; see Fig. 13. The shortest path problem is trivial in this case, and can be solved to finding the shortest path from node S to node T on the graph.

The penalty term P_c is a quantitative measure of the failure to assign customer c to any interval. Intuitively, this term must be large enough to ensure that if a potentially valid interval exists the customer will be assigned to this interval, yet small enough to ensure intervals far from the customer's service area are never chosen as valid service times. This measure is dependent on the length of the train. When a train is short we expect to see the cluster of GPS measurements which represent the service stop to be close to the customer service area. While a long train may result in the GPS readings of the service event occurring far from customer as the GPS device is housed in the locomotive of a train which may back into the customer. Intuitively, regardless of the train length, the penalty function should not exclude intervals very near to any customers. Thus for a train length less than 20 cars, the penalty term is fixed to always consider intervals within 420 m (the length of 20 rail cars and 3 locomotives) of the customer service area. Also, the penalty function is intuitively constant for very long train lengths. While a local train may be up to two miles in length, it would be very rare that a customer would be serviced by a train this length. A portion of the cars would likely be left elsewhere on the main line track and movements made with a train consisting of fewer cars. The penalty function was found experimentally and conforms to these intuitions. The penalty function used here is:

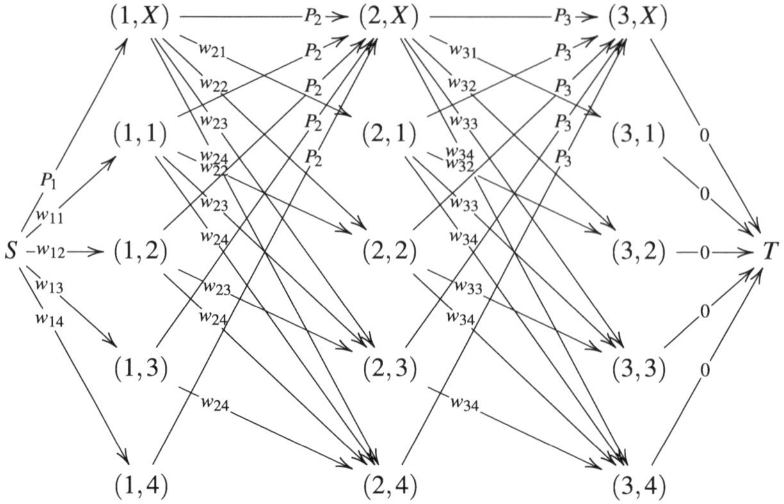

Fig. 14 An incorrectly formulated graph for solving the assignment as a shortest path problem for three customers to four potential service intervals

$$P_c = \begin{cases} e^{0.74194} & \text{if the train travels a short distance,} \\ 0.0024\delta^{-0.475}(D_{CSA}(c)+L)^{f(\delta)}, & \text{otherwise} \end{cases} \tag{2}$$

where $f(\delta) = 0.104\log_{10}\delta 1.35 - (\delta - 0.25)^2 - 85\delta$ and δ is the scaling term that was used to scale the distance values of the distance-versus-time graph between 0 and 10. If the work order shows that multiple customers were served at the same location, $D_{CSA}(c)$ is the maximum distance between any points of these overlapping service areas. $D_{CSA}(c)$ is zero if the customer c does not overlap with any other customers. The term L is the 420 m if the length of the train when visiting customer c is less than 20 and is equal to the total length of the train otherwise.

For a number of customers greater than one, consider repeating the same construction of $|I|+1$ nodes for the second customer. Label the nodes $(2,X)$ and $(2,j)$ for $j = 1,\ldots,|I|$. Based on our requirement that the sequence of customers on the work order is correct if the first customer will be assigned to interval i_j, the second customer must be assigned to some interval in $i_j+1,\ldots,|I|$ if it is assigned at all. Thus the arc weight $a_{(1,j),(2,k)}$ from node $(1,j)$ to node $(2,k)$ is

$$a_{(1,j),(2,k)} = \begin{cases} w_{2,k} & \text{if } j < k \\ \infty, & \text{otherwise} \end{cases} \tag{3}$$

for $j,k = 1,\ldots,|I|$. Also, each node $(1,j)$ for $j = 1,\ldots,|I|$ should also connect to $(2,X)$ with an arc of weight P_2 (the penalty for failing to assign customer 2); see Fig. 14. This logic can be repeated for all additional customers.

Unfortunately, the current construction of this graph is insufficient to solve the problem we require. Consider the case of attempting to assign three customers to four intervals (as shown in Fig. 14). Consider an optimal solution to the shortest path problem where customer 1 is assigned to interval 2 and no GPS pings are near customer 2 so it is unassigned. We wish to require that the customers are chosen in the order specified on the work order; however, customer 3 could be assigned to interval 1 instead of interval 3 if $w_{31} < w_{33}$. Such a scenario does in fact occur often in practice if a train stops near customer 3 on its way out to visit customer 1 and then returns to customer 3 along the same route.

To rectify the graph, for all customers other than the first, $|I|$ additional nodes are added to represent the null assignment for the subsequent customer. Label these $(c,X)_j$ for $j = 1,\ldots,|I|$ and $c = 2,\ldots,|C|$ which correspond to the case in which customer c is not assigned and the last customer assigned was assigned to interval i_j. In the graph only one arc enters node $(c,j)_X$ and its arc weight $a_{(c-1,j),(c,X)_j}$ from node $(c-1,j)$ to node $(c,2)_j$ is

$$a_{(c-1,j),(c,X)_j} = P_c \tag{4}$$

for all $j = 1,\ldots,|I|$ and $c = 2,\ldots,|C|$. The outgoing arcs from node $(c,X)_j$ have arc weights $a_{(c,X)_j,(c+1,k)}$ and connect node $(c,X)_j$ to node $(c+1,k)$ where

$$a_{(c,X)_j,(c+1,k)} = \begin{cases} w_{c+1,k} & \text{if } j < k \\ \infty, & \text{otherwise} \end{cases} \tag{5}$$

for all $j,k = 1,\ldots,|I|$ and $c = 2,\ldots,|C|$. The nodes (c,X) are used to represent the case where no customers have been assigned at all for customers $1,\ldots,c$; see Fig. 15. Solving a shortest path problem on this graph will provide us with the best assignment of customers to intervals.

After the best assignment of one (or fewer) stop interval to each customer has been found, additional stop intervals may still need to be assigned to a customer to account for the entire service time at that customer. For example, when a train arrives at a customer and stops to place cars on one track. After placing these cars the train moves over to a nearby track to pick up other cars for the same customer. The train then exits the customer's facility. The path of the train is characterized by the train moving to arrive, stopping or making small movements to drop off cars, moving again, stopping or making small movements to pick up cars, and finally, moving to depart. On the distance-versus-time graph the customer service time associated with dropping off and then picking up cars would most likely appear as two flat separate intervals. When solving the assignment problem, the stop interval with the lowest score will be assigned to the customer, leaving the other interval, with a presumably very similar score, unassigned. After the initial assignment, such unassigned, similar intervals will be assigned to customers in postprocessing.

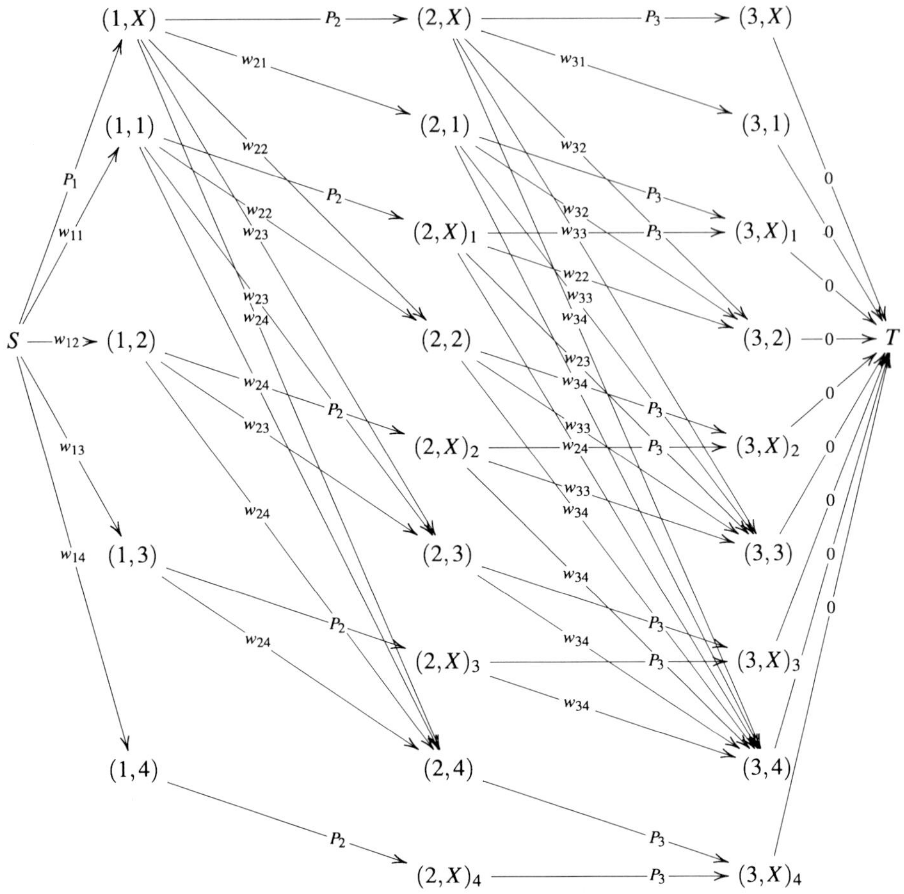

Fig. 15 The graph for solving the assignment as a shortest path problem for three customers to four potential service intervals

5 Results

While running the algorithm for numerous trains over a lengthy time span requires a very large amount of data, to run a single train for a single day requires relatively little. The number of GPS measurements per day for a single train will usually be less than one thousand records, and the work order and customer location information is minimal. Computationally, the polyline fitting procedure above consists of two parts: the splitting of the polyline and merging or smoothing to correct any potential over-fitting of the data. Each time the fit line is split requires $\mathcal{O}(n)$ and the polyline fit contained a maximum of 250 segments. The merging procedures used are $\mathcal{O}(n^2)$ in the worst case. The assignment of intervals to customers only requires solving a shortest path problem on a directed acyclic graph containing a number of nodes equal to the number of customers times the number of flat intervals. The total number of line segments possible in the fit graph is 250. However, in practice the merging procedures smooth the graph such that it would rarely contain such a large number of line segments. As the assignment problem only considers the flat intervals there will be a small number of nodes in the assignment problem. Because of this, for a single train on a single day the algorithm outlined above runs in less than one second on a desktop computer [64 bit, 2 Intel Xeon processors (2.27 GHz/2.26 GHz), 4 GB RAM]. The algorithm was implemented in C# as were the statistical tests (to avoid integrating a more robust statistical package). While in development, the algorithm did not access CSX's massive GPS databases; any required data was downloaded to a smaller development database which was accessed via Microsoft SQL Server.

To ensure a fully automated method that accurately approximated the service time at numerous unique customers, the model was designed and tested using the data of numerous trains and customers. Initially, 6 months of data from seven trains which serviced roughly 40 distinct customers was used. The output of the algorithm on each of these days which contained valid input data was examined visually. The GPS measurements for the day considered were overlaid on the CSX network map with customer information in the GPS software ArcGIS. This map was inspected and compared with expected plan information, the work order information and the assistance of those within the company who provided their extensive industry specific knowledge. The model was tested to provide the expected results based on all this information. The model was considered to deviate from these expected results if either the start or stop times deviated by more than 5 min. For most practical applications, such small deviations are certainly acceptable.

Figure 16 shows how the model compared with the expected behavior on five representative trains over a time frame of 6 months. The Over/Under Estimate count represents the number of times that the model's start or stop times appear to be off by greater or less than 5 min. The Expected Estimate values are the number of those experiments in which the algorithm provided the expected estimation, which was determined (as discussed above) by manually examining the GPS, work order, and algorithmic data for each date. The figure shows the algorithm's ability to automatically and accurately determine the service times for various trains. The occurrences of deviation are few, and among these the deviation is only marginal. All over or under estimates are less than 10 min. The relatively few representative points for the third train (Train 21) occur because this train is not often assigned to deliver to customers, but performs other work. This illustrates the large quantity of data which may be required to produce enough customer service time estimates for certain customers. If a customer is serviced very rarely, several months of data may be required in order to obtain a large enough sample of service events to determine the service time conditional on the numerous factors which could impact the work. Not all local trains run each day of the week, and certain customers may be served much less frequently.

As an additional method of comparison, as part of a customer service pilot, CSX had staff ride along the trains of this division for 3 days. These staff members recorded detailed notes of what the trains did at all times including when the train

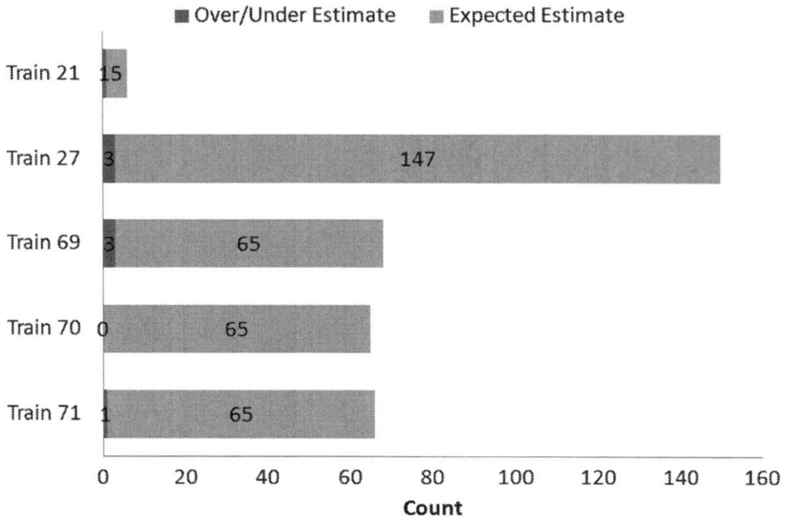

Fig. 16 Comparison of number of times the algorithm returned the expected result and the number of times the algorithm's start and stop times deviated by greater than 5 min from the expected result for five trains for 6 months of data

arrived at and departed from the customer, switches that were changed, number of cars handled, and any delays or train stops. This on-the-ground data provided a manually determined estimate of the customer service time. We compared the output of the model with these estimates. All customer service start and stop times were within 5 min each of the manually recorded times except for a single customer. After reviewing this customer it was determined that the customer's defined service area did not include certain tracks used to enter the customer's facility. After adjusting the customer's service area data to include these incoming tracks, and re-running the model for this customer, the model provided start/stop times within 5 min.

The current model works very well for numerous train and customer types and is very robust to changes up to a certain point. The model struggles in two areas: trains which travel over any area with a diameter less than four miles, and also for customers whose service areas/tracks overlap. For these two special cases alternative logic was developed which we will discuss in future work.

6 Conclusion

We presented an automated method for estimating customer service times on the rail network from GPS data given a work order sequence and minimal knowledge of the customer's tracks. The model performed well when compared with manually recorded customer service times and also when compared to the times expected by industry experts when observing the GPS measurements which described the motions of a train. The model discussed here provides useful information on the prior service behavior of trains at customers. In addition the service time estimates can be used to make characteristic predictions about future service events. The model can be applied to estimate the service times at customers for the large quantity of data which has been collected by CSX over several years. This provides multiple point estimates of the time required to serve a customer along with the type of work, number of cars handled, day of week, and train. Using statistical estimation techniques, a relationship between the service time estimates and the many possible influencing parameters can be found. The final goal of this statistical estimation techniques is an equation for each customer that can be used to predict future service times. Such techniques could provide service time estimates for several potential applications such as dynamic schedule creation and dynamic work-to-train assignments, real time service estimation, detailed profiles of service behavior for management, or real-time arrival time updates for customers slated to be served. Such automated methodology has high value for a large company wishing to monitor the customer service interactions of numerous trains resulting in millions of GPS records; clearly, manual monitoring is prohibitive on a network of this size.

Acknowledgments We wish to thank the CSX service design team (especially Shannon Slattery and Erik Henderson) and Jagadish Jampani and Dharma Acharya of the operations research department for their assistance.

References

1. ASAKURA, Y., AND HATO, E. Tracking survey for individual travel behaviour using mobile communication instruments. *Transportation Research Part C: Emerging Technologies 12* (2004), 273–291.
2. BIAGIONI, J., GERLICH, T., MERRIFIELD, T., AND ERIKSSON, J. Easytracker: automatic transit tracking, mapping, and arrival time prediction using smartphones. In *Proceedings of the 9th ACM Conference on Embedded Networked Sensor Systems* (2011), ACM, pp. 68–81.
3. BOX, G. E. P. Non-normality and tests on variances. *Biometrika 40* (1953), 318–335.
4. DOUGLAS, D., AND PEUCKER, T. Algorithms for the reduction of the number of points required to represent a digitized line or its caricature. *Cartographica: The International Journal for Geographic Information and Geovisualization 10*, 2 (1973), 112–122.
5. GERSTLE, D. Understanding bus travel time variation using AVL data. Master's thesis, Massachusetts Institute of Technology, 2012.
6. GUO, P., AND POLING, A. D. Geographic information systems/global positioning systems design for network travel time study. *Journal of the Transportation Research Board 1497* (1995), 135–139.
7. HARDING, J., SALWIN, A., D. S., AND GHAMAN, R. GPS applications for traffic engineering studies. In *Transportation Research Board Annual Meeting 1996* (preprint).
8. HECKBERT, P., AND GARLAND, M. Survey of polygonal surface simplification algorithms. Tech. rep., DTIC Document, 1997.
9. HERSHBERGER, J., AND SNOEYINK, J. *Speeding up the Douglas-Peucker line-simplification algorithm.* University of British Columbia, Department of Computer Science, 1992.
10. MARAIS, J., MEUNIER, B., AND BERBINEAU, M. Evaluation of GPS availability for train positioning along a railway line. In *IEEE Vehicular Technology Conference* (2000).

11. McCord, M., Goel, P., Brooks, C., Wallace, R., Dong, H., and Keefauver, D. E. Documenting truck activity times at international border crossings using redesigned geofences and existing onboard systems. *J Transportation Res. Board 2162* (2010), 81–89.
12. Morgan, W. A. A test for the significance of the difference between the two variances in a sample from a normal bivariate population. *Biometrika 31* (1939), 13–19.
13. Munson, J., and Gupta, V. Location-based notification as a general-purpose service. In *Proceedings of the 2nd international workshop on Mobile commerce* (2002), ACM, pp. 40–44.
14. Murakami, E., and Wagner, D. P. Can using global positioning system (GPS) improve trip reporting? *Transportation Research C 7* (1999), 149–165.
15. Pitman, E. J. A note on normal correlation. *Biometrika 31* (1939), 9–12.
16. Quiroga, C. A., and D. B. Travel time studies with global positioning and geographic information systems: an integrated methodology. *Transportation Research C 6* (1998), 101–127.
17. Reclus, F., and Drouard, K. Geofencing for fleet & freight management. In *Intelligent Transport Systems Telecommunications, (ITST), 2009 9th International Conference on* (2009), IEEE, pp. 353–356.
18. Scheffe, H. *The analysis of variance.* 1959.
19. Stopher, P., Bullock, P., and Jiang, Q. GPS, GIS and personal travel surveys: an exercise in visualisation. In *25th Australasian Transport Research Forum Incorporating the BTRE Transport Policy Colloquium* (2002).
20. Tsui, S., and Shalaby, A. Enhanced system for link and mode identification for personal travel surveys based on global positioning systems. *Transportation Research Record: Journal of the Transportation Research Board 1972* (2006), 38–45.
21. Wang, Y., and Potter, A. The application of real time tracking technologies in freight transport. In *Signal-Image Technologies and Internet-Based System, 2007. SITIS'07. Third International IEEE Conference on* (2007), IEEE, pp. 298–304.
22. Wilcox, R. R. Comparing the variances of two dependent groups. *Journal of Educational Statistics 15* (1990), 237–247.
23. Wilson, B., and Vincent, J. Estimating waste transfer station delays using GPS. *Waste Management 28* (2008), 1742–1750.
24. Wolf, J., Guensler, R., and Bachman, W. Elimination of the travel diary: Experiment to derive trip purpose from global positioning system travel data. *Journal of the Transportation Research Board 1768* (2001), 125–134.
25. Wolf, J., Loechl, M., Thompson, M., and Arce, C. Trip rate analysis in GPS-enhanced personal travel surveys. *Transport survey quality and innovation* (2003), 483–498.

A Risk-Averse Game-Theoretic Approach to Distributed Control

Khanh D. Pham and Meir Pachter

Abstract The research article gives a comprehensive presentation of the broad and still developing area of risk-averse decision-making approach to control of distributed stochastic systems. A distributed stochastic system considered here consists of the interconnection of two or more stochastic systems with the structural constraints of linear system dynamics, quadratic cost functionals, and additive stationary Wiener noises corrupting the system dynamics and measurements. Each system has an input from its incumbent agent or controller and an output to its local environment, in addition to links with the other neighboring systems. The problem of distributed control without communications between incumbent agents or controllers is formulated as a nonzero-sum stochastic differential game. Local best responses by each incumbent agents with risk-averse attitudes toward performance uncertainty are determined by a person-by-person equilibrium and subject to decentralized output-feedback information structures.

Keywords Distributed stochastic system • Distributed control • Self autonomy • Local best response • Performance risk • Output feedback • Risk-averse decision making • Person-by-person equilibrium

---

K.D. Pham (✉)
Air Force Research Laboratory, Space Vehicles Directorate, Kirtland Air Force Base, NM 87117, USA
e-mail: AFRL.RVSV@kirtland.af.mil

M. Pachter
Air Force Institute of Technology, Department of Electrical and Computer Engineering, Wright-Patterson Air Force Base, OH 45433, USA
e-mail: Meir.Pachter@afit.edu

A. Sorokin and P.M. Pardalos (eds.), *Dynamics of Information Systems: Algorithmic Approaches*, Springer Proceedings in Mathematics & Statistics 51,
DOI 10.1007/978-1-4614-7582-8_4,

1 Introduction

The purpose of this research investigation is to introduce to the readers the problem of control of distributed stochastic systems, to propose risk-averse decision making towards performance uncertainty, and to indicate emergent approaches for future research and development. The importance of broad flexibility and adaptability of the decision and control architectures of distributed control has spurred many large-scale applications such as military command and control hierarchies, spacecraft constellations, remotely piloted platform formations, and teams of humans and autonomous robots. where each member can be in best response to its neighbor actions and yet has no influence on other members to which it has no communication supports.

Despite the broad interest in distributed systems, there remain significant hurdles in applying them to practical problems of interest. Interplay between coalition objectives and individual member objectives can yield surprises and complex behaviors. Thus motivated, the main problem of the research herein is control of distributed systems via the game-theoretic framework with performance risk aversion. To the best knowledge of the authors, most studies, for instance, [1, 2] have mainly concentrated on the selection of open and/or closed-loop Nash strategy equilibria in accordance of expected utilities under the structural constraints of linear system dynamics, quadratic cost functionals, and additive independent white Gaussian noises corrupting the system dynamics and measurements. Very little work, if any, has been published on the subject of higher-order assessment of performance uncertainty and risks beyond expected performance.

For this reason attention in this research investigation is directed primarily towards a linear-quadratic class of nonzero-sum differential games which has linear system dynamics, quadratic cost functionals, and independent white zero-mean Gaussian noises additively corrupting the system dynamics and output measurements. Notice that, under these conditions, the quadratic cost functionals or outcomes associated with the game are random variables with the generalized chi-squared probability distributions. If a measure of uncertainty such as the variance of the possible outcome was used in addition to the expected outcome, the incumbent agents or controllers should be able to correctly order preferences for alternatives. This claim seems plausible, but it is not always correct. Various investigations have indicated that any evaluation scheme based on just the expected outcome and outcome variance would necessarily imply indifference between some courses of action; therefore, no criterion based solely on the two attributes of means and variances can correctly represent their preferences. See the works [3, 4] for more details.

Recent accounts by the first author [5, 6] have addressed risk aversion for performance uncertainty of cooperative and noncooperative large-scale stochastic systems, wherein the shape and functional form of an utility function tell a great deal about the basic attitudes of the agents or controllers toward the uncertain outcomes or performance risks. In particular, the new utility function or the so-called the generalized performance index, which is proposed therein as a linear manifold defined by a finite number of semi-invariants associated with a random quadratic cost functional, will provide a convenient allocation representation of apportioning performance robustness and reliability requirements into the multi-attribute requirement of qualitative characteristics of expected performance and performance risks.

The present research contributions are to extend the existing results in [7] toward some completely unexplored areas as such: (1) the design of decentralized filtering via constrained filters for self-directed agent subject to the linear-quadratic class of nonzero-sum stochastic differential games; (2) an efficient and tractable procedure that calculates exactly all the mathematical statistics associated with the generalized chi-squared performance measure for each self-directed agent; and (3) the risk-averse control and decision synthesis that is mostly via a person-by-person equilibrium for reliable performance.

Given the aforementioned background, the article is organized as follows. Section 2 contains the problem description in which basic assumptions related to the state-space model associated with each incumbent decision makers or controllers residing at distributed systems are discussed. In addition, the development of mathematical statistics for performance robustness whose the backward-in-time differential equations are characterized by making use of both compactness from the logics of the state-space representation and the quantitativity from a-priori knowledge of the underpinning probabilistic processes is further presented in detail. Subsequently, Sect. 3 provides the complete problem statements of statistical optimal decision making via the person-by-person equilibrium framework, unique notations, terminologies, definitions as well as the necessary and sufficient conditions for the existence of person-by-person equilibrium strategies. With regards to the theoretical constructs and design principles for distributed stochastic systems to include the requirements of performance reliability, decision making with risk consequences and emerging effects within the stochastic environment, the understanding of performance variations, risk-averse attitudes and the course correction required for realistic situations is determined and obtained in Sect. 4. Finally, conclusions pertaining to decisions with risk consequences and output feedback design for the linear-quadratic class of distributed stochastic linear systems with quadratic performance appraisals are presented in Sect. 5.

2 Mathematical Statistics for Performance Robustness

Before going into a formal presentation, it is necessary to consider some conceptual notations in this article. For instance, time t is modeled as continuous and the notation of the time interval is $[t_0, t_f]$. All random variables are defined on a probability space $(\Omega, \mathscr{F}, \mathscr{P})$ which is a triple consisting of a set Ω, a σ-algebra $\mathscr{F}$, and a probability measure $\mathscr{P} : \mathscr{F} \mapsto [0,1]$ and is equipped with a filtration

$\{\mathscr{F}_t : t \in [t_0, t_f]\}$. In addition, for a given Hilbert space X with norm $||\cdot||_X$, $1 \le p \le \infty$, a Banach space is defined as follows

$$\mathscr{L}^p_{\mathscr{F}}(t_0,t_f;X) \triangleq \left\{ \phi : [t_0,t_f] \times \Omega \mapsto X \text{ is an } X\text{-valued } \mathscr{F}_t\text{-measurable process} \right.$$
$$\left. \text{with } E\left\{ \int_{t_0}^{t_f} ||\phi(t,\omega)||_X^p dt \right\} < \infty \right\} \tag{1}$$

with norm

$$||\phi(\cdot)||_{\mathscr{F},p} \triangleq \left(E\left\{ \int_{t_0}^{t_f} ||\phi(t,\omega)||_X^p dt \right\} \right)^{1/p}. \tag{2}$$

Furthermore, the Banach space of X-valued continuous functionals on $[t_0,t_f]$ with the max-norm induced by $||\cdot||_X$ is denoted by $\mathscr{C}(t_0,t_f;X)$. The deterministic version of (1) and its associated norm (2) is written as $\mathscr{L}^p(t_0,t_f;X)$ and $||\cdot||_p$.

A distributed stochastic system that evolves over $[t_0,t_f]$ captures interactions among a finite number of incumbent systems. Each incumbent system that enters the distributed system is assigned a unique positive integer-valued index. The set of indices of incumbent systems is denoted by $\mathscr{I} \triangleq \{1,2,\ldots,N\}$ and a typical element by i. The set of immediate neighbors that have communication paths with an incumbent system i is denoted by $\mathscr{N}_i$, whereby the cardinality of $\mathscr{N}_i$ is notated as N_i. For concreteness, the heterogeneity of incumbent system i and $i \in \mathscr{I}$ is distinguished by an individual state that is governed by the stochastic differential equation with the known initial condition $x_i(t_0) = x_i^0$ and $t \in [t_0,t_f]$

$$dx_i(t) = \left(A_{ii}(t)x_i(t) + B_{ii}(t)u_i(t) + \sum_{j=1}^{N_i} B_{ij}(t)u_{ij}(t) \right) dt + G_{ii}(t)dw_i(t), \tag{3}$$

where the continuous-time coefficients $A_{ii} \in \mathscr{C}(t_0,t_f;\mathbb{R}^{n_i \times n_i})$, $B_{ii} \in \mathscr{C}(t_0,t_f;\mathbb{R}^{n_i \times m_i})$, $B_{ij} \in \mathscr{C}(t_0,t_f;\mathbb{R}^{n_i \times r_i})$ and $G_{ii} \in \mathscr{C}(t_0,t_f;\mathbb{R}^{n_i \times p_i})$ are deterministic matrix-valued functions. At time t, the recursive state of incumbent system i is denoted by $x_i \in \mathscr{L}^2_{\mathscr{F}_i}(t_0,t_f;\mathbb{R}^{n_i})$ with the initial state $x_i^0 \in \mathbb{R}^{n_i}$ known. The control policy from agent i to that system i is presented by $u_i \in \mathscr{L}^2_{\mathscr{F}_i}(t_0,t_f;\mathbb{R}^{m_i})$.

In addition, the interconnection inputs of that incumbent system i supported by the communication paths from immediate neighbors j and $j \in \mathscr{N}_i$ are viewed as the real-valued functions $u_{ij}(t)dt$ of the following random processes

$$u_{ij}(t)dt = (C_{ij}(t)x_j(t) + D_{ij}(t)u_j(t))dt + G_{ij}(t)dw_j(t), \quad j \in \mathscr{N}_i \tag{4}$$

where continuous-time coefficients $C_{ij} \in \mathscr{C}(t_0,t_f;\mathbb{R}^{r_i \times n_j})$, $D_{ij} \in \mathscr{C}(t_0,t_f;\mathbb{R}^{r_i \times m_j})$ and $G_{ij} \in \mathscr{C}(t_0,t_f;\mathbb{R}^{r_i \times p_j})$ are deterministic matrix-valued functions.

In the state-space representation (3) and (4) one postulates independent Wiener processes $w_i(t) \triangleq w_i(t,\omega_i) : [t_0,t_f] \times \Omega_i \mapsto \mathbb{R}^{p_i}$ and $w_j(t) \triangleq w_j(t,\omega_j) : [t_0,t_f] \times \Omega_j \mapsto \mathbb{R}^{p_j}$ defined by the underlying filtered probability spaces $(\Omega_i, \mathscr{F}_i, \{\mathscr{F}_i\}_t, \mathscr{P}_i)$ and $(\Omega_j, \mathscr{F}_j, \{\mathscr{F}_j\}_t, \mathscr{P}_j)$ with the correlations of independent increments

$$E\left\{[w_i(\tau_1) - w_i(\tau_2)][w_i(\tau_1) - w_i(\tau_2)]^T\right\} = W_i|\tau_1 - \tau_2|, \quad W_i > 0, \quad \tau_1, \tau_2 \in [t_0, t_f]$$

$$E\left\{[w_j(\tau_1) - w_j(\tau_2)][w_j(\tau_1) - w_j(\tau_2)]^T\right\} = W_j|\tau_1 - \tau_2|, \quad W_j > 0, \quad \tau_1, \tau_2 \in [t_0, t_f]$$

approximate the inherent design system uncertainty due to variability and lack of knowledge.

With the local agent dynamics (3) and the intertemporal interactions (4), the recursive dynamics of each interconnected systems that evolve over $[t_0,t_f]$ and capture direct interactions among incumbent agent i and its immediate neighbors j and $j \in \mathscr{N}_i$ are now given by

$$ds_i(t) = \left(A_i(t)s_i(t) + B_i(t)u_i(t) + \sum_{j=1, j\neq i}^{N_i} B_j(t)u_j(t)\right)dt + G_i(t)d\xi_i(t), \quad (5)$$

where for each incumbent agent i, the aggregate Wiener process $\xi_i \triangleq \begin{bmatrix} w_1^T & \dots & w_{N_i}^T \end{bmatrix}^T$ has the correlations of independent increments

$$E\left\{[\xi_i(\tau_1) - \xi_i(\tau_2)][\xi_i(\tau_1) - \xi_i(\tau_2)]^T\right\} = \Xi_i|\tau_1 - \tau_2|, \quad \forall \tau_1, \tau_2 \in [t_0, t_f], \quad \Xi_i > 0$$

whereas for each incumbent agent i, the augmented state variable s_i, its initial-valued condition $s_i(t_0) = s_i^0$, the local game coefficients and parameters are defined by

$$s_i(t) \triangleq \begin{bmatrix} x_1(t) \\ \vdots \\ x_{N_i}(t) \end{bmatrix}; s_i^0 \triangleq \begin{bmatrix} x_1^0 \\ \vdots \\ x_{N_i}^0 \end{bmatrix}; A_i \triangleq \begin{bmatrix} A_{11} & B_{12}C_{12} & \dots & B_{1N_i}C_{1N_i} \\ B_{21}C_{21} & A_{22} & \dots & B_{2N_i}C_{2N_i} \\ \vdots & \vdots & \ddots & \vdots \\ B_{N_i1}C_{N_i1} & \dots & B_{N_i(N_i-1)}C_{N_i(N_i-1)} & A_{N_iN_i} \end{bmatrix}$$

$$B_1 \triangleq \begin{bmatrix} B_{11} \\ B_{21}D_{21} \\ \vdots \\ B_{N_i1}D_{N_i1} \end{bmatrix}; \quad B_2 \triangleq \begin{bmatrix} B_{12}D_{12} \\ B_{22} \\ \vdots \\ B_{N_i2}D_{N_i2} \end{bmatrix}; \quad B_{N_i} \triangleq \begin{bmatrix} B_{1N_i}D_{1N_i} \\ \vdots \\ B_{(N_i-1)N_i}D_{(N_i-1)N_i} \\ B_{N_iN_i} \end{bmatrix}$$

$$G_i \triangleq \begin{bmatrix} G_{11} & B_{12}G_{12} & \dots & B_{1N_i}G_{1N_i} \\ B_{21}G_{21} & G_{22} & \dots & B_{2N_i}G_{2N_i} \\ \vdots & \vdots & \ddots & \vdots \\ B_{N_i1}G_{N_i1} & \dots & B_{N_i(N_i-1)}G_{N_i(N_i-1)} & G_{N_iN_i} \end{bmatrix}; \quad \Xi_i \triangleq \begin{bmatrix} W_1 & 0 & \dots & 0 \\ 0 & W_2 & \dots & 0 \\ \vdots & 0 & \ddots & \vdots \\ 0 & \dots & 0 & W_{N_i} \end{bmatrix}.$$

Practical situations where self-autonomy is possible require that each agent be able to possess the common knowledge of the parameters associated with potential noncooperative interactions (5). Viewed from the mutual influence of one agent to those of others, self-autonomy preferred by incumbent agent i is therefore described by a surrogate model with the initial value $z_i(t_0) = z_i^0 = s_i^0$

$$dz_i(t) = \left(A_i(t)z_i(t) + B_i(t)u_i(t) + \sum_{j=1, j\neq i}^{N_i} B_j(t)u_j(t) \right) dt + G_i(t)d\xi_i(t), \quad (6)$$

whereby each incumbent agent i and $i \in \mathscr{I}$ can presumably observe all interactions

$$\sum_{j=1, j\neq i}^{N_i} B_j(t)u_j(t)dt$$

from its immediate neighbors that are in turn corrupted by an uncorrelated stationary Wiener measurement noise process. Specifically, the following observations are locally available at incumbent agent i and $t \in [t_0, t_f]$

$$u_{-i}(t)dt = \sum_{j=1, j\neq i}^{N_i} B_j(t)u_j(t)dt + d\eta_i(t). \quad (7)$$

For the completely decentralizing information pattern, it is also assumed that the incomplete information structure available at each incumbent agent i consists of a linear transformation $C_i \in \mathscr{C}(t_0, t_f; \mathbb{R}^{q_i \times \sum_{j=1}^{N_i} n_j})$ of the states $z_i(t)$ through the local online data $\{y_i(\tau) : \tau \in [t_0, t]\}$

$$dy_i(t) = C_i(t)z_i(t)dt + dv_i(t). \quad (8)$$

Notice that all incumbent agents operate within the common and local environments modeled by the filtered probability spaces. Subsequently, they are then defined by the following uncorrelated stationary Wiener processes adapted for $[t_0, t_f]$ together with the correlations of independent increments for all $\tau_1, \tau_2 \in [t_0, t_f]$

$$E\left\{[\eta_i(\tau_1) - \eta_i(\tau_2)][\eta_i(\tau_1) - \eta_i(\tau_2)]^T\right\} = I_i|\tau_1 - \tau_2|$$
$$E\left\{[v_i(\tau_1) - v_i(\tau_2)][v_i(\tau_1) - v_i(\tau_2)]^T\right\} = V_i|\tau_1 - \tau_2|$$

whose a-priori second-order statistics I_i and $V_i > 0$ for $i = 1, \ldots, N$ are assumed known.

At this point, each decentralized filter associated with incumbent agent i and $i \in \mathscr{I}$, whose the output is the conditional mean estimate $\hat{z}_i(t)$ of the current state $z_i(t)$ and $t \in [t_0, t_f]$ has the form with the initial-value condition $\hat{z}_i(t_0) = z_i^0$

$$d\hat{z}_i(t) = (A_i(t)\hat{z}_i(t) + B_i(t)u_i(t) + u_{-i}(t))dt + L_i(t)[dy_i(t) - C_i(t)\hat{z}_i(t)dt], \quad (9)$$

whereby the decentralized filter gain $L_i(t)$ and $i \subset \mathscr{I}$ is given by

$$L_i(t) = \Sigma_i(t) C_i^T(t) V_i^{-1} \tag{10}$$

and is supported by the estimate-error covariance differential equation with the initial-value condition $\Sigma_i(t_0) = 0$

$$\dot{\Sigma}_i(t) = A_i(t)\Sigma_i(t) + \Sigma_i(t)A_i^T(t) + G_i(t)\Xi_i G_i^T(t) + I_i - \Sigma_i(t)C_i^T(t)V_i^{-1}C_i(t)\Sigma_i(t). \tag{11}$$

Using the definition for the estimate errors $\tilde{z}_i(t) \triangleq z_i(t) - \hat{z}_i(t)$, it can be shown that

$$d\tilde{z}_i(t) = (A_i(t) - L_i(t)C_i(t))\tilde{z}_i(t)dt + G_i(t)d\xi_i(t) - L_i(t)dv_i(t) - d\eta_i(t) \tag{12}$$

$$\tilde{z}_i(t_0) = 0$$

incumbent agent i, however, attempts to make risk-bearing decisions u_i from an admissible feedback policy set $\mathscr{U}_i \subset L^2_{\mathscr{F}^i_t}(t_0, t_f; \mathbb{R}^{m_i})$, which is the subset of Hilbert space of $\mathbb{R}^{m_i}$-valued square integrable processes on $[t_0, t_f]$ that are adapted to the σ-algebra $\mathscr{F}^i_t$ generated by $\{y_i(\tau) : \tau \in [t_0, t]\}$ for reliable attainments of payoffs or utilities. Associated with each admissible 2-tuple $(u_i(\cdot), u_{-i}(\cdot))$ is the generalized chi-squared random measure of performance

$$J_i(u_i, u_{-i}) = z_i^T(t_f)Q_f^i z_i(t_f) + \int_{t_0}^{t_f} [z_i^T(\tau)Q_i(\tau)z_i(\tau) + u_i^T(\tau)R_i(\tau)u_i(\tau) - u_{-i}^T(\tau)M_i(\tau)u_{-i}(\tau)]d\tau, \tag{13}$$

whereby the coefficients $Q_f^i \in \mathbb{R}^{\sum_{j=1}^{N_i} n_j \times \sum_{j=1}^{N_i} n_j}$, $Q_i \in \mathscr{C}(t_0, t_f; \mathbb{R}^{\sum_{j=1}^{N_i} n_j \times \sum_{j=1}^{N_i} n_j})$, $M_i \in \mathscr{C}(t_0, t_f; \mathbb{R}^{\sum_{j=1}^{N_i} n_j \times \sum_{j=1}^{N_i} n_j})$ and $R_i \in \mathscr{C}(t_0, t_f; \mathbb{R}^{m_i \times m_i})$ representing relative weightings for terminal and transient trade-offs between the regulatory of responses z_i, the effectiveness of the control and/or decision policy u_i and observable variations in the control and/or decision policies of all other neighbors u_{-i} are deterministic and positive semidefinite with $R_i(t)$ invertible.

Amongst some research issues for distributed control which are currently under investigation is how incumbent agent i for $i \in \mathscr{I}$ and its immediate neighbors j for $j \in \mathscr{N}_i$ carry out optimal control and decision synthesis for controlling of distributed stochastic systems. The approach to handle the problem with a tuple of two or more control laws or decision policies is to use the noncooperative game-theoretic paradigm. Particularly, an N_i-tuple policy $\{u_1^*, u_2^*, \ldots, u_{N_i}^*\}$ is said to constitute a person-by-person equilibrium solution for the distributed control problem (6) and performance measure (13) if

$$J_i^* \triangleq J_i(u_i^*, u_{-i}^*) \le J_i(u_i, u_{-i}^*), \qquad \forall i \in \mathscr{I}. \tag{14}$$

That is, none of the N_i agents can deviate unilaterally from the equilibrium policies and gain from doing so. The justification for the restriction to such an equilibrium is that the coalition effects u_{-i}^* being observed by incumbent agent i does not necessarily support its preference optimization. Therefore, they cannot do better than behave as if they strive for this equilibrium. It is reasonable to conclude that a person-by-person equilibrium of distributed control for incumbent agent i and its immediate neighbors $j \in \mathcal{N}_i$ is identical to the concept of a Nash equilibrium within a noncooperative game-theoretic setting.

Moreover, the N_i-tuple $(u_1^*, \ldots, u_{N_i}^*)$ of decision laws for incumbent agent i and its immediate neighbors j and $j \in \mathcal{N}_i$ that is satisfying the person-by-person equilibrium is also a minimal tuple of decision laws. The reasons being are that the input spaces u_i are continuous and criterion J_i are continuous, differentiable, and convex in the inputs u_i. Henceforth, a minimal tuple is obtained if incumbent agents individually optimize their criteria in a parallel fashion. See [8] for more details.

Next, the notion of admissible feedback policy sets is discussed. In the case of incomplete information, an admissible feedback policy u_i for local best response to all other immediate neighbors u_{-i}^* must be of the form, for some $\eth_i(\cdot,\cdot)$

$$u_i(t) = \eth_i(t, y_i(\tau)), \quad \tau \in [t_0, t]. \tag{15}$$

In general, the conditional density $p_i(z_i(t)|\mathcal{F}_t^i)$, which is the density of $z_i(t)$ conditioned on $\mathcal{F}_t^i$ (i.e., induced by the observation $\{y_i(\tau) : \tau \in [t_0, t]\}$) represents the sufficient statistics for describing the conditional stochastic effects of future feedback policy u_i. Under the Gaussian assumption the conditional density $p_i(z_i(t)|\mathcal{F}_t^i)$ is parameterized by the locally available conditional mean $\hat{z}_i(t) \triangleq E\{z_i(t)|\mathcal{F}_t^i\}$ and covariance $\Sigma_i(t) \triangleq E\{[z_i(t) - \hat{z}_i(t)][z_i(t) - \hat{z}_i(t)]^T|\mathcal{F}_t^i\}$ by incumbent agent i. With respect to the linear-Gaussian conditions, the covariance $\Sigma_i(t)$ is independent of feedback policy $u_i(t)$ and observations $\{y_i(\tau) : \tau \in [t_0, t]\}$. Therefore, to look for an optimal control and/or decision policy $u_i(t)$ of the form (15), it is only required that

$$u_i(t) = \gamma_i(t, \hat{z}_i(t)), \quad t \in [t_0, t_f].$$

Given the linear-quadratic properties of the surrogate system description (6)–(13), the search for an optimal feedback solution may be productively restricted to a linear time-varying feedback policy generated from the locally accessible state $\hat{z}_i(t)$ by

$$u_i(t) = K_i(t)\hat{z}_i(t), \quad t \in [t_0, t_f] \tag{16}$$

with $K_i \in \mathcal{C}(t_0, t_f; \mathbb{R}^{m_i \times \sum_{j=1}^{N_i} n_j})$ an admissible feedback form whose further defining properties will be stated shortly.

Hence, for the admissible pair (t_0, z_i^0), the observed knowledge about neighboring disturbances $u_{-i}^*(\cdot)$ and the admissible feedback policy (16), the aggregation of the dynamics (9) and (12) associated with incumbent agent i is described by the controlled stochastic differential equation

$$dz^i(t) = (F^i(t)z^i(t) + E^i(t)u^*_{-i}(t))dt + G^i(t)dw^i(t)\,, \quad z^i(t_0) = z^i_0 \tag{17}$$

with the performance measure (13) rewritten as follows

$$J_i(u_i, u^*_{-i}) = (z^i)^T(t_f)N^i_f z^i(t_f) + \int_{t_0}^{t_f} [(z^i)^T(\tau)N^i(\tau)z^i(\tau) - (u^*_{-i})^T(\tau)M_i(\tau)u^*_{-i}(\tau)]d\tau\,, \tag{18}$$

whereby for each incumbent agent i and $i \in \mathscr{I}$, the aggregate system states $z^i \triangleq \begin{bmatrix} (\hat{z}_i)^T & (\tilde{z}_i)^T \end{bmatrix}^T$, the stationary Wiener process noise $w^i \triangleq \begin{bmatrix} \xi_i^T & \eta_i^T & v_i^T \end{bmatrix}^T$ with the correlation of independent increments defined as

$$E\left\{[w^i(\tau_1) - w^i(\tau_2)][w^i(\tau_1) - w^i(\tau_2)]^T\right\} = W^i|\tau_1 - \tau_2|\,, \quad \forall \tau_1, \tau_2 \in [t_0, t_f]\,, W^i > 0$$

and the aggregate system coefficients are given by, for each $t \in [t_0, t_f]$

$$F^i(t) \triangleq \begin{bmatrix} A_i(t) + B_i(t)K_i(t) & L_i(t)C_i(t) \\ 0 & A_i(t) - L_i(t)C_i(t) \end{bmatrix}; \quad E^i(t) \triangleq \begin{bmatrix} I_{\sum_{j=1}^{N_i} n_j \times \sum_{j=1}^{N_i} n_j} \\ 0 \end{bmatrix}$$

$$G^i(t) \triangleq \begin{bmatrix} 0 & 0 & L_i(t) \\ G_i(t) & -I_{\sum_{j=1}^{N_i} n_j \times \sum_{j=1}^{N_i} n_j} & -L_i(t) \end{bmatrix}; \quad N^i_f \triangleq \begin{bmatrix} Q^i_f & Q^i_f \\ Q^i_f & Q^i_f \end{bmatrix}; \quad z^i_0 \triangleq \begin{bmatrix} z^0_i \\ 0 \end{bmatrix}$$

$$N^i(t) \triangleq \begin{bmatrix} Q_i(t) + K_i^T(t)R_i(t)K_i(t) & Q_i(t) \\ Q_i(t) & Q_i(t) \end{bmatrix}; \quad W^i \triangleq \begin{bmatrix} \Xi_i & 0 & 0 \\ 0 & I_i & 0 \\ 0 & 0 & V_i \end{bmatrix}.$$

Regarding the linear-quadratic structural constraints (17) and (18), the path-wise performance-measure (18), with which incumbent agent i is risk averse, is clearly a random variable of the generalized chi-squared type. Henceforth, the degree of uncertainty of the path-wise performance-measure (18) must be assessed via a complete set of higher-order statistics beyond the statistical mean or average. In an attempt to describe or model performance uncertainty, the essence of information about these higher-order performance-measure statistics is now considered as a source of information flow, which will affect perception of the problem and the environment at the risk-averse incumbent agent i.

Next, the question of how to characterize and influence performance information is answered by modeling and management of cumulants (also known as semi-invariants) associated with (18) as shown in the following result.

Theorem 1 (Cumulant-Generating Function). *Let each incumbent agent i and $i \in \mathscr{I}$ be associated with the state variable $z^i(\cdot)$ of the stochastic dynamics (17) and* *subject to the performance measure (18). Further, let initial states $z^i(\tau) \equiv z^i_\tau$ and $\tau \in [t_0, t_f]$ and the moment-generating function be denoted by*

$$\varphi^i\left(\tau, z^i_\tau, \theta\right) = \rho^i(\tau, \theta) \exp\left\{(z^i_\tau)^T \Upsilon^i(\tau, \theta) z^i_\tau + 2(z^i_\tau)^T \ell^i(\tau, \theta)\right\} \tag{19}$$

$$\upsilon^i(\tau, \theta) = \ln\{\rho^i(\tau, \theta)\}, \qquad \theta \in \mathbb{R}^+. \tag{20}$$

Then, the cumulant-generating function has the form of quadratic affine

$$\psi^i\left(\tau, z^i_\tau, \theta\right) = (z^i_\tau)^T \Upsilon^i(\tau, \theta) z^i_\tau + 2(z^i_\tau)^T \ell^i(\tau, \theta) + \upsilon^i(\tau, \theta), \tag{21}$$

where the scalar solution $\upsilon^i(\tau, \theta)$ solves the scalar-valued backward-in-time differential equation with the terminal-value condition $\upsilon^i\left(t_f, \theta\right) = 0$

$$\frac{d}{d\tau}\upsilon^i(\tau, \theta) = -\mathrm{Tr}\left\{\Upsilon^i(\tau, \theta) G^i(\tau) W^i (G^i)^T(\tau)\right\} + \theta (u^*_{-i})^T(\tau) M_i(\tau) u^*_{-i}(\tau) \tag{22}$$

whereas the matrix $\Upsilon^i(\tau, \theta)$ and vector $\ell^i(\tau, \theta)$ solutions satisfy the matrix and vector-valued backward-in-time differential equations

$$\begin{aligned}\frac{d}{d\tau}\Upsilon^i(\tau, \theta) &= -(F^i)^T(\tau)\Upsilon^i(\tau, \theta) - \Upsilon^i(\tau, \theta) F^i(\tau) \\ &\quad - 2\Upsilon^i(\tau, \theta) G^i(\tau) W^i (G^i)^T(\tau) \Upsilon^i(\tau, \theta) - \theta N^i(\tau), \quad \Upsilon^i(t_f, \theta) = \theta N^i_f\end{aligned} \tag{23}$$

$$\frac{d}{d\tau}\ell^i(\tau, \theta) = -\Upsilon^i(\tau, \theta) E^i(\tau) u^*_{-i}(\tau), \quad \ell^i(t_f, \theta) = 0. \tag{24}$$

Meanwhile, the scalar solution $\rho^i(\tau, \theta)$ satisfies the scalar-valued backward-in-time differential equation

$$\begin{aligned}\frac{d}{d\tau}\rho^i(\tau, \theta) &= -\rho^i(\tau, \theta)\left[\mathrm{Tr}\left\{\Upsilon^i(\tau, \theta) G^i(\tau) W^i (G^i)^T(\tau)\right\}\right. \\ &\quad \left. - \theta (u^*_{-i})^T(\tau) M_i(\tau) u^*_{-i}(\tau)\right], \quad \rho^i\left(t_f, \theta\right) = 1.\end{aligned} \tag{25}$$

Proof. For notional simplicity, it is convenient to have, for each $i \in \mathscr{I}$

$$\varpi^i\left(\tau, z^i_\tau, \theta\right) \triangleq \exp\left\{\theta J_i\left(\tau, z^i_\tau\right)\right\},$$

in which the performance measure (18) is rewritten as the cost-to-go function from an arbitrary state z^i_τ at a running time $\tau \in [t_0, t_f]$, that is,

$$\begin{aligned}J_i(\tau, z^i_\tau) &= (z^i)^T(t_f) N^i_f z^i(t_f) \\ &\quad + \int_\tau^{t_f} [(z^i)^T(t) N^i(t) z^i(t) - (u^*_{-i})^T(t) M_i(t) u^*_{-i}(t)] dt\end{aligned} \tag{26}$$

subject to

$$dz^i(t) = (F^i(t)z^i(t) + E^i(t)u^*_{-i}(t))dt + G^i(t)dw^i(t)\,, \quad z^i(\tau) = z^i_\tau\,. \tag{27}$$

By definition, the moment-generating function is

$$\varphi^i(\tau, z^i_\tau, \theta) \triangleq E\left\{\varpi^i\left(\tau, z^i_\tau, \theta\right)\right\}.$$

Thus, the total time derivative of $\varphi^i(\tau, z^i_\tau, \theta)$ is obtained as

$$\frac{d}{d\tau}\varphi^i\left(\tau, z^i_\tau, \theta\right) = -\theta[(z^i_\tau)^T N^i(\tau) z^i_\tau - (u^*_{-i})^T(\tau) M_i(\tau) u^*_{-i}(\tau)]\varphi^i\left(\tau, z^i_\tau, \theta\right).$$

Using the standard Ito's formula, it follows

$$\begin{aligned}
d\varphi^i\left(\tau, z^i_\tau, \theta\right) &= E\left\{d\varpi^i\left(\tau, z^i_\tau, \theta\right)\right\} \\
&= E\Big\{\varpi^i_\tau(\tau, z^i_\tau, \theta)d\tau + \varpi^i_{z^i_\tau}(\tau, z^i_\tau, \theta)dz^i_\tau \\
&\qquad + \frac{1}{2}\mathrm{Tr}\Big\{\varpi^i_{z^i_\tau z^i_\tau}(\tau, z^i_\tau, \theta)G^i(\tau)W^i(G^i)^T(\tau)\Big\}d\tau\Big\} \\
&= \varphi^i_\tau(\tau, z^i_\tau, \theta)d\tau + \varphi^i_{z^i_\tau}(\tau, z^i_\tau, \theta)(F^i(\tau)z^i_\tau + E^i(\tau)u^*_{-i}(\tau))d\tau \\
&\qquad + \frac{1}{2}\mathrm{Tr}\Big\{\varphi^i_{z^i_\tau z^i_\tau}(\tau, z^i_\tau, \theta)G^i(\tau)W^i(G^i)^T(\tau)\Big\}d\tau,
\end{aligned}$$

which under the definition of the moment-generating function; e.g.,

$$\varphi^i\left(\tau, z^i_\tau, \theta\right) = \rho^i(\tau, \theta)\exp\left\{(z^i_\tau)^T \Upsilon^i(\tau, \theta) z^i_\tau + 2(z^i_\tau)^T \ell^i(\tau, \theta)\right\}$$

and its partial derivatives lead to the result

$$\begin{aligned}
&-\theta[(z^i_\tau)^T N^i(\tau) z^i_\tau - (u^*_{-i})^T(\tau) M_i(\tau) u^*_{-i}(\tau)]\varphi^i(\tau, z_\tau, \theta) \\
&\quad = \Bigg\{\frac{\frac{d}{d\tau}\rho^i(\tau, \theta)}{\rho^i(\tau, \theta)} + (z^i_\tau)^T \frac{d}{d\tau}\Upsilon^i(\tau, \theta) z^i_\tau + 2(z^i_\tau)^T \frac{d}{d\tau}\ell^i(\tau, \theta) \\
&\qquad + (z^i_\tau)^T\left[(F^i)^T(\tau)\Upsilon^i(\tau, \theta) + \Upsilon^i(\tau, \theta)F^i(\tau)\right]z^i_\tau + 2(z^i_\tau)^T \Upsilon^i(\tau, \theta)E^i(\tau)u^*_{-i}(\tau) \\
&\qquad + 2(z^i_\tau)^T \Upsilon^i(\tau, \theta)G^i(\tau)W^i(G^i)^T(\tau)\Upsilon^i(\tau, \theta)z^i_\tau \\
&\qquad + \mathrm{Tr}\left\{\Upsilon^i(\tau, \theta)G^i(\tau)W^i(G^i)^T(\tau)\right\}\Bigg\}\varphi^i\left(\tau, z^i_\tau, \theta\right).
\end{aligned}$$

To have constant and quadratic terms be independent of arbitrary z^i_τ, it requires

$$\frac{d}{d\tau}\Upsilon^i(\tau,\theta) = -(F^i)^T(\tau)\Upsilon^i(\tau,\theta) - \Upsilon^i(\tau,\theta)F^i(\tau) - \theta N^i(\tau)$$
$$- 2\Upsilon^i(\tau,\theta)G^i(\tau)W^i(G^i)^T(\tau)\Upsilon^i(\tau,\theta)$$
$$\frac{d}{d\tau}\ell^i(\tau,\theta) = -\Upsilon^i(\tau,\theta)E^i(\tau)u^*_{-i}(\tau)$$
$$\frac{d}{d\tau}\rho^i(\tau,\theta) = -\rho^i(\tau,\theta)\left[\mathrm{Tr}\left\{\Upsilon^i(\tau,\theta)G^i(\tau)W^i(G^i)^T(\tau)\right\} - \theta(u^*_{-i})^T(\tau)M_i(\tau)u^*_{-i}(\tau)\right]$$

with the terminal-value conditions $\Upsilon^i(t_f,\theta) = \theta N^i_f$ and $\rho^i(t_f,\theta) = 1$. □

Finally, the backward-in-time differential equation satisfied by the scalar-valued solution $\upsilon^i(\tau,\theta)$ is obtained with the terminal-value condition $\upsilon^i(t_f,\theta) = 0$

$$\frac{d}{d\tau}\upsilon^i(\tau,\theta) = -\mathrm{Tr}\left\{\Upsilon^i(\tau,\theta)G^i(\tau)W^i(G^i)^T(\tau)\right\} + \theta(u^*_{-i})^T(\tau)M_i(\tau)u^*_{-i}(\tau),$$

which completes the proof.

As it turns out, all the higher-order characteristic distributions associated with performance uncertainty and risk are captured in the higher-order performance-measure statistics associated with (18). Subsequently, higher-order statistics that encapsulate the uncertain nature of (18) can now be generated via a MacLaurin series of the cumulant-generating function or the second characteristic function (21)

$$\psi^i\left(\tau,z^i_\tau,\theta\right) = \sum_{r=1}^{\infty} \frac{\partial^{(r)}}{\partial\theta^{(r)}}\psi^i(\tau,z^i_\tau,\theta)\bigg|_{\theta=0}\frac{\theta^r}{r!}, \tag{28}$$

from which all $\kappa^i_r \triangleq \frac{\partial^{(r)}}{\partial\theta^{(r)}}\psi^i(\tau,z^i_\tau,\theta)\Big|_{\theta=0}$ are defined as performance-measure statistics associated with incumbent agent i and $i \in \mathscr{I}$. In fact, the rth performance-measure statistic is determined by the series expansion coefficients; that is, it is obtained from the cumulant-generating function (21)

$$\kappa^i_r = \frac{\partial^{(r)}}{\partial\theta^{(r)}}\psi^i(\tau,z^i_\tau,\theta)\bigg|_{\theta=0} = (z^i)^T_\tau\frac{\partial^{(r)}}{\partial\theta^{(r)}}\Upsilon^i(\tau,\theta)\bigg|_{\theta=0} z^i_\tau$$
$$+2(z^i_\tau)^T\frac{\partial^{(r)}}{\partial\theta^{(r)}}\ell^i(\tau,\theta)\bigg|_{\theta=0} + \frac{\partial^{(r)}}{\partial\theta^{(r)}}\upsilon^i(\tau,\theta)\bigg|_{\theta=0}. \tag{29}$$

For notational convenience, the change of variables corresponding to each incumbent agent i and $i \in \mathscr{I}$

$$H^i_r(\tau) \triangleq \frac{\partial^{(r)}\Upsilon^i(\tau,\theta)}{\partial\theta^{(r)}}\bigg|_{\theta=0}, \quad \tau\in[t_0,t_f] \tag{30}$$

$$\breve{D}_r^i(\tau) \triangleq \left. \frac{\partial^{(r)} \ell^i(\tau,\theta)}{\partial \theta^{(r)}} \right|_{\theta=0} ; \quad D_r^i(\tau) \triangleq \left. \frac{\partial^{(r)} \upsilon^i(\tau,\theta)}{\partial \theta^{(r)}} \right|_{\theta=0} \tag{31}$$

is introduced so that the next theorem provides an effective and accurate capability for forecasting all the higher-order characteristics associated with performance uncertainty.

Theorem 2 (Performance-Measure Statistics). *Associate with each incumbent agent i and $i \in \mathcal{I}$ the decentralized stochastic system governed by (17) and (18), wherein the pairs (A_i, B_i) and (A_i, C_i) are uniformly stabilizable and detectable. For $k^i \in \mathbb{N}$ fixed, the k^ith cumulant of performance measure (18) concerned by incumbent agent i is given by*

$$\kappa_k^i = (z_0^i)^T H_{k^i}^i(t_0) z_0^i + 2(z_0^i)^T \breve{D}_{k^i}^i(t_0) + D_{k^i}^i(t_0) , \tag{32}$$

where the supporting variables $\{H_r^i(\tau)\}_{r=1}^{k^i}$, $\{\breve{D}_r^i(\tau)\}_{r=1}^{k^i}$ and $\{D_r^i(\tau)\}_{r=1}^{k^i}$ evaluated at $\tau = t_0$ satisfy the differential equations (with the dependence of $H_r^i(\tau)$, $\breve{D}_r^i(\tau)$ and $D_r(\tau)$ upon the admissible feedback policy gain $K_i(\tau)$ and $u_{-i}^*(\tau)$ suppressed)

$$\frac{d}{d\tau} H_1^i(\tau) = -(F^i)^T(\tau) H_1^i(\tau) - H_1^i(\tau) F^i(\tau) - N^i(\tau) \tag{33}$$

$$\frac{d}{d\tau} H_r^i(\tau) = -(F^i)^T(\tau) H_r^i(\tau) - H_r^i(\tau) F^i(\tau) - \sum_{s=1}^{r-1} \frac{2r!}{s!(r-s)!} H_s^i(\tau) G^i(\tau) W^i (G^i)^T(\tau) H_{r-s}^i(\tau) , \quad 2 \le r \le k^i \tag{34}$$

and

$$\frac{d}{d\tau} \breve{D}_r^i(\tau) = -H_r^i(\tau) E^i(\tau) u_{-i}^*(\tau) , \qquad 1 \le r \le k^i \tag{35}$$

and, finally,

$$\frac{d}{d\tau} D_1^i(\tau) = -\mathrm{Tr}\left\{ H_1^i(\tau) G^i(\tau) W^i (G^i)^T(\tau) \right\} + (u_{-i}^*)^T(\tau) M_i(\tau) u_{-i}^*(\tau) \tag{36}$$

$$\frac{d}{d\tau} D_r^i(\tau) = -\mathrm{Tr}\left\{ H_r^i(\tau) G^i(\tau) W^i (G^i)^T(\tau) \right\} , \quad 2 \le r \le k^i \tag{37}$$

whereby the terminal-value conditions $H_1^i(t_f) = N_f^i$, $H_r^i(t_f) = 0$ for $2 \le r \le k^i$, $\breve{D}_r^i(t_f) = 0$ for $1 \le r \le k^i$ and $D_r^i(t_f) = 0$ for $1 \le r \le k^i$.

Proof. The expression of performance-measure statistics described in (32) is readily justified by using result (29) and definition (30)–(31). What remains is to show that the solutions $H_r^i(\tau)$, $\breve{D}_r^i(\tau)$ and $D_r^i(\tau)$ for $1 \le r \le k^i$ indeed satisfy the dynamical equations (33)–(37). Notice that these backward-in-time equations (33)–(37) satisfied by the matrix-valued $H_r^i(\tau)$, vector-valued $\breve{D}_r^i(\tau)$, and scalar-valued $D_r^i(\tau)$ solutions are then obtained by successively taking derivatives with respect to θ of the supporting equations (22)–(24) and subject to the assumptions of (A_i, B_i) and (A_i, C_i) being uniformly stabilizable and detectable on $[t_0, t_f]$. □

3 Problem Statements

The purpose of this section is to make use of increased insight into the roles played by performance-measure statistics on the generalized chi-squared performance measure (18) for risk-averse Nash feedback strategies. The distributed optimization with Nash feedback policy here is distinguished by the fact that the evolution in time of all mathematical statistics (32) associated with the random performance measure (18) of the generalized chi-squared type is described by means of the matrix/vector/scalar-valued backward-in-time differential equations (33)–(37).

For such problems it is important to have a compact statement of the risk-averse decision and control optimization so as to aid mathematical manipulation. To make this more precise, one may think of the k^i-tuple state variables $\mathscr{H}^i(\cdot) \triangleq (\mathscr{H}_1^i(\cdot), \ldots, \mathscr{H}_{k^i}^i(\cdot))$, $\breve{\mathscr{D}}^i(\cdot) \triangleq (\breve{\mathscr{D}}_1^i(\cdot), \ldots, \breve{\mathscr{D}}_{k^i}^i(\cdot))$ and $\mathscr{D}^i(\cdot) \triangleq (\mathscr{D}_1^i(\cdot), \ldots, \mathscr{D}_{k^i}^i(\cdot))$ whose continuously differentiable states $\mathscr{H}_r^i \in \mathscr{C}^1(t_0, t_f; \mathbb{R}^{2\sum_{j=1}^{N_i} n_j \times 2\sum_{j=1}^{N_i} n_j})$, $\breve{\mathscr{D}}_r^i \in \mathscr{C}^1(t_0, t_f; \mathbb{R}^{2\sum_{j=1}^{N_i} n_j})$ and $\mathscr{D}_r^i \in \mathscr{C}^1(t_0, t_f; \mathbb{R})$ having the representations $\mathscr{H}_r^i(\cdot) \triangleq H_r^i(\cdot)$, $\breve{\mathscr{D}}_r^i(\cdot) \triangleq \breve{D}_r^i(\cdot)$ and $\mathscr{D}_r^i(\cdot) \triangleq D_r^i(\cdot)$ with the right members satisfying the dynamics (33)–(37) are defined on $[t_0, t_f]$. In the remainder of the development, the convenient mappings associated with incumbent agent i and $i \in \mathscr{I}$ are introduced as follows

$$\mathscr{F}_r^i : [t_0, t_f] \times (\mathbb{R}^{2\sum_{j=1}^{N_i} n_j \times 2\sum_{j=1}^{N_i} n_j})^{k^i} \times \mathbb{R}^{m_i \times \sum_{j=1}^{N_i} n_j} \mapsto \mathbb{R}^{2\sum_{j=1}^{N_i} n_j \times 2\sum_{j=1}^{N_i} n_j}$$

$$\breve{\mathscr{G}}_r^i : [t_0, t_f] \times (\mathbb{R}^{2\sum_{j=1}^{N_i} n_j})^{k^i} \mapsto \mathbb{R}^{2\sum_{j=1}^{N_i} n_j}$$

$$\mathscr{G}_r^i : [t_0, t_f] \times (\mathbb{R}^{2\sum_{j=1}^{N_i} n_j \times 2\sum_{j=1}^{N_i} n_j})^{k^i} \mapsto \mathbb{R},$$

where the rules of action are given by

$$\mathscr{F}_1^i(\tau, \mathscr{H}^i, K_i) \triangleq -(F^i)^T(\tau)\mathscr{H}_1^i(\tau) - \mathscr{H}_1^i(\tau)F^i(\tau) - N^i(\tau)$$

$$\mathscr{F}_r^i(\tau, \mathscr{H}^i, K_i) \triangleq -(F^i)^T(\tau)\mathscr{H}_r^i(\tau) - \mathscr{H}_r^i(\tau)F^i(\tau) - \sum_{s=1}^{r-1} \frac{2r!}{s!(r-s)!} \mathscr{H}_s^i(\tau)G^i(\tau)W^i(G^i)^T(\tau)\mathscr{H}_{r-s}^i(\tau), \quad 2 \le r \le k^i$$

$$\breve{\mathscr{G}}_r^i(\tau, \mathscr{H}^i) \triangleq -\mathscr{H}_r^i(\tau)E^i(\tau)u_{-i}^*(\tau), \quad 1 \le r \le k^i$$

$$\mathscr{G}_1^i(\tau, \mathscr{H}^i) \triangleq -\text{Tr}\left\{\mathscr{H}_r^1(\tau)G^i(\tau)W^i(G^i)^T(\tau)\right\} + (u_{-i}^*)^T(\tau)M_i(\tau)u_{-i}^*(\tau)$$
$$\mathscr{G}_r^i(\tau, \mathscr{H}^i) \triangleq -\text{Tr}\left\{\mathscr{H}_r^i(\tau)G^i(\tau)W^i(G^i)^T(\tau)\right\}, \quad 2 \le r \le k^i.$$

The product mappings that follow are necessary for a compact formulation; e.g.,

$$\mathscr{F}_1^i \times \cdots \times \mathscr{F}_{k^i}^i : [t_0, t_f] \times (\mathbb{R}^{2\sum_{j=1}^{N_i} n_j \times 2\sum_{j=1}^{N_i} n_j})^{k^i} \times \mathbb{R}^{m_i \times \sum_{j=1}^{N_i} n_j} \mapsto (\mathbb{R}^{2\sum_{j=1}^{N_i} n_j \times 2\sum_{j=1}^{N_i} n_j})^{k^i}$$
$$\breve{\mathscr{G}}_1^i \times \cdots \times \breve{\mathscr{G}}_{k^i}^i : [t_0, t_f] \times (\mathbb{R}^{2\sum_{j=1}^{N_i} n_j})^{k^i} \mapsto (\mathbb{R}^{2\sum_{j=1}^{N_i} n_j})^{k^i}$$
$$\mathscr{G}_1^i \times \cdots \times \mathscr{G}_{k^i}^i : [t_0, t_f] \times (\mathbb{R}^{2\sum_{j=1}^{N_i} n_j \times 2\sum_{j=1}^{N_i} n_j})^{k^i} \mapsto \mathbb{R}^{k^i}$$

whereby the corresponding notations

$$\mathscr{F}^i \triangleq \mathscr{F}_1^i \times \cdots \times \mathscr{F}_{k^i}^i$$
$$\breve{\mathscr{G}}^i \triangleq \breve{\mathscr{G}}_1^i \times \cdots \times \breve{\mathscr{G}}_{k^i}^i$$
$$\mathscr{G}^i \triangleq \mathscr{G}_1^i \times \cdots \times \mathscr{G}_{k^i}^i$$

are used. Thus, the dynamical equations (33)–(37) can be rewritten as follows

$$\frac{d}{d\tau}\mathscr{H}^i(\tau) = \mathscr{F}^i(\tau, \mathscr{H}^i(\tau), K_i(\tau)), \qquad \mathscr{H}^i(t_f) \equiv \mathscr{H}_f^i \tag{38}$$
$$\frac{d}{d\tau}\breve{\mathscr{D}}^i(\tau) = \breve{\mathscr{G}}^i\left(\tau, \mathscr{H}^i(\tau)\right), \qquad \breve{\mathscr{D}}^i(t_f) \equiv \breve{\mathscr{D}}_f^i \tag{39}$$
$$\frac{d}{d\tau}\mathscr{D}^i(\tau) = \mathscr{G}^i\left(\tau, \mathscr{H}^i(\tau)\right), \qquad \mathscr{D}^i(t_f) \equiv \mathscr{D}_f^i \tag{40}$$

whereby the k^i-tuple terminal-value conditions $\mathscr{H}_f^i \triangleq (N_f^i, 0, \ldots, 0)$, $\breve{\mathscr{D}}_f^i \triangleq (0, \ldots, 0)$ and $\mathscr{D}_f^i \triangleq (0, \ldots, 0)$.

Once immediate neighbors $j \in \mathscr{N}_i$ of incumbent agent i fix the control and decision parameters K_j^* of the person-by-person equilibrium strategies u_j^* and thus the interconnection effects u_{-i}^* underpinned by K_{-i}^*, incumbent agent i therefore obtains an optimal stochastic control problem with risk-averse performance considerations. The construction of agent i's person-by-person policy also involves the control and decision parameter K_i. In the sequel and elsewhere, when the dependence on K_i and K_{-i}^* is needed to be clear, then the notations

$$\mathscr{H}^i \equiv \mathscr{H}^i(\cdot, K_i, K_{-i}^*)$$
$$\breve{\mathscr{D}}^i \equiv \breve{\mathscr{D}}^i(\cdot, K_i, K_{-i}^*)$$
$$\mathscr{D}^i \equiv \mathscr{D}^i(\cdot, K_i, K_{-i}^*)$$

should be used to denote the solution trajectories of the dynamics (38)–(40) with the admissible 2-tuple (K_i, K_{-i}^*).

For the given terminal data $(t_f, \mathscr{H}_f^i, \breve{\mathscr{D}}_f^i, \mathscr{D}_f^i)$, the class of admissible feedback gains employed by incumbent agent i and $i \in \mathscr{I}$ is next defined.

Definition 1 (Admissible Feedback Policy Gains). Let compact subset $\overline{K}^i \subset \mathbb{R}^{m_i \times n}$ be the set of allowable feedback form values. For the given $k^i \in \mathbb{N}$ and sequence $\mu^i = \{\mu_r^i \geq 0\}_{r=1}^{k^i}$ with $\mu_1^i > 0$, the set of feedback gains $\mathscr{K}^i_{t_f, \mathscr{H}_f^i, \breve{\mathscr{D}}_f^i, \mathscr{D}_f^i; \mu^i}$ is assumed to be the class of $\mathscr{C}(t_0, t_f; \mathbb{R}^{m_i \times \sum_{j=1}^{N_i} n_j})$ with values $K_i(\cdot) \in \overline{K}^i$, for which the solutions to the dynamical equations (38)–(40) with the terminal-value conditions $\mathscr{H}^i(t_f) = \mathscr{H}_f^i$, $\breve{\mathscr{D}}^i(t_f) = \breve{\mathscr{D}}_f^i$ and $\mathscr{D}^i(t_f) = \mathscr{D}_f^i$ exist on the interval of optimization $[t_0, t_f]$.

One way to make sense of risk bearing existing at incumbent agent i is to identify performance vulnerability of (18) against all the sample-path realizations from the local environment and potential noncooperative influences u_{-i}^* from immediate neighbors j and $j \in \mathscr{N}_i$. The mechanism identified here that is under a finite set of selective weights associated with the mathematical statistics of (18) helps to unfold the complexity behind observed performance values and risks of person-by-person strategy dependence in the following formulation of a risk-value aware performance index. Notice that this custom set of design freedoms representing particular uncertainty aversions is hence different from the ones with aversion to risk captured in risk-sensitive optimal control [9, 10].

On $\mathscr{K}^i_{t_f, \mathscr{H}_f^i, \breve{\mathscr{D}}_f^i, \mathscr{D}_f^i; \mu^i}$ the performance index with risk-value considerations in risk-averse decision making is subsequently defined as follows.

Definition 2 (Risk-Value Aware Performance Index). Let incumbent agent i and $i \in \mathscr{I}$ select $k^i \in \mathbb{N}$ and the sequence of scalar coefficients $\mu^i = \{\mu_r^i \geq 0\}_{r=1}^{k^i}$ with $\mu_1^i > 0$. Then, the risk-value aware performance index

$$\phi_0^i : \{t_0\} \times (\mathbb{R}^{2\sum_{j=1}^{N_i} n_j \times 2\sum_{j=1}^{N_i} n_j})^{k^i} \times (\mathbb{R}^{2\sum_{j=1}^{N_i} n_j})^{k^i} \times \mathbb{R}^{k^i} \mapsto \mathbb{R}^+$$

pertaining to risk-averse decision making of the stochastic Nash game over $[t_0, t_f]$ is defined by

$$\begin{aligned}
\phi_0^i(t_0, \mathscr{H}^i(t_0), \breve{\mathscr{D}}^i(t_0), \mathscr{D}^i(t_0)) \triangleq{} & \underbrace{\mu_1^i \kappa_1^i}_{\text{Value Measure}} \underbrace{+ \mu_2^i \kappa_2^i + \cdots + \mu_{k^i}^i \kappa_{k^i}^i}_{\text{Risk Measures}} \\
={} & \sum_{r=1}^{k^i} \mu_r^i [(z_0^i)^T \mathscr{H}_r^i(t_0) z_0^i + 2(z_0^i)^T \breve{\mathscr{D}}_r^i(t_0) + \mathscr{D}_r^i(t_0)],
\end{aligned} \tag{41}$$

where additional design freedom by means of μ_r^i's utilized by incumbent agent i with risk-averse attitudes are sufficient to meet and exceed different levels of performance-based reliability requirements, for instance, mean (i.e., the average of performance measure), variance (i.e., the dispersion of values of performance measure around its mean), skewness (i.e., the anti-symmetry of the density of performance measure), kurtosis (i.e., the heaviness in the density tails of performance measure), etc., pertaining to closed-loop performance variations and uncertainties while the supporting solutions $\{\mathscr{H}_r^i(\tau)\}_{r=1}^{k^i}$, $\{\breve{\mathscr{D}}_r^i(\tau)\}_{r=1}^{k^i}$ and $\{\mathscr{D}_r^i(\tau)\}_{r=1}^{k^i}$ evaluated at $\tau = t_0$ satisfy the dynamical equations (38)–(40).

To specifically indicate the dependence of the risk-value aware performance index (41) expressed in Mayer form on K_i and the signaling effects u_{-i}^* or K_{-i}^* issued by all immediate neighbors j from $\mathscr{N}_i$, the multi-attribute utility function or performance index (41) for incumbent agent i is now rewritten explicitly as $\phi_0^i(K_i, K_{-i}^*)$.

Definition 3 (Nash Equilibrium Solution). An admissible set of feedback strategies $(K_i^*, \ldots, K_{N_i}^*)$ is a Nash equilibrium for the local N_i-person game, where each incumbent agent i and $i \in \mathscr{I}$ has the performance index $\phi_0^i(K_i, K_{-i}^*)$ of Mayer type, if for all admissible feedback strategies $(K_1, \ldots, K_{N_i})$ the inequalities hold

$$\phi_0^i(K_i^*, K_{-i}^*) \leq \phi_0^i(K_i, K_{-i}^*) .$$

For the sake of time consistency and subgame perfection, a Nash equilibrium solution is required to have an additional property that its restriction on the interval $[t_0, \tau]$ is also a Nash solution to the truncated version of the original problem, defined on $[t_0, \tau]$. With such a restriction so defined, the Nash equilibrium solution is now termed as a feedback Nash equilibrium solution, which is now free of any informational nonuniqueness, and thus whose derivation allows a dynamic programming type argument.

Definition 4 (Feedback Nash Equilibrium). Let K_i^* constitute a feedback Nash strategy which will be implemented by incumbent agent i such that

$$\phi_0^i(K_i^*, K_{-i}^*) \leq \phi_0^i(K_i, K_{-i}^*) , \qquad i \in \mathscr{I} \tag{42}$$

for all admissible $K_i \in \mathscr{K}^i_{t_f, \mathscr{H}_f^i, \breve{\mathscr{D}}_f^i, \mathscr{D}_f^i; \mu^i}$, upon which the solutions to the dynamical systems (38)–(40) exist on $[t_0, t_f]$.

Then, $\left(K_1^*, \ldots, K_{N_i}^*\right)$ when restricted to the interval $[t_0, \tau]$ is still an N_i-tuple feedback Nash equilibrium solution for the multiperson Nash decision problem with the appropriate terminal-value condition $(\tau, \mathscr{H}_*^i(\tau), \breve{\mathscr{D}}_*^i(\tau), \mathscr{D}_*^i(\tau))$ for all $\tau \in [t_0, t_f]$.

In conformity with the rigorous formulation of dynamic programming, the following development is important. Let the terminal time t_f and 3-tuple states $(\mathscr{H}_f^i, \breve{\mathscr{D}}_f^i, \mathscr{D}_f^i)$, the other end condition involved the initial time t_0 and 3-tuple states $(\mathscr{H}_0^i, \breve{\mathscr{D}}_0^i, \mathscr{D}_0^i)$ be specified by a target set requirement.

Definition 5 (Target Sets). $(t_0, \mathscr{H}_0^i, \breve{\mathscr{D}}_0^i, \mathscr{D}_0^i) \in \hat{\mathscr{M}}^i$, where the target set $\hat{\mathscr{M}}^i$ residing at incumbent agent i and $i \in \mathscr{I}$ is a closed subset of $[t_0, t_f] \times (\mathbb{R}^{2\sum_{j=1}^{N_i} n_j \times 2\sum_{j=1}^{N_i} n_j})^{k^i} \times (\mathbb{R}^{2\sum_{j=1}^{N_i} n_j})^{k^i} \times \mathbb{R}^{k^i}$.

Now, the decision optimization residing at incumbent agent i and $i \in \mathscr{I}$ is to minimize the risk-value aware performance index (41) over all admissible feedback strategies $K_i = K_i(\cdot)$ in $\mathscr{K}^i_{t_f, \mathscr{H}_f^i, \breve{\mathscr{D}}_f^i, \mathscr{D}_f^i; \mu^i}$ while subject to potential interferences from all immediate neighbors with the feedback Nash policies K_{-i}^*.

Definition 6 (Optimization of Mayer Problem). Given the sequence of scalars $\mu^i = \{\mu_r^i \geq 0\}_{r=1}^{k^i}$ with $\mu_1^i > 0$, the decision optimization on $[t_0, t_f]$ associated with incumbent agent i and $i \in \mathscr{I}$ is given by

$$\min_{K_i(\cdot) \in \mathscr{K}^i_{t_f, \mathscr{H}_f^i, \breve{\mathscr{D}}_f^i, \mathscr{D}_f^i; \mu^i}} \phi_0^i(K_i, K_{-i}^*), \tag{43}$$

subject to the dynamical equations (38)–(40) on $[t_0, t_f]$.

Notice that the optimization considered here is in Mayer form and can be solved by applying an adaptation of the Mayer form verification results as given in [11]. To embed this optimization facing incumbent agent i into a larger problem, the terminal time and states $(t_f, \mathscr{H}_f^i, \breve{\mathscr{D}}_f^i, \mathscr{D}_f^i)$ are parameterized as $(\varepsilon, \mathscr{Y}^i, \breve{\mathscr{Z}}^i, \mathscr{Z}^i)$, whereby $\mathscr{Y}^i \triangleq \mathscr{H}^i(\varepsilon)$, $\breve{\mathscr{Z}}^i \triangleq \breve{\mathscr{D}}^i(\varepsilon)$ and $\mathscr{Z}^i \triangleq \mathscr{D}^i(\varepsilon)$. Thus, the value function for this optimization problem is now depending on the parameterization of terminal-value conditions.

Definition 7 (Value Function). Suppose $(\varepsilon, \mathscr{Y}^i, \breve{\mathscr{Z}}^i, \mathscr{Z}^i) \in [t_0, t_f] \times (\mathbb{R}^{2\sum_{j=1} N_i n_j \times 2\sum_{j=1}^{N_i} n_j})^{k^i} \times (\mathbb{R}^{2\sum_{j=1}^{N_i} n_j})^{k^i} \times \mathbb{R}^{k^i}$ is given and fixed. Then, the value function $\mathscr{V}^i(\varepsilon, \mathscr{Y}^i, \breve{\mathscr{Z}}^i, \mathscr{Z}^i)$ and $i \in \mathscr{I}$ is defined by

$$\mathscr{V}^i(\varepsilon, \mathscr{Y}^i, \breve{\mathscr{Z}}^i, \mathscr{Z}^i) \triangleq \inf_{K_i(\cdot) \in \mathscr{K}^i_{\varepsilon, \mathscr{Y}^i, \breve{\mathscr{Z}}^i, \mathscr{Z}^i; \mu^i}} \phi_0^i(K_i, K_{-i}^*).$$

For convention, $\mathscr{V}^i(\varepsilon, \mathscr{Y}^i, \breve{\mathscr{Z}}^i, \mathscr{Z}^i) \triangleq \infty$ when $\mathscr{K}^i_{\varepsilon, \mathscr{Y}^i, \breve{\mathscr{Z}}^i, \mathscr{Z}^i; \mu^i}$ is empty. Next, some candidates for the value function are constructed with the help of the concept of reachable set.

Definition 8 (Reachable Sets). Let reachable set associated with incumbent agent i be $\mathscr{Q}^i \triangleq \{(\varepsilon, \mathscr{Y}^i, \breve{\mathscr{Z}}^i, \mathscr{Z}^i) \in [t_0, t_f] \times (\mathbb{R}^{2\sum_{j=1}^{N_i} n_j \times 2\sum_{j=1}^{N_i} n_j})^{k^i} \times (\mathbb{R}^{2\sum_{j=1}^{N_i} n_j})^{k^i} \times \mathbb{R}^{k^i}$ such that $\mathscr{K}^i_{\varepsilon, \mathscr{Y}^i, \breve{\mathscr{Z}}^i, \mathscr{Z}^i; \mu^i} \neq \emptyset\}$.

Moreover, it can be shown that the value function associated with incumbent agent i is satisfying a partial differential equation at interior points of $\mathscr{Q}^i$, at which it is differentiable.

Theorem 3 (Hamilton–Jacobi–Bellman (HJB) Equation-Mayer Problem). *Let $(\varepsilon, \mathscr{Y}^i, \breve{\mathscr{Z}}^i, \mathscr{Z}^i)$ be any interior point of the reachable set $\mathscr{Q}^i$ and $i \in \mathscr{I}$, at which the value function $\mathscr{V}^i(\varepsilon, \mathscr{Y}^i, \breve{\mathscr{Z}}^i, \mathscr{Z}^i)$ is differentiable. If there exists a feedback Nash strategy $K_i^* \in \mathscr{K}^i_{t_f, \mathscr{H}^i_f, \breve{\mathscr{D}}^i_f, \mathscr{D}^i_f; \mu^i}$, then the differential equation*

$$0 = \min_{K_i \in \overline{K}^i} \left\{ \frac{\partial}{\partial \varepsilon} \mathscr{V}^i(\varepsilon, \mathscr{Y}^i, \breve{\mathscr{Z}}^i, \mathscr{Z}^i) + \frac{\partial}{\partial \operatorname{vec}(\mathscr{Y}^i)} \mathscr{V}^i(\varepsilon, \mathscr{Y}^i, \breve{\mathscr{Z}}^i, \mathscr{Z}^i) \operatorname{vec}(\mathscr{F}^i(\varepsilon, \mathscr{Y}^i, K_i)) + \frac{\partial}{\partial \operatorname{vec}(\breve{\mathscr{Z}}^i)} \mathscr{V}^i(\varepsilon, \mathscr{Y}^i, \breve{\mathscr{Z}}^i, \mathscr{Z}^i) \operatorname{vec}(\breve{\mathscr{G}}^i(\varepsilon, \mathscr{Y}^i) + \frac{\partial}{\partial \operatorname{vec}(\mathscr{Z}^i)} \mathscr{V}^i(\varepsilon, \mathscr{Y}^i, \breve{\mathscr{Z}}^i, \mathscr{Z}^i) \operatorname{vec}(\mathscr{G}^i(\varepsilon, \mathscr{Y}^i) \right\} \quad (44)$$

is satisfied whereby $\mathscr{V}^i(t_0, \mathscr{Y}^i(t_0), \breve{\mathscr{Z}}^i(t_0), \mathscr{Z}^i(t_0)) = \phi_0^i(\mathscr{H}^i(t_0), \breve{\mathscr{D}}^i(t_0), \mathscr{D}^i(t_0))$.

Proof. By what have been shown in the recent results by the first author [12], the proof for the result herein is readily proven.

Finally, the following result gives the sufficient condition used to verify a feedback Nash strategy for incumbent agent i and $i \in \mathscr{I}$. □

Theorem 4 (Verification Theorem). *Let $\mathscr{W}^i(\varepsilon, \mathscr{Y}^i, \breve{\mathscr{Z}}^i, \mathscr{Z}^i)$ associated with incumbent agent i and $i \in \mathscr{I}$ be continuously differentiable solution of the HJB equation (44), which satisfies the following boundary condition*

$$\mathscr{W}^i(t_0, \mathscr{H}^i(t_0), \breve{\mathscr{D}}^i(t_0), \mathscr{D}^i(t_0)) = \phi_0^i(t_0, \mathscr{H}^i(t_0), \breve{\mathscr{D}}^i(t_0), \mathscr{D}^i(t_0)).$$

Let $(t_f, \mathscr{H}^i_f, \breve{\mathscr{D}}^i_f, \mathscr{D}^i_f) \in \mathscr{Q}^i$; let $K_i \in \mathscr{K}^i_{t_f, \mathscr{H}^i_f, \breve{\mathscr{D}}^i_f, \mathscr{D}^i_f; \mu^i}$; and let $(\mathscr{H}^i(\cdot), \breve{\mathscr{D}}^i(\cdot), \mathscr{D}^i(\cdot))$ be the trajectory solutions of the dynamical equations (38)–(40). Then, the scalar-valued function $\mathscr{W}^i(\tau, \mathscr{H}^i(\tau), \breve{\mathscr{D}}^i(\tau), \mathscr{D}^i(\tau))$ is time-backward increasing function of τ and $\tau \in [t_0, t_f]$.

If K_i^* is in $\mathscr{K}^i_{t_f, \mathscr{H}^i_f, \breve{\mathscr{D}}^i_f, \mathscr{D}^i_f; \mu^i}$ with the corresponding solutions $(\mathscr{H}^i_*(\cdot), \breve{\mathscr{D}}^i_*(\cdot), \mathscr{D}^i_*(\cdot))$ of the dynamical equations (38)–(40) such that, for $\tau \in [t_0, t_f]$

$$0 = \frac{\partial}{\partial \varepsilon} \mathscr{W}^i(\tau, \mathscr{H}^i_*(\tau), \breve{\mathscr{D}}^i_*(\tau), \mathscr{D}^i_*(\tau)) + \frac{\partial}{\partial \operatorname{vec}(\mathscr{Y}^i)} \mathscr{W}^i(\tau, \mathscr{H}^i_*(\tau), \breve{\mathscr{D}}^i_*(\tau), \mathscr{D}^i_*(\tau)) \operatorname{vec}(\mathscr{F}^i(\tau, \mathscr{H}^i_*(\tau), K_i^*(\tau))) + \frac{\partial}{\partial \operatorname{vec}(\breve{\mathscr{Z}}^i)} \mathscr{W}^i(\tau, \mathscr{H}^i_*(\tau), \breve{\mathscr{D}}^i_*(\tau), \mathscr{D}^i_*(\tau)) \operatorname{vec}(\breve{\mathscr{G}}^i(\tau, \mathscr{H}^i_*(\tau))) + \frac{\partial}{\partial \operatorname{vec}(\mathscr{Z}^i)} \mathscr{W}^i(\tau, \mathscr{H}^i_*(\tau), \breve{\mathscr{D}}^i_*(\tau), \mathscr{D}^i_*(\tau)) \operatorname{vec}(\mathscr{G}^i(\tau, \mathscr{H}^i_*(\tau))) \quad (45)$$

then, K_i^* is a feedback Nash strategy in $\mathscr{K}^i_{t_f,\mathscr{H}^i_f,\breve{\mathscr{D}}^i_f,\mathscr{D}^i_f;\mu^i}$,

$$\mathscr{W}^i(\varepsilon,\mathscr{Y}^i,\breve{\mathscr{Z}}^i,\mathscr{Z}^i) = \mathscr{V}^i(\varepsilon,\mathscr{Y}^i,\breve{\mathscr{Z}}^i,\mathscr{Z}^i), \tag{46}$$

where $\mathscr{V}^i(\varepsilon,\mathscr{Y}^i,\breve{\mathscr{Z}}^i,\mathscr{Z}^i)$ is the value function associated with incumbent agent i.

Proof. With the aid of the recent development in [12], the proof then follows for the verification theorem herein. □

4 Person-by-Person Equilibrium Strategies

The aim of the present section is to recognize the optimization problem of Mayer form existing at incumbent agent i and $i \in \mathscr{I}$, which can therefore be solved by an adaptation of the Mayer-form verification theorem. To this end the terminal time and states $(\varepsilon,\mathscr{H}^i_f,\breve{\mathscr{D}}^i_f,\mathscr{D}^i_f)$ of the dynamics (38)–(40) are now parameterized as $(\varepsilon,\mathscr{Y}^i,\breve{\mathscr{Z}}^i,\mathscr{Z}^i)$ for a broader family of optimization problems.

To apply properly the dynamic programming approach based on the HJB mechanism, together with the verification result, the solution procedure should be formulated as follows. For any given interior point $(\varepsilon,\mathscr{Y}^i,\breve{\mathscr{Z}}^i,\mathscr{Z}^i)$ of the reachable set $\mathscr{Q}^i$ and $i \in \mathscr{I}$, at which the following real-valued function is considered as a candidate solution $\mathscr{W}^i(\varepsilon,\mathscr{Y}^i,\breve{\mathscr{Z}}^i,\mathscr{Z}^i)$ to the HJB equation (44). Because the initial state z^i_0, which is arbitrarily fixed represents both quadratic and linear contributions to the performance index (41) of Mayer type, it is therefore concluded that the value function is linear and quadratic in z^i_0. Thus, a candidate function $\mathscr{W}^i \in \mathscr{C}^1(t_0,t_f;\mathbb{R})$ for the value function is of the form

$$\mathscr{W}^i(\varepsilon,\mathscr{Y}^i,\breve{\mathscr{Z}}^i,\mathscr{Z}^i) = (z^i_0)^T \sum_{r=1}^{k^i} \mu^i_r(\mathscr{Y}^i_r + \mathscr{E}^i_r(\varepsilon))\, z^i_0 \\ + 2(z^i_0)^T \sum_{r=1}^{k^i} \mu^i_r(\breve{\mathscr{Z}}^i_r + \breve{\mathscr{T}}^i_r(\varepsilon)) + \sum_{r=1}^{k^i} \mu^i_r(\mathscr{Z}^i_r + \mathscr{T}^i_r(\varepsilon)) \tag{47}$$

whereby the parametric functions of time $\mathscr{E}^i_r \in \mathscr{C}^1(t_0,t_f;\mathbb{R}^{2\sum_{j=1}^{N_i} n_j \times 2\sum_{j=1}^{N_i} n_j})$, $\breve{\mathscr{T}}^i_r \in \mathscr{C}^1(t_0,t_f;\mathbb{R}^{2\sum_{j=1}^{N_i} n_j})$, and $\mathscr{T}^i_r \in \mathscr{C}^1([t_0,t_f];\mathbb{R})$ are yet to be determined.

Moreover, it can be shown that the derivative of $\mathscr{W}^i(\varepsilon,\mathscr{Y}^i,\breve{\mathscr{Z}}^i,\mathscr{Z}^i)$ with respect to time ε is

$$\frac{d}{d\varepsilon}\mathscr{W}^i(\varepsilon,\mathscr{Y}^i,\breve{\mathscr{Z}}^i,\mathscr{Z}^i) = (z_0^i)^T\sum_{r=1}^{k^i}\mu_r^i[\mathscr{F}_r^i(\varepsilon,\mathscr{Y}^i,K_i)+\frac{d}{d\varepsilon}\mathscr{E}_r^i(\varepsilon)]z_0^i$$
$$+2(z_0^i)^T\sum_{r=1}^{k^i}\mu_r^i[\breve{\mathscr{G}}_r^i(\varepsilon,\mathscr{Y}^i)+\frac{d}{d\varepsilon}\breve{\mathscr{T}}_r^i(\varepsilon)]$$
$$+\sum_{r=1}^{k^i}\mu_r^i[\mathscr{G}_r^i(\varepsilon,\mathscr{Y}^i)+\frac{d}{d\varepsilon}\mathscr{T}_r^i(\varepsilon)]. \tag{48}$$

The substitution of this candidate (47) for the value function into the HJB equation (44) and making use of (48) yield

$$0=\min_{K_i\in\overline{K}^i}\Big\{(z_0^i)^T\sum_{r=1}^{k^i}\mu_r^i[\mathscr{F}_r^i(\varepsilon,\mathscr{Y}^i,K_i)+\frac{d}{d\varepsilon}\mathscr{E}_r^i(\varepsilon)]z_0^i$$
$$+2(z_0^i)^T\sum_{r=1}^{k^i}\mu_r^i[\breve{\mathscr{G}}_r^i(\varepsilon,\mathscr{Y}^i)+\frac{d}{d\varepsilon}\breve{\mathscr{T}}_r^i(\varepsilon)]+\sum_{r=1}^{k^i}\mu_r^i[\mathscr{G}_r^i(\varepsilon,\mathscr{Y}^i)+\frac{d}{d\varepsilon}\mathscr{T}_r^i(\varepsilon)]\Big\}. \tag{49}$$

Now the aggregate matrix coefficients $F^i(\cdot)$ and $N^i(\cdot)$ of the aggregate dynamics (17) are partitioned to conform with the n-dimensional structure of (6) by means of

$$I_0^T\triangleq\begin{bmatrix}I\ 0\end{bmatrix},\qquad I_1^T\triangleq\begin{bmatrix}0\ I\end{bmatrix},$$

where I is an $\sum_{j=1}^{N_i}n_j\times\sum_{j=1}^{N_i}n_j$ identity matrix and

$$F^i(\cdot)=I_0(A_i(\cdot)+B_i(\cdot)K_i(\cdot))I_0^T+I_0L_i(\cdot)C_i(\cdot)I_1^T+I_1(A_i(\cdot)-L_i(\cdot)C_i(\cdot))I_1^T \tag{50}$$
$$N^i(\cdot)=I_0(Q_i(\cdot)+K_i^T(\cdot)R_i(\cdot)K_i(\cdot))I_0^T+I_0Q_i(\cdot)I_1^T+I_1Q_i(\cdot)I_0^T+I_1Q_i(\cdot)I_1^T. \tag{51}$$

Taking the gradient with respect to K_i of the expression within the bracket of (49) yield the necessary conditions for an extremum of risk-value performance index (41) on the time interval $[t_0,\varepsilon]$

$$K_i=-R_i^{-1}(\varepsilon)B_i^T(\varepsilon)I_0^T\sum_{r=1}^{k^i}\hat{\mu}_r^i\mathscr{Y}_r^iI_0((I_0^TI_0)^{-1})^T,\quad i\in\mathscr{I} \tag{52}$$

where $\hat{\mu}_r^i\triangleq\mu_r^i/\mu_1^i$ for $\mu_1^i>0$. With the feedback Nash strategy (52) replaced in the expression of the bracket (49) and having $\{\mathscr{Y}_r^i\}_{r=1}^{k^i}$ evaluated on the optimal solution trajectories (38)–(40), the time-dependent functions $\mathscr{E}_r^i(\varepsilon)$, $\breve{\mathscr{T}}_r^i(\varepsilon)$ and $\mathscr{T}_r^i(\varepsilon)$ are therefore chosen such that the sufficient condition (45) in the verification theorem is satisfied in the presence of the arbitrary value of z_0^i; for example,

$$\frac{d}{d\varepsilon}\mathscr{E}_1^i(\varepsilon) = (F_*^i)^T(\varepsilon)\mathscr{H}_{1*}^i(\varepsilon) + \mathscr{H}_{1*}^i(\varepsilon)F_*^i(\varepsilon) + N_*^i(\varepsilon)$$

$$\frac{d}{d\varepsilon}\mathscr{E}_r^i(\varepsilon) = (F_*^i)^T(\varepsilon)\mathscr{H}_{r*}^i(\varepsilon) + \mathscr{H}_{r*}^i(\varepsilon)F_*^i(\varepsilon)$$

$$+\sum_{s=1}^{r-1}\frac{2r!}{s!(r-s)!}\mathscr{H}_{s*}^i(\varepsilon)G^i(\varepsilon)W^i(G^i)^T(\varepsilon)\mathscr{H}_{r-s*}^i(\varepsilon), \quad 2 \le r \le k^i$$

and

$$\frac{d}{d\varepsilon}\breve{\mathscr{T}}_r^i(\varepsilon) = \mathscr{H}_{r*}^i(\varepsilon)E^i(\varepsilon)u_{-i}^*(\varepsilon), \qquad 1 \le r \le k^i$$

and, finally

$$\frac{d}{d\varepsilon}\mathscr{T}_1^i(\varepsilon) = \mathrm{Tr}\left\{\mathscr{H}_{1*}^i(\varepsilon)G^i(\varepsilon)W^i(G^i)^T(\varepsilon)\right\} - (u_{-i}^*)^T(\varepsilon)M_i(\varepsilon)u_{-i}^*(\varepsilon)$$

$$\frac{d}{d\varepsilon}\mathscr{T}_r^i(\varepsilon) = \mathrm{Tr}\left\{\mathscr{H}_{r*}^i(\varepsilon)G^i(\varepsilon)W^i(G^i)^T(\varepsilon)\right\}, \quad 2 \le r \le k^i$$

with the initial-value conditions $\mathscr{E}_r^i(t_0) = 0$, $\breve{\mathscr{T}}_r^i(t_0) = 0$ and $\mathscr{T}_r^i(t_0) = 0$ for $1 \le r \le k^i$. Therefore, the sufficient condition (45) of the verification theorem is satisfied so that the extremizing feedback strategy (52) becomes optimal.

Therefore, the subsequent result for risk-bearing decisions is already proved and thus summarized for each incumbent agent i and $i \in \mathscr{I}$; who autonomously selects K_i^* for its person-by-person equilibrium (or equivalently, feedback Nash decision policy) strategy in presence of its immediate neighbors' feedback Nash policy parameters K_{-i}^*, as in Fig. 1.

Theorem 5 (Person-by-Person Equilibrium Policies for Distributed Control). *Consider the linear-quadratic class of distributed stochastic systems whose descriptions are governed by (6)–(13) and subject to the assumption of* (A_i, B_i) *and* (A_i, C_i) *for* $i \in \mathscr{I}$ *uniformly stablizable and detectable. Assume that incumbent systems or agents are constrained to admissible decision laws* $u_i(\cdot) = K_i(\cdot)\hat{z}_i(\cdot)$*, where the conditional mean estimates* $\hat{z}_i(\cdot)$ *are governed by the decentralized state-estimation dynamics (9). Further let incumbent agents* i *select* $k^i \in \mathbb{N}$ *and the sequence of nonnegative coefficients* $\mu^i = \{\mu_r^i \ge 0\}_{r=1}^{k^i}$ *with* $\mu_1^i > 0$*. Then, there exists a person-by-person equilibrium which strives to optimize the risk-value awareness performance indices (41); e.g.,*

$$u_*^i(t) = K_i^*(t)\hat{z}_i^*(t), \quad t \triangleq t_0 + t_f - \tau \tag{53}$$

$$K_i^*(\tau) = -R_i^{-1}(\tau)B_i^T(\tau)I_0^T\sum_{r=1}^{k^i}\hat{\mu}_r^i\mathscr{H}_{r*}^i(\tau)I_0((I_0^TI_0)^{-1})^T, \quad i \in \mathscr{I} \tag{54}$$

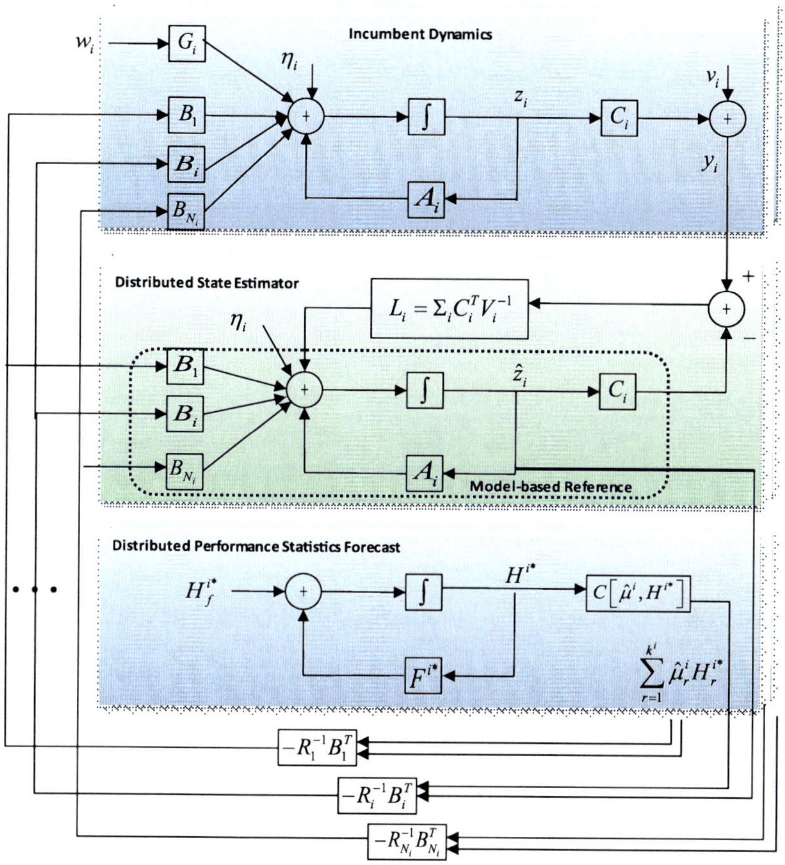

Fig. 1 Unified framework of measuring risk judgments and modeling choices and decisions

wherein the parametric design freedom through $\hat{\mu}_r^i$ represent the preferences toward specific summary statistical measures; e.g., mean, variance, skewness, etc. are chosen by incumbent agent i for performance reliability; whereas the optimal solutions $\mathscr{H}_{r*}^i(\cdot)$ satisfy the backward-in-time matrix-valued differential equations

$$\frac{d}{d\tau}\mathscr{H}_{1*}^i(\tau) = -(F_*^i)^T(\tau)\mathscr{H}_{1*}^i(\tau) - \mathscr{H}_{1*}^i(\tau)F_*^i(\tau) - N_*^i(\tau), \quad \mathscr{H}_{1*}^i(t_f) = N_f^i \tag{55}$$

$$\frac{d}{d\tau}\mathscr{H}_{r*}^i(\tau) = -(F_*^i)^T(\tau)\mathscr{H}_{r*}^i(\tau) - \mathscr{H}_{r*}^i(\tau)F_*^i(\tau)$$
$$-\sum_{s=1}^{r-1}\frac{2r!}{s!(r-s)!}\mathscr{H}_{s*}^i(\tau)G^i(\tau)W^i(G^i)^T(\tau)\mathscr{H}_{r-s*}^i(\tau),\ \mathscr{H}_{r*}^i(t_f)$$
$$= 0,\ 2 \le r \le k^i. \tag{56}$$

In addition, the decentralized state estimates $\hat{z}_i^*(t)$ associated with incumbent agent i and $i \in \mathcal{I}$ when the person-by-person equilibrium policy (53) are applied, are satisfying the forward-in-time vector-valued differential equation with $\hat{z}_i^*(t_0) = z_i^0$

$$d\hat{z}_i^*(t) = (A_i(t)\hat{z}_i^*(t) + B_i(t)u_i^*(t) + u_{-i}^*(t))dt + L_i(t)[dy_i^*(t) - C_i(t)\hat{z}_i^*(t)dt] \quad (57)$$

and

$$dz_i^*(t) = (A_i(t)z_i^*(t) + B_i(t)u_i^*(t) + u_{-i}^*(t)dt)dt + G_i(t)d\xi_i(t),\ z_i^*(t_0) = z_i^0 \quad (58)$$

$$u_{-i}^*(t)dt = \sum_{j=1, j\neq i}^{N_i} B_j(t)u_j^*(t)dt + d\eta_i(t) \quad (59)$$

$$dy_i^*(t) = C_i(t)z_i^*(t)dt + dv_i(t) \quad (60)$$

whereby the decentralized filter gain $L_i(t) = \Sigma_i(t)C_i^T(t)V_i^{-1}$ and the state-estimate error covariance $\Sigma_i(t)$ is determined by the forward-in-time matrix-valued differential equation with initial-value condition $\Sigma_i(t_0) = 0$

$$\frac{d}{dt}\Sigma_i(t) = A_i(t)\Sigma_i(t) + \Sigma_i(t)A_i^T(t) + G_i(t)\Xi_i G_i^T(t) + I_i - \Sigma_i(t)C_i^T(t)V_i^{-1}C_i(t)\Sigma_i(t).$$

Notice that to have the person-by-person equilibrium policy (53) of incumbent agent i be defined and continuous for all $\tau \in [t_0, t_f]$, the solutions $\mathcal{H}_{r*}^i(\tau)$ to the equations (55)–(56) when evaluated at $\tau = t_0$ must also exist. Thus, it is necessary that $\mathcal{H}_{r*}^i(\tau)$ are finite for all $\tau \in [t_0, t_f)$. Moreover, the solutions of (55)–(56) exist and are continuously differentiable in a neighborhood of t_f. Applying the result from [13], these solutions can further be extended to the left of t_f as long as $\mathcal{H}_{r*}^i(\tau)$ remain finite. Hence, the existence of unique and continuously differentiable solutions to (55)–(56) is certain if $\mathcal{H}_{r*}^i(\tau)$ are bounded for all $\tau \in [t_0, t_f)$. Subsequently, the candidate value functions $\mathcal{W}^i(\tau, \mathcal{H}^i, \breve{\mathcal{D}}^i, \mathcal{D}^i)$ are continuously differentiable.

5 Conclusions

The present research offers a theoretic lens and a novel approach that direct attention towards mathematical statistics of the chi-squared random performance measures concerned by incumbent agents of the class of distributed stochastic systems herein and thus provide new insights into complex dynamics of performance robustness and reliability. To account for mutual influence from immediate neighbors that give rise to interaction complexity such as potential noncooperation, each incumbent system or self-directed agent autonomously focuses on the search for a person-by-person equilibrium which is in turn locally supported by noisy state observations.

In views of performance risks, a new paradigm shift for understanding and building decentralized person-by-person equilibrium policies for the emergence of flexibly autonomous systems is obtained, with which the self-directed agents of incumbent systems, who are constrained to decentralized information processing and distributed decision making, are fully capable of implementing risk-bearing actions and local best responses in the furtherance of their own goals.

Acknowledgments The first author would like to acknowledge the Air Force Research Laboratory for financially supporting this research through the Collaborative Systems Control Strategic Technology Thrust Initiatives. The opinions expressed in this research article are those of the authors and do not necessarily represent, and should not be attributed to, the United States Air Force, the Department of Defense, or the United States Government.

References

1. Basar T and Olsder GJ (1999) Dynamic noncooperative game theory. 2nd Edn. Society for Industrial and Applied Mathematics
2. Engwerda JC (2005) LQ dynamic optimization and differential games. Wiley
3. Pollatsek A and Tversky A (1970) Theory of risk. Journal of Mathematical Psychology, 7:540–53
4. Luce RD (1980) Several possible measures of risk. Theory and Decision, 12:217–228
5. Pham KD (2008) New results in stochastic cooperative games: strategic coordination for multi-resolution performance robustness. In: Hirsch MJ, Pardalos PM, Murphey R, Grundel D (eds.) Optimization and Cooperative Control Strategies. Series Lecture Notes in Control and Information Sciences, 381:257–285. Springer Berlin: Heidelberg
6. Pham KD (2010) Performance-information analysis and distributed feedback stabilization in large-scale interconnected systems. Dynamics of Information Systems Theory and Applications Series: Springer Optimization and Its Applications, Hirsch, MJ; Pardalos, PM; Murphey R (Eds.), 40:45–81
7. Pham KD (2008) Cooperative outcomes for stochastic Nash games: decision strategies towards multi-attribute performance robustness. The 17th International Federation of Automatic Control World Congress, pp. 11750–11756
8. Marschak J and Radner R (1972) Economic theory of teams. New Haven: Yale University Press
9. Jacobson DH (1973) Optimal stochastic linear systems with exponential performance criteria and their relation to deterministic games. IEEE Transactions on Automatic Control, 18:124–131
10. Whittle P (1990) Risk sensitive optimal control. New York: John Wiley & Sons
11. Fleming WH and Rishel RW (1975) Deterministic and stochastic optimal control. Springer-Verlag
12. Pham KD (2011) Performance-reliability-aided decision-making in multiperson quadratic decision games against jamming and estimation confrontations. Journal of Optimization Theory and Applications, Edited by Giannessi F, 149(1):599–629
13. Dieudonne J (1960) Foundations of modern analysis. New York and London: Academic Press

Static Teams and Stochastic Games

Meir Pachter and Khanh Pham

Abstract Radner's solution of the static team decision problem is revisited. A careful and complete statement of the static decentralized optimization problem, also referred to as the team decision problem, is given. Decentralized optimization is considered in the framework of nonzero-sum game theory, and the impact of the partial information pattern on the structure of the optimal strategies is analyzed. The complete solution of the static decentralized multivariate Quadratic Gaussian (QG) optimization problem is obtained.

Keywords Static quadratic Gaussian team • Decentralized optimization

1 Introduction

A static stochastic decentralized optimization problem where a team consisting of two decision makers/players is at work is considered. The cost function is

$$J = J(u, v, \zeta) \tag{1}$$

where $u \in R^{m_u}$ and $v \in R^{m_v}$ are the two players' respective decision variables/controls and the state of nature, $\zeta \in R^n$, $n \geq 2$, is a random variable whose p.d.f. $f(\zeta)$ is known to both players. This is the players' prior information—it is

M. Pachter (✉)
Department of Electrical and Computer Engineering, Air Force Institute of Technology, Wright-Patterson Air Force Base, OH 45433, USA
e-mail: meir.pachter@afit.edu

K. Pham
Air Force Research Laboratory, Space Vehicles Directorate, Kirtland Air Force Base, NM 87117, USA
e-mail: kanh.pham@kirtland.af.mil

A. Sorokin and P.M. Pardalos (eds.), *Dynamics of Information Systems: Algorithmic Approaches*, Springer Proceedings in Mathematics & Statistics 51,
DOI 10.1007/978-1-4614-7582-8_5,

public information. The random variable ζ is partitioned

$$\zeta = (\zeta_1, \zeta_2)^T$$

and the information pattern is as follows. At decision time the component ζ_1 is known to the player whose control is u, the u-player, and the component ζ_2 is known to the player whose control is v, the v-player. Thus, both players have imperfect information. The u-player is oblivious of the ζ_2 component of the random variable, which is the v-player's private information, and consequently the strategy of the u-player is $u = u(\zeta_1)$. The v-player is oblivious of the ζ_1 component of the random variable, which is the u-player's private information, and consequently his or her strategy is $v = v(\zeta_2)$. The players have partial, or incomplete, information.

To obtain the *optimal* solution/strategies of the team/decentralized optimization problem, the following optimization problem in Hilbert space must be solved.

$$\begin{aligned} J^* &= \min_{u(\zeta_1), v(\zeta_2)} E_\zeta(J(u(\zeta_1), v(\zeta_2), \zeta)) \\ &= \min_{u(\zeta_1), v(\zeta_2)} \int_{\zeta_1} \int_{\zeta_2} J(u(\zeta_1), v(\zeta_2), (\zeta_1, \zeta_2)) f(\zeta_1, \zeta_2) d\zeta_1 d\zeta_2 \end{aligned} \tag{2}$$

The instance where the u-player is interested in minimizing the cost function (1) whereas the v-player strives to maximize the cost (1) calls for the formulation of a stochastic *zero-sum* game with incomplete information, where a saddle point in pure strategies, in Hilbert space, is sought: the value of the game, if it exists, is

$$\begin{aligned} J^* &= \min_{u(\zeta_1)} \max_{v(\zeta_2)} E_\zeta(J(u(\zeta_1), v(\zeta_2), \zeta)) \\ &= \min_{u(\zeta_1)} \max_{v(\zeta_2)} \int_{\zeta_1} \int_{\zeta_2} J(u(\zeta_1), v(\zeta_2), (\zeta_1, \zeta_2)) f(\zeta_1, \zeta_2) d\zeta_1 d\zeta_2 \end{aligned} \tag{3}$$

This static zero-sum game in Hilbert space is in normal form.

In both the decentralized optimization problem posed in (2) and in the zero-sum static game formulation (3), the u- and v-players have partial information. And in both the decentralized optimization problem and in the zero-sum game, the players decide on their respective strategies $u(\cdot)$ and $v(\cdot)$, knowing the type of information that will become available to them, but before the information is actually received. In (2) and (3), the players' strategies are of *prior commitment* type. This is the reason why, although the players have partial information and consequently it stands to reason that their respective costs are conditional on their private information and therefore they have different costs, the game (3) is nevertheless zero-sum. And for the same reason, the solution of the decentralized optimization problem (2) entails the minimization of just one cost functional.

The decentralized stochastic static optimization problem in Hilbert space (2), referred to as a team decision problem, was addressed by Radner in his pioneering paper [1]. The present work could aptly be named "variations on a *team* by Radner."

Since a strong interest in Witsenhausen's counterexample from 1968 [2] persists to this day, it is important to revisit Radner's 1962 paper. Indeed, after the appearance of Radner's paper and until the publication of Witsenhausen's counterexample, it was widely believed in the controls community that the linear quadratic Gaussian (LQG) paradigm is a guarantor of the applicability of the separation, or, certainty equivalence, principle, and, as in LQG optimal control, the state is Gaussian distributed so that the sufficient statistics are linear in the measurements/information and are provided by *linear* Kalman filters. Consequently, the players' optimal strategies will be linear in the sufficient statistic, and in particular, the linear state estimate. However, Radner showed in [1] that in the *static* Quadratic Gaussian (QG) optimization problem with incomplete information, although the players' optimal strategies are affine in the information, the separation, or, certainty equivalence, principle does not apply. And in [2] Witsenhausen showed that in the simplest decentralized *dynamic* LQG optimal control problem neither does the separation principle apply, nor are the optimal strategies linear in the measurements. The bottom line: Radner's paper [1] relates to Witsenhausen's paper [2] like the Statics and Dynamics fields in Mechanical Engineering. Thus, with a view to also obtaining a better understanding of Witsenhausen's counterexample, it is instructive to revisit Radner's work and closely examine the informational and game theoretic aspects of the decentralized static QG optimization problem/team decision problem.

The article is organized as follows. In Sect. 2 the decentralized optimization problem is analyzed using the concept of *delayed commitment* strategies and necessary conditions for the existence of a solution are obtained. The necessary conditions derived in Sect. 2 are used in Sect. 3 to directly obtain the solution of the decentralized static multivariate QG optimization problem. The applicability of the separation principle/certainty equivalence is discussed in Sect. 4. The necessary and sufficient conditions for the existence of a solution of the decentralized static multivariate QG optimization problem are discussed in Sect. 5. The solution of the decentralized static multivariate QG optimization problem using the concept of prior commitment strategies is presented in Sect. 6. It is shown that although in the static case the delayed commitment and prior commitment strategies are equivalent, when the concept of prior commitment strategies is used, the strategies are harder to derive. Finally, in Sect. 7 the decentralized static multivariate QG optimization problem where the players' information is asymmetric is solved. The structure of the optimal solutions for cases of extreme informational asymmetry yields interesting insights into decentralized optimal control. Conclusions are presented in Sect. 8.

2 Analysis

The solution of the static team/decentralized optimization problem pursued in this paper is based on the following approach. Rather than tackling the Hilbert space optimization problem (2) head on, we instead opt for a game theoretic analysis of the decision problem on hand.

Consider first the decision problem faced by the u-player after he has received the information ζ_1, but before anyone has acted. His or Her cost is evaluated as follows.

$$\begin{aligned} J^{(u)}(u,v(\cdot);\zeta_1) &= E_{\zeta}(\, J(u,v(\zeta_2),\zeta) \mid \zeta_1 \,) \\ &= E_{\zeta_2}(\, J(u,v(\zeta_2),(\zeta_1,\zeta_2)) \mid \zeta_1 \,) \\ &= \int_{\zeta_2} J(u,v(\zeta_2),(\zeta_1,\zeta_2)) f(\zeta_2 \mid \zeta_1) d\zeta_2 \;\rightarrow\; \min_u \end{aligned}$$

Similar considerations apply to the v-player: having received the ζ_2 information, the cost which the v-player strives to minimize is

$$\begin{aligned} J^{(v)}(u(\cdot),v;\zeta_2) &= E_{\zeta}(\, J(u(\zeta_1),v,\zeta) \mid \zeta_2 \,) \\ &= E_{\zeta_1}(\, J(u(\zeta_1),v,(\zeta_1,\zeta_2)) \mid \zeta_2 \,) \\ &= \int_{\zeta_1} J(u(\zeta_1),v,(\zeta_1,\zeta_2)) f(\zeta_1 \mid \zeta_2) d\zeta_1 \;\rightarrow\; \min_v \end{aligned}$$

Now, the u and v-players' strategies are of *delayed commitment* type. Consequently, although both players strive to minimize the cost function (1), since they have partial information, their expected costs will be conditional on their private information and will not be the same—each player minimizes his or her own cost functional. The static team problem/decentralized optimal control problem (2) has been reformulated as a stochastic nonzero-sum game with incomplete information. Hence, a Nash equilibrium is sought. Using *delayed commitment* type strategies has highlighted informational issues which are apparent in extensive-form games but are suppressed in normal form games.

If a solution to the team/decentralized control problem in the form of a Nash equilibrium exists, it can be obtained as follows.

The u-player's *value function* is

$$(J^{(u)}(\zeta_1;v^*(\cdot)))^* = \min_u \int_{\zeta_2} J(u,v^*(\zeta_2),(\zeta_1,\zeta_2)) f(\zeta_2 \mid \zeta_1) d\zeta_2 \tag{4}$$

and his or her optimal strategy is obtained as follows: the u-player calculates the vector in R^{m_u}

$$u^*(\zeta_1) = \arg\min_u \int_{\zeta_2} J(u,v^*(\zeta_2),(\zeta_1,\zeta_2)) f(\zeta_2 \mid \zeta_1) d\zeta_2 \;\; \forall\, \zeta_1$$

The v-player's *value function* is

$$(J^{(v)}(\zeta_2;u^*(\cdot)))^* = \min_v \int_{\zeta_1} J(u^*(\zeta_1),v,(\zeta_1,\zeta_2)) f(\zeta_1 \mid \zeta_2) d\zeta_1 \tag{5}$$

and his or her optimal strategy is obtained as follows: the v-player calculates the vector in R^{m_v}

$$v^*(\zeta_2) = \arg\min_v \int_{\zeta_1} J(u^*(\zeta_1), v, (\zeta_1, \zeta_2)) f(\zeta_1 \mid \zeta_2) d\zeta_1 \ \ \forall\, \zeta_2$$

Hence, in order to determine the players' optimal strategies, that is, the functions $u^*(\cdot)$ and $v^*(\cdot)$, the equation in $u \in R^{m_u}$,

$$\int \frac{\partial}{\partial u} J(u, v^*(\zeta_2), (\zeta_1, \zeta_2)) f(\zeta_2 \mid \zeta_1) d\zeta_2 = 0 \ \forall \zeta_1 \tag{6}$$

must be solved $\forall \zeta_1$, and in this way the u-player's strategy $u^*(\zeta_1)$ is obtained. At the same time the equation in $v \in R^{m_v}$

$$\int \frac{\partial}{\partial v} J(u^*(\zeta_1), v, (\zeta_1, \zeta_2)) f(\zeta_1 \mid \zeta_2) d\zeta_1 = 0 \ \forall \zeta_2 \tag{7}$$

must be solved $\forall \zeta_2$, and in this way the v-player's strategy $v^*(\zeta_2)$ is obtained. In addition, the following second-order conditions/inequalities must hold

$$\int \frac{\partial^2}{\partial u^2} J(u, v^*(\zeta_2), (\zeta_1, \zeta_2)) \mid_{u^*(\zeta_1)} f(\zeta_2 \mid \zeta_1) d\zeta_2 > 0 \ \forall \zeta_1 \tag{8}$$

and

$$\int \frac{\partial^2}{\partial v^2} J(u^*(\zeta_1), v, (\zeta_1, \zeta_2)) \mid_{v^*(\zeta_2)} f(\zeta_1 \mid \zeta_2) d\zeta_1 > 0 \ \forall \zeta_2 \tag{9}$$

A set of two coupled functional equations (6) and (7) has been derived whose solution, if it exists, yields the u- and v-players' respective Nash strategies $u^*(\zeta_1)$ and $v^*(\zeta_2)$. Evidently, the solution of static team/decentralized optimization problems and/or nonzero-sum stochastic games calls for the solution of a somewhat nonconventional mathematical problem, (6) and (7). The culprit is the partial information pattern.

At this juncture it is apparent that the solution concept advanced for the original team/decentralized control problem is a Nash equilibrium in the nonzero-sum stochastic game (4) and (5). Using delayed commitment strategies, a Person-By-Person Satisfactory (PBPS) minimization is pursued: the strategy $u^*(\cdot)$ of the u-player is best, given that the v-player uses the strategy $v^*(\cdot)$, and the strategy $v^*(\cdot)$ of the v-player must be best, given that the u-player uses the strategy $u^*(\cdot)$. Thus, the derived strategies $(u^*(\cdot), v^*(\cdot))$ are person-by-person minimal. This is so because the players' outcomes provided by $(u^*(\cdot), v^*(\cdot))$ cannot be improved by unilaterally changing, say, $u^*(\cdot)$ alone; and, vice versa, the strategy $(u^*(\cdot), v^*(\cdot))$ cannot be improved by changing $v^*(\cdot)$ alone—this being the essence of a Nash equilibrium. Now, in nonzero-sum games, the calculated Nash equilibrium better be unique, for the solution to be applicable. However, in the absence of conflict of interest, as is the case in our original team/decentralized optimization problem (2), uniqueness of the Nash equilibrium solution is not an issue and the players will naturally settle on that particular Nash equilibrium $(u^*(\cdot), v^*(\cdot))$ which yields the minimal expected cost—we here refer to the calculated expected cost (2), namely

$$J^* = J(u^*(\cdot), v^*(\cdot)) = E_\zeta(\, J(u^*(\zeta_1), v^*(\zeta_2), \zeta)\,)$$
$$= \int_{\zeta_1} \int_{\zeta_2} J(u^*(\zeta_1), v^*(\zeta_2), (\zeta_1, \zeta_2)) d\zeta_1 d\zeta_2 \qquad (10)$$

Uniqueness of the obtained Nash equilibrium follows if the cost function (1) is convex in u and in v. This is so because the weighted sum of convex functions is convex—see (4) and (5).

Clearly, the optimal solution of the original team/decentralized optimization problem (2), if it exists, is PBPS, that is, it is a Nash equilibrium. However, having found an even unique Nash equilibrium of the nonzero-sum stochastic game (4) and (5) does not guarantee optimality in the original team/decentralized control problem, where one is interested in the expected cost (2). To answer the question of the existence of an optimal solution of the original team/decentralized control problem, the optimization problem (2) must be considered in a Hilbert space setting, as in [1], and convexity in (u, v) of the cost function (1) is required.

In summary, if a solution of the team/decentralized optimization problem exists, the above outlined solution of the attendant nonzero-sum stochastic game (4) and (5) will yield its optimal solution. However, should the cost function (1) be convex in u and v, but not in (u, v), then, while a Nash equilibrium in the nonzero-sum game (4) and (5) might exist, a solution of the decentralized optimization problem (2) might not exist.

3 Static Quadratic Gaussian Team

Using the theory developed in Sect. 2, the complete solution of the multivariate QG team decision/decentralized optimization problem is now derived.

The payoff function (1) is quadratic:

$$J(u, v, \zeta) = -u^T R^{(u)} u - v^T R^{(v)} v + 2v^T R^{(u,v)} u + 2(u^T, v^T) \begin{pmatrix} \zeta_1 \\ \zeta_2 \end{pmatrix}$$

and the components of the random variable ζ are $\zeta_1 \in R^m$, $\zeta_2 \in R^{n-m}$. The u- and v-players' control variables are $u \in R^m$ and $v \in R^{n-m}$ and the respective controls' effort weighing matrices

$$R^{(u)} > 0, \;\; R^{(v)} > 0;$$

$R^{(u,v)}$ is an $(n-m) \times m$ coupling matrix.

We calculate the v-player's payoff

$$J^{(v)}(v,\zeta_2;u(\cdot)) = 2v^T\zeta_2 - v^T R^{(v)} v + 2v^T R^{(u,v)} E_{\zeta_1}(\ u(\zeta_1) \mid \zeta_2\ )$$
$$+E_{\zeta_1}(\ 2u^T(\zeta_1)\zeta_1 - u^T(\zeta_1) R^{(u)} u(\zeta_1) \mid \zeta_2\ ) \tag{11}$$

Differentiation in v yields the unique optimal control response to the u-player's strategy $u(\zeta_1)$,

$$v^*(\zeta_2) = (R^{(v)})^{-1}\zeta_2 + (R^{(v)})^{-1} R^{(u,v)} E_{\zeta_1}(\ u(\zeta_1) \mid \zeta_2\ )\ \forall \zeta_2 \tag{12}$$

The u-player's payoff is

$$J^{(u)}(u,\zeta_1;v(\cdot)) = 2u^T\zeta_1 - u^T R^{(u)} u + 2u^T (R^{(u,v)})^T E_{\zeta_2}(\ v(\zeta_2) \mid \zeta_1\ )$$
$$+E_{\zeta_2}(\ 2v^T(\zeta_2)\zeta_2 - v^T(\zeta_2) R^{(v)} v(\zeta_2) \mid \zeta_1\ ) \tag{13}$$

and differentiation in u yields the unique optimal control response to the v-player's strategy $v(\zeta_2)$,

$$u^*(\zeta_1) = (R^{(u)})^{-1}\zeta_1 + (R^{(u)})^{-1} (R^{(u,v)})^T E_{\zeta_2}(\ v(\zeta_2) \mid \zeta_1\ )\ \forall \zeta_1 \tag{14}$$

Furthermore, the positive definiteness of the controls' effort weighing matrices guarantees that the conditions (8) and (9) hold.

At this point we assume that the p.d.f. f of the random variable ζ is a multivariate normal distribution, that is,

$$f(\zeta) = \frac{1}{\sqrt{(2\pi)^n \mid \det(P) \mid}} \exp^{-\frac{1}{2}(\zeta-\overline{\zeta})^T P^{-1}(\zeta-\overline{\zeta})}$$

and the covariance matrix P is real, symmetric, and positive definite. In other words, the random variable

$$\zeta = \begin{pmatrix} \zeta_1 \\ \zeta_2 \end{pmatrix} \sim \mathscr{N}\left(\begin{pmatrix} \overline{\zeta}_1 \\ \overline{\zeta}_2 \end{pmatrix}, \begin{bmatrix} P_{1,1} & P_{1,2} \\ P_{1,2}^T & P_{2,2} \end{bmatrix} \right) \tag{15}$$

In the special case of a bivariate normal distribution with $\zeta_1, \zeta_2 \in R^1$,

$$\zeta \sim \mathscr{N}\left(\begin{pmatrix} \overline{\zeta}_1 \\ \overline{\zeta}_2 \end{pmatrix}, \begin{bmatrix} \sigma_1^2 & \rho\sigma_1\sigma_2 \\ \rho\sigma_1\sigma_2 & \sigma_2^2 \end{bmatrix} \right) \tag{16}$$

and $-1 < \rho < 1$.

The following is well known.

Lemma 1. *Consider the multivariate normal distribution (15). The distribution of ζ_1 conditional on ζ_2 is*

$$\zeta_1 \sim \mathcal{N}(\overline{\zeta}_1 + P_{1,2}P_{2,2}^{-1}(\zeta_2 - \overline{\zeta}_2), P_{1,1} - P_{1,2}P_{2,2}^{-1}P_{1,2}^T) \tag{17}$$

and the distribution of ζ_2 conditional on ζ_1 is

$$\zeta_2 \sim \mathcal{N}(\overline{\zeta}_2 + P_{1,2}^T P_{1,1}^{-1}(\zeta_1 - \overline{\zeta}_1), P_{2,2} - P_{1,2}^T P_{1,1}^{-1} P_{1,2}) \tag{18}$$

The marginal p.d.f.s $f_1(\zeta_1)$ and $f_2(\zeta_2)$ are also Gaussian, that is,

$$\zeta_1 \sim \mathcal{N}(\overline{\zeta}_1, P_{1,1}) \tag{19}$$

and

$$\zeta_2 \sim \mathcal{N}(\overline{\zeta}_2, P_{2,2}) \tag{20}$$

In the special case of a bivariate normal distribution (16), the distribution of ζ_1 conditional on ζ_2 is

$$\zeta_1 \sim \mathcal{N}\left(\overline{\zeta}_1 + \rho\frac{\sigma_1}{\sigma_2}(\zeta_2 - \overline{\zeta}_2), (1-\rho^2)\sigma_1^2\right) \tag{21}$$

and the distribution of ζ_2 conditional on ζ_1 is

$$\zeta_2 \sim \mathcal{N}\left(\overline{\zeta}_2 + \rho\frac{\sigma_2}{\sigma_1}(\zeta_1 - \overline{\zeta}_1), (1-\rho^2)\sigma_2^2\right) \tag{22}$$

The marginal p.d.f.s $f_1(\zeta_1)$ and $f_2(\zeta_2)$ are

$$\zeta_1 \sim \mathcal{N}(\overline{\zeta}_1, \sigma_1^2) \tag{23}$$

and

$$\zeta_2 \sim \mathcal{N}(\overline{\zeta}_2, \sigma_2^2) \tag{24}$$

Inserting (18) into (14) yields

$$\begin{aligned} u^*(\zeta_1) = {} & (R^{(u)})^{-1}\zeta_1 \\ & + (R^{(u)})^{-1}(R^{(u,v)})^T E_{w_1}(\, v(\overline{\zeta}_2 + P_{1,2}^T P_{1,1}^{-1}(\zeta_1 - \overline{\zeta}_1) + w_1)\,) \end{aligned}$$

where

$$w_1 \sim \mathcal{N}(0, P_{2,2} - P_{1,2}^T P_{1,1}^{-1} P_{1,2})$$

and inserting (17) into (12) yields

$$v^*(\zeta_2) = (R^{(v)})^{-1}\zeta_2 + (R^{(v)})^{-1}R^{(u,v)}E_{w_2}(\, u(\overline{\zeta}_1 + P_{1,2}P_{2,2}^{-1}(\zeta_2 - \overline{\zeta}_2) + w_2)\,)$$

where

$$w_2 \sim \mathcal{N}(0, P_{1,1} - P_{1,2}P_{2,2}^{-1}P_{1,2}^T)$$

Using the convolution notation we obtain

$$u^*(\zeta_1) = (R^{(u)})^{-1}\zeta_1 + (R^{(u)})^{-1}(R^{(u,v)})^T G_{P_{2,2}-P_{1,2}^T P_{1,1}^{-1}P_{1,2}} * v(P_{1,2}^T P_{1,1}^{-1}\zeta_1 + \overline{\zeta}_2 - P_{1,2}^T P_{1,1}^{-1}\overline{\zeta}_1)$$

where the function $G_{P_{2,2}-P_{1,2}^T P_{1,1}^{-1}P_{1,2}}$ is the p.d.f. of the Gaussian random variable w_1. Similarly

$$v^*(\zeta_2) = (R^{(v)})^{-1}\zeta_2 + (R^{(v)})^{-1}R^{(u,v)}G_{P_{1,1}-P_{1,2}P_{2,2}^{-1}P_{1,2}^T} * u(P_{1,2}P_{2,2}^{-1}\zeta_2 + \overline{\zeta}_1 - P_{1,2}P_{2,2}^{-1}\overline{\zeta}_2)$$

where the function $G_{P_{1,1}-P_{1,2}P_{2,2}^{-1}P_{1,2}^T}$ is the p.d.f. of the Gaussian random variable w_2. Hence, the optimal strategies satisfy the equations

$$u^*(\zeta_1) = (R^{(u)})^{-1}\zeta_1 + (R^{(u)})^{-1}(R^{(u,v)})^T G_{P_{2,2}-P_{1,2}^T P_{1,1}^{-1}P_{1,2}} * v^*(P_{1,2}^T P_{1,1}^{-1}\zeta_1 + \overline{\zeta}_2 - P_{1,2}^T P_{1,1}^{-1}\overline{\zeta}_1) \tag{25}$$

and

$$v^*(\zeta_2) = (R^{(v)})^{-1}\zeta_2 + (R^{(v)})^{-1}R^{(u,v)}G_{P_{1,1}-P_{1,2}P_{2,2}^{-1}P_{1,2}^T} * u^*(P_{1,2}P_{2,2}^{-1}\zeta_2 + \overline{\zeta}_1 - P_{1,2}P_{2,2}^{-1}\overline{\zeta}_2) \tag{26}$$

Equations (25) and (26) constitute a linear system of two convolution-type Fredholm integral equations of the second kind with Gaussian kernels, in the unknown functions/optimal strategies $u^*(\cdot)$ and $v^*(\cdot)$. Moreover, the forcing functions are linear in their arguments. In view of these observations, we apply

Ansatz 2. The u- and v-players' optimal strategies are affine, that is,

$$u^*(\zeta_1) = K^{(u)}\zeta_1 + c^{(u)} \tag{27}$$

and

$$v^*(\zeta_2) = K^{(v)}\zeta_2 + c^{(v)} \tag{28}$$

□

Inserting the strategies (27) and (28) into the respective (25) and (26), we calculate

$$\begin{aligned} K^{(v)}\zeta_2 + c^{(v)} &= (R^{(v)})^{-1}\zeta_2 + (R^{(v)})^{-1}R^{(u,v)}K^{(u)}(P_{1,2}P_{2,2}^{-1}\zeta_2 + \overline{\zeta}_1 - P_{1,2}P_{2,2}^{-1}\overline{\zeta}_2) \\ &\quad + (R^{(v)})^{-1}R^{(u,v)}c^{(u)} \ \forall\ \zeta_2 \end{aligned}$$

and

$$\begin{aligned} K^{(u)}\zeta_1 + c^{(u)} &= (R^{(u)})^{-1}\zeta_1 + (R^{(u)})^{-1}(R^{(u,v)})^T K^{(v)}(P_{1,2}^T P_{1,1}^{-1}\zeta_1 + \overline{\zeta}_2 - P_{1,2}^T P_{1,1}^{-1}\overline{\zeta}_1) \\ &\quad + (R^{(u)})^{-1}(R^{(u,v)})^T c^{(v)} \ \forall\ \zeta_1 \end{aligned}$$

We conclude that the following four *linear* equations in the four unknowns $K^{(u)}_{m\times m}$, $K^{(v)}_{(n-m)\times(n-m)}$, $c^{(u)} \in R^m$ and $c^{(v)} \in R^{n-m}$ hold:

$$K^{(v)} = (R^{(v)})^{-1}(I + R^{(u,v)}K^{(u)}P_{1,2}P_{2,2}^{-1})\,, \tag{29}$$

$$K^{(u)} = (R^{(u)})^{-1}(I + (R^{(u,v)})^T K^{(v)}P_{1,2}^T P_{1,1}^{-1})\,, \tag{30}$$

$$c^{(v)} = (R^{(v)})^{-1}R^{(u,v)}K^{(u)}(\overline{\zeta}_1 - P_{1,2}P_{2,2}^{-1}\overline{\zeta}_2) + (R^{(v)})^{-1}R^{(u,v)}c^{(u)}\,, \tag{31}$$

and

$$c^{(u)} = (R^{(u)})^{-1}(R^{(u,v)})^T K^{(v)}(\overline{\zeta}_2 - P_{1,2}^T P_{1,1}^{-1}\overline{\zeta}_1) + (R^{(u)})^{-1}(R^{(u,v)})^T c^{(v)} \tag{32}$$

Combining (29) and (30) yields the respective linear, Lyapunov type, matrix equations for $K^{(u)}$ and $K^{(v)}$,

$$\begin{aligned} &R^{(u)}K^{(u)}P_{1,1} - (R^{(u,v)})^T(R^{(v)})^{-1}R^{(u,v)}K^{(u)}P_{1,2}P_{2,2}^{-1}P_{1,2}^T \\ &\quad = P_{1,1} + (R^{(u,v)})^T(R^{(v)})^{-1}P_{1,2}^T \end{aligned} \tag{33}$$

and

$$R^{(v)}K^{(v)}P_{2,2} - R^{(u,v)}(R^{(u)})^{-1}(R^{(u,v)})^T K^{(v)}P_{1,2}^T P_{1,1}^{-1}P_{1,2} = P_{2,2} + R^{(u,v)}(R^{(u)})^{-1}P_{1,2} \tag{34}$$

Solving the linear Lyapunov-type matrix equations (33) and (34) yields the optimal gains $K^{(u)}$ and $K^{(v)}$, whereupon the constant vectors $c^{(u)} \in R^{m_u}$ and $c^{(v)} \in R^{m_v}$ are

$$\begin{pmatrix} c^{(u)} \\ c^{(v)} \end{pmatrix} = \begin{bmatrix} R^{(u)} & -(R^{(u,v)})^T \\ -R^{(u,v)} & R^{(v)} \end{bmatrix}^{-1} \begin{pmatrix} (R^{(u,v)})^T K^{(v)}(\overline{\zeta}_2 - P_{1,2}^T P_{1,1}^{-1}\overline{\zeta}_1) \\ R^{(u,v)} K^{(u)}(\overline{\zeta}_1 - P_{1,2} P_{2,2}^{-1}\overline{\zeta}_2) \end{pmatrix}$$

Concerning the calculation of the intercepts $c^{(u)}$ and $c^{(v)}$, the following holds.

A necessary condition for the existence of a solution to the multivariate decentralized QG optimization problem is that the Schur complements $R^{(u)} - (R^{(u,v)})^T (R^{(v)})^{-1} R^{(u,v)}$ and $R^{(v)} - R^{(u,v)}(R^{(u)})^{-1}(R^{(u,v)})^T$ are nonsingular.

In the special case where the controls are scalars and the p.d.f. of the random variable ζ is the bivariate normal distribution (16), the optimal gains are

$$K^{(u)} = \frac{R^{(v)} + \rho \frac{\sigma_2}{\sigma_1} R^{(u,v)}}{R^{(u)} R^{(v)} - \rho^2 (R^{(u,v)})^2} \tag{35}$$

and

$$K^{(v)} = \frac{R^{(u)} + \rho \frac{\sigma_1}{\sigma_2} R^{(u,v)}}{R^{(u)} R^{(v)} - \rho^2 (R^{(u,v)})^2} \tag{36}$$

The intercepts are the solution of the linear system

$$\begin{aligned} \begin{bmatrix} R^{(u)} & -R^{(u,v)} \\ R^{(u,v)} & -R^{(v)} \end{bmatrix} \begin{pmatrix} c^{(u)} \\ c^{(v)} \end{pmatrix} &= R^{(u,v)} \begin{pmatrix} (\overline{\zeta}_2 - \rho \frac{\sigma_2}{\sigma_1} \overline{\zeta}_1) K^{(v)} \\ -(\overline{\zeta}_1 - \rho \frac{\sigma_1}{\sigma_2} \overline{\zeta}_2) K^{(u)} \end{pmatrix} \\ &= \frac{R^{(u,v)}}{R^{(u)} R^{(v)} - \rho^2 (R^{(u,v)})^2} \\ &\quad \begin{pmatrix} (\overline{\zeta}_2 - \rho \frac{\sigma_2}{\sigma_1} \overline{\zeta}_1)(R^{(u)} + \rho \frac{\sigma_1}{\sigma_2} R^{(u,v)}) \\ -(\overline{\zeta}_1 - \rho \frac{\sigma_1}{\sigma_2} \overline{\zeta}_2)(R^{(v)} + \rho \frac{\sigma_2}{\sigma_1} R^{(u,v)}) \end{pmatrix} \end{aligned}$$

so that

$$\begin{aligned} c^{(u)} &= \frac{R^{(u,v)}}{(R^{(u)} R^{(v)} - \rho^2 (R^{(u,v)})^2)((R^{(u,v)})^2 - R^{(u)} R^{(v)})} \\ &\quad \left\{ \left[(\rho^2 - 1) R^{(u,v)} R^{(v)} - \rho \frac{\sigma_2}{\sigma_1} ((R^{(u,v)})^2 \right.\right. \\ &\quad \left.\left. - R^{(u)} R^{(v)}) \right] \overline{\zeta}_1 + [\rho^2 (R^{(u,v)})^2 - R^{(u)} R^{(v)}] \overline{\zeta}_2 \right\} \end{aligned} \tag{37}$$

and

$$c^{(v)} = \frac{R^{(u,v)}}{(R^{(u)}R^{(v)} - \rho^2(R^{(u,v)})^2)((R^{(u,v)})^2 - R^{(u)}R^{(v)})} \Big\{\Big[(\rho^2-1)R^{(u,v)}R^{(u)} - \rho\frac{\sigma_1}{\sigma_2}((R^{(u,v)})^2 - R^{(u)}R^{(v)})\Big]\overline{\zeta}_2 + [\rho^2(R^{(u,v)})^2 - R^{(u)}R^{(v)}]\overline{\zeta}_1\Big\} \tag{38}$$

The following holds.

Proposition 3. *The necessary and sufficient conditions for the existence of a solution of the scalar decentralized QG optimization problem using delayed commitment strategies are*

$$R^{(u)} > 0,$$

$$R^{(v)} > 0,$$

$$R^{(u)}R^{(v)} \neq (R^{(u,v)})^2,$$

and

$$R^{(u)}R^{(v)} \neq \rho^2(R^{(u,v)})^2$$

The u- and v-players' optimal strategies are specified in (35)–(38) and are determined by the scalar problem parameters $R^{(u)}$, $R^{(v)}$, $R^{(u,v)}$, $\overline{\zeta}_1$, $\overline{\zeta}_2$, σ_1, σ_2, and ρ. The optimal solution (35)–(38) is symmetric. □

Corollary 4. *In the special scalar case where the random variable's components ζ_1 and ζ_2 are uncorrelated and $\rho = 0$, the optimal strategies are*

$$u^*(\zeta_1) = \frac{1}{R^{(u)}}\zeta_1 + \frac{R^{(u,v)}}{R^{(u)}R^{(v)} - (R^{(u,v)})^2}\left(\frac{R^{(u,v)}}{R^{(u)}}\overline{\zeta}_1 + \overline{\zeta}_2\right)$$

and

$$v^*(\zeta_2) = \frac{1}{R^{(v)}}\zeta_2 + \frac{R^{(u,v)}}{R^{(u)}R^{(v)} - (R^{(u,v)})^2}\left(\frac{R^{(u,v)}}{R^{(v)}}\overline{\zeta}_2 + \overline{\zeta}_1\right)$$

Also, in the special case where in the quadratic cost function there is no coupling and $R^{(u,v)} = 0$, the optimal strategies are linear:

$$u^*(\zeta_1) = \frac{1}{R^{(u)}}\zeta_1 \tag{39}$$

and

$$v^*(\zeta_2) = \frac{1}{R^{(v)}}\zeta_2 \tag{40}$$

□

4 Certainty Equivalence

We briefly digress and first examine the centralized static QG optimization problem.

4.1 Centralized QG Optimization Problem

In the centralized static QG optimization problem where both players have complete knowledge of the state of nature $(\zeta_1, \zeta_2)^T$, a necessary and sufficient condition for the existence of an optimal solution is

$$M \equiv \begin{bmatrix} R^{(u)} & -(R^{(u,v)})^T \\ -R^{(u,v)} & R^{(v)} \end{bmatrix} > 0$$

and the optimal controls $(u^*, v^*)^T$ are

$$\begin{pmatrix} u^* \\ v^* \end{pmatrix} = \begin{bmatrix} R^{(u)} & -(R^{(u,v)})^T \\ -R^{(u,v)} & R^{(v)} \end{bmatrix}^{-1} \begin{pmatrix} \zeta_1 \\ \zeta_2 \end{pmatrix}$$

We shall require the following.

Lemma 5. *Consider the blocked symmetric matrix*

$$M = \begin{bmatrix} M_{1,1} & M_{1,2} \\ M_{1,2}^T & M_{2,2} \end{bmatrix}$$

and let

$$N \equiv M^{-1}$$

Assuming the required matrix inverses exit, the inverse matrix

$$N = \begin{bmatrix} N_{1,1} & N_{1,2} \\ N_{1,2}^T & N_{2,2} \end{bmatrix}$$

where the blocks

$$N_{1,1} = M_{1,1}^{-1}[I + M_{1,2}(M_{2,2} - M_{1,2}^T M_{1,1}^{-1} M_{1,2})^{-1} M_{1,2}^T M_{1,1}^{-1}]$$

$$N_{1,2} = M_{1,1}^{-1} M_{1,2}(M_{1,2}^T M_{1,1}^{-1} M_{1,2} - M_{2,2})^{-1}$$

$$N_{2,2} = -(M_{1,2}^T M_{1,1}^{-1} M_{1,2} - M_{2,2})^{-1}$$

An alternative representation in blocked form of the inverse matrix N is

$$N_{1,1} = (M_{1,1} - M_{1,2} M_{2,2}^{-1} M_{1,2}^T)^{-1}$$

$$N_{1,2} = (M_{1,2} M_{2,2}^{-1} M_{1,2}^T - M_{1,1})^{-1} M_{1,2} M_{2,2}^{-1}$$

$$N_{2,2} = M_{2,2}^{-1} + M_{2,2}^{-1} M_{1,2}^T (M_{1,1} - M_{1,2} M_{2,2}^{-1} M_{1,2}^T)^{-1} M_{1,2} M_{2,2}^{-1}$$

Proof. By inspection, and the application of the Matrix Inversion Lemma. □

We shall also require

Lemma 6. *The real symmetric matrix*

$$M = \begin{bmatrix} R^{(u)} & -(R^{(u,v)})^T \\ -R^{(u,v)} & R^{(v)} \end{bmatrix}$$

is positive definite iff the matrices $R^{(v)} > 0$, $R^{(u)} > 0$ and their respective Schur complements are positive definite, that is,

$$R^{(u)} - (R^{(u,v)})^T (R^{(v)})^{-1} R^{(u,v)} > 0$$

$$R^{(v)} - R^{(u,v)} (R^{(u)})^{-1} (R^{(u,v)})^T > 0$$

□

In view of Lemmas 5 and 6, the following holds.

$$\begin{bmatrix} R^{(u)} & -(R^{(u,v)})^T \\ -R^{(u,v)} & R^{(v)} \end{bmatrix}^{-1} = \begin{bmatrix} N_{1,1} & N_{1,2} \\ N_{1,2}^T & N_{2,2} \end{bmatrix}$$

where

$$N_{1,1} = (R^{(u)})^{-1}[I + (R^{(u,v)})^T (R^{(v)} - R^{(u,v)} (R^{(u)})^{-1} (R^{(u,v)})^T)^{-1} R^{(u,v)} (R^{(u)})^{-1}]$$

$$N_{1,2} = (R^{(u)})^{-1} (R^{(u,v)})^T (R^{(v)} - R^{(u,v)} (R^{(u)})^{-1} (R^{(u,v)})^T)^{-1}$$

$$N_{2,2} = (R^{(v)} - R^{(u,v)} (R^{(u)})^{-1} (R^{(u,v)})^T)^{-1}$$

or, alternatively,

$$N_{1,1} = (R^{(u)} - (R^{(u,v)})^T (R^{(v)})^{-1} R^{(u,v)})^{-1}$$
$$N_{1,2} = (R^{(u)} - (R^{(u,v)})^T (R^{(v)})^{-1} R^{(u,v)})^{-1} (R^{(u,v)})^T (R^{(v)})^{-1}$$
$$N_{2,2} = (R^{(v)})^{-1} + (R^{(v)})^{-1} R^{(u,v)} (R^{(u)} - (R^{(u,v)})^T (R^{(v)})^{-1} R^{(u,v)})^{-1} (R^{(u,v)})^T (R^{(v)})^{-1}$$

Hence, in the centralized scenario the explicit formulae for the optimal controls are

$$u^*(\zeta_1, \zeta_2) = (R^{(u)} - (R^{(u,v)})^T (R^{(v)})^{-1} R^{(u,v)})^{-1} (\zeta_1 + (R^{(u,v)})^T (R^{(v)})^{-1} \zeta_2) \tag{41}$$

and

$$v^*(\zeta_1, \zeta_2) = (R^{(v)} - R^{(u,v)} (R^{(u)})^{-1} (R^{(u,v)})^T)^{-1} (R^{(u,v)} (R^{(u)})^{-1} \zeta_1 + \zeta_2) \tag{42}$$

Corollary 7. *In the special case where the controls are scalars, the necessary and sufficient conditions for the existence of an optimal solution are*

$$R^{(u)} > 0,$$
$$R^{(v)} > 0,$$

and

$$R^{(u)} R^{(v)} > (R^{(u,v)})^2$$

The optimal controls are linear and the solution is symmetric:

$$u^*(\zeta_1, \zeta_2) = \frac{1}{R^{(u)} R^{(v)} - (R^{(u,v)})^2} (R^{(v)} \zeta_1 + R^{(u,v)} \zeta_2) \tag{43}$$
$$v^*(\zeta_1, \zeta_2) = \frac{1}{R^{(u)} R^{(v)} - (R^{(u,v)})^2} (R^{(u,v)} \zeta_1 + R^{(u)} \zeta_2) \tag{44}$$

□

4.2 Separation Principle

We now return to the decentralized QG optimization problem and ascertain the applicability of *certainty equivalence*, a.k.a., the separation principle. We confine our attention to the scalar case and a bivariate Gaussian random variable (16).

When the information available to the u-player is restricted to the ζ_1 component of the state of nature, then, according to Lemma 1, his or her Maximum Likelihood (ML) estimate of the ζ_2 component of the state of nature will be

$$\hat{\zeta}_2 = \overline{\zeta}_2 + \rho\frac{\sigma_2}{\sigma_1}(\zeta_1 - \overline{\zeta}_1)$$

Similarly, when the information available to the v-player is restricted to the ζ_2 component of the state of nature, then, according to Lemma 1, his or her ML estimate of the ζ_1 component of the state of nature will be

$$\hat{\zeta}_1 = \overline{\zeta}_1 + \rho\frac{\sigma_1}{\sigma_2}(\zeta_2 - \overline{\zeta}_2)$$

Replacing ζ_2 in the centralized solution given by Corollary 7, (43), by the u-player's ML estimate $\hat{\zeta}_2$ of ζ_2 yields the u-player's certainty equivalence-based affine strategy

$$\begin{aligned} u^*(\zeta_1) &= \frac{1}{R^{(u)}R^{(v)} - (R^{(u,v)})^2}\left\{R^{(v)}\zeta_1 + R^{(u,v)}\left[\overline{\zeta}_2 + \rho\frac{\sigma_2}{\sigma_1}(\zeta_1 - \overline{\zeta}_1)\right]\right\} \\ &= \frac{1}{R^{(u)}R^{(v)} - (R^{(u,v)})^2}\left[\left(R^{(v)} + \rho\frac{\sigma_2}{\sigma_1}R^{(u,v)}\right)\zeta_1 + R^{(u,v)}\left(\overline{\zeta}_2 - \rho\frac{\sigma_2}{\sigma_1}\overline{\zeta}_1\right)\right] \end{aligned} \tag{45}$$

and replacing ζ_1 in the centralized solution given by Corollary 7, (44), by the v-player's ML estimate $\hat{\zeta}_1$ of ζ_1 yields the v-player's affine strategy

$$\begin{aligned} v^*(\zeta_2) &= \frac{1}{R^{(u)}R^{(v)} - (R^{(u,v)})^2}\left\{R^{(u,v)}\left[\overline{\zeta}_1 + \rho\frac{\sigma_1}{\sigma_2}(\zeta_2 - \overline{\zeta}_2)\right] + R^{(u)}\zeta_2\right\} \\ &= \frac{1}{R^{(u)}R^{(v)} - (R^{(u,v)})^2}\left[\left(R^{(u)} + \rho\frac{\sigma_1}{\sigma_2}R^{(u,v)}\right)\zeta_2 + R^{(u,v)}\left(\overline{\zeta}_1 - \rho\frac{\sigma_1}{\sigma_2}\overline{\zeta}_2\right)\right] \end{aligned} \tag{46}$$

In the special case where the random variable's components ζ_1 and ζ_2 are not correlated, that is, $\rho = 0$, the players' certainty equivalence-based affine strategies are

$$u(\zeta_1) = \frac{1}{R^{(u)}R^{(v)} - (R^{(u,v)})^2}(R^{(v)}\zeta_1 + R^{(u,v)}\overline{\zeta}_2)$$

and

$$v(\zeta_2) = \frac{1}{R^{(u)}R^{(v)} - (R^{(u,v)})^2}(R^{(u)}\zeta_2 + R^{(u,v)}\overline{\zeta}_1)$$

In the special case where there is no coupling in the quadratic payoff function and $R^{(u,v)} = 0$, the players' certainty equivalence-based strategies are (39) and (40).

5 Discussion

Similar to the optimal strategies in the decentralized control problem, also the certainty equivalence-based strategies (45) and (46) are affine and symmetric. However, a comparison of the u-player's optimal strategy which is specified in (35) and (37), and his or her certainty equivalence-based strategy (45), and similarly, a comparison of the v-player's optimal strategy which is specified in (36) and (38), and his or her certainty equivalence-based strategy (46), leads one to conclude that certainty equivalence does *not* hold. This is so even when there is no correlation and the parameter $\rho = 0$. Certainty equivalence holds only in the special case where there is no coupling in the quadratic payoff function and $R^{(u,v)} = 0$. This state of affairs is attributable to the partial information pattern.

It is also interesting to contrast the conditions for the existence of a solution of the centralized QG optimization problem and the conditions for the existence of a solution of the decentralized QG optimization problem. We note that the solution (41) and (42) of the centralized optimization problem can be formally derived using the PBPS solution concept. For this we need

$$R^{(u)} > 0$$
$$R^{(v)} > 0$$

and the Schur complements must be nonsingular, that is,

$$\det(R^{(u)} - (R^{(u,v)})^T (R^{(v)})^{-1} R^{(u,v)}) \neq 0$$

and

$$\det(R^{(v)} - R^{(u,v)} (R^{(u)})^{-1} (R^{(u,v)}) T) \neq 0$$

At the same time, we know that an optimal solution of the centralized optimization problem exists *iff* the matrix M is positive definite. Hence, in view of Lemma 6, we conclude that the positive definiteness of the Schur complements of the positive definite matrices $R^{(u)}$ and $R^{(v)}$ is a necessary condition for the existence of an optimal solution of the centralized optimization problem. At the same time, the invertibility of the Schur complements, while not sufficient to guarantee the existence of a solution of the centralized optimal control problem, is sufficient to allow a solution which conforms to the PBPS solution concept-based decentralized optimization problem—we have obtained a unique Nash solution and in the scalar case the respective u and v-players' Nash strategies are determined by (35), (37), and (36), (38), respectively.

Now, in view of [1], the positive definiteness of M is sufficient for the existence of an optimal solution of the decentralized optimization problem (2): the necessary and sufficient condition for the existence of a solution of the centralized optimization problem is a sufficient condition for the existence of an optimal solution of the decentralized problem, and moreover, the u- and v-players' Nash strategies determined by (35), (37), and (36), (38), respectively, are then optimal. However, if the matrix M is not positive definite but the matrices $R^{(u)}$ and $R^{(v)}$ are positive definite and their Schur complements are nonsingular, then while an optimal solution to the centralized optimization problem does not exist, in the decentralized control problem a PBPS solution concept-based unique Nash equilibrium exists.

6 Decentralized Static Quadratic Gaussian Optimization Problem

The original formulation of the decentralized optimization problem with a quadratic payoff functional, as formulated by Radner, (2), is considered in the special context of the multivariate QG optimization problem:

$$
\begin{aligned}
J(u(\zeta_1), v(\zeta_2), \zeta) = \int_{\zeta_1}\int_{\zeta_2} \Bigg[& -u^T(\zeta_1)R^{(u)}u(\zeta_1) - v^T(\zeta_2)R^{(v)}v(\zeta_2) \\
& +2v^T(\zeta_2)R^{(u,v)}u(\zeta_1) \\
& +2(u^T(\zeta_1), v^T(\zeta_2)) \begin{pmatrix} \zeta_1 \\ \zeta_2 \end{pmatrix} \Bigg] f(\zeta_1, \zeta_2)d\zeta_1 d\zeta_2 \\
= & \int_{\zeta_1} [-u^T(\zeta_1)R^{(u)}u(\zeta_1) + 2u^T(\zeta_1)\zeta_1] f_1(\zeta_1)d\zeta_1 \\
& + \int_{\zeta_2} [-v^T(\zeta_2)R^{(v)}v(\zeta_2) + 2v^T(\zeta_2)\zeta_2] f_2(\zeta_2)d\zeta_2 \\
& +2\int_{\zeta_1}\int_{\zeta_2} v^T(\zeta_2)R^{(u,v)}u(\zeta_1)f(\zeta_1,\zeta_2)d\zeta_1 d\zeta_2
\end{aligned} \tag{47}
$$

From [1] we know that optimal prior commitment strategies $u^*(\cdot)$ and $v^*(\cdot)$ exist and they are affine, provided the quadratic cost function is convex, that is, the matrix M is positive definite. Thus, the u- and v-players' strategies are parameterized as follows:

$$
u(\zeta_1) = K_p^{(u)}\zeta_1 + c_p^{(u)} \tag{48}
$$

and

$$
v(\zeta_2) = K_p^{(v)}\zeta_2 + c_p^{(v)} \tag{49}
$$

The subscript $_p$ indicates that now the strategies are of the *prior commitment* type.

Inserting the expressions (48) and (49) into (47) yields

$$\begin{aligned} J(K_p^{(u)}, K_p^{(v)}, c_p^{(u)}, c_p^{(v)}) = & -E_{\zeta_1}(\,(K_p^{(u)}\zeta_1 + c_p^{(u)})^T R^{(u)}(K_p^{(u)}\zeta_1 + c_p^{(u)}) \\ & +2(K_p^{(u)}\zeta_1 + c_p^{(u)})^T \zeta_1\,) \\ & -E_{\zeta_2}(\,(K_p^{(v)}\zeta_2 + c_p^{(v)})^T R^{(v)}(K_p^{(v)}\zeta_2 + c_p^{(v)}) \\ & +2(K_p^{(v)}\zeta_2 + c_p^{(v)})^T \zeta_2\,) \\ & +2E_{\zeta}(\,(K_p^{(v)}\zeta_2 + c_p^{(v)})^T R^{(u,v)}(K_p^{(u)}\zeta_1 + c_p^{(u)})) \end{aligned} \quad (50)$$

The payoff (50) is a function of the parameters $K_p^{(u)}$, $K_p^{(v)}$, $c_p^{(u)}$, and $c_p^{(v)}$.

The payoff function is differentiated in the parameters and the derivatives are set equal to zero. We can interchange the order of integration and differentiation. We shall use the notation.

e_i is the unit vector in the Euclidian spaces R^m or R^{n-m}, all of whose entries are zeroes except entry number i.

The following calculations are needed.

Lemma 8.

$$\frac{\partial}{\partial (K_p^{(u)})_{i,j}}((K_p^{(u)}\zeta_1 + c_p^{(u)})^T R^{(u)}(K_p^{(u)}\zeta_1 + c_p^{(u)}))$$
$$= 2\zeta_1^T e_j e_i^T R^{(u)} K_p^{(u)}\zeta_1 + 2\zeta_1^T e_j e_i^T R^{(u)} c_p^{(u)}$$

and consequently, using the properties of the Trace operator and the fact that the marginal p.d.f. of ζ_1 is Gaussian with expectation $\overline{\zeta}_1$ and covariance $P_{1,1}$, we calculate

$$\begin{aligned} E_{\zeta_1}\left(\frac{\partial}{\partial (K_p^{(u)})_{i,j}}((K_p^{(u)}\zeta_1 + c_p^{(u)})^T R^{(u)}(K_p^{(u)}\zeta_1 + c_p^{(u)}))\right) & = 2e_i^T R^{(u)} K_p^{(u)} P_{1,1} e_j \\ & \quad +2\overline{\zeta}_1^T e_j e_i^T R^{(u)} K_p^{(u)}\overline{\zeta}_1 \\ & \quad +2\overline{\zeta}_1^T e_j e_i^T R^{(u)} c_p^{(u)} \\ & = 2e_i^T R^{(u)} K_p^{(u)} P_{1,1} e_j \\ & \quad +2e_j^T\overline{\zeta}_1 \cdot e_i^T R^{(u)} K_p^{(u)}\overline{\zeta}_1 \\ & \quad +2e_j^T\overline{\zeta}_1 \cdot e_i^T R^{(u)} c_p^{(u)}, \\ & i = 1,\ldots,m,\ j = 1,\ldots,m \end{aligned}$$

Similarly,

$$\frac{\partial}{\partial (K_p^{(v)})_{i,j}}((K_p^{(v)}\zeta_2 + c_p^{(v)})^T R^{(v)} (K_p^{(v)}\zeta_2 + c_p^{(v)}))$$
$$= 2\zeta_2^T e_j e_i^T R^{(v)} K_p^{(v)} \zeta_2 + 2\zeta_2^T e_j e_i^T R^{(v)} c_p^{(v)}$$

and consequently

$$\begin{aligned} E_{\zeta_2}\left(\frac{\partial}{\partial (K_p^{(v)})_{i,j}}(K_p^{(v)}\zeta_2 + c_p^{(v)})^T R^{(v)} (K_p^{(v)}\zeta_2 + c_p^{(v)})\right) &= 2e_i^T R^{(v)} K_p^{(v)} P_{2,2} e_j \\ &\quad + 2\overline{\zeta}_2^T e_j e_i^T R^{(v)} K_p^{(v)} \overline{\zeta}_2 \\ &\quad + 2\overline{\zeta}_2^T e_j e_i^T R^{(v)} c_p^{(v)} \\ &= 2e_i^T R^{(v)} K_p^{(v)} P_{2,2} e_j \\ &\quad + 2e_j^T \overline{\zeta}_2 \cdot e_i^T R^{(v)} K_p^{(v)} \overline{\zeta}_2 \\ &\quad + 2e_j^T \overline{\zeta}_2 \cdot e_i^T R^{(v)} c_p^{(v)}, \\ i = 1, \ldots, n-m,\ & j = 1, \ldots, n-m \end{aligned}$$

In addition

$$\frac{\partial}{\partial (K_p^{(u)})_{i,j}}(\zeta_1^T (K_p^{(u)}\zeta_1 + c_p^{(u)})) = \zeta_1^T e_i e_j^T \zeta_1$$

and consequently

$$\begin{aligned} E_{\zeta_1}\left(\frac{\partial}{\partial (K_p^{(u)})_{i,j}}(\zeta_1^T (K_p^{(u)}\zeta_1 + c_p^{(u)}))\right) &= e_j^T P_{1,1} e_i + e_i^T \overline{\zeta}_1 \cdot e_j^T \overline{\zeta}_1, \\ i = 1, \ldots, m,\ & j = 1, \ldots, m \end{aligned}$$

Similarly,

$$\frac{\partial}{\partial (K_p^{(v)})_{i,j}}(\zeta_2^T (K_p^{(v)}\zeta_2 + c_p^{(v)})) = e_i^T \zeta_2 \cdot e_j^T \zeta_2$$

and consequently

$$\begin{aligned} E_{\zeta_2}\left(\frac{\partial}{\partial (K_p^{(v)})_{i,j}}(\zeta_2^T (K_p^{(v)}\zeta_2 + c_p^{(v)}))\right) &= e_j^T P_{2,2} e_i + e_i^T \overline{\zeta}_2 \cdot e_j^T \overline{\zeta}_2, \\ i = 1, \ldots, n-m,\ & j = 1, \ldots, n-m \end{aligned}$$

Also,

$$\frac{\partial}{\partial (K_p^{(u)})_{i,j}}((K_p^{(v)}\zeta_2 + c_p^{(v)})^T R^{(u,v)}(K_p^{(u)}\zeta_1 + c_p^{(u)})) = (K_p^{(v)}\zeta_2 + c_p^{(v)})^T R^{(u,v)} e_i \cdot e_j^T \zeta_1$$

and consequently

$$\begin{aligned} E_\zeta\left(\frac{\partial}{\partial (K_p^{(u)})_{i,j}} ((K_p^{(v)}\zeta_2 + c_p^{(v)})^T R^{(u,v)}(K_p^{(u)}\zeta_1 + c_p^{(u)}))\right) &= e_j^T \overline{\zeta}_1 \cdot e_i^T (R^{(u,v)})^T c_p^{(v)} \\ &+ e_j^T \overline{\zeta}_1 \cdot e_i^T (R^{(u,v)})^T K_p^{(v)} \overline{\zeta}_2 \\ &+ e_i^T (R^{(u,v)})^T K_p^{(v)} P_{2,1} e_j\ , \\ i = 1, \ldots, m,\ j &= 1, \ldots, m \end{aligned}$$

Similarly,

$$\frac{\partial}{\partial (K_p^{(v)})_{i,j}}((K_p^{(v)}\zeta_2 + c_p^{(v)})^T R^{(u,v)}(K_p^{(u)}\zeta_1 + c_p^{(u)})) = (K_p^{(u)}\zeta_1 + c_p^{(u)})^T R^{(u,v)} e_i e_j^T \zeta_2$$

and consequently

$$\begin{aligned} E_\zeta\left(\frac{\partial}{\partial (K_p^{(v)})_{i,j}} ((K_p^{(v)}\zeta_2 + c_p^{(v)})^T R^{(u,v)}(K_p^{(u)}\zeta_1 + c_p^{(u)}))\right) &= \overline{\zeta}_2^T e_j e_i^T (R^{(u,v)})^T c_p^{(u)} \\ &+ \overline{\zeta}_2^T e_j e_i^T (R^{(u,v)})^T K_p^{(u)} \overline{\zeta}_1 \\ &+ e_i^T (R^{(u,v)})^T K_p^{(u)} P_{1,2} e_j \\ &= e_j^T \overline{\zeta}_2 \cdot e_i^T (R^{(u,v)})^T c_p^{(u)} \\ &+ e_j^T \overline{\zeta}_2 \cdot e_i^T (R^{(u,v)})^T K_p^{(u)} \overline{\zeta}_1 \\ &+ e_i^T (R^{(u,v)})^T K_p^{(u)} P_{1,2} e_j\ , \\ i = 1, \ldots, n-m,\ j &= 1, \ldots, n-m \end{aligned}$$

Furthermore,

$$\frac{\partial}{\partial (c_p^{(u)})}((K_p^{(u)}\zeta_1 + c_p^{(u)})^T R^{(u)}(K_p^{(u)}\zeta_1 + c_p^{(u)})) = 2R^{(u)} c_p^{(u)} + 2R^{(u)} K_p^{(u)} \zeta_1$$

and consequently

$$E_{\zeta_1}\left(\frac{\partial}{\partial c_p^{(u)}}((K_p^{(u)}\zeta_1 + c_p^{(u)})^T R^{(u)}(K_p^{(u)}\zeta_1 + c_p^{(u)}))\right) = 2R^{(u)} c_p^{(u)} + 2R^{(u)} K_p^{(u)} \overline{\zeta}_1$$

Similarly,

$$\frac{\partial}{\partial(c_p^{(v)}}((K_p^{(v)}\zeta_2+c_p^{(v)})^T R^{(v)}(K_p^{(v)}\zeta_2+c_p^{(v)}))=2R^{(v)}c_p^{(v)}+2R^{(v)}K_p^{(v)}\zeta_2$$

and consequently

$$E_{\zeta_2}\left(\frac{\partial}{\partial c_p^{(v)}}((K_p^{(v)}\zeta_2+c_p^{(v)})^T R^{(v)}(K_p^{(v)}\zeta_2+c_p^{(v)}))\right)=2R^{(v)}c_p^{(v)}+2R^{(v)}K_p^{(v)}\overline{\zeta}_2$$

In addition,

$$\frac{\partial}{\partial c_p^{(u)}}((K_p^{(u)}\zeta_1+c_p^{(u)})^T\zeta_1)=\zeta_1$$

and consequently

$$E_{\zeta_1}\left(\frac{\partial c_p^{(u)}}{\partial}((K_p^{(u)}\zeta_1+c_p^{(u)})^T\zeta_1)\right)=\overline{\zeta}_1$$

Similarly,

$$\frac{\partial}{\partial c_p^{(v)}}((K_p^{(v)}\zeta_2+c_p^{(v)})^T\zeta_2)=\zeta_2$$

and consequently

$$E_{\zeta_2}\left(\frac{\partial}{\partial c_p^{(v)}}((K_p^{(v)}\zeta_2+c_p^{(v)})^T\zeta_2)\right)=\overline{\zeta}_2$$

Finally,

$$\frac{\partial}{\partial c_p^{(u)}}((K_p^{(v)}\zeta_2+c_p^{(v)})^T R^{(u,v)}(K_p^{(u)}\zeta_1+c_p^{(u)}))=(R^{(u,v)})^T(K_p^{(v)}\zeta_2+c_p^{(v)})$$

and consequently

$$E_{\zeta}\left(\frac{\partial}{\partial c_p^{(u)}}((K_p^{(v)}\zeta_2+c_p^{(v)})^T R^{(u,v)}(K_p^{(u)}\zeta_1+c_p^{(u)}))\right)=(R^{(u,v)})^T K_p^{(v)}\overline{\zeta}_2+(R^{(u,v)})^T c_p^{(v)}$$

Similarly,

$$\frac{\partial}{\partial c_p^{(v)}}((K_p^{(v)}\zeta_2 + c_p^{(v)})^T R^{(u,v)}(K_p^{(u)}\zeta_1 + c_p^{(u)})) = R^{(u,v)}(K_p^{(u)}\zeta_1 + c_p^{(u)})$$

and consequently

$$E_\zeta\left(\frac{\partial}{\partial c_p^{(v)}}((K_p^{(v)}\zeta_2 + c_p^{(v)})^T R^{(u,v)}(K_p^{(u)}\zeta_1 + c_p^{(u)}))\right) = R^{(u,v)}K_p^{(u)}\overline{\zeta}_1 + R^{(u,v)}c_p^{(u)}$$

□

The optimality conditions and Lemma 8 yield the system of $n(n+1) - 2m(n-m)$ linear equations

$$\begin{aligned} & e_i^T R^{(u)} K_p^{(u)} P_{1,1} e_j + (e_j^T \overline{\zeta}_1) \cdot e_i^T R^{(u)} K_p^{(u)} \overline{\zeta}_1 \\ & \quad + (e_j^T \overline{\zeta}_1) \cdot e_i^T R^{(u)} c_p^{(u)} - (e_j^T \overline{\zeta}_1) \cdot e_i^T (R^{(u,v)})^T c_p^{(v)} \\ & \quad - (e_j^T \overline{\zeta}_1) \cdot e_i^T (R^{(u,v)})^T K_p^{(v)} \overline{\zeta}_2 - e_i^T (R^{(u,v)})^T K_p^{(v)} P_{2,1} e_j \\ & = e_j^T P_{1,1} e_i + (e_i^T \overline{\zeta}_1) \cdot (e_j^T \overline{\zeta}_1) \end{aligned} \tag{51}$$

where $e_i, e_j \in R^m$ and $i = 1, \ldots, m,\ j = 1, \ldots, m,$

$$\begin{aligned} & e_i^T R^{(v)} K_p^{(v)} P_{2,2} e_j + (e_j^T \overline{\zeta}_2) \cdot e_i^T R^{(v)} K_p^{(v)} \overline{\zeta}_2 \\ & \quad + (e_j^T \overline{\zeta}_2) \cdot e_i^T R^{(v)} c_p^{(v)} - (e_j^T \overline{\zeta}_2) \cdot e_i^T R^{(u,v)} c_p^{(u)} \\ & \quad - (e_j^T \overline{\zeta}_2) \cdot e_i^T R^{(u,v)} K_p^{(u)} \overline{\zeta}_1 - e_i^T R^{(u,v)} K_p^{(u)} P_{1,2} e_j \\ & = e_j^T P_{2,2} e_i + (e_i^T \overline{\zeta}_2) \cdot (e_j^T \overline{\zeta}_2) \end{aligned} \tag{52}$$

where $e_i, e_j \in R^{n-m}$ and $i = 1, \ldots, n-m,\ j = 1, \ldots, n-m,$

$$(R^{(u,v)})^T K_p^{(v)} \overline{\zeta}_2 + (R^{(u,v)})^T c_p^{(v)} + \overline{\zeta}_1 = R^{(u)} c_p^{(u)} + R^{(u)} K_p^{(u)} \overline{\zeta}_1, \tag{53}$$

and

$$R^{(u,v)} K_p^{(u)} \overline{\zeta}_1 + R^{(u,v)} c_p^{(u)} + \overline{\zeta}_2 = R^{(v)} c_p^{(v)} + R^{(v)} K_p^{(v)} \overline{\zeta}_2 \tag{54}$$

The unknowns are $K_p^{(u)}$, an $m \times m$ matrix, $K_p^{(v)}$, an $(n-m) \times (n-m)$ matrix, $c_p^{(u)} \in R^m$ and $c_p^{(v)} \in R^{n-m}$, a total of $n(n+1) - 2m(n-m)$ unknowns.

Using (53) and (54) we express the intercepts $c_p^{(u)}$ and $c_p^{(v)}$ as linear functions of $K_p^{(u)}$ and $K_p^{(v)}$:

$$\begin{pmatrix} c_p^{(u)} \\ c_p^{(v)} \end{pmatrix} = \begin{bmatrix} R^{(u)} & -(R^{(u,v)})^T \\ -R^{(u,v)} & R^{(v)} \end{bmatrix}^{-1} \begin{pmatrix} \overline{\zeta}_1 + (R^{(u,v)})^T K_p^{(v)}\overline{\zeta}_2 - R^{(u)}K_p^{(u)}\overline{\zeta}_1 \\ \overline{\zeta}_2 + R^{(u,v)}K_p^{(u)}\overline{\zeta}_1 - R^{(v)}K_p^{(v)}\overline{\zeta}_2 \end{pmatrix}$$

Hence,

$$\begin{aligned} c_p^{(u)} &= (R^{(u)} - (R^{(u,v)})^T (R^{(v)})^{-1} R^{(u,v)})^{-1} [\overline{\zeta}_1 + (R^{(u,v)})^T K_p^{(v)}\overline{\zeta}_2 - R^{(u)}K_p^{(u)}\overline{\zeta}_1] \\ &\quad + (R^{(u)} - (R^{(u,v)})^T (R^{(v)})^{-1} R^{(u,v)})^{-1} (R^{(u,v)})^T (R^{(v)})^{-1} \\ &\quad [\overline{\zeta}_2 + R^{(u,v)}K_p^{(u)}\overline{\zeta}_1 - R^{(v)}K_p^{(v)}\overline{\zeta}_2]\,, \\ c_p^{(v)} &= (R^{(v)})^{-1} R^{(u,v)} (R^{(u)} - (R^{(u,v)})^T (R^{(v)})^{-1} R^{(u,v)})^{-1} \\ &\quad [\overline{\zeta}_1 + (R^{(u,v)})^T K_p^{(v)}\overline{\zeta}_2 - R^{(u)}K_p^{(u)}\overline{\zeta}_1] \\ &\quad + (R^{(v)} - R^{(u,v)} (R^{(u)})^{-1} (R^{(u,v)})^T)^{-1} [\overline{\zeta}_2 + R^{(u,v)}K_p^{(u)}\overline{\zeta}_1 - R^{(v)}K_p^{(v)}\overline{\zeta}_2] \end{aligned}$$

Substituting these expressions into (51) and (52) yields a reduced linear system of $n^2 - 2m(n-m)$ equations in the $n^2 - 2m(n-m)$ unknowns which populate the matrices $K_p^{(u)}$ and $K_p^{(v)}$. Note that if $\overline{\zeta}_1 = 0$ and $\overline{\zeta}_2 = 0$, $c_p^{(u)} = 0$, $c_p^{(v)} = 0$ and the equations for $K_p^{(u)}$ and $K_p^{(v)}$ are

$$\begin{aligned} & e_i^T R^{(u)} K_p^{(u)} P_{1,1} e_j + (e_j^T \overline{\zeta}_1) \cdot e_i^T R^{(u)} K_p^{(u)} \overline{\zeta}_1 - (e_j^T \overline{\zeta}_1) \cdot e_i^T (R^{(u,v)})^T K_p^{(v)} \overline{\zeta}_2 \\ & \quad - e_i^T (R^{(u,v)})^T K_p^{(v)} P_{2,1} e_j = e_j^T P_{1,1} e_i + (e_i^T \overline{\zeta}_1) \cdot (e_j^T \overline{\zeta}_1) \\ & e_i^T R^{(v)} K_p^{(v)} P_{2,2} e_j + (e_j^T \overline{\zeta}_2) \cdot e_i^T R^{(v)} K_p^{(v)} \overline{\zeta}_2 - (e_j^T \overline{\zeta}_2) \cdot e_i^T R^{(u,v)} K_p^{(u)} \overline{\zeta}_1 \\ & \quad - e_i^T R^{(u,v)} K_p^{(u)} P_{1,2} e_j = e_j^T P_{2,2} e_i + (e_i^T \overline{\zeta}_2) \cdot (e_j^T \overline{\zeta}_2) \end{aligned}$$

Example. In the special case of scalar controls and a bivariate normal distribution we obtain a system of four linear equations for the four scalar unknowns $K_p^{(u)}$, $K_p^{(v)}$, $c_p^{(u)}$, and $c_p^{(v)}$:

$$\begin{aligned} & (\sigma_1^2 + \overline{\zeta}_1^2) R^{(u)} K_p^{(u)} - (\rho\sigma_1\sigma_2 + \overline{\zeta}_1\overline{\zeta}_2) R^{(u,v)} K_p^{(v)} \\ & \quad + \overline{\zeta}_1 R^{(u)} c_p^{(u)} - \overline{\zeta}_1 R^{(u,v)} c_p^{(v)} = \sigma_1^2 + \overline{\zeta}_1^2 \end{aligned} \tag{55}$$

$$\begin{aligned} & (\sigma_2^2 + \overline{\zeta}_2^2) R^{(v)} K_p^{(v)} - (\rho\sigma_1\sigma_2 + \overline{\zeta}_1\overline{\zeta}_2) R^{(u,v)} K_p^{(u)} \\ & \quad + \overline{\zeta}_2 R^{(v)} c_p^{(v)} - \overline{\zeta}_2 R^{(u,v)} c_p^{(u)} = \sigma_2^2 + \overline{\zeta}_2^2 \end{aligned} \tag{56}$$

$$c_p^{(u)} = \frac{1}{R^{(u)}R^{(v)} - (R^{(u,v)})^2}(R^{(v)}\overline{\zeta}_1 + R^{(u,v)}\zeta_2) - \overline{\zeta}_1 K_p^{(u)} \tag{57}$$

$$c_p^{(v)} = \frac{1}{R^{(u)}R^{(v)} - (R^{(u,v)})^2}(R^{(u,v)}\overline{\zeta}_1 + R^{(u)}\overline{\zeta}_2) - \overline{\zeta}_2 K_p^{(v)} \tag{58}$$

Compare the optimal prior commitment strategies specified in (55)–(58) and the delayed commitment strategies explicitly specified in (35)–(38). The optimization problem is static and therefore the prior commitment and delayed commitment strategies are all the same:

$$K_p^{(u)} = K^{(u)},\ K_p^{(v)} = K^{(v)},\ c_p^{(u)} = c^{(u)},\ c_p^{(v)} = c^{(v)}$$

So the two sets of formulae (35)–(38) and (55)–(58) give rise to interesting identities. In particular, in the multivariate case new matrix identities will be obtained.

Taking a game theoretic approach naturally leads to the concept of delayed commitment strategies. Although the prior commitment strategies and delayed commitment strategies are equivalent, the above example illustrates that it is much easier to calculate the latter.

7 Asymmetric Players

Scenarios where one team member is strongly informationally disadvantaged relative to the second team member are investigated.

7.1 *Asymmetric Players: Case 1*

Assume the u-player has perfect information, that is, he is privy to the state of nature $\zeta = (\zeta_1, \zeta_2)$, whereas the v-player has access to ζ_2 only. At the same time, the u-player knows that the v-player has the prior information $\overline{\zeta}_1$, $\overline{\zeta}_2$, ρ, σ_1, and σ_2; in fact, and in the best tradition of Bayesian games, it is tacitly assumed that both players are simultaneously presented the prior information before the game starts—the prior information is public information.

In this case the u-player's payoff is

$$\begin{aligned} J^{(u)}(u, v(\cdot); \zeta) &= E_\zeta(J(u, v(\zeta_2); \zeta) \mid \zeta) \\ &= J(u, v(\zeta_2); \zeta), \end{aligned}$$

that is, in the case of perfect information the u-player need not calculate an expectation; $v(\zeta_2)$ is the unknown input of the v-player.

If the payoff function J is quadratic,

$$J^{(u)}(u,v(\cdot);\zeta) = -u^T R^{(u)} u - v^T(\zeta_2) R^{(v)} v(\zeta_2) + 2v^T(\zeta_2) R^{(u,v)} u + 2u^T \zeta_1 + 2v^T(\zeta_2)\zeta_2$$

and differentiation in u yields the relationship

$$u^*(\zeta_1,\zeta_2) = (R^{(u)})^{-1}[(R^{(u,v)})^T v^*(\zeta_2) + \zeta_1]$$

The v-player's payoff function is

$$\begin{aligned} J^{(v)}(u(\cdot),v;\zeta) &= -v^T R^{(v)} v + 2v^T \zeta_2 + E_\zeta(-u^T(\zeta) R^{(u)} u(\zeta) \\ &\quad + 2v^T R^{(u,v)} u(\zeta_2) + 2u^T \zeta_1 \mid \zeta_2) \end{aligned}$$

and differentiating it in v yields the relationship

$$\begin{aligned} R^{(v)} v^*(\zeta_2) &= \zeta_2 + R^{(u,v)} E_\zeta(u^*(\zeta) \mid \zeta_2) \\ &= \zeta_2 + R^{(u,v)} E_\zeta((R^{(u)})^{-1}[(R^{(u,v)})^T v^*(\zeta_2) + \zeta_1] \mid \zeta_2) \\ &= \zeta_2 + R^{(u,v)}(R^{(u)})^{-1}(R^{(u,v)})^T v^*(\zeta_2) + R^{(u,v)}(R^{(u)})^{-1} E_\zeta(\zeta_1 \mid \zeta_2) \\ &= \zeta_2 + R^{(u,v)}(R^{(u)})^{-1}(R^{(u,v)})^T v^*(\zeta_2) + R^{(u,v)}(R^{(u)})^{-1} E_{\zeta_1}(\zeta_1 \mid \zeta_2) \\ &= \zeta_2 + R^{(u,v)}(R^{(u)})^{-1}(R^{(u,v)})^T v^*(\zeta_2) + R^{(u,v)}(R^{(u)})^{-1} \\ &\quad (\overline{\zeta}_1 + P_{1,2}P_{2,2}^{-1}(\zeta_2 - \overline{\zeta}_2)) \end{aligned}$$

Hence,

$$\begin{aligned} v^*(\zeta_2) &= [R^{(v)} - R^{(u,v)}(R^{(u)})^{-1}(R^{(u,v)})^T]^{-1}[I + P_{1,2}P_{2,2}^{-1} R^{(u,v)}(R^{(u)})^{-1}]\zeta_2 \\ &\quad + [R^{(v)} - R^{(u,v)}(R^{(u)})^{-1}(R^{(u,v)})^T]^{-1} R^{(u,v)}(R^{(u)})^{-1} \left(\overline{\zeta}_1 - P_{1,2}P_{2,2}^{-1}\overline{\zeta}_2\right) \end{aligned}$$

In the special case of scalar inputs and a bivariate normal distribution (16), the optimal strategy of the v-player is

$$v^*(\zeta_2) = \frac{R^{(u)} + \rho\frac{\sigma_1}{\sigma_2}R^{(u,v)}}{R^{(u)}R^{(v)} - (R^{(u,v)})^2}\zeta_2 + \frac{R^{(u,v)}}{R^{(u)}R^{(v)} - (R^{(u,v)})^2}\left(\overline{\zeta}_1 - \rho\frac{\sigma_1}{\sigma_2}\overline{\zeta}_2\right),$$

provided that $R^{(u,v)}$ is not the geometric mean of $R^{(u)}$ and $R^{(v)}$—which is the case if the quadratic payoff function is concave in the control variable (u,v), whereupon $R^{(u)}R^{(v)} - (R^{(u,v)})^2 > 0$. The optimal strategy of the u-player is

$$u^*(\zeta_1,\zeta_2) = \frac{1}{R^{(u)}}\zeta_1 + R^{(u,v)}\frac{1+\rho\frac{\sigma_1}{\sigma_2}\frac{R^{(u,v)}}{R^{(u)}}}{R^{(u)}R^{(v)}-(R^{(u,v)})^2}\zeta_2$$
$$+\frac{\frac{(R^{(u,v)})^2}{R^{(u)}}}{R^{(u)}R^{(v)}-(R^{(u,v)})^2}\left(\overline{\zeta}_1-\rho\frac{\sigma_1}{\sigma_2}\overline{\zeta}_2\right)$$

Interestingly, although the u-player has complete state of nature information, his or her optimal strategy is affine and he also uses the public prior information. Concerning the strategy of the informationally disadvantaged v-player: certainty equivalence holds.

7.2 Asymmetric Players: Case 2

As in Sects. 1–6, the private information of the u-player is the ζ_1 component of the state of nature vector ζ. However, we now assume the v-player has no private information and he is totally dependent on the public prior information. As in Sect. 7.1, the u-player is aware that the public information is available to the v-player and he also knows that the v-player is "blind."

The v-player's payoff is

$$J^{(v)}(u(\cdot),v) = 2v^T E_\zeta(\zeta_2) - v^T R^{(v)} v + 2v^T R^{(u,v)} E_\zeta(u(\zeta_1))$$
$$+E_{\zeta_1}(2u^T(\zeta_1)\zeta_1 - u^T(\zeta_1)R^{(u)}u(\zeta_1))$$

and differentiation in v yields the unique optimal control response to the u-player's strategy $u(\zeta_1)$,

$$v^* = (R^{(v)})^{-1}E_\zeta(\zeta_2) + (R^{(v)})^{-1}R^{(u,v)}E_\zeta(u(\zeta_1)) \tag{59}$$

The expectation $E_\zeta(\zeta_2)$ in (59) is calculated as follows.

$$E_\zeta(\zeta_2) = \int_{-\infty}^{\infty}\int_{-\infty}^{\infty}\zeta_2 f(\zeta_1,\zeta_2)d\zeta_1 d\zeta_2$$
$$= \int_{-\infty}^{\infty}\zeta_2\left(\int_{-\infty}^{\infty}f(\zeta_1,\zeta_2)d\zeta_1\right)d\zeta_2$$
$$= \int_{-\infty}^{\infty}\zeta_2 f_{\mathrm{m}}(\zeta_2)d\zeta_2$$
$$= \overline{\zeta}_2,$$

where $f(\zeta_1,\zeta_2)$ is the p.d.f. of the state of nature Gaussian random variable ζ and $f_{\mathrm{m}}(\zeta_2)$ is a marginal Gaussian p.d.f. of $f(\zeta_1,\zeta_2)$. Recall that to obtain the marginal distribution over a subset of the components of a multivariate normal random variable, one only needs to drop the irrelevant variables (the variables that one wants to marginalize out) from the mean vector and the covariance matrix. For example, in the bivariate normal case the (Gaussian) marginal p.d.f. $f_{\mathrm{m}}(\zeta_1)$ is characterized by the parameters $(\overline{\zeta}_1, \sigma_1)$ and the marginal p.d.f. $f_{\mathrm{m}}(\zeta_2)$ is characterized by the parameters $(\overline{\zeta}_2, \sigma_2)$. Similarly,

$$E_\zeta(u(\zeta_1)) = \int_{-\infty}^{\infty} u(\zeta_1) f_{\mathrm{m}}(\zeta_1) d\zeta_1$$

Thus,

$$v^* = (R^{(v)})^{-1}\overline{\zeta}_2 + (R^{(v)})^{-1} R^{(u,v)} \int_{-\infty}^{\infty} u(\zeta_1) f_{\mathrm{m}}(\zeta_1) d\zeta_1 \tag{60}$$

The u-player's payoff is

$$J^{(u)}(u, v; \zeta_1) = 2u^T \zeta_1 - u^T R^{(u)} u + 2u^T (R^{(u,v)})^T v - v^T R^{(v)} v + 2v^T E_{\zeta_2}(\zeta_2 \mid \zeta_1)$$

Note: Now, as far as the u-player is concerned, the v-player does not employ a strategy, therefore the v-player's input v is no longer a random variable and one need not compute an expectation: the u-player knows that the v-player is "blind."

Differentiation in u yields the unique optimal control response to the v-player's input v

$$u^*(\zeta_1) = (R^{(u)})^{-1}\zeta_1 + (R^{(u)})^{-1}(R^{(u,v)})^T v \tag{61}$$

Combining (60) and (61) yields the relationship

$$v^* = (R^{(v)})^{-1}\overline{\zeta}_2 + (R^{(v)})^{-1} R^{(u,v)}[(R^{(u)})^{-1}\overline{\zeta}_1 + (R^{(u)})^{-1}(R^{(u,v)})^T v^*],$$

that is, the v-player's optimal control is

$$v^* = [R^{(v)} - R^{(u,v)}(R^{(u)})^{-1}(R^{(u,v)})^T]^{-1}[R^{(u,v)}(R^{(u)})^{-1}\overline{\zeta}_1 + \overline{\zeta}_2]$$

and the u-player's optimal strategy is

$$u^*(\zeta_1) = (R^{(u)})^{-1}\zeta_1 + (R^{(u)})^{-1}(R^{(u,v)})^T [R^{(v)} - R^{(u,v)}(R^{(u)})^{-1}(R^{(u,v)})^T]^{-1} [R^{(u,v)}(R^{(u)})^{-1}\overline{\zeta}_1 + \overline{\zeta}_2]$$

If the controls are scalars,

$$u^*(\zeta_1) = \frac{1}{R^{(u)}}\zeta_1 + \frac{R^{(u,v)}}{R^{(u)}R^{(v)} - (R^{(u,v)})^2}\left(\frac{R^{(u,v)}}{R^{(u)}}\overline{\zeta}_1 + \overline{\zeta}_2\right)$$

and

$$v^* = \frac{1}{R^{(u)}R^{(v)} - (R^{(u,v)})^2}(R^{(u,v)}\overline{\zeta}_1 + R^{(u)}\overline{\zeta}_2)$$

In conclusion, in the case where the v-player is "blind," the strategy of the u-player is as if there would be no correlation, that is, the parameter $\rho = 0$—as in Corollary 4. As far as the v-player is concerned, certainty equivalence holds. A little bit of thought will convince the reader that these results are expected.

8 Conclusion

The static decentralized decision problem has been analyzed. Special attention is given to the multivariate Quadratic Gaussian (QG) payoff function. The optimization problem is static, yet the players have partial information and as such, this is a small step away from the celebrated LQG paradigm. Informational issues, prior commitment strategies vs. delayed commitment strategies, as well as Nash equilibria solution concepts, are discussed. Necessary and sufficient conditions for the existence of a solution are provided and the optimal strategies are calculated. Extreme cases of informational asymmetry are also explored. This work lays the groundwork for gaining a better understanding of optimization problems with partial information where also dynamics are at play.

References

1. Radner, R.: "Team Decision Problems", Annals of Mathematical Statistics, Vol. 33, No. 3, September 1962, pp. 857–881.
2. H. S. Witsenhausen: "A Counterexample in Stochastic Optimal Control", SIAM Journal of Control, Vol. 6, No 1, pp. 131–147, 1968.

A Framework for Coordination in Distributed Stochastic Systems: Output Feedback and Performance Risk Aversion

Khanh D. Pham

Abstract This research article considers a class of distributed stochastic systems where interconnected systems closely keep track of reference signals issued by a coordinator. Much of the existing literature concentrates on conducting decisions and control synthesis based solely on expected utilities and averaged performance. However, research in psychology and behavioral decision theory suggests that performance risk plays an important role in shaping preferences in decisions under uncertainty. Thus motivated, a new equilibrium concept, called "person-by-person equilibrium" for local best responses is proposed for analyzing signaling effects and mutual influences between an incumbent system, its coordinator, and immediate neighbors. Individual member objectives are defined by the multi-attribute utility functions that capture both performance expectation and risk measures to model the satisfaction associated with local best responses with risk-averse attitudes. The problem class and approach of coordination control of distributed stochastic systems proposed here are applicable to and exemplified in military organizations and flexibly autonomous systems.

Keywords Distributed stochastic system • Coordination control • Mutual influences • Signaling effects • Performance measure • Performance risk • Output feedback • Risk-value aware performance index • Person-by-person equilibrium

K.D. Pham (✉)
Air Force Research Laboratory, Space Vehicles Directorate, 3550 Aberdeen Ave. SE, Kirtland Air Force Base, NM 87117, USA
e-mail: AFRL.RVSV@kirtland.af.mil

A. Sorokin and P.M. Pardalos (eds.), *Dynamics of Information Systems: Algorithmic Approaches*, Springer Proceedings in Mathematics & Statistics 51,
DOI 10.1007/978-1-4614-7582-8_6,

1 Introduction

Control and coordination of distributed stochastic systems offers a framework to analyzing intertemporal strategic interactions between individual agents or controllers, one for each interconnected systems and based on local observations. The importance of evaluating approaches in a dynamic setting and the broad flexibility and adaptability of the decision and control architectures of distributed control with communications has spurred many large-scale applications such as military command and control hierarchies, spacecraft constellations, remotely piloted platform formations, teams of humans and autonomous robots, etc. where each member can be in best response to its neighbor actions and yet has no influence on other members to which it has no communication supports.

Despite the broad interest in distributed systems, there remain significant hurdles in applying them to practical problems of interest. Interplay between common team objectives and individual member objectives can yield surprises and complex behaviors. Hence, a form of coordination control that helps balance between cooperative goals and adversarial behavior in addition to fundamentals for team and individual decisions, is necessarily required.

Thus motivated, this research article proposes a new framework and analysis to study risk-averse control of a distributed stochastic system, in particular coordination control with risk-averse attitudes toward performance uncertainty and robustness. The approach of noncooperative game-theoretic decision making and optimization is suited to coordination control, where a distributed stochastic system is distinguished into a coordinator (also known as dominant player) with significant reference signals and incumbent systems (also known as nondominant players) with fringe couplings. To account for uncertainty in inherent design problem and in preference assessment, a multi-attribute utility function that enables incumbent systems' decision makers or controllers to select the best risk-averse strategy for the attribute trade-offs between performance expectation and risks is therefore considered. Notice that this dominant/nondominant game structure is also prevalent in both economics [1] and social sciences [2].

The game-theoretic model of mixed player behaviors considered herein is particularly related to the research [3] that has extended the large population linear-quadratic-Gaussian games to include a major player and a large number of minor players. As such, minor players are more sensitive to variations in the behavior of major player than those of individual minor players. To overcome the curse of dimensionality, computational concerns have typically resorted the analysis to the so-called Nash certainty equivalence method, where the key idea is to break the large population game into a family of limiting two-player games. The synthesis of decentralized strategies is obtained via a set of aggregate quantities giving the mean field approximation. In contrast with such existing literature, this appealing research representing the interplay between stochastics, statistics, and dynamics as well as the extension of the recent accounts [4, 5] investigates: (1) a stochastic dynamic game model of behavior where nondominant players not only keep track closely of the large impact by the dominant player but also monitor rivals from the peers in a less detailed way and (2) a tractable paradigm of performance assessment uncertainty forecast for which sufficient statistics summarize all performance relevant information and thus are used in the person-by-person equilibrium strategies by nondominant players.

In summary, the proposed game-theoretic framework is prevalent in distributed stochastic systems with a dominant/fringe coordination structure, capturing the attributes that are important to inherent design problem and preference assessment uncertainties, their trade-off behavior over these attributes and their risk attitude. The rest of this article is organized as follows. Section 2 introduces a new computationally tractable model for distributed stochastic systems with state-space representations of a dominant coordinator and many nondominant systems. In addition, the preliminary results on sufficient mathematical statistics that summarize all performance measure or utility relevant history and for which the person-by-person equilibrium strategies are optimal for nondominant systems are discussed in great detail. Section 3 contains precise problem statements for coordination control analysis and decision optimization for the person-by-person equilibrium or feedback Nash strategy concerned by autonomous agents and incumbent systems. The construction of person-by-person strategies is established in Sect. 4, while some conclusions and future research directions are drawn in Sect. 5.

2 Problem Formulation

Before going into a formal presentation, it is necessary to consider some conceptual notations in this article. For instance, time t is modeled as continuous and the notation of the time interval is $[t_0, t_f]$. All random variables are defined on a probability space $(\Omega, \mathscr{F}, \mathscr{P})$ which is a triple consisting of a set Ω, a σ-algebra $\mathscr{F}$, and a probability measure $\mathscr{P} : \mathscr{F} \mapsto [0,1]$ and is equipped with a filtration $\{\mathscr{F}_t : t \in [t_0, t_f]\}$. In addition, for a given Hilbert space X with norm $||\cdot||_X$, $1 \le p \le \infty$, a Banach space is defined as follows

$$\mathscr{L}^p_{\mathscr{F}}(t_0, t_f; X) \triangleq \Bigg\{ \phi : [t_0, t_f] \times \Omega \mapsto X \text{ is an } X\text{-valued } \mathscr{F}_t\text{-measurable process with } E\left\{ \int_{t_0}^{t_f} ||\phi(t,\omega)||^p_X dt \right\} < \infty \Bigg\} \tag{1}$$

with norm

$$||\phi(\cdot)||_{\mathscr{F},p} \triangleq \left(E\left\{ \int_{t_0}^{t_f} ||\phi(t,\omega)||^p_X \, dt \right\} \right)^{1/p}. \tag{2}$$

Furthermore, the Banach space of X-valued continuous functionals on $[t_0,t_f]$ with the max-norm induced by $||\cdot||_X$ is denoted by $\mathscr{C}(t_0,t_f;X)$. The deterministic version of (1) and its associated norm (2) is written as $\mathscr{L}^p(t_0,t_f;X)$ and $||\cdot||_p$.

A distributed stochastic system that evolves over $[t_0,t_f]$ captures interactions among a coordinator and finite number of incumbent systems. Each incumbent system that enters the distributed system is assigned a unique positive integer-valued index. The set of indices of incumbent systems is denoted by $\overline{I} \triangleq \{1,2,\ldots,N\}$ and a typical element by i. The set of immediate neighbors associated with an incumbent system i is denoted by N_i. For concreteness, the heterogeneity of incumbent system i and $i \in \overline{I}$ is distinguished by an individual state; that is governed by the stochastic differential equation with the initial-value condition $x_i(t_0) = x_i^0$

$$dx_i(t) = (A_{ii}(t)x_i(t) + B_{ii}(t)u_i(t) + C_{ii}(t)z_i(t) + \sum_{j=1}^{N_i} B_{ij}(t)u_{ij}(t))dt + G_i(t)dw_i(t) \quad (3)$$

$$dy_i(t) = C_i(t)x_i(t)dt + dv_i(t), \quad (4)$$

where the continuous-time coefficients $A_{ii} \in \mathscr{C}(t_0,t_f;\mathbb{R}^{n_i \times n_i})$, $B_{ii} \in \mathscr{C}(t_0,t_f;\mathbb{R}^{n_i \times m_i})$, $C_{ii} \in \mathscr{C}(t_0,t_f;\mathbb{R}^{n_i \times q_i})$, $B_{ij} \in \mathscr{C}(t_0,t_f;\mathbb{R}^{n_i \times r_i})$, $G_i \in \mathscr{C}(t_0,t_f;\mathbb{R}^{n_i \times p_i})$ as well as $C_i \in \mathscr{C}(t_0,t_f;\mathbb{R}^{r_i \times n_i})$ are deterministic matrix-valued functions. At time t, the recursive state and output of incumbent system i are denoted by $x_i \in \mathscr{L}^2_{\mathscr{F}_i}(t_0,t_f;\mathbb{R}^{n_i})$ and $y_i \in \mathscr{L}^2_{\mathscr{F}_i}(t_0,t_f;\mathbb{R}^{r_i})$ with the initial state $x_i^0 \in \mathbb{R}^{n_i}$ known. The control policies from agent i to that system i are presented by $u_i \in \mathscr{L}^2_{\mathscr{F}_i}(t_0,t_f;\mathbb{R}^{m_i})$ and $z_i \in \mathscr{L}^2_{\mathscr{F}_i}(t_0,t_f;\mathbb{R}^{q_i})$. In addition, the interconnection inputs and linkage effects of that incumbent system i supported by the communication paths from immediate neighbors j and $j \in N_i$ are viewed as the real-valued functions $u_{ij}(t)dt$ of the following random processes

$$du_{ij}(t) \triangleq u_{ij}(t)dt = (C_{ij}(t)x_j(t) + D_{ij}(t)u_j(t))dt + dv_j(t), \quad j \in N_i \quad (5)$$

where continuous-time coefficients $C_{ij} \in \mathscr{C}(t_0,t_f;\mathbb{R}^{r_i \times n_j})$ and $D_{ij} \in \mathscr{C}(t_0,t_f;\mathbb{R}^{r_i \times m_j})$ are deterministic matrix-valued functions. As the number of incumbent systems grows large, it is unrealistic to believe that binding agents i associated with incumbent systems i and $i \in \overline{I}$ are capable of monitoring the evolution of their immediate neighbors. Instead, it is reasonable to assume that incumbent systems only keep track of actual interactions or signaling references provided by coordinator c and $c \in \overline{I}_c$, where the set of partaking coordinators is predetermined and does not change over time.

A challenging task for all multiscale modeling and coordination control is to transfer the knowledge gained from one resolution to another. As such, in coordination control there is an ongoing need for a coordinator c issuing reference signals to two or more incumbent systems i and $i \in \overline{I}$ such that

$$z_{ic}(t)dt = (A_{ic}(t)x_c(t) + B_{ic}(t)u_c(t))dt + G_{ic}(t)dv_c(t) \quad (6)$$

but the incumbent systems i do not directly send signals to the coordinator c. In practice, it is further desirable to have decentralized decision making without intensive communication overheads. A potential alternative therefore involves the selection of a crude model of reduced order for the interactions among coordinator c and binding agents i associated with incumbent systems i. The actual reference signals imposed by coordinator c are now approximated by an explicit model-following of the type

$$dz_{ic}(t) = (A_{ic}(t)z_{ic}(t) + B_{ic}(t)u_c(t))dt + G_{ic}(t)dw_{ic}(t), \quad z_{ic}(t_0) = 0 \tag{7}$$

$$dy_{ic}(t) = C_{ic}(t)z_{ic}(t)dt + dv_{ic}(t) \tag{8}$$

whereby continuous-time coefficients $A_{ic} \in \mathscr{C}(t_0,t_f;\mathbb{R}^{q_i \times q_i})$, $B_{ic} \in \mathscr{C}(t_0,t_f;\mathbb{R}^{q_i \times m_c})$, $G_{ic} \in \mathscr{C}(t_0,t_f;\mathbb{R}^{q_i \times p_{ic}})$ and $C_{ic} \in \mathscr{C}(t_0,t_f;\mathbb{R}^{r_{ic} \times q_i})$ are deterministic matrix-valued function and potentially come from a structural decomposition of a monolithic distributed system with centralized dynamics. In this exposition, $z_{ic} \in \mathscr{L}^2_{\mathscr{F}_i}(t_0,t_f;\mathbb{R}^{q_i \times q_i})$ is the coordinator state, $u_c \in \mathscr{L}^2_{\mathscr{F}_i}(t_0,t_f;\mathbb{R}^{m_c})$ is the coordinator control input and $y_{ic} \in \mathscr{L}^2_{\mathscr{F}_i}(t_0,t_f;\mathbb{R}^{r_{ic}})$ is the coordinator output.

In the state-space representations (3)–(4) and (7)–(8) one postulates uncorrelated Wiener processes $w_i(t) \triangleq w_i(t,\omega_i) : [t_0,t_f] \times \Omega_i \mapsto \mathbb{R}^{p_i}$, $v_i(t) \triangleq v_i(t,\omega_i) : [t_0,t_f] \times \Omega_i \mapsto \mathbb{R}^{r_i}$, $w_{ic}(t) \triangleq w_{ic}(t,\omega_{ic}) : [t_0,t_f] \times \Omega_{ic} \mapsto \mathbb{R}^{p_{ic}}$ and $v_{ic}(t) \triangleq v_{ic}(t,\omega_{ic}) : [t_0,t_f] \times \Omega_{ic} \mapsto \mathbb{R}^{r_{ic}}$ defined by the underlying filtered probability spaces $(\Omega_i, \mathscr{F}_i, \{\mathscr{F}_i\}_t, \mathscr{P}_i)$ and $(\Omega_{ic}, \mathscr{F}_{ic}, \{\mathscr{F}_{ic}\}_t, \mathscr{P}_{ic})$ with the correlations of independent increments

$$E\left\{[w_i(\tau_1) - w_i(\tau_2)][w_i(\tau_1) - w_i(\tau_2)]^T\right\} = W_i|\tau_1 - \tau_2|, \quad W_i > 0;\ \tau_1,\tau_2 \in [t_0,t_f]$$

$$E\left\{[v_i(\tau_1) - v_i(\tau_2)][v_i(\tau_1) - v_i(\tau_2)]^T\right\} = V_i|\tau_1 - \tau_2|, \quad V_i > 0$$

$$E\left\{[w_{ic}(\tau_1) - w_{ic}(\tau_2)][w_{ic}(\tau_1) - w_{ic}(\tau_2)]^T\right\} = W_{ic}|\tau_1 - \tau_2|, \quad W_{ic} > 0$$

$$E\left\{[v_{ic}(\tau_1) - v_{ic}(\tau_2)][v_{ic}(\tau_1) - v_{ic}(\tau_2)]^T\right\} = V_{ic}|\tau_1 - \tau_2|, \quad V_{ic} > 0$$

which now approximate the inherent design system uncertainty due to variability and lack of knowledge.

Furthermore, the model primitives of the state recursion (3) in the absence of links from the immediate neighbors and environmental disturbances are also assumed to be uniformly exponentially stable. For instance, there exist positive constants η_1 and η_2 such that the pointwise matrix norm of the closed-loop state transition matrix associated with incumbent system (3) satisfies the inequality

$$||\Phi_i(t,\tau)|| \le \eta_1 e^{-\eta_2(t-\tau)} \qquad \forall\, t \ge \tau \ge t_0\ .$$

The pair $(A_{ii}(t), [B_{ii}(t), C_{ii}(t)])$ is pointwise stabilizable if there exist bounded matrix-valued functions $K_{x_i}(t)$ and $K_{z_i}(t)$ so that the closed-loop system $dx_i(t) = (A_{ii}(t) + B_{ii}(t)K_{x_i}(t) + C_{ii}(t)K_{z_i}(t))\,x_i(t)dt$ is uniformly exponentially stable.

With the local agent dynamics (3) considered herein, each agent i associated with incumbent system i only plays a local dynamical game with its immediate neighbors $j \in N_i$. Mutual influence controlled by the control policies from the immediate neighbors of agent i is defined by $u_{-i} \triangleq \{u_{ij} : j \in N_i\}$. Assuming its coalition N_i conveys mutual influence information u_{-i}, agent i selects, at each time instant, a tuple of control policies to optimize its multi-attribute utility function. The tuple of control laws is defined by the control processes u_i and z_i, of which z_i is supposed to follow the prediction process z_{ic} for the reference signals from coordinator c. Thus, the subsequent states of agent i is determined by its current individual states x_i and z_{ic}, its chosen action (u_i, z_i) and the coalition effects u_{-i}. In fact, the selected action (u_i, z_i) will depend on agent i's individual states x_i and z_{ic} as well as the coalition effects u_{-i}.

To further illustrate the applicability of the coordination control framework as proposed here, the classes of admissible control policies associated with (3) are defined by $U_i \times Z_i \subset \mathscr{L}^2_{\mathscr{F}_i}(t_0, t_f; \mathbb{R}^{m_i}) \times \mathscr{L}^2_{\mathscr{F}_{mi}}(t_0, t_f; \mathbb{R}^{q_i})$. For any given coalition effects u_{-i}, the 3-tuple $(x_i(\cdot), u_i(\cdot), z_i(\cdot))$ shall be therefore referred to as an admissible 3-tuple if $x_i(\cdot) \in \mathscr{L}^2_{\mathscr{F}_i}(t_0, t_f; \mathbb{R}^{n_i})$ is the solution trajectory of the stochastic differential equation (3) when $u_i(\cdot) \in U_i$ and $z_i(\cdot) \in Z_i$.

In the subsequent analysis, the problem of observation and/or estimation in the distributed stochastic system is investigated with a major emphasis on the design of a set of locally optimal decision and control policies for incumbent agent i and $i \in \overline{I}$ in a completely decentralized environment with interconnection patterns. More precisely, since (A_{ii}, C_i) are detectable, it is possible to construct the local observers

$$
\begin{aligned}
d\hat{x}_i(t) &= (A_{ii}(t)\hat{x}_i(t) + B_{ii}(t)u_i(t) + C_{ii}(t)z_i(t) + \sum_{j=1, j\neq i}^{N_i} B_{ij}(t)u_{ij}(t))dt \\
&\quad + L_i(t)(dy_i(t) - C_i(t)\hat{x}_i(t)dt), \quad \hat{x}_i(t_0) = x_i^0
\end{aligned} \tag{9}
$$

whereby $\hat{x}_i(t) \in \mathbb{R}^{n_i}$ is the state estimate of $x_i(t)$ for incumbent agent i and $i \in \overline{I}$ and $L_i(t) \in \mathbb{R}^{n_i \times r_i}$ are the decentralized filtering gains determined by suitably modifying the dynamics of the local observers; for example

$$
L_i(t) = \Sigma_i(t) C_i^T(t) V_i^{-1} \tag{10}
$$

$$
\begin{aligned}
\frac{d}{dt}\Sigma_i(t) &= A_{ii}(t)\Sigma_i(t) + \Sigma_i(t)A_{ii}^T(t) - \Sigma_i(t)C_i^T(t)V_i^{-1}C_i(t)\Sigma_i(t) + G_{ii}(t)W_i G_{ii}^T(t) \\
&\quad + \sum_{j=1, j\neq i}^{N_i} B_{ij}(t)W_j \sum_{j=1, j\neq i}^{N_i} B_{ij}^T(t), \quad \Sigma_i(t_0) = 0.
\end{aligned} \tag{11}
$$

It is readily evident that the decentralized observation scheme developed in (9)–(11) incorporates the knowledge of the interconnection functions or the outputs of the other immediate neighbors of agent i and $i \in \overline{I}$.

In similar to the state-regulation case, independent decentralized optimal estimators may be designed hereafter for certain compensating signals from coordinators c to agent i; e.g.,

$$d\hat{z}_{ic}(t) = (A_{ic}(t)\hat{z}_{ic}(t) + B_{ic}(t)u_c(t))dt + L_{ic}(t)(dy_{ic}(t) - C_{ic}(t)\hat{z}_{ic}(t)dt) \qquad (12)$$
$$\hat{z}_{ic}(t_0) = 0$$

whereby $L_{ic}(t) \in \mathbb{R}^{q_i \times r_{ic}}$ is given by $L_{ic}(t) = \Sigma_{ic}(t)C_{ic}^T(t)V_{ic}^{-1}$ and $\Sigma_{ic}(t) \in \mathbb{R}^{q_i \times q_i}$ is the covariance of the error process $\tilde{z}_{ic}(t) = z_{ic}(t) - \hat{z}_{ic}(t)$, satisfying the forward-in-time differential equation

$$\frac{d}{dt}\Sigma_{ic}(t) = A_{ic}(t)\Sigma_{ic}(t) + \Sigma_{ic}(t)A_{ic}^T(t)$$
$$- \Sigma_{ic}(t)C_{ic}^T(t)V_{ic}^{-1}C_{ic}(t)\Sigma_{ic}(t) + G_{ic}(t)W_{ic}G_{ic}^T(t), \quad \Sigma_{ic}(t_0) = 0. \quad (13)$$

In terms of the observation errors $\tilde{x}_i(t) = x_i(t) - \hat{x}_i(t)$ and $\tilde{z}_{ic}(t) = z_{ic}(t) - \hat{z}_{ic}(t)$, it follows from (3), (4), (7), and (8) that, for $\tilde{x}_i(t_0) = 0$ and $\tilde{z}_{ic}(t_0) = 0$

$$d\tilde{x}_i(t) = (A_{ii}(t) - L_i(t)C_i(t))\tilde{x}_i(t)dt + G_{ii}(t)dw_i(t) - L_i(t)dv_i(t) \qquad (14)$$
$$d\tilde{z}_{ic}(t) = (A_{ic}(t) - L_{ic}(t)C_{ic}(t))\tilde{z}_{ic}(t)dt + G_{ic}(t)dw_{ic}(t) - L_{ic}(t)dv_{ic}(t). \qquad (15)$$

Indeed, the system (14)–(15) will function as observers for the system (3) and (7) if the design parameters $L_i(t)$ and $L_{ic}(t)$ can be selected such that the local observers (9) and (12) are asymptotically stable.

Next, agent i evaluates its performance and makes control policies that are consistent with its preferences. There are performance trade-offs among the closeness of locally accessible states $\hat{x}_i$ from desired states ζ_i, the size of local actions u_i and the closeness of interaction enforcements between local efforts z_i and local estimates $\hat{z}_{ic}$ of reference signals imposed by coordinator c. Henceforth, agent i must carefully balance the three in order to achieve its local performance measure. Mathematically, there assumes existence of an integral-quadratic form (IQF) performance-measure $J_i : U_i \times Z_i \mapsto \mathbb{R}_+$

$$J_i(u_i, z_i; u_{-i}) = [\hat{x}_i(t_f) - \zeta_i(t_f)]^T Q_i^f [\hat{x}_i(t_f) - \zeta_i(t_f)]$$
$$+ \int_{t_0}^{t_f} \{\hat{x}_i^T(\tau)Q_{ii}(\tau)\hat{x}_i(\tau) + [\hat{x}_i(\tau) - \zeta_i(\tau)]^T Q_i[\hat{x}_i(\tau) - \zeta_i(\tau)]\}d\tau$$
$$+ \int_{t_0}^{t_f} \{u_i^T(\tau)R_{ii}(\tau)u_i(\tau) + [z_i(\tau) - \hat{z}_{ic}(\tau)]^T R_{zi}(\tau)[z_i(\tau) - \hat{z}_{ic}(\tau)]\}d\tau, \qquad (16)$$

where the deterministic matrix-valued functions $Q_i^f \in \mathbb{R}^{n_i \times n_i}$, $Q_{ii} \in \mathscr{C}(t_0, t_f; \mathbb{R}^{n_i \times n_i})$, $Q_i \in \mathscr{C}(t_0, t_f; \mathbb{R}^{n_i \times n_i})$ $R_{ii} \in \mathscr{C}(t_0, t_f; \mathbb{R}^{m_i \times m_i})$ and $R_{zi} \in \mathscr{C}(t_0, t_f; \mathbb{R}^{q_i \times q_i})$ representing design parameters for terminal states, transient state estimates for regulation and tracking, regulating efforts and coordination effort mismatches are positive semidefinite with $R_{ii}(t)$ and $R_{zi}(t)$ invertible.

Control of (collective and aggregated) distributed stochastic systems on coordination levels is a major challenge and research theme. The approach to handling the problem with a tuple of two or more control laws is to use the noncooperative game-theoretic paradigm. Particularly, an N-tuple policy $\{(u_1^*, z_1^*), (u_2^*, z_2^*), \ldots, (u_N^*, z_N^*)\}$ is said to constitute a person-by-person equilibrium solution for the coordination control problem (3) and performance measure (16) if

$$J_i^* \triangleq J_i(u_i^*, z_i^*; u_{-i}^*) \leq J_i(u_i, z_i; u_{-i}^*), \qquad \forall i \in \bar{I}. \tag{17}$$

That is, none of the N agents can deviate unilaterally from the equilibrium policies and gain from doing so. The justification for the restriction to such an equilibrium is that the coalition effects u_{-i}^* sent to agent i does not necessarily support its preference optimization. Therefore, they cannot do better than behave as if they strive for this equilibrium. It is reasonable to conclude that a person-by-person equilibrium of distributed control is identical to the concept of a Nash equilibrium within a noncooperative game-theoretic setting.

Because admissible feedback policy sets for agent i are not discussed, the determination of a person-by-person equilibrium for the distributed stochastic system is still not straightforward. Therefore, a further restriction is imposed next. For the moment, it will suffice to say that in the case of incomplete information, an admissible 2-tuple feedback policy (u_i, z_i) for local best responses to all other immediate neighbors u_{-i}^* must be of the form, for some $\eth_i(\cdot,\cdot)$ and $\hbar_i(\cdot,\cdot)$

$$u_i(t) = \eth_i(t, y_i(\tau)), \quad \tau \in [t_0, t] \tag{18}$$

$$z_i(t) = \hbar_i(t, y_i(\tau)). \tag{19}$$

In general, the conditional density $p_i(x_i(t)|\mathscr{F}_t^i)$, which is the density of $x_i(t)$ conditioned on $\mathscr{F}_t^i$ (i.e., induced by the observation $\{y_i(\tau) : \tau \in [t_0, t]\}$) represents the sufficient statistics for describing the conditional stochastic effects of future 2-tuple feedback policies (u_i, z_i). It is natural that under the Gaussian assumption, the conditional density $p_i(x_i(t)|\mathscr{F}_t^i)$ is parameterized by the locally available conditional mean $\hat{x}_i(t) \triangleq E\{x_i(t)|\mathscr{F}_t^i\}$ and error-estimate covariance $\Sigma_i(t) \triangleq E\{[x_i(t) - \hat{x}_i(t)][x_i(t) - \hat{x}_i(t)]^T|\mathscr{F}_t^i\}$ by incumbent agent i. With respect to the linear-Gaussian conditions, the error-estimate covariances $\Sigma_i(t)$ are independent of feedback policies $u_i(t)$ and $z_i(t)$ and observations $\{y_i(\tau) : \tau \in [t_0, t]\}$. Hereafter, to look for observer-based optimal control and/or decision policies $u_i(t)$ and $z_i(t)$ of the form (18) and (19), it is only required that

$$u_i(t) = \gamma_i(t, \hat{x}_i(t)), \quad t \in [t_0, t_f]$$

$$z_i(t) = \wp_i(t, \hat{x}_i(t)).$$

In view of the linear-quadratic properties of the state-space description (3) and (16), the search for linear time-varying feedback policies generated from the locally accessible state $\hat{x}_i(t)$ is now proceeded to consider

$$u_i(t) = K_{x_i}(t)\hat{x}_i(t) + p_{x_i}(t) \tag{20}$$

$$z_i(t) = K_{z_i}(t)\hat{x}_i(t) + p_{z_i}(t), \quad t \in [t_0, t_f] \tag{21}$$

with the feedback policy parameters $K_{x_i} \in \mathscr{C}(t_0, t_f; \mathbb{R}^{m_i \times n_i})$, $K_{z_i} \in \mathscr{C}(t_0, t_f; \mathbb{R}^{q_i \times n_i})$, $p_{x_i} \in \mathscr{C}(t_0, t_f; \mathbb{R}^{m_i})$ and $p_{z_i} \in \mathscr{C}(t_0, t_f; \mathbb{R}^{q_i})$ admissible feedback policy parameters whose further defining properties will be stated shortly.

For the given (t_0, x^0_{ai}) and subject to the feedback control policies (20)–(21), agent i forms a local awareness of its state recursion (3) and (7) as follows

$$dx_{ai}(t) = (A_{ai}(t)x_{ai}(t) + l_{ai}(t))dt + G_{ai}(t)dw_{ai}(t), \quad x_{ai}(t_0) = x^0_{ai} \tag{22}$$

in which the aggregate Wiener process w_{ai} has the correlations of independent increments, for all $\tau_1, \tau_2 \in [t_0, t_f]$ and $W_{ai} > 0$

$$E\left\{[w_{ai}(\tau_1) - w_{ai}(\tau_2)][w_{ai}(\tau_1) - w_{ai}(\tau_2)]^T\right\} = W_{ai}|\tau_1 - \tau_2|,$$

whereas the augmented state variable x_{ai}, its initial-valued condition x^0_{ai}, the system coefficients and parameters are defined by

$$x_{ai} \triangleq \begin{bmatrix} \hat{x}_i \\ \tilde{x}_i \\ \hat{z}_{ic} \\ \tilde{z}_{ic} \end{bmatrix}; \quad x^0_{ai} \triangleq \begin{bmatrix} x^0_i \\ 0 \\ 0 \\ 0 \end{bmatrix}; \quad w_{ai} \triangleq \begin{bmatrix} w_i \\ v_i \\ w_{ic} \\ v_{ic} \end{bmatrix}; \quad G_{ai} \triangleq \begin{bmatrix} 0 & L_i & 0 & 0 \\ G_i & -L_i & 0 & 0 \\ 0 & 0 & 0 & L_{ic} \\ 0 & 0 & G_{ic} & -L_{ic} \end{bmatrix}$$

$$A_{ai} \triangleq \begin{bmatrix} A_{ii} + B_{ii}K_{x_i} + C_{ii}K_{z_i} & L_iC_i & 0 & 0 \\ 0 & A_{ii} - L_iC_i & 0 & 0 \\ 0 & 0 & A_{ic} & L_{ic}C_{ic} \\ 0 & 0 & 0 & A_{ic} - L_{ic}C_{ic} \end{bmatrix}; \quad W_{ai} \triangleq \begin{bmatrix} W_i & 0 & 0 & 0 \\ 0 & V_i & 0 & 0 \\ 0 & 0 & W_{ic} & 0 \\ 0 & 0 & 0 & V_{ic} \end{bmatrix}$$

$$l_{ai} \triangleq \begin{bmatrix} B_{ii}p_{x_i} + C_{ii}p_{z_i} + \sum_{j=1}^{N_i} B_{ij}u^*_{ij} \\ 0 \\ B_{ic}u_c \\ 0 \end{bmatrix}.$$

Moreover, the sample-path function of the random performance measure (16) is now rewritten as below

$$\begin{aligned} J_i(K_{x_i}, p_{x_i}; K_{z_i}, p_{z_i}) &= x^T_{ai}(t_f)Q^f_{ai}x_{ai}(t_f) + 2x^T_{ai}(t_f)S^f_{ai} + \zeta^T_i(t_f)Q^f_i\zeta_i(t_f) \\ &+ \int_{t_0}^{t_f} [x^T_{ai}(\tau)Q_{ai}(\tau)x_{ai}(\tau) + 2x^T_{ai}(\tau)S_{ai}(\tau) + \zeta^T_i(\tau)Q_i(\tau)\zeta_i(\tau) \\ &+ p^T_{x_i}(\tau)R_{ii}(\tau)p_{x_i}(\tau) + p^T_{z_i}(\tau)R_{zi}(\tau)p_{z_i}(\tau)]d\tau \end{aligned} \tag{23}$$

whereby the corresponding weightings are given by

$$Q^f_{ai} \triangleq \begin{bmatrix} Q^f_i & 0 & 0 & 0 \\ 0 & 0 & 0 & 0 \\ 0 & 0 & 0 & 0 \\ 0 & 0 & 0 & 0 \end{bmatrix}; \quad S^f_{ai} \triangleq \begin{bmatrix} -Q^f_i \zeta_i(t_f) \\ 0 \\ 0 \\ 0 \end{bmatrix}; \quad S_{ai} \triangleq \begin{bmatrix} K^T_{x_i} R_{ii} p_{x_i} + K^T_{z_i} R_{zi} p_{z_i} - Q_i \zeta_i \\ 0 \\ -R_{zi} p_{z_i} \\ 0 \end{bmatrix}$$

$$Q_{ai} \triangleq \begin{bmatrix} Q_{ii} + Q_i + K^T_{x_i} R_{ii} K_{x_i} + K^T_{z_i} R_{zi} K_{z_i} & 0 & -2K^T_{z_i} R_{zi} & 0 \\ 0 & 0 & 0 & 0 \\ 0 & 0 & R_{zi} & 0 \\ 0 & 0 & 0 & 0 \end{bmatrix}.$$

In views of the linear-quadratic structure of the problem (22) and (23), the performance measure (23) is clearly a random variable with chi-squared type. To account for performance uncertainty, a methodology that enables agent i to select robust decisions under uncertainty from a Pareto front which is acquired using envelopes of a finite set of higher-order statistics associated with (23). This methodology assists the preferences by agent i to be captured perfectly; i.e., what performance attributes that are important to agent i, their trade-off behavior over these attributes and their risk attitude. Recently, the research [6, 7] show how performance uncertainty affects different aspects of risk-averse decision making which can now serve as a starting point for such a knowledge extraction in terms of performance-measure statistics hereafter.

Theorem 1 (Performance-Measure Statistics). *Let the pairs* (A_{ii}, B_{ii}) *and* (A_{ii}, C_{ii}) *be uniformly stabilizable on* $[t_0, t_f]$ *in the incumbent system* i *and* $i \in \bar{I}$ *governed by (22) and (23). Then for the given initial condition* (t_0, x^0_i)*, incumbent agent* i *obtains the* k_i*-th cumulant associated with (23)*

$$\kappa^i_{k_i} = (x^0_{ai})^T H_i(t_0, k_i) x^0_{ai} + 2(x^0_{ai})^T \breve{D}_i(t_0, k_i) + D_i(t_0, k_i), \qquad k_i \in \mathbb{N} \tag{24}$$

whereby the supporting variables $\{H_i(s,r)\}^{k_i}_{r=1}$, $\{\breve{D}_i(s,r)\}^{k_i}_{r=1}$ *and* $\{D_i(s,r)\}^{k_i}_{r=1}$ *satisfy the time-backward differential equations (with the dependence of* $H_i(s,r)$, $\breve{D}_i(s,r)$ *and* $D_i(s,r)$ *upon the admissible* K_{x_i}, K_{z_i}, p_{x_i} *and* p_{z_i} *suppressed)*

$$\frac{d}{ds} H_i(s,1) = -A^T_{ai}(s) H_i(s,1) - H_i(s,1) A_{ai}(s) - Q_{ai}(s) \tag{25}$$

$$\frac{d}{ds} H_i(s,r) = -A^T_{ai}(s) H_i(s,r) - H_i(s,r) A_{ai}(s) - \sum_{v=1}^{r-1} \frac{2r!}{v!(r-v)!} H_i(s,v) G_{ai}(s) W_{ai} G^T_{ai}(s) H_i(s,r-v), \quad 2 \le r \le k_i \tag{26}$$

$$\frac{d}{ds} \breve{D}_i(s,1) = -A^T_{ai}(s) \breve{D}_i(s,1) - H_i(s,1) l_{ai}(s) - S_{ai}(s) \tag{27}$$

$$\frac{d}{ds} \breve{D}_i(s,r) = -A^T_{ai}(s) \breve{D}_i(s,r) - H_i(s,r) l_{ai}(s), \quad 2 \le r \le k_i \tag{28}$$

$$\frac{d}{ds}D_i(s,1) = -\mathrm{Tr}\{H_i(s,1)G_{ai}(s)W_{ai}G_{ai}^T(s)\} - 2\breve{D}_i^T(s,1)l_{ai}(s)$$
$$- p_{x_i}^T(s)R_{ii}(s)p_{x_i}(s) - p_{z_i}^T(s)R_{zi}(s)p_{z_i}(s) - \zeta_i^T(s)Q_i(s)\zeta_i(s) \quad (29)$$

$$\frac{d}{ds}D_i(s,r) = -\mathrm{Tr}\{H_i(s,r)G_{ai}(s)W_{ai}G_{ai}^T(s)\} - 2\breve{D}_i^T(s,r)l_{ai}(s),\ 2 \le r \le k_i \quad (30)$$

whereby the terminal-value conditions $H_i(t_f,1) = Q_{ai}^f$, $H_i(t_f,r) = 0$ *for* $2 \le r \le k_i$; $\breve{D}_i(t_f,1) = S_{ai}^f$, $\breve{D}_i(t_f,r) = 0$ *for* $2 \le r \le k_i$; *and* $D_i(t_f,1) = \zeta_i^T(t_f)Q_i^f\zeta_i(t_f)$, $D_i(t_f,r) = 0$ *for* $2 \le r \le k_i$.

Proof. A key challenge of the problem at hand is to come up with a tractable way to handle performance uncertainty such that its probabilistic nature is manageable. Therefore, only its statistics can be optimized. Most researchers find it easier to understand or describe a random variable through both moment and cumulant generating functions.

Precisely stated, it is necessary to parameterize the initial condition (t_0, x_{ai}^0) as any arbitrary pair (s, x_{ai}^s). Then, for the given admissible affine inputs p_{x_i} and p_{z_i} in addition with admissible feedback gains K_{x_i} and K_{z_i}, the "running" version of performance measure (23) is introduced as follows

$$J_i(s,x_{ai}^s) = x_{ai}^T(t_f)Q_{ai}^f x_{ai}(t_f) + 2x_{ai}^T(t_f)S_{ai}^f + \zeta_i^T(t_f)Q_i^f\zeta_i(t_f)$$
$$+ \int_s^{t_f} [x_{ai}^T(\tau)Q_{ai}(\tau)x_{ai}(\tau) + 2x_{ai}^T(\tau)S_{ai}(\tau) + \zeta_i^T(\tau)Q_i(\tau)\zeta_i(\tau)$$
$$+ p_{x_i}^T(\tau)R_{ii}(\tau)p_{x_i}(\tau) + p_{z_i}^T(\tau)R_{zi}(\tau)p_{z_i}(\tau)]d\tau, \quad i \in \bar{I}. \quad (31)$$

The moment-generating function associated with agent i of (31) is defined by

$$\varphi_i(s,x_{ai}^s;\theta_i) \triangleq E\{\exp(\theta_i J_i(s,x_{ai}^s))\}, \quad (32)$$

for some small parameters θ_i in an open interval about 0. Thus, the cumulant-generating function immediately follows

$$\psi_i(s,x_{ai}^s;\theta_i) \triangleq \ln\{\varphi_i(s,x_{ai}^s;\theta_i)\}, \quad (33)$$

for some θ_i in some (possibly smaller) open interval about 0 while $\ln\{\cdot\}$ denotes the natural logarithmic transformation.

For notational simplicity, it is convenient to define $\varpi_i(s,x_{ai}^s;\theta_i) \triangleq \exp\{\theta_i J_i(s,x_{ai}^s)\}$ and $\varphi_i(s,x_{ai}^s;\theta_i) \triangleq E\{\varpi_i(s,x_{ai}^s;\theta_i)\}$ together with the time derivative of

$$\frac{d}{ds}\varphi_i(s,x_{ai}^s;\theta_i) = -\theta_i\Big\{(x_{ai}^s)^T Q_{ai}(s)x_{ai}^s + 2(x_{ai}^s)^T S_{ai}(s)$$
$$+ \zeta_i^T(s)Q_i(s)\zeta_i(s) + p_{x_i}^T(s)R_{ii}(s)p_{x_i}(s) + p_{z_i}^T(s)R_{zi}(s)p_{z_i}(s)\Big\}$$
$$\varphi_i(s,x_{ai}^s;\theta_i). \quad (34)$$

Using the standard Ito's formula, it yields

$$
\begin{aligned}
d\varphi_i(s, x^s_{ai}; \theta_i) &= E\{d\varpi_i(s, x^s_{ai}; \theta_i)\}, \\
&= \varphi_{i,s}(s, x^s_{ai}; \theta_i)\, ds + \varphi_{i,x^s_{ai}}(s, x^s_{ai}; \theta_i)\left[A_{ai}(s) x^s_{ai} + l_{ai}(s)\right] ds \\
&\quad + \frac{1}{2}\mathrm{Tr}\{\varphi_{i,x^s_{ai}x^s_{ai}}(s, x^s_{ai}; \theta_i)\, G_{ai}(s)\, W_{ai} G^T_{ai}(s)\} ds.
\end{aligned}
$$

Furthermore, the moment-generating function of (31) can also be expressed by

$$
\varphi_i(s, x^s_{ai}; \theta_i) \triangleq \rho_i(s; \theta_i) \exp\left\{(x^s_{ai})^T \Upsilon_i(s; \theta_i) x^s_{ai} + 2(x^s_{ai})^T \eta_i(s; \theta_i)\right\} \tag{35}
$$

whereby all the supporting entities are going to be determined in the sequel. In particular, the partial derivatives of (35) results in

$$
\begin{aligned}
\frac{d}{ds}\varphi_i(s, x^s_{ai}; \theta_i) = \Bigg\{ & \frac{\frac{d}{ds}\rho_i(s; \theta_i)}{\rho_i(s, \theta_i)} + (x^s_{ai})^T \frac{d}{ds}\Upsilon_i(s; \theta_i) x^s_{ai} \\
& + 2(x^s_{ai})^T \frac{d}{ds}\eta_i(s; \theta_i) \\
& + (x^s_{ai})^T A^T_{ai}(s) \Upsilon_i(s; \theta_i) x^s_{ai} + (x^s_{ai})^T \Upsilon_i(s; \theta_i) A_{ai}(s) x^s_{ai} \\
& + 2(x^s_{ai})^T A^T_{ai}(s) \eta_i(s; \theta_i) \\
& + 2(x^s_{ai})^T \Upsilon_i(s; \theta_i) l_{ai}(s) + 2\eta^T_i(s; \theta_i) l_{ai}(s) \\
& + \mathrm{Tr}\{\Upsilon_i(s; \theta_i) G_{ai}(s) W_{ai} G^T_{ai}(s)\} \\
& + 2(x^s_{ai})^T \Upsilon_i(s; \theta_i) G_{ai}(s) W_{ai} G^T_{ai}(s) \Upsilon_i(s; \theta_i) x^s_{ai} \Bigg\} \varphi_i(s, x^s_{ai}; \theta_i).
\end{aligned} \tag{36}
$$

Equating the expression (34) with that of (36) and having both linear and quadratic terms independent of x^s_{ai} yield the following results

$$
\begin{aligned}
\frac{d}{ds}\Upsilon_i(s; \theta_i) &= -A^T_{ai}(s)\Upsilon_i(s; \theta_i) - \Upsilon_i(s; \theta_i) A_{ai}(s) \\
&\quad - 2\Upsilon_i(s; \theta_i) G_{ai}(s) W_{ai} G^T_{ai}(s) \Upsilon_i(s; \theta_i) - \theta_i Q_{ai}(s)
\end{aligned} \tag{37}
$$

$$
\frac{d}{ds}\eta_i(s; \theta_i) = -A^T_{ai}(s)\eta_i(s; \theta_i) - \Upsilon_i(s; \theta_i) l_{ai}(s) - \theta_i S_{ai}(s) \tag{38}
$$

$$
\begin{aligned}
\frac{d}{ds}\upsilon_i(s; \theta_i) &= -\mathrm{Tr}\left\{\Upsilon_i(s; \theta_i) G_{ai}(s) W_{ai} G^T_{ai}(s)\right\} - 2\eta^T_i(s; \theta_i) l_{ai}(s) - \theta_i \zeta^T_i(s) Q_i(s) \zeta_i(s) \\
&\quad - \theta_i p^T_{x_i}(s) R_{ii}(s) p_{x_i}(s) - \theta_i p^T_{z_i}(s) R_{zi}(s) p_{z_i}(s)
\end{aligned} \tag{39}
$$

wherein $\upsilon_i(s; \theta_i) \triangleq \ln\{\rho_i(s; \theta_i)\}$. At the final time $s = t_f$, it follows that

$$\varphi_i(t_f, x_{ai}(t_f); \theta_i) = \rho_i(t_f; \theta_i) \exp\left\{x_{ai}^T(t_f)\Upsilon_i(t_f; \theta_i)x_{ai}(t_f) + 2x_{ai}^T(t_f)\eta_i(t_f; \theta_i)\right\}$$
$$= E\left\{\exp\left\{\theta_i[x_{ai}^T(t_f)Q_{ai}^f x_{ai}(t_f) + 2x_{ai}^T(t_f)S_{ai}^f + \zeta_i^T(t_f)Q_i^f\zeta_i(t_f)]\right\}\right\}$$

which in turn yields the terminal-value conditions as $\Upsilon_i(t_f; \theta_i) = \theta_i Q_{ai}^f$; $\eta_i(t_f; \theta_i) = \theta_i S_{ai}^f$; and $\upsilon_i(t_f; \theta_i) = \theta_i \zeta_i^T(t_f)Q_i^f\zeta_i(t_f)$.

Hereafter, all the higher-order performance-measure statistics associated with the chi-squared random performance measure (31) will be utilized to generate a Pareto front with which incumbent agent i and $i \in \bar{I}$ is enabled to choose one or more trade-offs between multiple performance attributes and risk attitude. In views of the expression (35) and the definition of (33), the cumulant-generating function or second-order characteristic function of (31) is rewritten as follows

$$\psi_i(s, x_{ai}^s; \theta_i) = (x_{ai}^s)^T\Upsilon_i(s; \theta_i)x_{ai}^s + 2(x_{ai}^s)^T\eta_i(s; \theta_i) + \upsilon_i(s; \theta_i). \tag{40}$$

Subsequently, higher-order statistics of the random performance measure (31) that depict the performance uncertainty can now be determined by a Maclaurin series expansion of the cumulant-generating function (40); e.g.,

$$\psi_i(s, x_{ai}^s; \theta_i) = \sum_{r=1}^{\infty} \frac{\partial^{(r)}}{\partial\theta_i^{(r)}}\psi_i(s, x_{ai}^s; \theta_i)\bigg|_{\theta_i=0} \frac{\theta_i^r}{r!}, \tag{41}$$

from which all $\kappa_r \triangleq \frac{\partial^{(r)}}{\partial\theta_i^{(r)}}\psi_i(s, x_{ai}^s; \theta_i)\bigg|_{\theta_i=0}$ are known as the mathematical statistics or cumulants of the performance measure (31).

Moreover, the series expansion coefficients are computed by using the cumulant-generating function (40)

$$\frac{\partial^{(r)}}{\partial\theta_i^{(r)}}\psi_i(s, x_{ai}^s; \theta_i)\bigg|_{\theta_i=0} = (x_{ai}^s)^T \frac{\partial^{(r)}}{\partial\theta_i^{(r)}}\Upsilon_i(s; \theta_i)\bigg|_{\theta_i=0} x_{ai}^s + 2(x_{ai}^s)^T \frac{\partial^{(r)}}{\partial\theta_i^{(r)}}\eta_i(s; \theta_i)\bigg|_{\theta_i=0} + \frac{\partial^{(r)}}{\partial\theta_i^{(r)}}\upsilon_i(s; \theta_i)\bigg|_{\theta_i=0}. \tag{42}$$

In view of the definition (41), the rth performance-measure statistic is given by

$$\kappa_r = (x_{ai}^s)^T \frac{\partial^{(r)}}{\partial\theta_i^{(r)}}\Upsilon_i(s; \theta_i)\bigg|_{\theta_i=0} x_{ai}^s + 2(x_{ai}^s)^T \frac{\partial^{(r)}}{\partial\theta_i^{(r)}}\eta_i(s; \theta_i)\bigg|_{\theta_i=0} + \frac{\partial^{(r)}}{\partial\theta_i^{(r)}}\upsilon_i(s; \theta_i)\bigg|_{\theta_i=0} \tag{43}$$

for any finite $1 \leq r < \infty$. For notational convenience, the change of notations

$$H_i(s,r) \triangleq \left.\frac{\partial^{(r)} \Upsilon_i(s;\theta_i)}{\partial \theta_i^{(r)}}\right|_{\theta_i=0} ; \breve{D}_i(s,r) \triangleq \left.\frac{\partial^{(r)} \eta_i(s;\theta_i)}{\partial \theta_i^{(r)}}\right|_{\theta_i=0} ; D_i(s,r) \triangleq \left.\frac{\partial^{(r)} \upsilon_i(s;\theta_i)}{\partial \theta_i^{(r)}}\right|_{\theta_i=0}$$

is introduced. What remains is to show that the solutions $H_i(s,r)$, $\breve{D}_i(s,r)$, and $D_i(s,r)$ for $1 \leq r \leq k_i$ and $k_i \in \mathbb{N}$ indeed satisfy the time-backward matrix, vector, and scalar-valued differential equations (25)–(30). Notice that these differential equations (25)–(30) are readily obtained by successively taking derivatives with respect to θ_i of the cumulant-supporting equations (37)–(39) under the assumption of (A_{ii}, B_{ii}) and (A_{ii}, C_{ii}) uniformly stabilizable on the interval $[t_0, t_f]$. □

Furthermore, some attractive properties of the solutions to the cumulant-generating equations (25)–(30), for which the problem of coordination control with risk-averse performance of the class of distributed stochastic systems considered here is therefore well posed, are presented as follows.

Theorem 2 (Existence of Solutions for Performance-Measure Statistics). *Let the pairs $(A_{ii}(\cdot), B_{ii}(\cdot))$ and $(A_{ii}(\cdot), C_{ii}(\cdot))$ be uniformly stabilizable. Then, for any given $k_i \in \mathbb{N}$, the cumulant-generating equations (25)–(30) admit unique and bounded solutions $\{H_i(\cdot,r)\}_{r=1}^{k_i}$, $\{\breve{D}_i(\cdot,r)\}_{r=1}^{k_i}$ and $\{D_i(\cdot,r)\}_{r=1}^{k_i}$ on $[t_0, t_f]$.*

Proof. Under the assumption of stabilizability, there always exist some feedback parameters $K_{x_i}(\cdot)$ and $K_{z_i}(\cdot)$ such that the continuous-time aggregate state matrix $A_{ai}(\cdot)$ is exponentially stable on $[t_0, t_f]$. According to the results in [8], the state transition matrix $\Phi_{ai}(t,t_0)$, associated with the continuous-time composite state matrix $A_{ai}(\cdot)$, has the following properties

$$\frac{d}{dt}\Phi_{ai}(t,t_0) = A_{ai}(t)\Phi_{ai}(t,t_0), \qquad \Phi_{ai}(t_0,t_0) = I,$$

$$\lim_{t_f \to \infty} ||\Phi_{ai}(t_f,\tau)|| = 0, \qquad \lim_{t_f \to \infty} \int_{t_0}^{t_f} ||\Phi_{ai}(t_f,\tau)||^2 d\tau < \infty.$$

By the matrix variation of constant formula, the unique solutions to the time-backward matrix differential equations (25)–(30) together with the terminal-value conditions are then written as follows

$$H_i(s,1) = \Phi_{ai}^T(t_f,s) Q_{ai}^f \Phi_{ai}(t_f,s) + \int_s^{t_f} \Phi_{ai}^T(\tau,s) Q_{ai}(\tau) \Phi_{ai}(\tau,s) d\tau$$

$$H_i(s,r) = \int_s^{t_f} \Phi_{ai}^T(\tau,s) \sum_{v=1}^{r-1} \frac{2r!}{v!(r-v)!} H_i(\tau,v) G_{ai}(\tau) W_{ai} G_{ai}^T(\tau) H_i(\tau,r-v) \Phi_{ai}(\tau,s) d\tau$$

$$\breve{D}_i(s,1) = -\Phi_{ai}^T(t_f,s) Q_i^f \zeta_i(t_f) + \int_s^{t_f} \Phi_{ai}^T(\tau,s) \{H_i(\tau,1) l_{ai}(\tau) + S_{ai}(\tau)\} d\tau$$

$$\breve{D}_i(s,r) = \int_s^{t_f} \Phi_{ai}^T(\tau,s) H_i(\tau,r) l_{ai}(\tau) d\tau\,, \quad 2 \le r \le k_i$$

$$D_i(s,1) = \zeta_i^T(t_f) Q_i^f \zeta_i(t_f) + \int_s^{t_f} \{\mathrm{Tr}\{H_i(\tau,1) G_{ai}(\tau) W_{ai} G_{ai}^T(\tau)\} + 2\breve{D}_i^T(\tau,1) l_{ai}(\tau)$$
$$+ p_{x_i}^T(\tau) R_{ii}(\tau) p_{x_i}(\tau) + p_{z_i}^T(\tau) R_{zi}(\tau) p_{z_i}(\tau) + \zeta_i^T(\tau) Q_i(\tau) \zeta_i(\tau)\} d\tau$$

$$D_i(s,r) = \int_s^{t_f} \{\mathrm{Tr}\{H_i(\tau,r) G_{ai}(\tau) W_{ai} G_{ai}^T(\tau)\} + 2\breve{D}_i^T(\tau,r) l_{ai}(\tau)\} d\tau, \quad 2 \le r \le k_i\,.$$

As long as the growth rates of the integrals are not faster than those of exponentially decreasing $\Phi_{ai}(\cdot,\cdot)$ and $\Phi_{ai}^T(\cdot,\cdot)$ factors, it is therefore concluded that there exist upper bounds on the nonnegative and monotically increasing solutions $H_i(\cdot,r)$, $\breve{D}_i(\cdot,r)$ and $D_i(\cdot,r)$ for any time interval $[t_0,t_f]$. □

3 Problem Statements

The problem of adapting to performance uncertainty is now addressed by leveraging increased insight into the roles played by performance-measure statistics (24). It is interesting to note that all the performance-measure statistics (24) are functions of time-backward evolutions and do not depend on intermediate recursive state values $x_{ai}(t)$ governed by the state-space representation (22)–(23) for incumbent agent i at each point of time $t \in [t_0,t_f]$. Henceforth, these time-backward evolutions (25)–(30) of which the admissible decision variables K_{x_i}, K_{z_i}, p_{x_i}, and p_{z_i} from the 2-tuple person-by-person equilibrium strategy (20)–(21) are embedded, are therefore considered as the new dynamical equations with the associated state variables $H_i(\cdot,r)$, $\breve{D}_i(\cdot,r)$ and $D_i(\cdot,r)$, not the traditional system states $x_{ai}(\cdot)$.

To properly develop the problem statements within the concept of the person-by-person equilibrium strategy for agent i and $i \in \bar{I}$, the new dynamics (25)–(30) based upon the performance-measure statistics of (24) is rewritten in accordance with the following matrix partitions, for $1 \le r \le k_i$ and $k_i \in \mathbb{N}$

$$H_i(\cdot,r) \triangleq \begin{bmatrix} (H_i^r)_{11}(\cdot) & (H_i^r)_{12}(\cdot) & (H_i^r)_{13}(\cdot) & (H_i^r)_{14}(\cdot) \\ (H_i^r)_{21}(\cdot) & (H_i^r)_{22}(\cdot) & (H_i^r)_{23}(\cdot) & (H_i^r)_{24}(\cdot) \\ (H_i^r)_{31}(\cdot) & (H_i^r)_{32}(\cdot) & (H_i^r)_{33}(\cdot) & (H_i^r)_{34}(\cdot) \\ (H_i^r)_{41}(\cdot) & (H_i^r)_{42}(\cdot) & (H_i^r)_{43}(\cdot) & (H_i^r)_{44}(\cdot) \end{bmatrix}, \quad \breve{D}_i(\cdot,r) \triangleq \begin{bmatrix} (\breve{D}_i^r)_{11}(\cdot) \\ (\breve{D}_i^r)_{21}(\cdot) \\ (\breve{D}_i^r)_{31}(\cdot) \\ (\breve{D}_i^r)_{41}(\cdot) \end{bmatrix}.$$

For notational simplicity, it is now useful to denote the right members of the dynamics (25)–(30) as the mappings

$$(\mathscr{F}_i^r)_{11} : [t_0,t_f] \times (\mathbb{R}^{4n_i \times 4n_i})^{k_i} \times \mathbb{R}^{m_i \times n_i} \times \mathbb{R}^{q_i \times n_i} \mapsto \mathbb{R}^{n_i \times n_i}$$

$$(\mathscr{F}_i^r)_{12} : [t_0,t_f] \times (\mathbb{R}^{4n_i \times 4n_i})^{k_i} \times \mathbb{R}^{m_i \times n_i} \times \mathbb{R}^{q_i \times n_i} \mapsto \mathbb{R}^{n_i \times n_i}$$

$$(\mathscr{F}_i^r)_{13} : [t_0,t_f] \times (\mathbb{R}^{4n_i \times 4n_i})^{k_i} \times \mathbb{R}^{m_i \times n_i} \times \mathbb{R}^{q_i \times n_i} \mapsto \mathbb{R}^{n_i \times n_i}$$

$$(\mathscr{F}_i^r)_{14} : [t_0,t_f] \times (\mathbb{R}^{4n_i \times 4n_i})^{k_i} \times \mathbb{R}^{m_i \times n_i} \times \mathbb{R}^{q_i \times n_i} \mapsto \mathbb{R}^{n_i \times n_i}$$

$$(\mathscr{F}_i^r)_{21} : [t_0,t_f] \times (\mathbb{R}^{4n_i \times 4n_i})^{k_i} \times \mathbb{R}^{m_i \times n_i} \times \mathbb{R}^{q_i \times n_i} \mapsto \mathbb{R}^{n_i \times n_i}$$

$$(\mathscr{F}_i^r)_{22} : [t_0,t_f] \times (\mathbb{R}^{4n_i \times 4n_i})^{k_i} \mapsto \mathbb{R}^{n_i \times n_i}$$

$$(\mathscr{F}_i^r)_{23} : [t_0,t_f] \times (\mathbb{R}^{4n_i \times 4n_i})^{k_i} \mapsto \mathbb{R}^{n_i \times n_i}$$

$$(\mathscr{F}_i^r)_{24} : [t_0,t_f] \times (\mathbb{R}^{4n_i \times 4n_i})^{k_i} \mapsto \mathbb{R}^{n_i \times n_i}$$

$$(\mathscr{F}_i^r)_{31} : [t_0,t_f] \times (\mathbb{R}^{4n_i \times 4n_i})^{k_i} \times \mathbb{R}^{m_i \times n_i} \times \mathbb{R}^{q_i \times n_i} \mapsto \mathbb{R}^{n_i \times n_i}$$

$$(\mathscr{F}_i^r)_{32} : [t_0,t_f] \times (\mathbb{R}^{4n_i \times 4n_i})^{k_i} \mapsto \mathbb{R}^{n_i \times n_i}$$

$$(\mathscr{F}_i^r)_{33} : [t_0,t_f] \times (\mathbb{R}^{4n_i \times 4n_i})^{k_i} \mapsto \mathbb{R}^{n_i \times n_i}$$

$$(\mathscr{F}_i^r)_{34} : [t_0,t_f] \times (\mathbb{R}^{4n_i \times 4n_i})^{k_i} \mapsto \mathbb{R}^{n_i \times n_i}$$

$$(\mathscr{F}_i^r)_{41} : [t_0,t_f] \times (\mathbb{R}^{4n_i \times 4n_i})^{k_i} \times \mathbb{R}^{m_i \times n_i} \times \mathbb{R}^{q_i \times n_i} \mapsto \mathbb{R}^{n_i \times n_i}$$

$$(\mathscr{F}_i^r)_{42} : [t_0,t_f] \times (\mathbb{R}^{4n_i \times 4n_i})^{k_i} \mapsto \mathbb{R}^{n_i \times n_i}$$

$$(\mathscr{F}_i^r)_{43} : [t_0,t_f] \times (\mathbb{R}^{4n_i \times 4n_i})^{k_i} \mapsto \mathbb{R}^{n_i \times n_i}$$

$$(\mathscr{F}_i^r)_{44} : [t_0,t_f] \times (\mathbb{R}^{4n_i \times 4n_i})^{k_i} \mapsto \mathbb{R}^{n_i \times n_i}$$

$$(\breve{\mathscr{G}}_i^r)_{11} : [t_0,t_f] \times (\mathbb{R}^{4n_i \times 4n_i})^{k_i} \times (\mathbb{R}^{4n_i})^{k_i} \times \mathbb{R}^{m_i \times n_i} \times \mathbb{R}^{q_i \times n_i} \times \mathbb{R}^{m_i} \times \mathbb{R}^{q_i} \mapsto \mathbb{R}^{n_i}$$

$$(\breve{\mathscr{G}}_i^r)_{21} : [t_0,t_f] \times (\mathbb{R}^{4n_i \times 4n_i})^{k_i} \times (\mathbb{R}^{4n_i})^{k_i} \times \mathbb{R}^{m_i} \times \mathbb{R}^{q_i} \mapsto \mathbb{R}^{n_i}$$

$$(\breve{\mathscr{G}}_i^r)_{31} : [t_0,t_f] \times (\mathbb{R}^{4n_i \times 4n_i})^{k_i} \times (\mathbb{R}^{4n_i})^{k_i} \times \mathbb{R}^{m_i} \times \mathbb{R}^{q_i} \mapsto \mathbb{R}^{n_i}$$

$$(\breve{\mathscr{G}}_i^r)_{41} : [t_0,t_f] \times (\mathbb{R}^{4n_i \times 4n_i})^{k_i} \times (\mathbb{R}^{4n_i})^{k_i} \times \mathbb{R}^{m_i} \times \mathbb{R}^{q_i} \mapsto \mathbb{R}^{n_i}$$

$$\mathscr{G}_i^r : [t_0,t_f] \times (\mathbb{R}^{4n_i \times 4n_i})^{k_i} \times (\mathbb{R}^{4n_i})^{k_i} \times \mathbb{R}^{m_i} \times \mathbb{R}^{q_i} \mapsto \mathbb{R}$$

with the rules of action

$$\begin{aligned}(\mathscr{F}_i^1)_{11}(s,\mathscr{H}_i,K_{x_i},K_{z_i}) \triangleq & -[A_{ii}(s)+B_{ii}(s)K_{x_i}(s)+C_{ii}(s)K_{z_i}(s)]^T(\mathscr{H}_i^1)_{11}(s)\\ & -(\mathscr{H}_i^1)_{11}(s)[A_{ii}(s)+B_{ii}(s)K_{x_i}(s)+C_{ii}(s)K_{z_i}(s)]\\ & -Q_{ii}(s)-Q_i(s)-K_{x_i}^T(s)R_{ii}(s)K_{x_i}(s)-K_{z_i}^T(s)R_{zi}(s)K_{z_i}(s)\end{aligned}$$

$$\begin{aligned}(\mathscr{F}_i^r)_{11}(s,\mathscr{H}_i,K_{x_i},K_{z_i}) \triangleq & -[A_{ii}(s)+B_{ii}(s)K_{x_i}(s)+C_{ii}(s)K_{z_i}(s)]^T(\mathscr{H}_i^r)_{11}(s)\\ & -(\mathscr{H}_i^r)_{11}(s)[A_{ii}(s)+B_{ii}(s)K_{x_i}(s)+C_{ii}(s)K_{z_i}(s)]\\ & -\sum_{v=1}^{r-1}\frac{2r!}{v!(r-v)!}\Big\{(\mathscr{H}_i^v)_{11}(s)L_i(s)V_iL_i^T(s)(\mathscr{H}_i^{r-v})_{11}(s)\\ & -(\mathscr{H}_i^v)_{12}(s)L_i(s)V_iL_i^T(s)(\mathscr{H}_i^{r-v})_{11}(s)-(\mathscr{H}_i^v)_{11}(s)L_i(s)V_iL_i^T(s)(\mathscr{H}_i^{r-v})_{21}(s)\end{aligned}$$

$$+(\mathscr{H}_i^v)_{12}(s)G_i(s)W_iG_i^T(s)\mathscr{H}_i^{r-v})_{21}(s)+(\mathscr{H}_i^v)_{12}(s)L_i(s)V_iL_i^T(s)(\mathscr{H}_i^{r-v})_{21}(s)$$
$$+(\mathscr{H}_i^v)_{13}(s)L_{ic}(s)V_{ic}L_{ic}^T(s)(\mathscr{H}_i^{r-v})_{31}(s)-(\mathscr{H}_i^v)_{14}(s)L_{ic}(s)V_{ic}L_{ic}^T(s)(\mathscr{H}_i^{r-v})_{31}(s)$$
$$-(\mathscr{H}_i^v)_{13}(s)L_{ic}(s)V_{ic}L_{ic}^T(s)(\mathscr{H}_i^{r-v})_{41}(s)+(\mathscr{H}_i^v)_{14}(s)G_{ic}(s)W_{ic}G_{ic}^T(s)(\mathscr{H}_i^{r-v})_{41}(s)$$
$$+(\mathscr{H}_i^v)_{14}(s)L_{ic}(s)L_{ic}L_{ic}^T(s)(\mathscr{H}_i^{r-v})_{41}(s)\Big\}, \qquad 2\le r\le k_i$$

$$(\mathscr{F}_i^1)_{12}(s,\mathscr{H}_i,K_{x_i},K_{z_i}) \triangleq -[A_{ii}(s)+B_{ii}(s)K_{x_i}(s)+C_{ii}(s)K_{z_i}(s)]^T(\mathscr{H}_i^1)_{12}(s)$$
$$-(\mathscr{H}_i^1)_{11}(s)L_i(s)C_i(s)-(\mathscr{H}_i^1)_{12}(s)(A_{ii}(s)-L_i(s)C_i(s))$$

$$(\mathscr{F}_i^r)_{12}(s,\mathscr{H}_i,K_{x_i},K_{z_i}) \triangleq -[A_{ii}(s)+B_{ii}(s)K_{x_i}(s)+C_{ii}(s)K_{z_i}(s)]^T(\mathscr{H}_i^r)_{12}(s)$$
$$-(\mathscr{H}_i^r)_{11}(s)L_i(s)C_i(s)-(\mathscr{H}_i^r)_{12}(A_{ii}(s)-L_i(s)C_i(s))$$
$$-\sum_{v=1}^{r-1}\frac{2r!}{v!(r-v)!}\Big\{(\mathscr{H}_i^v)_{11}(s)L_i(s)V_iL_i^T(s)(\mathscr{H}_i^{r-v})_{12}(s)$$
$$-(\mathscr{H}_i^v)_{12}(s)L_i(s)V_iL_i^T(s)(\mathscr{H}_i^{r-v})_{12}(s)-(\mathscr{H}_i^v)_{11}(s)L_i(s)V_iL_i^T(s)(\mathscr{H}_i^{r-v})_{22}(s)$$
$$+(\mathscr{H}_i^v)_{12}(s)G_i(s)W_iG_i^T(s)(\mathscr{H}_i^{r-s})_{22}(s)+(\mathscr{H}_i^v)_{12}(s)L_i(s)W_iL_i^T(s)(\mathscr{H}_i^{r-s})_{22}(s)$$
$$+(\mathscr{H}_i^v)_{13}(s)L_{ic}(s)V_{ic}L_{ic}^T(s)(\mathscr{H}_i^{r-v})_{32}(s)-(\mathscr{H}_i^v)_{14}(s)L_{ic}(s)V_{ic}L_{ic}^T(s)(\mathscr{H}_i^{r-v})_{32}(s)$$
$$-(\mathscr{H}_i^v)_{13}(s)L_{ic}(s)V_{ic}L_{ic}^T(s)(\mathscr{H}_i^{r-v})_{42}(s)+(\mathscr{H}_i^v)_{14}(s)G_{ic}(s)W_{ic}G_{ic}^T(s)(\mathscr{H}_i^{r-v})_{42}(s)$$
$$+(\mathscr{H}_i^v)_{14}(s)L_{ic}(s)V_{ic}L_{ic}^T(s)(\mathscr{H}_i^{r-v})_{42}(s)\Big\}, \qquad 2\le r\le k_i$$

$$(\mathscr{F}_i^1)_{13}(s,\mathscr{H}_i,K_{x_i},K_{z_i}) \triangleq -[A_{ii}(s)+B_{ii}(s)K_{x_i}(s)+C_{ii}(s)K_{z_i}(s)]^T(\mathscr{H}_i^1)_{13}(s)$$
$$-(\mathscr{H}_i^1)_{13}(s)A_{ic}(s)+2K_{z_i}^T(s)R_{zi}(s)$$

$$(\mathscr{F}_i^r)_{13}(s,\mathscr{H}_i,K_{x_i},K_{z_i}) \triangleq -[A_{ii}(s)+B_{ii}(s)K_{x_i}(s)+C_{ii}(s)K_{z_i}(s)]^T(\mathscr{H}_i^r)_{13}(s)$$
$$-(\mathscr{H}_i^r)_{13}(s)A_{ic}(s)-\sum_{v=1}^{r-1}\frac{2r!}{v!(r-v)!}\Big\{(\mathscr{H}_i^v)_{11}(s)L_i(s)V_iL_i^T(s)(\mathscr{H}_i^{r-v})_{13}(s)$$
$$-(\mathscr{H}_i^v)_{12}(s)L_i(s)V_iL_i^T(s)(\mathscr{H}_i^{r-v})_{13}(s)-(\mathscr{H}_i^v)_{11}(s)L_i(s)V_iL_i^T(s)(\mathscr{H}_i^{r-v})_{23}(s)$$
$$+(\mathscr{H}_i^v)_{12}(s)G_i(s)W_iG_i^T(s)(\mathscr{H}_i^{r-v})_{23}(s)+(\mathscr{H}_i^v)_{12}(s)L_i(s)V_iL_i^T(s)(\mathscr{H}_i^{r-v})_{23}(s)$$

$$+ (\mathscr{H}_i^v)_{13}(s)L_{ic}(s)V_{ic}L_{ic}^T(s)(\mathscr{H}_i^{r-v})_{33}(s) - (\mathscr{H}_i^v)_{14}(s)L_{ic}(s)V_{ic}L_{ic}^T(s)(\mathscr{H}_i^{r-v})_{33}(s)$$
$$- (\mathscr{H}_i^v)_{13}(s)L_{ic}(s)V_{ic}L_{ic}^T(s)(\mathscr{H}_i^{r-v})_{43}(s) + (\mathscr{H}_i^v)_{14}(s)G_{ic}(s)W_{ic}G_{ic}^T(s)(\mathscr{H}_i^{r-v})_{43}(s)$$
$$+ (\mathscr{H}_i^v)_{14}(s)L_{ic}(s)V_{ic}L_{ic}^T(s)(\mathscr{H}_i^{r-v})_{43}(s)\Big\}, \qquad 2 \le r \le k_i$$

$$(\mathscr{F}_i^1)_{14}(s,\mathscr{H}_i,K_{x_i},K_{z_i}) \triangleq -[A_{ii}(s)+B_{ii}(s)K_{x_i}(s)+C_{ii}(s)K_{z_i}(s)]^T(\mathscr{H}_i^1)_{14}(s)$$
$$- (\mathscr{H}_i^1)_{13}(s)L_i(s)C_i(s) - (\mathscr{H}_i^1)_{14}(s)(A_{ic}(s)-L_{ic}(s)C_{ic}(s))$$

$$(\mathscr{F}_i^r)_{14}(s,\mathscr{H}_i,K_{x_i},K_{z_i}) \triangleq -[A_{ii}(s)+B_{ii}(s)K_{x_i}(s)+C_{ii}(s)K_{z_i}(s)]^T(\mathscr{H}_i^r)_{14}(s)$$
$$- (\mathscr{H}_i^r)_{13}(s)L_i(s)C_i(s) - (\mathscr{H}_i^r)_{14}(s)(A_{ic}(s)-L_{ic}(s)C_{ic}(s))$$
$$- \sum_{v=1}^{r-1} \frac{2r!}{v!(r-v)!}\Big\{(\mathscr{H}_i^v)_{11}(s)L_i(s)V_iL_i^T(s)(\mathscr{H}_i^{r-v})_{14}(s)$$
$$- (\mathscr{H}_i^v)_{12}(s)L_i(s)V_iL_i^T(s)(\mathscr{H}_i^{r-v})_{14}(s) - (\mathscr{H}_i^v)_{11}(s)L_i(s)V_iL_i^T(s)(\mathscr{H}_i^{r-v})_{24}(s)$$
$$+ (\mathscr{H}_i^v)_{12}(s)G_i(s)W_iG_i^T(s)(\mathscr{H}_i^{r-v})_{24}(s) + (\mathscr{H}_i^v)_{12}(s)L_i(s)V_iL_i^T(s)(\mathscr{H}_i^{r-v})_{24}(s)$$
$$+ (\mathscr{H}_i^v)_{13}(s)L_{ic}(s)V_{ic}L_{ic}^T(s)(\mathscr{H}_i^{r-v})_{34}(s) - (\mathscr{H}_i^v)_{14}(s)L_{ic}(s)V_{ic}L_{ic}^T(s)(\mathscr{H}_i^{r-v})_{34}(s)$$
$$- (\mathscr{H}_i^v)_{13}(s)L_{ic}(s)V_{ic}L_{ic}^T(s)(\mathscr{H}_i^{r-v})_{44}(s) + (\mathscr{H}_i^v)_{14}(s)G_{ic}(s)W_{ic}G_{ic}^T(s)(\mathscr{H}_i^{r-v})_{44}(s)$$
$$+ (\mathscr{H}_i^v)_{14}(s)L_{ic}(s)V_{ic}L_{ic}^T(s)(\mathscr{H}_i^{r-v})_{44}(s)\Big\}, \qquad 2 \le r \le k_i$$

$$(\mathscr{F}_i^1)_{21}(s,\mathscr{H}_i,K_{x_i},K_{z_i}) \triangleq -(\mathscr{H}_i^1)_{21}(s)[A_{ii}(s)+B_{ii}(s)K_{x_i}(s)+C_{ii}(s)K_{z_i}(s)]$$
$$- (L_i(s)C_i(s))^T(\mathscr{H}_i^1)_{11}(s) - (A_{ii}(s)-L_i(s)C_i(s))^T(\mathscr{H}_i^1)_{21}(s)$$

$$(\mathscr{F}_i^r)_{21}(s,\mathscr{H}_i,K_{x_i},K_{z_i}) \triangleq -(\mathscr{H}_i^r)_{21}(s)[A_{ii}(s)+B_{ii}(s)K_{x_i}(s)+C_{ii}(s)K_{z_i}(s)]$$
$$- (L_i(s)C_i(s))^T(\mathscr{H}_i^r)_{11}(s) - (A_{ii}(s)-L_i(s)C_i(s))^T(\mathscr{H}_i^r)_{21}(s)$$
$$- \sum_{v=1}^{r-1} \frac{2r!}{v!(r-v)!}\Big\{(\mathscr{H}_i^v)_{21}(s)L_i(s)V_iL_i^T(s)(\mathscr{H}_i^{r-v})_{11}(s)$$
$$- (\mathscr{H}_i^v)_{22}(s)L_i(s)V_iL_i^T(s)(\mathscr{H}_i^{r-v})_{11}(s) - (\mathscr{H}_i^v)_{21}(s)L_i(s)V_iL_i^T(s)(\mathscr{H}_i^{r-v})_{21}(s)$$
$$+ (\mathscr{H}_i^v)_{22}(s)G_i(s)W_iG_i^T(s)(\mathscr{H}_i^{r-v})_{21}(s) + (\mathscr{H}_i^v)_{22}(s)L_i(s)V_iL_i^T(s)(\mathscr{H}_i^{r-v})_{21}(s)$$

$$+ (\mathscr{H}_i^v)_{23}(s)L_i(s)V_iL_i^T(s)(\mathscr{H}_i^{r-v})_{31}(s) - (\mathscr{H}_i^v)_{24}(s)L_{ic}(s)V_{ic}L_{ic}^T(s)(\mathscr{H}_i^{r-v})_{31}(s)$$
$$- (\mathscr{H}_i^v)_{23}(s)L_{ic}(s)V_{ic}L_{ic}^T(s)(\mathscr{H}_i^{r-v})_{41}(s) + (\mathscr{H}_i^v)_{24}(s)G_{ic}(s)W_{ic}G_{ic}^T(s)(\mathscr{H}_i^{r-v})_{41}(s)$$
$$+ (\mathscr{H}_i^v)_{24}(s)L_{ic}(s)V_{ic}L_{ic}^T(s)(\mathscr{H}_i^{r-v})_{41}(s)\Big\}, \qquad 2 \le r \le k_i$$

$$(\mathscr{F}_i^1)_{22}(s,\mathscr{H}_i) \triangleq -(A_{ii}(s) - L_i(s)C_i(s))^T(\mathscr{H}_i^1)_{22}(s) - (L_i(s)C_i(s))^T(\mathscr{H}_i^1)_{12}(s)$$
$$- (\mathscr{H}_i^1)_{22}(s)(A_{ii}(s) - L_i(s)C_i(s)) - (\mathscr{H}_i^1)_{21}(s)L_i(s)C_i(s)$$

$$(\mathscr{F}_i^r)_{22}(s,\mathscr{H}_i) \triangleq -(A_{ii}(s) - L_i(s)C_i(s))^T(\mathscr{H}_i^r)_{22}(s) - (L_i(s)C_i(s))^T(\mathscr{H}_i^r)_{12}(s)$$
$$- (\mathscr{H}_i^r)_{22}(s)(A_{ii}(s) - L_i(s)C_i(s)) - (\mathscr{H}_i^r)_{21}(s)L_i(s)C_i(s)$$
$$- \sum_{v=1}^{r-1} \frac{2r!}{v!(r-v)!}\Big\{(\mathscr{H}_i^v)_{21}(s)L_i(s)V_iL_i^T(s)(\mathscr{H}_i^{r-v})_{12}(s)$$
$$- (\mathscr{H}_i^v)_{22}(s)L_i(s)V_iL_i^T(s)(\mathscr{H}_i^{r-v})_{12}(s) - (\mathscr{H}_i^v)_{21}(s)L_i(s)V_iL_i^T(s)(\mathscr{H}_i^{r-v})_{22}(s)$$
$$+ (\mathscr{H}_i^v)_{22}(s)G_i(s)W_iG_i^T(s)(\mathscr{H}_i^{r-v})_{22}(s) + (\mathscr{H}_i^v)_{22}(s)L_i(s)V_iL_i^T(s)(\mathscr{H}_i^{r-v})_{22}(s)$$
$$+ (\mathscr{H}_i^v)_{23}(s)L_{ic}(s)V_{ic}L_{ic}^T(s)(\mathscr{H}_i^{r-v})_{32}(s) - (\mathscr{H}_i^v)_{24}(s)L_{ic}(s)V_{ic}L_{ic}^T(s)(\mathscr{H}_i^{r-v})_{32}(s)$$
$$- (\mathscr{H}_i^v)_{23}(s)L_{ic}(s)V_{ic}L_{ic}^T(s)(\mathscr{H}_i^{r-v})_{42}(s) + (\mathscr{H}_i^v)_{24}(s)G_{ic}(s)W_{ic}G_{ic}^T(s)(\mathscr{H}_i^{r-v})_{42}(s)$$
$$+ (\mathscr{H}_i^v)_{24}(s)L_{ic}(s)V_{ic}L_{ic}^T(s)(\mathscr{H}_i^{r-v})_{42}(s)\Big\}, \qquad 2 \le r \le k_i$$

$$(\mathscr{F}_i^1)_{23}(s,\mathscr{H}_i) \triangleq -(A_{ii}(s) - L_i(s)C_i(s))^T(\mathscr{H}_i^1)_{23}(s) - (\mathscr{H}_i^1)_{23}(s)A_{ic}(s)$$
$$- (L_i(s)C_i(s))^T(\mathscr{H}_i^1)_{13}(s)$$

$$(\mathscr{F}_i^r)_{23}(s,\mathscr{H}_i) \triangleq -(A_{ii}(s) - L_i(s)C_i(s))^T(\mathscr{H}_i^r)_{23}(s) - (\mathscr{H}_i^r)_{23}(s)A_{ic}(s)$$
$$- (L_i(s)C_i(s))^T(\mathscr{H}_i^r)_{13}(s) - \sum_{v=1}^{r-1} \frac{2r!}{v!(r-v)!}\Big\{(\mathscr{H}_i^v)_{21}(s)L_i(s)V_iL_i^T(s)(\mathscr{H}_i^{r-v})_{13}(s)$$
$$- (\mathscr{H}_i^v)_{22}(s)L_i(s)V_iL_i^T(s)(\mathscr{H}_i^{r-v})_{13}(s) - (\mathscr{H}_i^v)_{21}(s)L_i(s)V_iL_i^T(s)(\mathscr{H}_i^{r-v})_{23}(s)$$
$$+ (\mathscr{H}_i^v)_{22}(s)G_i(s)W_iG_i^T(s)(\mathscr{H}_i^{r-v})_{23}(s) + (\mathscr{H}_i^v)_{22}(s)L_i(s)V_iL_i^T(s)(\mathscr{H}_i^{r-v})_{23}(s)$$
$$+ (\mathscr{H}_i^v)_{23}(s)L_{ic}(s)V_{ic}L_{ic}^T(s)(\mathscr{H}_i^{r-v})_{33}(s) - (\mathscr{H}_i^v)_{24}(s)L_{ic}(s)V_{ic}L_{ic}^T(s)(\mathscr{H}_i^{r-v})_{33}(s)$$

$$
- (\mathscr{H}_i^{v})_{23}(s)L_{ic}(s)V_{ic}L_{ic}^T(s)(\mathscr{H}_i^{r-v})_{43}(s) + (\mathscr{H}_i^{v})_{24}(s)G_{ic}(s)W_{ic}G_{ic}^T(s)(\mathscr{H}_i^{r-v})_{43}(s)
+ (\mathscr{H}_i^{v})_{24}(s)L_{ic}(s)V_{ic}L_{ic}^T(s)(\mathscr{H}_i^{r-v})_{43}(s)\Big\}, \qquad 2 \le r \le k_i
$$

$$
(\mathscr{F}_i^1)_{24}(s,\mathscr{H}_i) \triangleq -(A_{ii}(s)-L_i(s)C_i(s))^T(\mathscr{H}_i^1)_{24}(s) - (\mathscr{H}_i^1)_{23}(s)L_{ic}(s)C_{ic}(s)
- (L_{ic}(s)C_{ic}(s))^T(\mathscr{H}_i^1)_{14}(s) - (\mathscr{H}_i^1)_{24}(s)(A_{ii}(s)-L_i(s)C_i(s))
$$

$$
(\mathscr{F}_i^r)_{24}(s,\mathscr{H}_i) \triangleq -(A_{ii}(s)-L_i(s)C_i(s))^T(\mathscr{H}_i^r)_{24}(s) - (\mathscr{H}_i^r)_{23}(s)L_{ic}(s)C_{ic}(s)
- (L_{ic}(s)C_{ic}(s))^T(\mathscr{H}_i^r)_{14}(s) - (\mathscr{H}_i^r)_{24}(s)(A_{ii}(s)-L_i(s)C_i(s))
- \sum_{v=1}^{r-1}\frac{2r!}{v!(r-v)!}\Big\{(\mathscr{H}_i^{v})_{21}(s)L_i(s)V_iL_i^T(s)(\mathscr{H}_i^{r-v})_{14}(s)
- (\mathscr{H}_i^{v})_{22}(s)L_i(s)V_iL_i^T(s)(\mathscr{H}_i^{r-v})_{14}(s) - (\mathscr{H}_i^{v})_{21}(s)L_i(s)V_iL_i^T(s)(\mathscr{H}_i^{r-v})_{24}(s)
+ (\mathscr{H}_i^{v})_{22}(s)G_i(s)W_iG_i^T(s)(\mathscr{H}_i^{r-v})_{24}(s) + (\mathscr{H}_i^{v})_{22}(s)L_i(s)V_iL_i^T(s)(\mathscr{H}_i^{r-v})_{24}(s)
+ (\mathscr{H}_i^{v})_{23}(s)L_{ic}(s)V_{ic}L_{ic}^T(s)(\mathscr{H}_i^{r-v})_{34}(s) - (\mathscr{H}_i^{v})_{24}(s)L_{ic}(s)V_{ic}L_{ic}^T(s)(\mathscr{H}_i^{r-v})_{34}(s)
- (\mathscr{H}_i^{v})_{23}(s)L_{ic}(s)V_{ic}L_{ic}^T(s)(\mathscr{H}_i^{r-v})_{44}(s) + (\mathscr{H}_i^{v})_{24}(s)G_{ic}(s)W_{ic}G_{ic}^T(s)(\mathscr{H}_i^{r-v})_{44}(s)
+ (\mathscr{H}_i^{v})_{24}(s)L_{ic}(s)V_{ic}L_{ic}^T(s)(\mathscr{H}_i^{r-v})_{44}(s)\Big\}, \qquad 2 \le r \le k_i
$$

$$
(\mathscr{F}_i^1)_{31}(s,\mathscr{H}_i,K_{x_i},K_{z_i}) \triangleq -A_{ic}^T(s)(\mathscr{H}_i^1)_{31}(s)
- (\mathscr{H}_i^1)_{31}(s)[A_{ii}(s)+B_{ii}(s)K_{x_i}(s)+C_{ii}(s)K_{z_i}(s)]
$$

$$
(\mathscr{F}_i^r)_{31}(s,\mathscr{H}_i,K_{x_i},K_{z_i}) \triangleq -(\mathscr{H}_i^r)_{31}(s)[A_{ii}(s)+B_{ii}(s)K_{x_i}(s)+C_{ii}(s)K_{z_i}(s)]
- A_{ic}^T(s)(\mathscr{H}_i^r)_{31}(s) - \sum_{v=1}^{r-1}\frac{2r!}{v!(r-v)!}\Big\{(\mathscr{H}_i^{v})_{31}(s)L_i(s)V_iL_i^T(s)(\mathscr{H}_i^{r-v})_{11}(s)
- (\mathscr{H}_i^{v})_{32}(s)L_i(s)V_iL_i^T(s)(\mathscr{H}_i^{r-v})_{11}(s) - (\mathscr{H}_i^{v})_{31}(s)L_i(s)V_iL_i^T(s)(\mathscr{H}_i^{r-v})_{21}(s)
+ (\mathscr{H}_i^{v})_{32}(s)G_i(s)W_iG_i^T(s)(\mathscr{H}_i^{r-v})_{21}(s) + (\mathscr{H}_i^{v})_{32}(s)L_i(s)V_iL_i^T(s)(\mathscr{H}_i^{r-v})_{21}(s)
+ (\mathscr{H}_i^{v})_{33}(s)L_{ic}(s)V_{ic}L_{ic}^T(s)(\mathscr{H}_i^{r-v})_{31}(s) - (\mathscr{H}_i^{v})_{34}(s)L_{ic}(s)V_{ic}L_{ic}^T(s)(\mathscr{H}_i^{r-v})_{31}(s)
- (\mathscr{H}_i^{v})_{33}(s)L_{ic}(s)V_{ic}L_{ic}^T(s)(\mathscr{H}_i^{r-v})_{41}(s) + (\mathscr{H}_i^{v})_{34}(s)G_{ic}(s)W_{ic}G_{ic}^T(s)(\mathscr{H}_i^{r-v})_{41}(s)
+ (\mathscr{H}_i^{v})_{34}(s)L_{ic}(s)V_{ic}L_{ic}^T(s)(\mathscr{H}_i^{r-v})_{41}(s)\Big\}, \qquad 2 \le r \le k_i
$$

$$(\mathscr{F}_i^1)_{32}(s,\mathscr{H}_i) \triangleq -A_{ic}^T(s)(\mathscr{H}_i^1)_{32}(s) - (\mathscr{H}_i^1)_{31}(s)L_i(s)C_i(s) - (\mathscr{H}_i^1)_{32}(s)(A_{ii}(s) - L_i(s)C_i(s))$$

$$\begin{aligned}(\mathscr{F}_i^r)_{32}(s,\mathscr{H}_i) \triangleq{} & -A_{ic}^T(s)(\mathscr{H}_i^r)_{32}(s) - (\mathscr{H}_i^r)_{32}(s)(A_{ii}(s) - L_i(s)C_i(s)) \\ & - (\mathscr{H}_i^r)_{31}(s)L_i(s)C_i(s) - \sum_{v=1}^{r-1}\frac{2r!}{v!(r-v)!}\Big\{(\mathscr{H}_i^v)_{31}(s)L_i(s)V_iL_i^T(s)(\mathscr{H}_i^{r-v})_{12}(s) \\ & - (\mathscr{H}_i^v)_{32}(s)L_i(s)V_iL_i^T(s)(\mathscr{H}_i^{r-v})_{12}(s) - (\mathscr{H}_i^v)_{31}(s)L_i(s)V_iL_i^T(s)(\mathscr{H}_i^{r-v})_{22}(s) \\ & + (\mathscr{H}_i^v)_{32}(s)G_i(s)W_iG_i^T(s)(\mathscr{H}_i^{r-v})_{22}(s) + (\mathscr{H}_i^v)_{32}(s)L_i(s)V_iL_i^T(s)(\mathscr{H}_i^{r-v})_{22}(s) \\ & + (\mathscr{H}_i^v)_{33}(s)L_{ic}(s)V_{ic}L_{ic}^T(s)(\mathscr{H}_i^{r-v})_{32}(s) - (\mathscr{H}_i^v)_{34}(s)L_{ic}(s)V_{ic}L_{ic}^T(s)(\mathscr{H}_i^{r-v})_{32}(s) \\ & - (\mathscr{H}_i^v)_{33}(s)L_{ic}(s)V_{ic}L_{ic}^T(s)(\mathscr{H}_i^{r-v})_{42}(s) + (\mathscr{H}_i^v)_{34}(s)G_{ic}(s)W_{ic}G_{ic}^T(s)(\mathscr{H}_i^{r-v})_{42}(s) \\ & + (\mathscr{H}_i^v)_{34}(s)L_{ic}(s)V_{ic}L_{ic}^T(s)(\mathscr{H}_i^{r-v})_{42}(s)\Big\}, \qquad 2 \le r \le k_i\end{aligned}$$

$$(\mathscr{F}_i^1)_{33}(s,\mathscr{H}_i) \triangleq -A_{ic}^T(s)(\mathscr{H}_i^1)_{33}(s) - (\mathscr{H}_i^1)_{33}(s)A_{ic}(s) - R_{zi}(s)$$

$$\begin{aligned}(\mathscr{F}_i^r)_{33}(s,\mathscr{H}_i) \triangleq{} & -A_{ic}^T(s)(\mathscr{H}_i^r)_{33}(s) - (\mathscr{H}_i^r)_{33}(s)A_{ic}(s) \\ & - \sum_{v=1}^{r-1}\frac{2r!}{v!(r-v)!}\Big\{(\mathscr{H}_i^v)_{31}(s)L_i(s)V_iL_i^T(s)(\mathscr{H}_i^{r-v})_{13}(s) \\ & - (\mathscr{H}_i^v)_{32}(s)L_i(s)V_iL_i^T(s)(\mathscr{H}_i^{r-v})_{13}(s) - (\mathscr{H}_i^v)_{31}(s)L_i(s)V_iL_i^T(s)(\mathscr{H}_i^{r-v})_{23}(s) \\ & + (\mathscr{H}_i^v)_{32}(s)G_i(s)W_iG_i^T(s)(\mathscr{H}_i^{r-v})_{23}(s) + (\mathscr{H}_i^v)_{32}(s)L_i(s)V_iL_i^T(s)(\mathscr{H}_i^{r-v})_{23}(s) \\ & + (\mathscr{H}_i^v)_{33}(s)L_{ic}(s)V_{ic}L_{ic}^T(s)(\mathscr{H}_i^{r-v})_{33}(s) - (\mathscr{H}_i^v)_{34}(s)L_{ic}(s)V_{ic}L_{ic}^T(s)(\mathscr{H}_i^{r-v})_{33}(s) \\ & - (\mathscr{H}_i^v)_{33}(s)L_{ic}(s)V_{ic}L_{ic}^T(s)(\mathscr{H}_i^{r-v})_{43}(s) + (\mathscr{H}_i^v)_{34}(s)G_{ic}(s)W_{ic}G_{ic}^T(s)(\mathscr{H}_i^{r-v})_{43}(s) \\ & + (\mathscr{H}_i^v)_{34}(s)L_{ic}(s)V_{ic}L_{ic}^T(s)(\mathscr{H}_i^{r-v})_{43}(s)\Big\}, \qquad 2 \le r \le k_i\end{aligned}$$

$$(\mathscr{F}_i^1)_{34}(s,\mathscr{H}_i) \triangleq -A_{ic}^T(s)(\mathscr{H}_i^1)_{34}(s) - (\mathscr{H}_i^1)_{34}(s)(A_{ic}(s) - L_{ic}(s)C_{ic}(s)) - (\mathscr{H}_i^1)_{33}(s)L_{ic}(s)C_{ic}(s)$$

$$
\begin{aligned}
(\mathscr{F}_i^r)_{34}(s,\mathscr{H}_i) &\triangleq -A_{ic}^T(s)(\mathscr{H}_i^r)_{34}(s) - (\mathscr{H}_i^r)_{34}(s)(A_{ic}(s) - L_{ic}(s)C_{ic}(s)) \\
&- (\mathscr{H}_i^r)_{33}(s)L_{ic}(s)C_{ic}(s) - \sum_{v=1}^{r-1}\frac{2r!}{v!(r-v)!}\Big\{(\mathscr{H}_i^v)_{31}(s)L_i(s)V_iL_i^T(s)(\mathscr{H}_i^{r-v})_{14}(s) \\
&- (\mathscr{H}_i^v)_{32}(s)L_i(s)V_iL_i^T(s)(\mathscr{H}_i^{r-v})_{14}(s) - (\mathscr{H}_i^v)_{31}(s)L_i(s)V_iL_i^T(s)(\mathscr{H}_i^{r-v})_{24}(s) \\
&+ (\mathscr{H}_i^v)_{32}(s)G_i(s)W_iG_i^T(s)(\mathscr{H}_i^{r-v})_{24}(s) + (\mathscr{H}_i^v)_{32}(s)L_i(s)V_iL_i^T(s)(\mathscr{H}_i^{r-v})_{24}(s) \\
&+ (\mathscr{H}_i^v)_{33}(s)L_{ic}(s)V_{ic}L_{ic}^T(s)(\mathscr{H}_i^{r-v})_{34}(s) - (\mathscr{H}_i^v)_{34}(s)L_{ic}(s)V_{ic}L_{ic}^T(s)(\mathscr{H}_i^{r-v})_{34}(s) \\
&- (\mathscr{H}_i^v)_{33}(s)L_{ic}(s)V_{ic}L_{ic}^T(s)(\mathscr{H}_i^{r-v})_{44}(s) + (\mathscr{H}_i^v)_{34}(s)G_{ic}(s)W_{ic}G_{ic}^T(s)(\mathscr{H}_i^{r-v})_{44}(s) \\
&+ (\mathscr{H}_i^v)_{34}(s)L_{ic}(s)V_{ic}L_{ic}^T(s)(\mathscr{H}_i^{r-v})_{44}(s)\Big\}, \qquad 2 \le r \le k_i
\end{aligned}
$$

$$
\begin{aligned}
(\mathscr{F}_i^1)_{41}(s,\mathscr{H}_i,K_{x_i},K_{z_i}) &\triangleq -(A_{ii}(s) - L_i(s)C_i(s))^T(\mathscr{H}_i^1)_{41}(s) \\
&- (L_i(s)C_i(s))^T(\mathscr{H}_i^1)_{31}(s) - (\mathscr{H}_i^1)_{41}(s)[A_{ii}(s) + B_{ii}(s)K_{x_i}(s) + C_{ii}(s)K_{z_i}(s)]
\end{aligned}
$$

$$
\begin{aligned}
(\mathscr{F}_i^r)_{41}(s,\mathscr{H}_i,K_{x_i},K_{z_i}) &\triangleq -(A_{ic}(s) - L_{ic}(s)C_{ic}(s))^T(\mathscr{H}_i^r)_{41}(s) \\
&- (L_{ic}(s)C_{ic}(s))^T(\mathscr{H}_i^r)_{31}(s) - (\mathscr{H}_i^r)_{41}(s)[A_{ii}(s) + B_{ii}(s)K_{x_i}(s) + C_{ii}(s)K_{z_i}(s)] \\
&- \sum_{v=1}^{r-1}\frac{2r!}{v!(r-v)!}\Big\{(\mathscr{H}_i^v)_{41}(s)L_i(s)V_iL_i^T(s)(\mathscr{H}_i^{r-v})_{11}(s) \\
&- (\mathscr{H}_i^v)_{42}(s)L_i(s)V_iL_i^T(s)(\mathscr{H}_i^{r-v})_{11}(s) - (\mathscr{H}_i^v)_{41}(s)L_i(s)V_iL_i^T(s)(\mathscr{H}_i^{r-v})_{21}(s) \\
&+ (\mathscr{H}_i^v)_{42}(s)G_i(s)W_iG_i^T(s)(\mathscr{H}_i^{r-v})_{21}(s) + (\mathscr{H}_i^v)_{42}(s)L_i(s)V_iL_i^T(s)(\mathscr{H}_i^{r-v})_{21}(s) \\
&+ (\mathscr{H}_i^v)_{43}(s)L_{ic}(s)V_{ic}L_{ic}^T(s)(\mathscr{H}_i^{r-v})_{31}(s) - (\mathscr{H}_i^v)_{44}(s)L_{ic}(s)V_{ic}L_{ic}^T(s)(\mathscr{H}_i^{r-v})_{31}(s) \\
&- (\mathscr{H}_i^v)_{43}(s)L_{ic}(s)V_{ic}L_{ic}^T(s)(\mathscr{H}_i^{r-v})_{41}(s) + (\mathscr{H}_i^v)_{44}(s)G_{ic}(s)W_{ic}G_{ic}^T(s)(\mathscr{H}_i^{r-v})_{41}(s) \\
&+ (\mathscr{H}_i^v)_{44}(s)L_{ic}(s)V_{ic}L_{ic}^T(s)(\mathscr{H}_i^{r-v})_{41}(s)\Big\}, \qquad 2 \le r \le k_i
\end{aligned}
$$

$$
\begin{aligned}
(\mathscr{F}_i^1)_{42}(s,\mathscr{H}_i) &\triangleq -(A_{ii}(s) - L_i(s)C_i(s))^T(\mathscr{H}_i^1)_{42}(s) - (L_i(s)C_i(s))^T(\mathscr{H}_i^1)_{32}(s) \\
&- (\mathscr{H}_i^1)_{41}(s)L_i(s)C_i(s) - (\mathscr{H}_i^1)_{42}(s)(A_{ii}(s) - L_i(s)C_i(s))
\end{aligned}
$$

$$
\begin{aligned}
(\mathscr{F}_i^r)_{42}(s,\mathscr{H}_i) &\triangleq -(A_{ii}(s)-L_i(s)C_i(s))^T(\mathscr{H}_i^r)_{42}(s) - (L_i(s)C_i(s))^T(\mathscr{H}_i^r)_{32}(s) \\
&\quad - (\mathscr{H}_i^r)_{41}(s)L_i(s)C_i(s) - (\mathscr{H}_i^r)_{42}(s)(A_{ii}(s)-L_i(s)C_i(s)) \\
&\quad - \sum_{v=1}^{r-1}\frac{2r!}{v!(r-v)!}\Big\{(\mathscr{H}_i^v)_{41}(s)L_i(s)V_iL_i^T(s)(\mathscr{H}_i^{r-v})_{12}(s) \\
&\quad - (\mathscr{H}_i^v)_{42}(s)L_i(s)V_iL_i^T(s)(\mathscr{H}_i^{r-v})_{12}(s) - (\mathscr{H}_i^v)_{41}(s)L_i(s)V_iL_i^T(s)(\mathscr{H}_i^{r-v})_{22}(s) \\
&\quad + (\mathscr{H}_i^v)_{42}(s)G_i(s)W_iG_i^T(s)(\mathscr{H}_i^{r-v})_{22}(s) + (\mathscr{H}_i^v)_{42}(s)L_i(s)V_iL_i^T(s)(\mathscr{H}_i^{r-v})_{22}(s) \\
&\quad + (\mathscr{H}_i^v)_{43}(s)L_{ic}(s)V_{ic}L_{ic}^T(s)(\mathscr{H}_i^{r-v})_{32}(s) - (\mathscr{H}_i^v)_{44}(s)L_{ic}(s)V_{ic}L_{ic}^T(s)(\mathscr{H}_i^{r-v})_{32}(s) \\
&\quad - (\mathscr{H}_i^v)_{43}(s)L_{ic}(s)V_{ic}L_{ic}^T(s)(\mathscr{H}_i^{r-v})_{42}(s) + (\mathscr{H}_i^v)_{44}(s)G_{ic}(s)W_{ic}G_{ic}^T(s)(\mathscr{H}_i^{r-v})_{42}(s) \\
&\quad + (\mathscr{H}_i^v)_{44}(s)L_{ic}(s)V_{ic}L_{ic}^T(s)(\mathscr{H}_i^{r-v})_{42}(s)\Big\}, \qquad 2 \le r \le k_i
\end{aligned}
$$

$$
\begin{aligned}
(\mathscr{F}_i^1)_{43}(s,\mathscr{H}_i) &\triangleq -(A_{ii}(s)-L_i(s)C_i(s))^T(\mathscr{H}_i^1)_{43}(s) - (L_i(s)C_i(s))^T(\mathscr{H}_i^1)_{33}(s) \\
&\quad - (\mathscr{H}_i^1)_{43}(s)A_{ic}(s)
\end{aligned}
$$

$$
\begin{aligned}
(\mathscr{F}_i^r)_{43}(s,\mathscr{H}_i) &\triangleq -(A_{ii}(s)-L_i(s)C_i(s))^T(\mathscr{H}_i^r)_{43}(s) - (L_i(s)C_i(s))^T(\mathscr{H}_i^r)_{33}(s) \\
&\quad - (\mathscr{H}_i^r)_{43}(s)A_{ic}(s) - \sum_{v=1}^{r-1}\frac{2r!}{v!(r-v)!}\Big\{(\mathscr{H}_i^v)_{41}(s)L_i(s)V_iL_i^T(s)(\mathscr{H}_i^{r-v})_{13}(s) \\
&\quad - (\mathscr{H}_i^v)_{42}(s)L_i(s)V_iL_i^T(s)(\mathscr{H}_i^{r-v})_{13}(s) - (\mathscr{H}_i^v)_{41}(s)L_i(s)V_iL_i^T(s)(\mathscr{H}_i^{r-v})_{23}(s) \\
&\quad + (\mathscr{H}_i^v)_{42}(s)G_i(s)W_iG_i^T(s)(\mathscr{H}_i^{r-v})_{23}(s) + (\mathscr{H}_i^v)_{42}(s)L_i(s)V_iL_i^T(s)(\mathscr{H}_i^{r-v})_{23}(s) \\
&\quad + (\mathscr{H}_i^v)_{43}(s)L_{ic}(s)V_{ic}L_{ic}^T(s)(\mathscr{H}_i^{r-v})_{33}(s) - (\mathscr{H}_i^v)_{44}(s)L_{ic}(s)V_{ic}L_{ic}^T(s)(\mathscr{H}_i^{r-v})_{33}(s) \\
&\quad - (\mathscr{H}_i^v)_{43}(s)L_{ic}(s)V_{ic}L_{ic}^T(s)(\mathscr{H}_i^{r-v})_{43}(s) + (\mathscr{H}_i^v)_{44}(s)G_{ic}(s)W_{ic}G_{ic}^T(s)(\mathscr{H}_i^{r-v})_{43}(s) \\
&\quad + (\mathscr{H}_i^v)_{44}(s)L_{ic}(s)V_{ic}L_{ic}^T(s)(\mathscr{H}_i^{r-v})_{43}(s)\Big\}, \qquad 2 \le r \le k_i
\end{aligned}
$$

$$
\begin{aligned}
(\mathscr{F}_i^1)_{44}(s,\mathscr{H}_i) &\triangleq -(A_{ii}(s)-L_i(s)C_i(s))^T(\mathscr{H}_i^1)_{44}(s) - (L_i(s)C_i(s))^T(\mathscr{H}_i^1)_{34}(s) \\
&\quad - (\mathscr{H}_i^1)_{43}(s)L_{ic}(s)C_{ic}(s) - (\mathscr{H}_i^1)_{44}(s)(A_{ic}(s)-L_{ic}(s)C_{ic}(s))
\end{aligned}
$$

$$
\begin{aligned}
(\mathscr{F}_i^r)_{44}(s,\mathscr{H}_i) &\triangleq -(A_{ii}(s)-L_i(s)C_i(s))^T(\mathscr{H}_i^r)_{44}(s)-(L_i(s)C_i(s))^T(\mathscr{H}_i^r)_{34}(s) \\
&-(\mathscr{H}_i^r)_{43}(s)L_{ic}(s)C_{ic}(s)-(\mathscr{H}_i^r)_{44}(s)(A_{ic}(s)-L_{ic}(s)C_{ic}(s)) \\
&-\sum_{v=1}^{r-1}\frac{2r!}{v!(r-v)!}\Big\{(\mathscr{H}_i^v)_{41}(s)L_i(s)V_iL_i^T(s)(\mathscr{H}_i^{r-v})_{14}(s) \\
&-(\mathscr{H}_i^v)_{44}(s)L_{ic}(s)V_{ic}L_{ic}^T(s)(\mathscr{H}_i^{r-v})_{14}(s)-(\mathscr{H}_i^v)_{41}(s)L_i(s)V_iL_i^T(s)(\mathscr{H}_i^{r-v})_{24}(s) \\
&+(\mathscr{H}_i^v)_{42}(s)G_i(s)W_iG_i^T(s)(\mathscr{H}_i^{r-v})_{24}(s)+(\mathscr{H}_i^v)_{42}(s)L_i(s)V_iL_i^T(s)(\mathscr{H}_i^{r-v})_{24}(s) \\
&+(\mathscr{H}_i^v)_{43}(s)L_{ic}(s)V_{ic}L_{ic}^T(s)(\mathscr{H}_i^{r-v})_{34}(s)-(\mathscr{H}_i^v)_{44}(s)L_{ic}(s)V_{ic}L_{ic}^T(s)(\mathscr{H}_i^{r-v})_{34}(s) \\
&-(\mathscr{H}_i^v)_{43}(s)L_{ic}(s)V_{ic}L_{ic}^T(s)(\mathscr{H}_i^{r-v})_{44}(s)+(\mathscr{H}_i^v)_{44}(s)G_{ic}(s)W_{ic}G_{ic}^T(s)(\mathscr{H}_i^{r-v})_{44}(s) \\
&+(\mathscr{H}_i^v)_{44}(s)L_{ic}(s)V_{ic}L_{ic}^T(s)(\mathscr{H}_i^{r-v})_{44}(s)\Big\}, \qquad 2\le r\le k_i
\end{aligned}
$$

$$
\begin{aligned}
(\breve{\mathscr{G}}_i^1)_{11}(s,\mathscr{H}_i,\breve{\mathscr{D}}_i,K_{x_i},K_{z_i},p_{x_i},p_{z_i}) &\triangleq -(\mathscr{H}_i^1)_{13}(s)B_{ic}(s)u_c(s)-K_{x_i}^T(s)R_{ii}(s)p_{x_i}(s) \\
&-K_{z_i}^TR_{zi}(s)p_{z_i}(s)-[A_{ii}(s)+B_{ii}(s)K_{x_i}(s)+C_{ii}(s)K_{z_i}(s)]^T(\breve{\mathscr{D}}_i^1)_{11}(s) \\
&-(\mathscr{H}_i^1)_{11}(s)\Big[B_{ii}(s)p_{x_i}(s)+C_{ii}(s)p_{z_i}(s)+\sum_{j=1}^{N_i}B_{ij}(s)u_{ij}^*(s)\Big]+Q_i(s)\zeta_i(s)
\end{aligned}
$$

$$
\begin{aligned}
(\breve{\mathscr{G}}_i^r)_{11}(s,\mathscr{H}_i,\breve{\mathscr{D}}_i,K_{x_i},K_{z_i},p_{x_i},p_{z_i}) &\triangleq -(\mathscr{H}_i^r)_{13}(s)B_{ic}(s)u_c(s) \\
&-[A_{ii}(s)+B_{ii}(s)K_{x_i}(s)+C_{ii}(s)K_{z_i}(s)]^T(\breve{\mathscr{D}}_i^r)_{11}(s) \\
&-(\mathscr{H}_i^r)_{11}(s)\Big[B_{ii}(s)p_{x_i}(s)+C_{ii}(s)p_{z_i}(s)+\sum_{j=1}^{N_i}B_{ij}(s)u_{ij}^*(s)\Big]
\end{aligned}
$$

$$
\begin{aligned}
(\breve{\mathscr{G}}_i^r)_{21}(s,\mathscr{H}_i,\breve{\mathscr{D}}_i,p_{x_i},p_{z_i}) &\triangleq -(A_{ii}(s)-L_i(s)C_i(s))^T(\breve{\mathscr{D}}_i^r)_{21}(s) \\
&-(L_i(s)C_i(s))^T(s)(\breve{\mathscr{D}}_i^r)_{11}(s)-(\mathscr{H}_i^r)_{23}(s)B_{ic}(s)u_c(s) \\
&-(\mathscr{H}_i^r)_{21}(s)\Big[B_{ii}(s)p_{x_i}(s)+C_{ii}(s)p_{z_i}(s)+\sum_{j=1}^{N_i}B_{ij}(s)u_{ij}^*(s)\Big], \qquad 1\le r\le k_i
\end{aligned}
$$

$$(\breve{\mathscr{G}}_i^1)_{31}(s,\mathscr{H}_i,\breve{\mathscr{D}}_i,p_{x_i},p_{z_i}) \triangleq -A_{ic}^T(s)(\breve{\mathscr{D}}_i^1)_{31}(s)+R_{zi}(s)p_{z_i}(s)$$
$$-(\mathscr{H}_i^1)_{33}(s)B_{ic}(s)u_c(s)-(\mathscr{H}_i^1)_{31}(s)\left[B_{ii}(s)p_{x_i}(s)+C_{ii}(s)p_{z_i}(s)+\sum_{j=1}^{N_i}B_{ij}(s)u_{ij}^*(s)\right]$$

$$(\breve{\mathscr{G}}_i^r)_{31}(s,\mathscr{H}_i,\breve{\mathscr{D}}_i,p_{x_i},p_{z_i}) \triangleq -A_{ic}^T(s)(\breve{\mathscr{D}}_i^r)_{31}(s)-(\mathscr{H}_i^r)_{33}(s)B_{ic}(s)u_c(s)$$
$$-(\mathscr{H}_i^r)_{31}(s)\left[B_{ii}(s)p_{x_i}(s)+C_{ii}(s)p_{z_i}(s)+\sum_{j=1}^{N_i}B_{ij}(s)u_{ij}^*(s)\right], \qquad 2\le r\le k_i$$

$$(\breve{\mathscr{G}}_i^r)_{41}(s,\mathscr{H}_i,\breve{\mathscr{D}}_i,p_{x_i},p_{z_i}) \triangleq -(L_{ic}(s)C_{ic}(s))^T(\breve{\mathscr{D}}_i^r)_{31}(s)$$
$$-(A_{ic}(s)-L_{ic}(s)C_{ic}(s))^T(\breve{\mathscr{D}}_i^r)_{41}(s)-(\mathscr{H}_i^r)_{43}(s)B_{ic}(s)u_c(s)$$
$$-(\mathscr{H}_i^r)_{41}(s)\left[B_{ii}(s)p_{x_i}(s)+C_{ii}(s)p_{z_i}(s)+\sum_{j=1}^{N_i}B_{ij}(s)u_{ij}^*(s)\right], \qquad 1\le r\le k_i$$

$$\mathscr{G}_i^1(s,\mathscr{H}_i,\breve{\mathscr{D}}_i,p_{x_i},p_{z_i}) \triangleq -\mathrm{Tr}\{(\mathscr{H}_i^1)_{11}(s)L_i(s)V_iL_i^T(s)-(\mathscr{H}_i^1)_{12}(s)L_i(s)V_iL_i^T(s)\}$$
$$-\mathrm{Tr}\{-(\mathscr{H}_i^1)_{21}(s)L_i(s)V_iL_i^T(s)+(\mathscr{H}_i^1)_{22}(s)(G_i(s)W_iG_i^T(s)+L_i(s)V_iL_i^T(s))\}$$
$$-\mathrm{Tr}\{(\mathscr{H}_i^1)_{33}(s)L_{ic}(s)V_{ic}L_{ic}^T(s)-(\mathscr{H}_i^1)_{34}(s)L_{ic}(s)V_{ic}L_{ic}^T(s)\}$$
$$-\mathrm{Tr}\{-(\mathscr{H}_i^1)_{43}(s)L_{ic}(s)V_{ic}L_{ic}^T(s)+(\mathscr{H}_i^1)_{44}(s)(G_{ic}W_{ic}G_{ic}^T(s)+L_{ic}(s)V_{ic}L_{ic}^T(s))\}$$
$$-2(\breve{\mathscr{D}}_i^1)_{11}^T(s)\left[B_{ii}(s)p_{x_i}(s)+C_{ii}(s)p_{z_i}(s)+\sum_{j=1}^{N_i}B_{ij}(s)u_{ij}^*(s)\right]-2(\breve{\mathscr{D}}_i^1)_{31}^T(s)B_{ic}(s)u_c(s)$$
$$-\zeta_i^T(s)Q_i(s)\zeta_i(s)-p_{x_i}^T(s)R_{ii}(s)p_{x_i}(s)-p_{z_i}^T(s)R_{zi}(s)p_{z_i}(s)$$

$$\mathscr{G}_i^r(s,\mathscr{H}_i,\breve{\mathscr{D}}_i,p_{x_i},p_{z_i}) \triangleq -\mathrm{Tr}\{(\mathscr{H}_i^r)_{11}(s)L_i(s)V_iL_i^T(s)-(\mathscr{H}_i^r)_{12}(s)L_i(s)V_iL_i^T(s)\}$$
$$-\mathrm{Tr}\{-(\mathscr{H}_i^r)_{21}(s)L_i(s)V_iL_i^T(s)+(\mathscr{H}_i^r)_{22}(s)(G_i(s)W_iG_i^T(s)+L_i(s)V_iL_i^T(s))\}$$
$$-\mathrm{Tr}\{(\mathscr{H}_i^r)_{33}(s)L_{ic}(s)V_{ic}L_{ic}^T(s)-(\mathscr{H}_i^r)_{34}(s)L_{ic}(s)V_{ic}L_{ic}^T(s)\}$$
$$-\mathrm{Tr}\{-(\mathscr{H}_i^r)_{43}(s)L_{ic}(s)V_{ic}L_{ic}^T(s)+(\mathscr{H}_i^r)_{44}(s)(G_{ic}W_{ic}G_{ic}^T(s)+L_{ic}(s)V_{ic}L_{ic}^T(s))\}$$
$$-2(\breve{\mathscr{D}}_i^r)_{11}^T(s)\left[B_{ii}(s)p_{x_i}(s)+C_{ii}(s)p_{z_i}(s)+\sum_{j=1}^{N_i}B_{ij}(s)u_{ij}^*(s)\right]-2(\breve{\mathscr{D}}_i^r)_{31}^T(s)B_{ic}(s)u_c(s)$$

whereby the k_i-tuple $\mathscr{H}_i$, k_i-tuple $\breve{\mathscr{D}}_i$, and k_i-tuple $\mathscr{D}_i$ variables are defined by

$$\begin{aligned}
\mathscr{H}_i &\triangleq ((\mathscr{H}_i^1)_{11}, \ldots, (\mathscr{H}_i^{k_i})_{11}, (\mathscr{H}_i^1)_{12}, \ldots, (\mathscr{H}_i^{k_i})_{12}, (\mathscr{H}_i^1)_{13}, \ldots, (\mathscr{H}_i^{k_i})_{13},\\
&\quad (\mathscr{H}_i^1)_{14}, \ldots, (\mathscr{H}_i^{k_i})_{14}, (\mathscr{H}_i^1)_{21}, \ldots, (\mathscr{H}_i^{k_i})_{21}, (\mathscr{H}_i^1)_{22}, \ldots, (\mathscr{H}_i^{k_i})_{22},\\
&\quad (\mathscr{H}_i^1)_{23}, \ldots, (\mathscr{H}_i^{k_i})_{23}, (\mathscr{H}_i^1)_{24}, \ldots, (\mathscr{H}_i^{k_i})_{24}, (\mathscr{H}_i^1)_{31}, \ldots, (\mathscr{H}_i^{k_i})_{31},\\
&\quad (\mathscr{H}_i^1)_{32}, \ldots, (\mathscr{H}_i^{k_i})_{32}, (\mathscr{H}_i^1)_{33}, \ldots, (\mathscr{H}_i^{k_i})_{33}, (\mathscr{H}_i^1)_{34}, \ldots, (\mathscr{H}_i^{k_i})_{34},\\
&\quad (\mathscr{H}_i^1)_{41}, \ldots, (\mathscr{H}_i^{k_i})_{41}, (\mathscr{H}_i^1)_{42}, \ldots, (\mathscr{H}_i^{k_i})_{42}, (\mathscr{H}_i^1)_{43}, \ldots, (\mathscr{H}_i^{k_i})_{43},\\
&\quad (\mathscr{H}_i^1)_{44}, \ldots, (\mathscr{H}_i^{k_i})_{44})\\
&\equiv ((H_i^1)_{11}, \ldots, (H_i^{k_i})_{11}, (H_i^1)_{12}, \ldots, (H_i^{k_i})_{12}, (H_i^1)_{13}, \ldots, (H_i^{k_i})_{13},\\
&\quad (H_i^1)_{14}, \ldots, (H_i^{k_i})_{14}, (H_i^1)_{21}, \ldots, (H_i^{k_i})_{21}, (H_i^1)_{22}, \ldots, (H_i^{k_i})_{22},\\
&\quad (H_i^1)_{23}, \ldots, (H_i^{k_i})_{23}, (H_i^1)_{24}, \ldots, (H_i^{k_i})_{24}, (H_i^1)_{31}, \ldots, (H_i^{k_i})_{31},\\
&\quad (H_i^1)_{32}, \ldots, (H_i^{k_i})_{32}, (H_i^1)_{33}, \ldots, (H_i^{k_i})_{33}, (H_i^1)_{34}, \ldots, (H_i^{k_i})_{34},\\
&\quad (H_i^1)_{41}, \ldots, (H_i^{k_i})_{41}, (H_i^1)_{42}, \ldots, (H_i^{k_i})_{42}, (H_i^1)_{43}, \ldots, (H_i^{k_i})_{43},\\
&\quad (H_i^1)_{44}, \ldots, (H_i^{k_i})_{44})
\end{aligned}$$

$$\begin{aligned}
\breve{\mathscr{D}}_i &\triangleq ((\breve{\mathscr{D}}_i^1)_{11}, \ldots, (\breve{\mathscr{D}}_i^{k_i})_{11}, (\breve{\mathscr{D}}_i^1)_{21}, \ldots, (\breve{\mathscr{D}}_i^{k_i})_{21}, (\breve{\mathscr{D}}_i^1)_{31}, \ldots, (\breve{\mathscr{D}}_i^{k_i})_{31},\\
&\quad (\breve{\mathscr{D}}_i^1)_{41}, \ldots, (\breve{\mathscr{D}}_i^{k_i})_{41})\\
&\equiv ((\breve{D}_i^1)_{11}, \ldots, \breve{D}_i^{k_i})_{11}, (\breve{D}_i^1)_{21}, \ldots, \breve{D}_i^{k_i})_{21}, (\breve{D}_i^1)_{31}, \ldots, \breve{D}_i^{k_i})_{31},\\
&\quad (\breve{D}_i^1)_{41}, \ldots, \breve{D}_i^{k_i})_{41})\\
\mathscr{D}_i &\triangleq (\mathscr{D}_i^1, \ldots, \mathscr{D}_i^{k_i}) \equiv (D_i^1, \ldots, D_i^{k_i}).
\end{aligned}$$

Hence, the product system of dynamical equations in coordination control of the problem class with performance risk aversion becomes

$$\frac{d}{ds}\mathscr{H}_i(s) = \mathscr{F}_i(s, \mathscr{H}_i(s), K_{x_i}(s), K_{z_i}(s)), \qquad \mathscr{H}_i(t_f) = \mathscr{H}_i^f, \tag{44}$$

$$\frac{d}{ds}\breve{\mathscr{D}}_i(s) = \breve{\mathscr{G}}_i(s, \mathscr{H}_i(s), \mathscr{D}_i(s), K_{x_i}(s), K_{z_i}(s), p_{x_i}(s), p_{z_i}(s)), \quad \breve{\mathscr{D}}_i(t_f) = \breve{\mathscr{D}}_i^f, \tag{45}$$

$$\frac{d}{ds}\breve{\mathscr{D}}_i(s) = \breve{\mathscr{G}}_i(s, \mathscr{H}_i(s), \mathscr{D}_i(s), p_{x_i}(s), p_{z_i}(s)), \quad \breve{\mathscr{D}}_i(t_f) = \breve{\mathscr{D}}_i^f, \tag{46}$$

whereby

$$\mathscr{F}_i \triangleq (\mathscr{F}_i^1)_{11} \times \cdots \times (\mathscr{F}_i^{k_i})_{11} \times \cdots \times (\mathscr{F}_i^1)_{44} \times \cdots \times (\mathscr{F}_i^{k_i})_{44}$$

$$\breve{\mathscr{G}}_i \triangleq (\breve{\mathscr{G}}_i^1)_{11} \times \cdots \times (\breve{\mathscr{G}}_i^{k_i})_{11} \times \cdots \times (\breve{\mathscr{G}}_i^1)_{41} \times \cdots \times (\breve{\mathscr{G}}_i^{k_i})_{41}$$

$$\mathscr{G}_i \triangleq \mathscr{G}_i^1 \times \cdots \times \mathscr{G}_i^{k_i}$$

in addition to the product system of the terminal-value conditions

$$\mathscr{H}_i^f \triangleq Q_i^f \times \underbrace{0 \times \cdots \times 0}_{(16k_i-1)\text{-times}} \; ; \quad \breve{\mathscr{D}}_i^f \triangleq -Q_i^f \zeta_i(t_f) \times \underbrace{0 \times \cdots \times 0}_{(4k_i-1)\text{-times}}$$

$$\mathscr{D}_i^f \triangleq \zeta_i^T(t_f) Q_i^f \zeta_i(t_f) \times \underbrace{0 \times \cdots \times 0}_{(k_i-1)\text{-times}} \; ; \quad i \in \bar{I}.$$

Once immediate neighbors $j \in N_i$ of agent i fix the corresponding person-by-person equilibrium strategies u_j^* and thus the signaling or coordination effects u_{-i}^*, agent i then obtains an optimal stochastic control problem with risk-averse performance considerations. The construction of agent i's person-by-person policy now involves the 4-tuple $(K_{x_i}, K_{z_i}, p_{x_i}, p_{z_i})$. Furthermore, the solutions of the equations (44)–(46) also depend on the admissible feedback gains K_{x_i} and K_{z_i}, in addition to the affine inputs p_{x_i} and p_{z_i}. In the sequel and elsewhere, when this dependence is needed to be clear, then the notations $\mathscr{H}_i(s, K_{x_i}, K_{z_i}; u_{-i}^*)$, $\breve{\mathscr{D}}_i(s, K_{x_i}, K_{z_i}, p_{x_i}, p_{z_i}; u_{-i}^*)$ and $\mathscr{D}_i(s, K_{x_i}, K_{z_i}, p_{x_i}, p_{z_i}; u_{-i}^*)$ should be used to denote the solution trajectories of the dynamics (44)–(46) with the admissible 5-tuple $(K_{x_i}, K_{z_i}, p_{x_i}, p_{z_i}; u_{-i}^*)$.

For the given terminal data $(t_f, \mathscr{H}_i^f, \breve{\mathscr{D}}_i^f, \mathscr{D}_i^f)$, the theoretical framework for risk-averse control of the distributed stochastic system with possibly noncooperative u_{-i}^*, is then analyzed by a class of admissible feedback policies employed by agent i.

Definition 1 (Admissible Feedback Policies). Let compact subsets $\overline{K}^{x_i} \subset \mathbb{R}^{m_i \times n_i}$, $\overline{K}^{z_i} \subset \mathbb{R}^{q_i \times n_i}$, $\overline{P}^{x_i} \subset \mathbb{R}^{m_i}$, and $\overline{P}^{z_i} \subset \mathbb{R}^{q_i}$ be the sets of allowable feedback form values available at agent i and $i \in \bar{I}$. For the given $k_i \in \mathbb{N}$ and sequence $\mu_i = \{\mu_r^i \geq 0\}_{r=1}^{k_i}$ with $\mu_1^i > 0$, the set of feedback gains $\mathscr{K}^{x_i}_{t_f, \mathscr{H}_i^f, \breve{\mathscr{D}}_i^f, \mathscr{D}_i^f; \mu_i}$, $\mathscr{K}^{z_i}_{t_f, \mathscr{H}_i^f, \breve{\mathscr{D}}_i^f, \mathscr{D}_i^f; \mu_i}$, $\mathscr{P}^{x_i}_{t_f, \mathscr{H}_i^f, \breve{\mathscr{D}}_i^f, \mathscr{D}_i^f; \mu_i}$ and $\mathscr{P}^{z_i}_{t_f, \mathscr{H}_i^f, \breve{\mathscr{D}}_i^f, \mathscr{D}_i^f; \mu_i}$ are assumed to be the classes of $\mathscr{C}(t_0, t_f; \mathbb{R}^{m_i \times n_i})$, $\mathscr{C}(t_0, t_f; \mathbb{R}^{q_i \times n_i})$, $\mathscr{C}(t_0, t_f; \mathbb{R}^{m_i})$ and $\mathscr{C}(t_0, t_f; \mathbb{R}^{q_i})$ with values $K_{x_i}(\cdot) \in \overline{K}^{x_i}$, $K_{z_i}(\cdot) \in \overline{K}^{z_i}$, $p_{x_i}(\cdot) \in \overline{P}^{x_i}$ and $p_{z_i}(\cdot) \in \overline{P}^{z_i}$, for which the solutions to the dynamic equations (44)–(46) with the terminal-value conditions $\mathscr{H}_i(t_f) = \mathscr{H}_i^f$, $\breve{\mathscr{D}}_i(t_f) = \breve{\mathscr{D}}_i^f$ and $\mathscr{D}_i(t_f) = \mathscr{D}_i^f$ exist on the interval of optimization $[t_0, t_f]$.

To determine agent i's the person-by-person equilibrium strategy with risk bearing so as to minimize its performance vulnerability of (23) against all the sample-path realizations from uncertain environments w_{ai} and noncooperative coordination u_{-i}^* from immediate neighbors j and $j \in N_i$, performance risks are henceforth interpreted as worries and fears about certain undesirable characteristics of performance distributions of (23) and thus are proposed to manage through a finite set of selective weights. This custom set of design freedoms representing particular uncertainty aversions is hence different from the ones with aversion to risk captured in risk-sensitive optimal control [9, 10].

On $\mathcal{K}^{x_i}_{t_f,\mathcal{H}_i^f,\breve{\mathcal{D}}_i^f,\mathcal{D}_i^f;\mu_i}$, $\mathcal{K}^{z_i}_{t_f,\mathcal{H}_i^f,\breve{\mathcal{D}}_i^f,\mathcal{D}_i^f;\mu_i}$, $\mathcal{P}^{x_i}_{t_f,\mathcal{H}_i^f,\breve{\mathcal{D}}_i^f,\mathcal{D}_i^f;\mu_i}$ and $\mathcal{P}^{z_i}_{t_f,\mathcal{H}_i^f,\breve{\mathcal{D}}_i^f,\mathcal{D}_i^f;\mu_i}$, the performance index with risk-value considerations in risk-averse decision making is subsequently defined as follows.

Definition 2 (Risk-Value Aware Performance Index). Let incumbent agent i and $i \in \overline{I}$ select $k_i \in \mathbb{N}$ and the sequence of scalar coefficients $\mu_i = \{\mu_r^i \geq 0\}_{r=1}^{k_i}$ with $\mu_1^i > 0$. Then for the given initial condition (t_0, x_i^0), the risk-value aware performance index, $\phi_i^0 : \{t_0\} \times (\mathbb{R}^{n_i \times n_i})^{k_i} \times (\mathbb{R}^{n_i})^{k_i} \times \mathbb{R}^{k_i} \mapsto \mathbb{R}^+$ pertaining to risk-averse decision making of agent i over $[t_0, t_f]$ is defined by

$$\phi_i^0(t_0, \mathcal{H}_i(t_0), \breve{\mathcal{D}}_i(t_0), \mathcal{D}_i(t_0)) \triangleq \underbrace{\mu_1^i \kappa_1^i}_{\text{Value Measure}} + \underbrace{\mu_2^i \kappa_2^i + \cdots + \mu_{k_i}^i \kappa_{k_i}^i}_{\text{Risk Measures}}$$
$$= \sum_{r=1}^{k_i} \mu_r^i [(x_i^0)^T \mathcal{H}_i^r(t_0) x_i^0 + 2(x_i^0)^T \breve{\mathcal{D}}_i^r(t_0) + \mathcal{D}_i^r(t_0)], \tag{47}$$

wherein the additional design freedom by means of μ_r^i's utilized by agent i with risk-averse attitudes are sufficient to meet and exceed different levels of performance-based reliability requirements, for instance, mean (i.e., the average of performance measure), variance (i.e., the dispersion of values of performance measure around its mean), skewness (i.e., the anti-symmetry of the density of performance measure), kurtosis (i.e., the heaviness in the density tails of performance measure), etc., pertaining to closed-loop performance variations and uncertainties while the supporting solutions $\{\mathcal{H}_i^r(s)\}_{r=1}^{k_i}$, $\{\breve{\mathcal{D}}_i^r(s)\}_{r=1}^{k_i}$ and $\{\mathcal{D}_i^r(s)\}_{r=1}^{k_i}$ evaluated at $s = t_0$ satisfy the dynamical equations (44)–(46).

To specifically indicate the dependence of the risk-value aware performance index (47) expressed in Mayer form on $(K_{x_i}, K_{z_i}, p_{x_i}, p_{z_i})$ and the signaling effects u_{-i}^* issued by all immediate neighbors j from N_i, the multi-attribute utility function or performance index (47) for agent i is now rewritten explicitly as $\phi_i^0(K_{x_i}, K_{z_i}, p_{x_i}, p_{z_i}; u_{-i}^*)$.

Definition 3 (Nash Equilibrium Solution). An N-tuple of policies $\{(K_{x_1}^*, K_{z_1}^*, p_{x_1}^*, p_{z_1}^*), \ldots, (K_{x_N}^*, K_{z_N}^*, p_{x_N}^*, p_{z_N}^*)\}$ is said to constitute a Nash equilibrium solution for the distributed N-agent stochastic game if, for all $i \in \overline{N}$, the Nash inequality condition holds

$$\phi_i^0(K_{x_1}^*, K_{z_1}^*, p_{x_1}^*, p_{z_1}^*; u_{-i}^*) \leq \phi_i^0(K_{x_1}, K_{z_1}, p_{x_1}, p_{z_1}; u_{-i}^*). \tag{48}$$

For the sake of time consistency and subgame perfection, a Nash equilibrium solution is required to have an additional property that its restriction on the interval $[t_0, \tau]$ is also a Nash solution to the truncated version of the original problem, defined on $[t_0, \tau]$. With such a restriction so defined, the Nash equilibrium solution is now termed as a feedback Nash equilibrium solution, which is now free of any informational nonuniqueness, and thus whose derivation allows a dynamic programming type argument.

Definition 4 (Feedback Nash Equilibrium). Let $(K_{x_i}^*, K_{z_i}^*, p_{x_i}^*, p_{z_i}^*)$ constitute a feedback Nash strategy for agent i such that

$$\phi_i^0(K_{x_i}^*, K_{z_i}^*, p_{x_i}^*, p_{z_i}^*; u_{-i}^*) \leq \phi_i^0(K_{x_i}, K_{z_i}, p_{x_i}, p_{z_i}; u_{-i}^*), \qquad i \in \bar{I} \tag{49}$$

for admissible $K_{x_i} \in \mathscr{K}^{x_i}_{t_f, \mathscr{H}_i^f, \breve{\mathscr{D}}_i^f, \mathscr{D}_i^f; \mu_i}$, $K_{z_i} \in \mathscr{K}^{z_i}_{t_f, \mathscr{H}_i^f, \breve{\mathscr{D}}_i^f, \mathscr{D}_i^f; \mu_i}$, $p_{x_i} \in \mathscr{P}^{x_i}_{t_f, \mathscr{H}_i^f, \breve{\mathscr{D}}_i^f, \mathscr{D}_i^f; \mu_i}$ and $p_{z_i} \in \mathscr{P}^{z_i}_{t_f, \mathscr{H}_i^f, \breve{\mathscr{D}}_i^f, \mathscr{D}_i^f; \mu_i}$, upon which the solutions to the dynamical systems (44)–(46) exist on $[t_0, t_f]$.

Then, $\{(K_{x_1}^*, K_{z_1}^*, p_{x_1}^*, p_{z_1}^*), \ldots, (K_{x_N}^*, K_{z_N}^*, p_{x_N}^*, p_{z_N}^*)\}$ when restricted to the interval $[t_0, \tau]$ is still an N-tuple feedback Nash equilibrium solution for the multiperson Nash decision problem with the appropriate terminal-value condition $(\tau, \mathscr{H}_i^*(\tau), \breve{\mathscr{D}}_i^*(\tau), \mathscr{D}_i^*(\tau))$ for all $\tau \in [t_0, t_f]$.

In conformity with the rigorous formulation of dynamic programming, the following development is important. Let the terminal time t_f and 3-tuple states $(\mathscr{H}_i^f, \breve{\mathscr{D}}_i^f, \mathscr{D}_i^f)$, the other end condition involved the initial time t_0 and 3-tuple states $(\mathscr{H}_i^0, \breve{\mathscr{D}}_i^0, \mathscr{D}_i^0)$ be specified by a target set requirement.

Definition 5 (Target Sets). $(t_0, \mathscr{H}_i^0, \breve{\mathscr{D}}_i^0, \mathscr{D}_i^0) \in \mathscr{M}_i$, where the *target set* $\mathscr{M}_i$ is a closed subset of $[t_0, t_f] \times (\mathbb{R}^{n_i \times n_i})^{k_i} \times (\mathbb{R}^{n_i})^{k_i} \times \mathbb{R}^{k_i}$.

Now, the decision optimization residing at incumbent agent i is to minimize the risk-value aware performance index (47) over admissible feedback strategies composed by $K_{x_i} \equiv K_{x_i}(\cdot) \in \mathscr{K}^{x_i}_{t_f, \mathscr{H}_i^f, \breve{\mathscr{D}}_i^f, \mathscr{D}_i^f; \mu_i}$, $K_{z_i} \equiv K_{z_i}(\cdot) \in \mathscr{K}^{z_i}_{t_f, \mathscr{H}_i^f, \breve{\mathscr{D}}_i^f, \mathscr{D}_i^f; \mu_i}$, $p_{x_i} \equiv p_{x_i}(\cdot) \in \mathscr{P}^{x_i}_{t_f, \mathscr{H}_i^f, \breve{\mathscr{D}}_i^f, \mathscr{D}_i^f; \mu_i}$ and $p_{z_i} \equiv p_{z_i}(\cdot) \in \mathscr{P}^{z_i}_{t_f, \mathscr{H}_i^f, \breve{\mathscr{D}}_i^f, \mathscr{D}_i^f; \mu_i}$ while subject to interconnection links from all immediate neighbors with corresponding feedback Nash policies u_{-i}^*.

Definition 6 (Optimization of Mayer Problem). Given the sequence of scalars $\mu_i = \{\mu_r^i \geq 0\}_{r=1}^{k^i}$ with $\mu_1^i > 0$, the decision optimization over $[t_0, t_f]$ is given by

$$\min_{K_{x_i}, K_{z_i}, p_{x_i}, p_{z_i}} \phi_i^0(K_{x_i}, K_{z_i}, p_{x_i}, p_{z_i}; u_{-i}^*) \tag{50}$$

subject to the dynamical equations (44)–(46) on $[t_0, t_f]$.

Notice that the optimization considered here is in Mayer form and can be solved by applying an adaptation of the Mayer form verification results as given in [11].

To embed this optimization facing agent i into a larger problem, the terminal time and states $(t_f, \mathscr{H}_i^f, \breve{\mathscr{D}}_i^f, \mathscr{D}_i^f)$ are parameterized as $(\varepsilon, \mathscr{Y}_i, \breve{\mathscr{Z}}_i, \mathscr{Z}_i)$, whereby $\mathscr{Y}_i \triangleq \mathscr{H}_i(\varepsilon)$, $\breve{\mathscr{Z}}_i \triangleq \breve{\mathscr{D}}_i(\varepsilon)$ and $\mathscr{Z}_i \triangleq \mathscr{D}_i(\varepsilon)$. Thus, the value function for this optimization problem is now depending on the parameterization of terminal-value conditions.

Definition 7 (Value Function). Suppose $(\varepsilon, \mathscr{Y}_i, \breve{\mathscr{Z}}_i, \mathscr{Z}_i) \in [t_0, t_f] \times (\mathbb{R}^{n_i \times n_i})^{k_i} \times (\mathbb{R}^{n_i})^{k_i} \times \mathbb{R}^{k_i}$ is given and fixed. Then, the value function $\mathscr{V}_i(\varepsilon, \mathscr{Y}_i, \breve{\mathscr{Z}}_i, \mathscr{Z}_i)$ is defined by

$$\mathscr{V}_i(\varepsilon, \mathscr{Y}_i, \breve{\mathscr{Z}}_i, \mathscr{Z}_i) \triangleq \inf_{K_{x_i}, K_{z_i}, p_{x_i}, p_{z_i}} \phi_i^0(K_{x_i}, K_{z_i}, p_{x_i}, p_{z_i}; u_{-i}^*).$$

For convention, $\mathscr{V}_i(\varepsilon, \mathscr{Y}_i, \breve{\mathscr{Z}}_i, \mathscr{Z}_i) \triangleq \infty$ when $\mathscr{K}^{x_i}_{t_f, \mathscr{H}_i^f, \breve{\mathscr{D}}_i^f, \mathscr{D}_i^f; \mu_i} \times \mathscr{K}^{z_i}_{t_f, \mathscr{H}_i^f, \breve{\mathscr{D}}_i^f, \mathscr{D}_i^f; \mu_i} \times \mathscr{P}^{x_i}_{t_f, \mathscr{H}_i^f, \breve{\mathscr{D}}_i^f, \mathscr{D}_i^f; \mu_i} \times \mathscr{P}^{z_i}_{t_f, \mathscr{H}_i^f, \breve{\mathscr{D}}_i^f, \mathscr{D}_i^f; \mu_i}$ is empty. Next, some candidates for the value function are constructed with the help of the concept of reachable set.

Definition 8 (Reachable Sets). Let a reachable set be defined by $\mathscr{Q}_i \triangleq \Big\{(\varepsilon, \mathscr{Y}_i, \breve{\mathscr{Z}}_i, \mathscr{Z}_i) \in [t_0, t_f] \times (\mathbb{R}^{n_i \times n_i})^{k_i} \times (\mathbb{R}^{n_i})^{k_i} \times \mathbb{R}^{k_i}$ such that the Cartesian product $\mathscr{K}^{x_i}_{t_f, \mathscr{H}_i^f, \breve{\mathscr{D}}_i^f, \mathscr{D}_i^f; \mu_i} \times \mathscr{K}^{z_i}_{t_f, \mathscr{H}_i^f, \breve{\mathscr{D}}_i^f, \mathscr{D}_i^f; \mu_i} \times \mathscr{P}^{x_i}_{t_f, \mathscr{H}_i^f, \breve{\mathscr{D}}_i^f, \mathscr{D}_i^f; \mu_i} \times \mathscr{P}^{z_i}_{t_f, \mathscr{H}_i^f, \breve{\mathscr{D}}_i^f, \mathscr{D}_i^f; \mu_i} \neq \emptyset\Big\}$.

Moreover, it can be shown that the value function associated with agent i is satisfying a partial differential equation at interior points of $\mathscr{Q}_i$, at which it is differentiable.

Theorem 3 (Hamilton–Jacobi–Bellman Equation–Mayer Problem). *Let $(\varepsilon, \mathscr{Y}_i, \breve{\mathscr{Z}}_i, \mathscr{Z}_i)$ be any interior point of the reachable set $\mathscr{Q}_i$, at which the value function $\mathscr{V}_i(\varepsilon, \mathscr{Y}_i, \breve{\mathscr{Z}}_i, \mathscr{Z}_i)$ is differentiable. If there exists a feedback Nash strategy which is supported by $K_{x_i}^*(\cdot) \in \mathscr{K}^{x_i}_{t_f, \mathscr{H}_i^f, \breve{\mathscr{D}}_i^f, \mathscr{D}_i^f; \mu_i}$, $K_{z_i}^*(\cdot) \in \mathscr{K}^{z_i}_{t_f, \mathscr{H}_i^f, \breve{\mathscr{D}}_i^f, \mathscr{D}_i^f; \mu_i}$, $p_{x_i}^*(\cdot) \in \mathscr{P}^{x_i}_{t_f, \mathscr{H}_i^f, \breve{\mathscr{D}}_i^f, \mathscr{D}_i^f; \mu_i}$ and $p_{z_i}^*(\cdot) \in \mathscr{P}^{z_i}_{t_f, \mathscr{H}_i^f, \breve{\mathscr{D}}_i^f, \mathscr{D}_i^f; \mu_i}$, then the differential equation*

$$\begin{aligned} 0 = \min_{K_{x_i} \in \overline{K}^{x_i}, K_{z_i} \in \overline{K}^{z_i}, p_{x_i} \in \overline{P}^{x_i}, p_{z_i} \in \overline{P}^{z_i}} & \Big\{ \frac{\partial}{\partial \varepsilon} \mathscr{V}_i(\varepsilon, \mathscr{Y}_i, \breve{\mathscr{Z}}_i, \mathscr{Z}_i) \\ & + \frac{\partial}{\partial \operatorname{vec}(\mathscr{Y}_i)} \mathscr{V}_i(\varepsilon, \mathscr{Y}_i, \breve{\mathscr{Z}}_i, \mathscr{Z}_i) \operatorname{vec}(\mathscr{F}_i(\varepsilon, \mathscr{Y}_i, K_{x_i}, K_{z_i})) \\ & + \frac{\partial}{\partial \operatorname{vec}(\breve{\mathscr{Z}}_i)} \mathscr{V}_i(\varepsilon, \mathscr{Y}_i, \breve{\mathscr{Z}}_i, \mathscr{Z}_i) \operatorname{vec}(\breve{\mathscr{G}}_i(\varepsilon, \mathscr{Y}_i, \breve{\mathscr{Z}}_i, K_{x_i}, K_{z_i}, p_{x_i}, p_{z_i}) \\ & + \frac{\partial}{\partial \operatorname{vec}(\mathscr{Z}_i)} \mathscr{V}_i(\varepsilon, \mathscr{Y}_i, \breve{\mathscr{Z}}_i, \mathscr{Z}_i) \operatorname{vec}(\mathscr{G}_i(\varepsilon, \mathscr{Y}_i, \breve{\mathscr{Z}}_i, p_{x_i}, p_{z_i}) \Big\} \end{aligned} \tag{51}$$

is satisfied whereby $\mathscr{V}_i(t_0, \mathscr{Y}_i(t_0), \breve{\mathscr{Z}}_i(t_0), \mathscr{Z}_i(t_0)) = \phi_i^0(\mathscr{H}_i(t_0), \breve{\mathscr{D}}_i(t_0), \mathscr{D}_i(t_0))$.

Proof. By what have been shown in the recent results by the author [7], the proof for the result herein is readily proven. □

Finally, the following result gives the sufficient condition used to verify a feedback Nash strategy for incumbent agent i and $i \in \bar{I}$.

Theorem 4 (Verification Theorem). *Let $\mathscr{W}_i(\varepsilon, \mathscr{Y}_i, \breve{\mathscr{Z}}_i, \mathscr{Z}_i)$ be continuously differentiable solution of the Hamilton–Jacobi–Bellman (HJB) equation (51), which satisfies the boundary condition*

$$\mathscr{W}_i(t_0, \mathscr{H}_i(t_0), \breve{\mathscr{D}}_i(t_0), \mathscr{D}_i(t_0)) = \phi_i^0(t_0, \mathscr{H}_i(t_0), \breve{\mathscr{D}}_i(t_0), \mathscr{D}_i(t_0)) \ . \tag{52}$$

Let $(t_f, \mathscr{H}_i^f, \breve{\mathscr{D}}_i^f, \mathscr{D}_i^f)$ be a 4-tuple point in $\mathscr{Q}_i$; let $K_{x_i} \in \mathscr{K}^{x_i}_{t_f, \mathscr{H}_i^f, \breve{\mathscr{D}}_i^f, \mathscr{D}_i^f; \mu_i}$, $K_{z_i} \in \mathscr{K}^{z_i}_{t_f, \mathscr{H}_i^f, \breve{\mathscr{D}}_i^f, \mathscr{D}_i^f; \mu_i}$, $p_{x_i} \in \mathscr{P}^{x_i}_{t_f, \mathscr{H}_i^f, \breve{\mathscr{D}}_i^f, \mathscr{D}_i^f; \mu_i}$, $p_{z_i} \in \mathscr{P}^{z_i}_{t_f, \mathscr{H}_i^f, \breve{\mathscr{D}}_i^f, \mathscr{D}_i^f; \mu_i}$; and let $\mathscr{H}_i(\cdot)$, $\breve{\mathscr{D}}_i(\cdot)$ and $\mathscr{D}_i(\cdot)$ be the corresponding solutions of the equations of motion (44)–(46). Then, $\mathscr{W}_i(s, \mathscr{H}_i(s), \breve{\mathscr{D}}_i(s), \mathscr{D}_i(s))$ is time-backward increasing function of s.

If $K_{x_i}^* \in \mathscr{K}^{x_i}_{t_f, \mathscr{H}_i^f, \breve{\mathscr{D}}_i^f, \mathscr{D}_i^f; \mu_i}$, $K_{z_i}^* \in \mathscr{K}^{z_i}_{t_f, \mathscr{H}_i^f, \breve{\mathscr{D}}_i^f, \mathscr{D}_i^f; \mu_i}$, $p_{x_i}^* \in \mathscr{P}^{x_i}_{t_f, \mathscr{H}_i^f, \breve{\mathscr{D}}_i^f, \mathscr{D}_i^f; \mu_i}$ and $p_{z_i}^* \in \mathscr{P}^{z_i}_{t_f, \mathscr{H}_i^f, \breve{\mathscr{D}}_i^f, \mathscr{D}_i^f; \mu_i}$ defining a person-by-person equilibrium or feedback Nash strategy for agent i with the corresponding solutions $\mathscr{H}_i^*(\cdot)$, $\breve{\mathscr{D}}_i^*(\cdot)$ and $\mathscr{D}_i^*(\cdot)$ of the dynamical equations (44)–(46) such that, for $s \in [t_0, t_f]$

$$\begin{aligned}
0 = {} & \frac{\partial}{\partial \varepsilon} \mathscr{W}_i(s, \mathscr{H}_i^*(s), \breve{\mathscr{D}}_i^*(s), \mathscr{D}_i^*(s)) + \frac{\partial}{\partial \operatorname{vec}(\mathscr{Y}_i)} \mathscr{W}_i(s, \mathscr{H}_i^*(s), \breve{\mathscr{D}}_i^*(s), \mathscr{D}_i^*(s)) \\
& \cdot \operatorname{vec}(\mathscr{F}_i(s, \mathscr{Y}_i^*(s), K_{x_i}^*(s), K_{z_i}^*(s))) + \frac{\partial}{\partial \operatorname{vec}(\breve{\mathscr{Z}}_i)} \mathscr{W}_i(s, \mathscr{H}_i^*(s), \breve{\mathscr{D}}_i^*(s), \mathscr{D}_i^*(s)) \\
& \cdot \operatorname{vec}(\breve{\mathscr{G}}_i(s, \mathscr{H}_i^*(s), \breve{\mathscr{D}}_i^*(s), K_{x_i}^*(s), K_{z_i}^*(s), p_{x_i}^*(s), p_{z_i}^*(s)) \\
& + \frac{\partial}{\partial \operatorname{vec}(\mathscr{Z}_i)} \mathscr{W}_i(s, \mathscr{H}_i^*(s), \breve{\mathscr{D}}_i^*(s), \mathscr{D}_i^*(s)) \operatorname{vec}(\mathscr{G}_i(s, \mathscr{H}_i^*(s), \breve{\mathscr{D}}_i^*(s), p_{x_i}^*(s), p_{z_i}^*(s))
\end{aligned} \tag{53}$$

then $(K_{x_i}^*, K_{z_i}^*, p_{x_i}^*, p_{z_i}^*)$ results in a feedback Nash strategy or person-by-person equilibrium for agent i in $\mathscr{K}^{x_i}_{t_f, \mathscr{H}_i^f, \breve{\mathscr{D}}_i^f, \mathscr{D}_i^f; \mu_i} \times \mathscr{K}^{z_i}_{t_f, \mathscr{H}_i^f, \breve{\mathscr{D}}_i^f, \mathscr{D}_i^f; \mu_i} \times \mathscr{P}^{x_i}_{t_f, \mathscr{H}_i^f, \breve{\mathscr{D}}_i^f, \mathscr{D}_i^f; \mu_i} \times \mathscr{P}^{z_i}_{t_f, \mathscr{H}_i^f, \breve{\mathscr{D}}_i^f, \mathscr{D}_i^f; \mu_i}$. Furthermore, it follows that

$$\mathscr{W}_i(\varepsilon, \mathscr{Y}_i, \breve{\mathscr{Z}}_i, \mathscr{Z}_i) = \mathscr{V}_i(\varepsilon, \mathscr{Y}_i, \breve{\mathscr{Z}}_i, \mathscr{Z}_i), \tag{54}$$

whereby $\mathscr{V}_i(\varepsilon, \mathscr{Y}_i, \breve{\mathscr{Z}}_i, \mathscr{Z}_i)$ is the value function associated with incumbent agent i.

Proof. With the aid of the recent development [7], the proof then follows for the verification theorem herein. □

4 Distributed Person-by-Person Equilibrium Strategies

Reflecting on the Mayer-form optimization problem of the person-by-person equilibrium strategy concerned by incumbent agent i and $i \in \overline{I}$, the technical approach is to apply an adaptation of the Mayer-form verification theorem of dynamic programming as given in [11]. In the framework of dynamic programming, it is often required to denote the terminal time and states of a family of optimization problems as $(\varepsilon, \mathscr{Y}_i, \breve{\mathscr{Z}}_i, \mathscr{Z}_i)$ rather than $(t_f, \mathscr{H}_i^f, \breve{\mathscr{D}}_i^f, \mathscr{D}_i^f)$. Stating precisely, for $\varepsilon \in [t_0, t_f]$ and $1 \le r \le k_i$, the states of the performance robustness (44)–(46) defined on the interval $[t_0, \varepsilon]$ have the terminal values denoted by $\mathscr{H}_i(\varepsilon) \equiv \mathscr{Y}_i$, $\breve{\mathscr{D}}_i(\varepsilon) \equiv \breve{\mathscr{Z}}_i$ and $\mathscr{D}_i(\varepsilon) \equiv \mathscr{Z}_i$.

Since the performance index (47) is quadratic affine in terms of arbitrarily fixed x_i^0, the resulting insight suggests a solution to the adapted Hamilton–Jacobi–Bellman equation (51) is of the form as follows. It is assumed that $(\varepsilon, \mathscr{Y}_i, \breve{\mathscr{Z}}_i, \mathscr{Z}_i)$ is any interior point of the reachable set $\mathscr{Q}_i$ at which the real-valued function

$$\mathscr{W}_i(\varepsilon, \mathscr{Y}_i, \breve{\mathscr{Z}}_i, \mathscr{Z}_i) = (x_i^0)^T \sum_{r=1}^{k_i} \mu_r^i (\mathscr{Y}_i^r + \mathscr{E}_i^r(\varepsilon)) x_i^0 + 2(x_i^0)^T \sum_{r=1}^{k_i} \mu_r^i (\breve{\mathscr{Z}}_i^r + \breve{\mathscr{T}}_i^r(\varepsilon)) + \sum_{r=1}^{k_i} \mu_r^i (\mathscr{Z}_i^r + \mathscr{T}_i^r(\varepsilon)) \quad (55)$$

is differentiable. The parametric functions of time $\mathscr{E}_i^r \in \mathscr{C}^1(t_0, t_f; \mathbb{R}^{n_i \times n_i})$, $\breve{\mathscr{T}}_i^r \in \mathscr{C}^1(t_0, t_f; \mathbb{R}_i^n)$ and $\mathscr{T}_i^r \in \mathscr{C}^1(t_0, t_f; \mathbb{R})$ are yet to be determined. Furthermore, the time derivative of $\mathscr{W}_i(\varepsilon, \mathscr{Y}_i, \breve{\mathscr{Z}}_i, \mathscr{Z}_i)$ can be shown to be

$$\begin{aligned}\frac{d}{d\varepsilon}\mathscr{W}_i(\varepsilon, \mathscr{Y}_i, \breve{\mathscr{Z}}_i, \mathscr{Z}_i) &= (x_i^0)^T \sum_{r=1}^{k_i} \mu_r^i \Big(\mathscr{F}_i^r(\varepsilon, \mathscr{Y}_i, K_{x_i}, K_{z_i}) + \frac{d}{d\varepsilon}\mathscr{E}_i^r(\varepsilon)\Big) x_i^0 \\ &\quad + 2(x_i^0)^T \sum_{r=1}^{k_i} \mu_r^i \Big(\breve{\mathscr{G}}_i^r(\varepsilon, \mathscr{Y}_i, \breve{\mathscr{Z}}_i, K_{x_i}, K_{z_i}, p_{x_i}, p_{z_i}) + \frac{d}{d\varepsilon}\breve{\mathscr{T}}_i^r(\varepsilon)\Big) \\ &\quad + \sum_{r=1}^{k_i} \mu_r^i \Big(\mathscr{G}_i^r(\varepsilon, \mathscr{Y}_i, \breve{\mathscr{Z}}_i, p_{x_i}, p_{z_i}) + \frac{d}{d\varepsilon}\mathscr{T}_i^r(\varepsilon)\Big). \end{aligned} \quad (56)$$

The substitution of this hypothesized solution (55) into the HJB equation (51) and making use of (56) results in

$$0 \equiv \min_{K_{x_i} \in \overline{K}^{x_i}, K_{z_i} \in \overline{K}^{z_i}, p_{x_i} \in \overline{P}^{x_i}, p_{z_i} \in \overline{P}^{z_i}} \left\{ (x_i^0)^T \sum_{r=1}^{k_i} \mu_r^i \frac{d}{d\varepsilon} \mathscr{E}_i^r(\varepsilon) x_i^0 + 2(x_i^0)^T \sum_{r=1}^{k_i} \mu_r^i \frac{d}{d\varepsilon} \breve{\mathscr{T}}_i^r(\varepsilon) \right.$$
$$+ \sum_{r=1}^{k_i} \mu_r^i \frac{d}{d\varepsilon} \mathscr{T}_i^r(\varepsilon) + 2(x_i^0)^T \sum_{r=1}^{k_i} \mu_r^i \breve{\mathscr{G}}_i^r(\varepsilon, \mathscr{Y}_i, \breve{\mathscr{Z}}_i, K_{x_i}, K_{z_i}, p_{x_i}, p_{z_i})$$
$$\left. + (x_i^0)^T \sum_{r=1}^{k_i} \mu_r^i \mathscr{F}_i^r(\varepsilon, \mathscr{Y}_i, K_{x_i}, K_{z_i}) x_i^0 + \sum_{r=1}^{k_i} \mu_r^i \mathscr{G}_i^r(\varepsilon, \mathscr{Y}_i, \breve{\mathscr{Z}}_i, p_{x_i}, p_{z_i}) \right\}. \quad (57)$$

Differentiating the expression within the bracket of (57) with respect to K_{x_i}, K_{z_i}, p_{x_i} and p_{z_i} yields the necessary conditions for an extremum of (51) on $[t_0, \varepsilon]$,

$$0 \equiv \left[B_{ii}^T(\varepsilon) \sum_{r=1}^{k_i} \mu_r^i \mathscr{Y}_i^r + \mu_i^1 R_{ii}(\varepsilon) K_{x_i} \right] x_i^0 (x_i^0)^T + \left[B_{ii}^T(\varepsilon) \sum_{r=1}^{k_i} \mu_i^r \breve{\mathscr{Z}}_i^r + \mu_i^1 R_{ii}(\varepsilon) p_{x_i} \right] (x_i^0)^T$$

$$0 \equiv \left[C_{ii}^T(\varepsilon) \sum_{r=1}^{k_i} \mu_i^r \mathscr{Y}_i^r + \mu_i^1 R_{zi}(\varepsilon) K_{z_i} \right] x_i^0 (x_i^0)^T + \left[C_{ii}^T(\varepsilon) \sum_{r=1}^{k_i} \mu_i^r \breve{\mathscr{Z}}_i^r + \mu_i^1 R_{zi}(\varepsilon) p_{z_i} \right] (x_i^0)^T$$

$$0 \equiv \left[B_{ii}^T(\varepsilon) \sum_{r=1}^{k_i} \mu_i^r \mathscr{Y}_i^r + \mu_i^1 R_{ii}(\varepsilon) K_{x_i} \right] x_i^0 + \left[B_{ii}^T(\varepsilon) \sum_{r=1}^{k_i} \mu_i^r \breve{\mathscr{Z}}_i^r + \mu_i^1 R_{ii}(\varepsilon) p_{x_i} \right]$$

$$0 \equiv \left[C_{ii}^T(\varepsilon) \sum_{r=1}^{k_i} \mu_i^r \mathscr{Y}_i^r + \mu_i^1 R_{zi}(\varepsilon) K_{z_i} \right] x_i^0 + \left[C_{ii}^T(\varepsilon) \sum_{r=1}^{k_i} \mu_i^r \breve{\mathscr{Z}}_i^r + \mu_i^1 R_{zi}(\varepsilon) p_{z_i} \right].$$

Because all $x_i^0 (x_i^0)^T$, $(x_i^0)^T$ and x_i^0 have arbitrary ranks of one, it must be true that

$$K_{x_i} = -(\mu_i^1 R_{ii}(\varepsilon))^{-1} B_{ii}^T(\varepsilon) \sum_{r=1}^{k_i} \mu_i^r \mathscr{Y}_i^r, \quad (58)$$

$$K_{z_i} = -(\mu_i^1 R_{zi}(\varepsilon))^{-1} C_{ii}^T(\varepsilon) \sum_{r=1}^{k_i} \mu_i^r \mathscr{Y}_i^r, \quad (59)$$

$$p_{x_i} = -(\mu_i^1 R_{ii}(\varepsilon))^{-1} B_{ii}^T(\varepsilon) \sum_{r=1}^{k_i} \mu_i^r \breve{\mathscr{Z}}_i^r, \quad (60)$$

$$p_{z_i} = -(\mu_i^1 R_{zi}(\varepsilon))^{-1} C_{ii}^T(\varepsilon) \sum_{r=1}^{k_i} \mu_i^r \breve{\mathscr{Z}}_i^r. \quad (61)$$

Replacing these results (58)–(61) into the right member of the HJB equation (51) yields the value of the minimum whose mathematical details are omitted herein for the purpose of brevity.

For each agent i and $i \in \overline{I}$, it is necessary to exhibit $\{\mathscr{E}_i^r(\cdot)\}_{r=1}^{k_i}$, $\{\breve{\mathscr{T}}_i^r(\cdot)\}_{r=1}^{k_i}$ and $\{\mathscr{T}_i^r(\cdot)\}_{r=1}^{k_i}$ which render the left side of the HJB equation (51) equal to zero for

$\varepsilon \in [t_0, t_f]$, when $\{\mathscr{Y}_i^r\}_{r=1}^{k_i}$, $\{\breve{\mathscr{Z}}_i^r\}_{r=1}^{k_i}$ and $\{\mathscr{Z}_i^r\}_{r=1}^{k_i}$ are evaluated along the solution trajectories of the dynamical equations (44)–(46). With a careful examination of the expression (57), it reveals that

$$\begin{aligned}
\frac{d}{d\varepsilon}\mathscr{E}_i^1(\varepsilon) = \Big[& A_{ii}(\varepsilon) - B_{ii}(\varepsilon)R_{ii}^{-1}(\varepsilon)B_{ii}^T(\varepsilon)\sum_{s=1}^{k_i}\frac{\mu_i^s}{\mu_i^1}(\mathscr{H}_i^s)_{11}(\varepsilon) \\
& - C_{ii}(\varepsilon)R_{zi}^{-1}(\varepsilon)C_{ii}^T(\varepsilon)\sum_{s=1}^{k_i}\frac{\mu_i^s}{\mu_i^1}(\mathscr{H}_i^s)_{11}(\varepsilon)\Big]^T(\mathscr{H}_i^1)_{11}(\varepsilon) \\
& + (\mathscr{H}_i^1)_{11}(\varepsilon)\Big[A_{ii}(\varepsilon) - B_{ii}(\varepsilon)R_{ii}^{-1}(\varepsilon)B_{ii}^T(\varepsilon)\sum_{s=1}^{k_i}\frac{\mu_i^s}{\mu_i^1}(\mathscr{H}_i^s)_{11}(\varepsilon) \\
& - C_{ii}(\varepsilon)R_{zi}^{-1}(\varepsilon)C_{ii}^T(\varepsilon)\sum_{s=1}^{k_i}\frac{\mu_i^s}{\mu_i^1}\mathscr{H}_i^s(\varepsilon)\Big] + Q_{ii}(\varepsilon) + Q_i(\varepsilon) \\
& + \sum_{s=1}^{k_i}\frac{\mu_i^s}{\mu_i^1}(\mathscr{H}_i^s)_{11}(\varepsilon)B_{ii}(\varepsilon)R_{ii}^{-1}(\varepsilon)B_{ii}^T(\varepsilon)\sum_{r=1}^{k_i}\frac{\mu_i^r}{\mu_i^1}(\mathscr{H}_i^r)_{11}(\varepsilon) \\
& + \sum_{s=1}^{k_i}\frac{\mu_i^s}{\mu_i^1}(\mathscr{H}_i^s)_{11}(\varepsilon)C_{ii}(\varepsilon)R_{zi}^{-1}(\varepsilon)C_{ii}^T(\varepsilon)\sum_{r=1}^{k_i}\frac{\mu_i^r}{\mu_i^1}(\mathscr{H}_i^r)_{11}(\varepsilon) \qquad (62)
\end{aligned}$$

$$\begin{aligned}
\frac{d}{d\varepsilon}\mathscr{E}_i^r(\varepsilon) = \Big[& A_{ii}(\varepsilon) - B_{ii}(\varepsilon)R_{ii}^{-1}(\varepsilon)B_{ii}^T(\varepsilon)\sum_{s=1}^{k_i}\frac{\mu_i^s}{\mu_i^1}(\mathscr{H}_i^s)_{11}(\varepsilon) \\
& - C_{ii}(\varepsilon)R_{zi}^{-1}(\varepsilon)C_{ii}^T(\varepsilon)\sum_{s=1}^{k_i}\frac{\mu_i^s}{\mu_i^1}(\mathscr{H}_i^s)_{11}(\varepsilon)\Big]^T(\mathscr{H}_i^r)_{11}(\varepsilon) + (\mathscr{H}_i^r)_{11}(\varepsilon)\Big[A_{ii}(\varepsilon) \\
& - B_{ii}(\varepsilon)R_{ii}^{-1}(\varepsilon)B_{ii}^T(\varepsilon)\sum_{s=1}^{k_i}\frac{\mu_i^s}{\mu_i^1}(\mathscr{H}_i^s)_{11}(\varepsilon) - C_{ii}(\varepsilon)R_{zi}^{-1}(\varepsilon)C_{ii}^T(\varepsilon)\sum_{s=1}^{k_i}\frac{\mu_i^s}{\mu_i^1}(\mathscr{H}_i^s)_{11}(\varepsilon)\Big] \\
& + \sum_{v=1}^{r-1}\frac{2r!}{v!(r-v)!}\Big\{(\mathscr{H}_i^v)_{11}(\varepsilon)L_i(\varepsilon)V_iL_i^T(\varepsilon)(\mathscr{H}_i^{r-v})_{11}(\varepsilon) \\
& - (\mathscr{H}_i^v)_{12}(\varepsilon)L_i(\varepsilon)V_iL_i^T(\varepsilon)(\mathscr{H}_i^{r-v})_{11}(\varepsilon) - (\mathscr{H}_i^v)_{11}(\varepsilon)L_i(\varepsilon)V_iL_i^T(\varepsilon)(\mathscr{H}_i^{r-v})_{21}(\varepsilon) \\
& + (\mathscr{H}_i^v)_{12}(\varepsilon)G_i(\varepsilon)W_iG_i^T(\varepsilon)\mathscr{H}_i^{r-v})_{21}(\varepsilon) + (\mathscr{H}_i^v)_{12}(\varepsilon)L_i(\varepsilon)V_iL_i^T(\varepsilon)(\mathscr{H}_i^{r-v})_{21}(\varepsilon) \\
& + (\mathscr{H}_i^v)_{13}(\varepsilon)L_{ic}(\varepsilon)V_{ic}L_{ic}^T(\varepsilon)(\mathscr{H}_i^{r-v})_{31}(\varepsilon) - (\mathscr{H}_i^v)_{14}(\varepsilon)L_{ic}(\varepsilon)V_{ic}L_{ic}^T(\varepsilon)(\mathscr{H}_i^{r-v})_{31}(\varepsilon) \\
& - (\mathscr{H}_i^v)_{13}(\varepsilon)L_{ic}(\varepsilon)V_{ic}L_{ic}^T(\varepsilon)(\mathscr{H}_i^{r-v})_{41}(\varepsilon) + (\mathscr{H}_i^v)_{14}(\varepsilon)G_{ic}(\varepsilon)W_{ic}G_{ic}^T(\varepsilon)(\mathscr{H}_i^{r-v})_{41}(\varepsilon) \\
& + (\mathscr{H}_i^v)_{14}(\varepsilon)L_{ic}(\varepsilon)L_{ic}L_{ic}^T(\varepsilon)(\mathscr{H}_i^{r-v})_{41}(\varepsilon)\Big\}, \quad 2 \le r \le k_i \qquad (63)
\end{aligned}$$

$$\frac{d}{d\varepsilon}\breve{\mathscr{T}}_i^1(\varepsilon) = \Big[A_{ii}(\varepsilon) - B_{ii}(\varepsilon)R_{ii}^{-1}(\varepsilon)B_{ii}^T(\varepsilon)\sum_{s=1}^{k_i}\frac{\mu_i^s}{\mu_i^1}(\mathscr{H}_i^s)_{11}(\varepsilon)$$

$$- C_{ii}(\varepsilon)R_{zi}^{-1}(\varepsilon)C_{ii}^T(\varepsilon)\sum_{s=1}^{k_i}\frac{\mu_i^s}{\mu_i^1}(\mathscr{H}_i^s)_{11}(\varepsilon)\Big]^T(\mathscr{D}_i^1)_{11}(\varepsilon)$$

$$+ (\mathscr{H}_i^1)_{11}(\varepsilon)\Big[-B_{ii}(\varepsilon)R_{ii}^{-1}(\varepsilon)B_{ii}^T(\varepsilon)\sum_{s=1}^{k_i}\frac{\mu_i^s}{\mu_i^1}(\breve{\mathscr{D}}_i^s)_{11}(\varepsilon)$$

$$- C_{ii}(\varepsilon)R_{zi}^{-1}(\varepsilon)C_{ii}^T(\varepsilon)\sum_{s=1}^{k_i}\frac{\mu_i^s}{\mu_i^1}(\breve{\mathscr{D}}_i^s)_{11}(\varepsilon) + \sum_{j=1}^{N_i}B_{ij}(\varepsilon)u_{ij}^*(\varepsilon)\Big] + (\mathscr{H}_i^1)_{13}(\varepsilon)B_{ic}(\varepsilon)u_c(\varepsilon)$$

$$+ \sum_{s=1}^{k_i}\frac{\mu_i^s}{\mu_i^1}(\mathscr{H}_i^s)_{11}(\varepsilon)B_{ii}(\varepsilon)R_{ii}^{-1}(\varepsilon)B_{ii}^T(\varepsilon)\sum_{r=1}^{k_i}\frac{\mu_i^r}{\mu_i^1}(\breve{\mathscr{D}}_i^r)_{11}(\varepsilon)$$

$$+ \sum_{s=1}^{k_i}\frac{\mu_i^s}{\mu_i^1}(\mathscr{H}_i^s)_{11}(\varepsilon)C_{ii}(\varepsilon)R_{zi}^{-1}(\varepsilon)C_{ii}^T(\varepsilon)\sum_{r=1}^{k_i}\frac{\mu_i^r}{\mu_i^1}(\breve{\mathscr{D}}_i^r)_{11}(\varepsilon) - Q_i(\varepsilon)\zeta_i(\varepsilon) \quad (64)$$

$$\frac{d}{d\varepsilon}\breve{\mathscr{T}}_i^r(\varepsilon) = \Big[A_{ii}(\varepsilon) - B_{ii}(\varepsilon)R_{ii}^{-1}(\varepsilon)B_{ii}^T(\varepsilon)\sum_{s=1}^{k_i}\frac{\mu_i^s}{\mu_i^1}(\mathscr{H}_i^s)_{11}(\varepsilon)$$

$$- C_{ii}(\varepsilon)R_{zi}^{-1}(\varepsilon)C_{ii}^T(\varepsilon)\sum_{s=1}^{k_i}\frac{\mu_i^s}{\mu_i^1}(\mathscr{H}_i^s)_{11}(\varepsilon)\Big]^T(\mathscr{D}_i^r)_{11}(\varepsilon) + (\mathscr{H}_i^r)_{13}(\varepsilon)B_{ic}(\varepsilon)u_c(\varepsilon)$$

$$+ (\mathscr{H}_i^r)_{11}(\varepsilon)\Big[-B_{ii}(\varepsilon)R_{ii}^{-1}(\varepsilon)B_{ii}^T(\varepsilon)\sum_{s=1}^{k_i}\frac{\mu_i^s}{\mu_i^1}(\breve{\mathscr{D}}_i^s)_{11}(\varepsilon)$$

$$- C_{ii}(\varepsilon)R_{zi}^{-1}(\varepsilon)C_{ii}^T(\varepsilon)\sum_{s=1}^{k_i}\frac{\mu_i^s}{\mu_i^1}(\breve{\mathscr{D}}_i^s)_{11}(\varepsilon) + \sum_{j=1}^{N_i}B_{ij}(\varepsilon)u_{ij}^*(\varepsilon)\Big], \qquad 2 \le r \le k_i \quad (65)$$

$$\frac{d}{d\varepsilon}\mathscr{T}_i^1(\varepsilon) = \mathrm{Tr}\{(\mathscr{H}_i^1)_{11}(\varepsilon)L_i(\varepsilon)V_iL_i^T(\varepsilon) - (\mathscr{H}_i^1)_{12}(\varepsilon)L_i(\varepsilon)V_iL_i^T(\varepsilon)\}$$

$$+ \mathrm{Tr}\{-(\mathscr{H}_i^1)_{21}(\varepsilon)L_i(\varepsilon)V_iL_i^T(\varepsilon) + (\mathscr{H}_i^1)_{22}(\varepsilon)(G_i(\varepsilon)W_iG_i^T(\varepsilon) + L_i(\varepsilon)V_iL_i^T(\varepsilon))\}$$

$$+ \mathrm{Tr}\{(\mathscr{H}_i^1)_{33}(\varepsilon)L_{ic}(\varepsilon)V_{ic}L_{ic}^T(\varepsilon) - (\mathscr{H}_i^1)_{34}(\varepsilon)L_{ic}(\varepsilon)V_{ic}L_{ic}^T(\varepsilon)\}$$

$$+ \mathrm{Tr}\{(\mathscr{H}_i^1)_{44}(\varepsilon)(G_{ic}(\varepsilon)W_{ic}G_{ic}^T(\varepsilon) + L_{ic}(\varepsilon)V_{ic}L_{ic}^T(\varepsilon)) - (\mathscr{H}_i^1)_{43}(\varepsilon)L_{ic}(\varepsilon)V_{ic}L_{ic}^T(\varepsilon)\}$$

$$+ 2(\breve{\mathscr{D}}_i^1)_{11}^T(\varepsilon)\Big[-B_{ii}(\varepsilon)R_{ii}^{-1}(\varepsilon)B_{ii}^T(\varepsilon)\sum_{s=1}^{k_i}\frac{\mu_i^s}{\mu_i^1}(\breve{\mathscr{D}}_i^s)_{11}(\varepsilon)$$

$$-C_{ii}(\varepsilon)R_{zi}^{-1}(\varepsilon)C_{ii}^T(\varepsilon)\sum_{s=1}^{k_i}\frac{\mu_i^s}{\mu_i^1}(\breve{\mathscr{D}}_i^s)_{11}(\varepsilon)+\sum_{j=1}^{N_i}B_{ij}(\varepsilon)u_{ij}^*(\varepsilon)\Big]+2(\breve{\mathscr{D}}_i^1)_{13}^T(\varepsilon)B_{ic}(\varepsilon)u_c(\varepsilon)$$

$$+\zeta_i^T(\varepsilon)Q_i(\varepsilon)\zeta_i(\varepsilon)+\sum_{r=1}^{k_i}\frac{\mu_i^r}{\mu_i^1}(\breve{\mathscr{D}}_i^r)_{11}^T(\varepsilon)B_{ii}(\varepsilon)R_{ii}^{-1}(\varepsilon)B_{ii}^T(\varepsilon)\sum_{s=1}^{k_i}\frac{\mu_i^s}{\mu_i^1}(\breve{\mathscr{D}}_i^s)_{11}(\varepsilon)$$

$$+\sum_{r=1}^{k_i}\frac{\mu_i^r}{\mu_i^1}(\breve{\mathscr{D}}_i^r)_{11}^T(\varepsilon)C_{ii}(\varepsilon)R_{zi}^{-1}(\varepsilon)C_{ii}^T(\varepsilon)\sum_{s=1}^{k_i}\frac{\mu_i^s}{\mu_i^1}(\breve{\mathscr{D}}_i^s)_{11}(\varepsilon) \quad (66)$$

$$\frac{d}{d\varepsilon}\mathscr{T}_i^r(\varepsilon)=\mathrm{Tr}\{(\mathscr{H}_i^r)_{11}(\varepsilon)L_i(\varepsilon)V_iL_i^T(\varepsilon)-(\mathscr{H}_i^r)_{12}(\varepsilon)L_i(\varepsilon)V_iL_i^T(\varepsilon)\}$$

$$+\mathrm{Tr}\{-(\mathscr{H}_i^r)_{21}(\varepsilon)L_i(\varepsilon)V_iL_i^T(\varepsilon)+(\mathscr{H}_i^r)_{22}(\varepsilon)(G_i(\varepsilon)W_iG_i^T(\varepsilon)+L_i(\varepsilon)V_iL_i^T(\varepsilon))\}$$

$$+\mathrm{Tr}\{(\mathscr{H}_i^r)_{33}(\varepsilon)L_{ic}(\varepsilon)V_{ic}L_{ic}^T(\varepsilon)-(\mathscr{H}_i^r)_{34}(\varepsilon)L_{ic}(\varepsilon)V_{ic}L_{ic}^T(\varepsilon)\}$$

$$+\mathrm{Tr}\{(\mathscr{H}_i^r)_{44}(\varepsilon)(G_{ic}(\varepsilon)W_{ic}G_{ic}^T(\varepsilon)+L_{ic}(\varepsilon)V_{ic}L_{ic}^T(\varepsilon))-(\mathscr{H}_i^r)_{43}(\varepsilon)L_{ic}(\varepsilon)V_{ic}L_{ic}^T(\varepsilon)\}$$

$$+2(\mathscr{D}_i^r)_{13}^T(\varepsilon)B_{ic}(\varepsilon)u_c(\varepsilon)+2(\mathscr{D}_i^r)_{11}^T(\varepsilon)\Big[-B_{ii}(\varepsilon)R_{ii}^{-1}(\varepsilon)B_{ii}^T(\varepsilon)\sum_{s=1}^{k_i}\frac{\mu_i^s}{\mu_i^1}(\breve{\mathscr{D}}_i^s)_{11}(\varepsilon)$$

$$-C_{ii}(\varepsilon)R_{zi}^{-1}(\varepsilon)C_{ii}^T(\varepsilon)\sum_{s=1}^{k_i}\frac{\mu_i^s}{\mu_i^1}(\breve{\mathscr{D}}_i^s)_{11}(\varepsilon)+\sum_{j=1}^{N_i}B_{ij}(\varepsilon)u_{ij}^*(\varepsilon)\Big], \quad 2\le r\le k_i \quad (67)$$

will work. Furthermore, the boundary condition associated with the verification theorem requires that

$$(x_i^0)^T\sum_{r=1}^{k_i}\mu_i^r((\mathscr{H}_i^r)_{11}(t_0)+\mathscr{E}_i^r(t_0))x_i^0+2(x_i^0)^T\sum_{r=1}^{k_i}\mu_i^r((\breve{\mathscr{D}}_i^r)_{11}(t_0)+\breve{\mathscr{T}}_i^r(t_0))$$

$$+\sum_{r=1}^{k_i}\mu_i^r(\mathscr{D}_i^r(t_0)+\mathscr{T}_i^r(t_0))$$

$$=(x_i^0)^T\sum_{r=1}^{k_i}\mu_i^r(\mathscr{H}_i^r)_{11}(t_0)x_i^0+2(x_i^0)^T\sum_{r=1}^{k_i}\mu_i^r(\breve{\mathscr{D}}_i^r)_{11}(t_0)+\sum_{r=1}^{k_i}\mu_i^r\mathscr{D}_i^r(t_0).$$

Thus, matching the boundary condition yields the initial value conditions $\mathscr{E}_i^r(t_0)=0$, $\breve{\mathscr{T}}_i^r(t_0)=0$ and $\mathscr{T}_i^r(t_0)=0$ for the forward-in-time differential equations (62)–(67).

Applying the 4-tuple $(K_{x_i},K_{z_i},p_{x_i},p_{z_i})$ in (58)–(61) that is defining the person-by-person equilibrium for each agent i and $i\in\bar{I}$ along the solution trajectories of the backward-in-time differential equations (44)–(46), these equations become the backward-in-time Riccati-type differential equations

$$
\begin{aligned}
\frac{d}{d\varepsilon}(\mathscr{H}_i^1)_{11}(\varepsilon) &= -\Big[A_{ii}(\varepsilon) - B_{ii}(\varepsilon)R_{ii}^{-1}(\varepsilon)B_{ii}^T(\varepsilon)\sum_{s=1}^{k_i}\frac{\mu_i^s}{\mu_i^1}(\mathscr{H}_i^s)_{11}(\varepsilon) \\
&\quad - C_{ii}(\varepsilon)R_{zi}^{-1}(\varepsilon)C_{ii}^T(\varepsilon)\sum_{s=1}^{k_i}\frac{\mu_i^s}{\mu_i^1}(\mathscr{H}_i^s)_{11}(\varepsilon)\Big]^T(\mathscr{H}_i^1)_{11}(\varepsilon) \\
&\quad - (\mathscr{H}_i^1)_{11}(\varepsilon)\Big[A_{ii}(\varepsilon) - B_{ii}(\varepsilon)R_{ii}^{-1}(\varepsilon)B_{ii}^T(\varepsilon)\sum_{s=1}^{k_i}\frac{\mu_i^s}{\mu_i^1}(\mathscr{H}_i^s)_{11}(\varepsilon) \\
&\quad - C_{ii}(\varepsilon)R_{zi}^{-1}(\varepsilon)C_{ii}^T(\varepsilon)\sum_{s=1}^{k_i}\frac{\mu_i^s}{\mu_i^1}\mathscr{H}_i^s(\varepsilon)\Big] - Q_{ii}(\varepsilon) - Q_i(\varepsilon) \\
&\quad - \sum_{s=1}^{k_i}\frac{\mu_i^s}{\mu_i^1}(\mathscr{H}_i^s)_{11}(\varepsilon)B_{ii}(\varepsilon)R_{ii}^{-1}(\varepsilon)B_{ii}^T(\varepsilon)\sum_{r=1}^{k_i}\frac{\mu_i^r}{\mu_i^1}(\mathscr{H}_i^r)_{11}(\varepsilon) \\
&\quad - \sum_{s=1}^{k_i}\frac{\mu_i^s}{\mu_i^1}(\mathscr{H}_i^s)_{11}(\varepsilon)C_{ii}(\varepsilon)R_{zi}^{-1}(\varepsilon)C_{ii}^T(\varepsilon)\sum_{r=1}^{k_i}\frac{\mu_i^r}{\mu_i^1}(\mathscr{H}_i^r)_{11}(\varepsilon) \qquad (68)
\end{aligned}
$$

$$
\begin{aligned}
\frac{d}{d\varepsilon}(\mathscr{H}_i^r)_{11}(\varepsilon) &= -\Big[A_{ii}(\varepsilon) - B_{ii}(\varepsilon)R_{ii}^{-1}(\varepsilon)B_{ii}^T(\varepsilon)\sum_{s=1}^{k_i}\frac{\mu_i^s}{\mu_i^1}(\mathscr{H}_i^s)_{11}(\varepsilon) \\
&\quad - C_{ii}(\varepsilon)R_{zi}^{-1}(\varepsilon)C_{ii}^T(\varepsilon)\sum_{s=1}^{k_i}\frac{\mu_i^s}{\mu_i^1}(\mathscr{H}_i^s)_{11}(\varepsilon)\Big]^T(\mathscr{H}_i^r)_{11}(\varepsilon) - (\mathscr{H}_i^r)_{11}(\varepsilon)\Big[A_{ii}(\varepsilon) \\
&\quad - B_{ii}(\varepsilon)R_{ii}^{-1}(\varepsilon)B_{ii}^T(\varepsilon)\sum_{s=1}^{k_i}\frac{\mu_i^s}{\mu_i^1}(\mathscr{H}_i^s)_{11}(\varepsilon) - C_{ii}(\varepsilon)R_{zi}^{-1}(\varepsilon)C_{ii}^T(\varepsilon)\sum_{s=1}^{k_i}\frac{\mu_i^s}{\mu_i^1}(\mathscr{H}_i^s)_{11}(\varepsilon)\Big] \\
&\quad - \sum_{v=1}^{r-1}\frac{2r!}{v!(r-v)!}\Big\{(\mathscr{H}_i^v)_{11}(\varepsilon)L_i(\varepsilon)V_iL_i^T(\varepsilon)(\mathscr{H}_i^{r-v})_{11}(\varepsilon) \\
&\quad - (\mathscr{H}_i^v)_{12}(\varepsilon)L_i(\varepsilon)V_iL_i^T(\varepsilon)(\mathscr{H}_i^{r-v})_{11}(\varepsilon) - (\mathscr{H}_i^v)_{11}(\varepsilon)L_i(\varepsilon)V_iL_i^T(\varepsilon)(\mathscr{H}_i^{r-v})_{21}(\varepsilon) \\
&\quad + (\mathscr{H}_i^v)_{12}(\varepsilon)G_i(\varepsilon)W_iG_i^T(\varepsilon)\mathscr{H}_i^{r-v})_{21}(\varepsilon) + (\mathscr{H}_i^v)_{12}(\varepsilon)L_i(\varepsilon)V_iL_i^T(\varepsilon)(\mathscr{H}_i^{r-v})_{21}(\varepsilon) \\
&\quad + (\mathscr{H}_i^v)_{13}(\varepsilon)L_{ic}(\varepsilon)V_{ic}L_{ic}^T(\varepsilon)(\mathscr{H}_i^{r-v})_{31}(\varepsilon) - (\mathscr{H}_i^v)_{14}(\varepsilon)L_{ic}(\varepsilon)V_{ic}L_{ic}^T(\varepsilon)(\mathscr{H}_i^{r-v})_{31}(\varepsilon) \\
&\quad - (\mathscr{H}_i^v)_{13}(\varepsilon)L_{ic}(\varepsilon)V_{ic}L_{ic}^T(\varepsilon)(\mathscr{H}_i^{r-v})_{41}(\varepsilon) + (\mathscr{H}_i^v)_{14}(\varepsilon)G_{ic}W_{ic}G_{ic}^T(\varepsilon)(\mathscr{H}_i^{r-v})_{41}(\varepsilon) \\
&\quad + (\mathscr{H}_i^v)_{14}(\varepsilon)L_{ic}(\varepsilon)L_{ic}L_{ic}^T(\varepsilon)(\mathscr{H}_i^{r-v})_{41}(\varepsilon)\Big\}, \qquad 2 \le r \le k_i \qquad (69)
\end{aligned}
$$

$$\frac{d}{d\varepsilon}(\breve{\mathscr{D}}_i^1)_{11}(\varepsilon) = -\Big[A_{ii}(\varepsilon) - B_{ii}(\varepsilon)R_{ii}^{-1}(\varepsilon)B_{ii}^T(\varepsilon)\sum_{s=1}^{k_i}\frac{\mu_i^s}{\mu_i^1}(\mathscr{H}_i^s)_{11}(\varepsilon)$$
$$-C_{ii}(\varepsilon)R_{zi}^{-1}(\varepsilon)C_{ii}^T(\varepsilon)\sum_{s=1}^{k_i}\frac{\mu_i^s}{\mu_i^1}(\mathscr{H}_i^s)_{11}(\varepsilon)\Big]^T(\mathscr{D}_i^1)_{11}(\varepsilon)$$
$$-(\mathscr{H}_i^1)_{11}(\varepsilon)\Big[-B_{ii}(\varepsilon)R_{ii}^{-1}(\varepsilon)B_{ii}^T(\varepsilon)\sum_{s=1}^{k_i}\frac{\mu_i^s}{\mu_i^1}(\breve{\mathscr{D}}_i^s)_{11}(\varepsilon)$$
$$-C_{ii}(\varepsilon)R_{zi}^{-1}(\varepsilon)C_{ii}^T(\varepsilon)\sum_{s=1}^{k_i}\frac{\mu_i^s}{\mu_i^1}(\breve{\mathscr{D}}_i^s)_{11}(\varepsilon) + \sum_{j=1}^{N_i}B_{ij}(\varepsilon)u_{ij}^*(\varepsilon)\Big] - (\mathscr{H}_i^1)_{13}(\varepsilon)B_{ic}(\varepsilon)u_c(\varepsilon)$$
$$-\sum_{s=1}^{k_i}\frac{\mu_i^s}{\mu_i^1}(\mathscr{H}_i^s)_{11}(\varepsilon)B_{ii}(\varepsilon)R_{ii}^{-1}(\varepsilon)B_{ii}^T(\varepsilon)\sum_{r=1}^{k_i}\frac{\mu_i^r}{\mu_i^1}(\breve{\mathscr{D}}_i^r)_{11}(\varepsilon)$$
$$-\sum_{s=1}^{k_i}\frac{\mu_i^s}{\mu_i^1}(\mathscr{H}_i^s)_{11}(\varepsilon)C_{ii}(\varepsilon)R_{zi}^{-1}(\varepsilon)C_{ii}^T(\varepsilon)\sum_{r=1}^{k_i}\frac{\mu_i^r}{\mu_i^1}(\breve{\mathscr{D}}_i^r)_{11}(\varepsilon) + Q_i(\varepsilon)\zeta_i(\varepsilon) \quad (70)$$

$$\frac{d}{d\varepsilon}(\breve{\mathscr{D}}_i^r)_{11}(\varepsilon) = -\Big[A_{ii}(\varepsilon) - B_{ii}(\varepsilon)R_{ii}^{-1}(\varepsilon)B_{ii}^T(\varepsilon)\sum_{s=1}^{k_i}\frac{\mu_i^s}{\mu_i^1}(\mathscr{H}_i^s)_{11}(\varepsilon)$$
$$-C_{ii}(\varepsilon)R_{zi}^{-1}(\varepsilon)C_{ii}^T(\varepsilon)\sum_{s=1}^{k_i}\frac{\mu_i^s}{\mu_i^1}(\mathscr{H}_i^s)_{11}(\varepsilon)\Big]^T(\mathscr{D}_i^r)_{11}(\varepsilon) - (\mathscr{H}_i^r)_{13}(\varepsilon)B_{ic}(\varepsilon)u_c(\varepsilon)$$
$$-(\mathscr{H}_i^r)_{11}(\varepsilon)\Big[-B_{ii}(\varepsilon)R_{ii}^{-1}(\varepsilon)B_{ii}^T(\varepsilon)\sum_{s=1}^{k_i}\frac{\mu_i^s}{\mu_i^1}(\breve{\mathscr{D}}_i^s)_{11}(\varepsilon)$$
$$-C_{ii}(\varepsilon)R_{zi}^{-1}(\varepsilon)C_{ii}^T(\varepsilon)\sum_{s=1}^{k_i}\frac{\mu_i^s}{\mu_i^1}(\breve{\mathscr{D}}_i^s)_{11}(\varepsilon) + \sum_{j=1}^{N_i}B_{ij}(\varepsilon)u_{ij}^*(\varepsilon)\Big], \qquad 2 \le r \le k_i \quad (71)$$

$$\frac{d}{d\varepsilon}\mathscr{D}_i^1(\varepsilon) = -\mathrm{Tr}\{(\mathscr{H}_i^1)_{11}(\varepsilon)L_i(\varepsilon)V_iL_i^T(\varepsilon) - (\mathscr{H}_i^1)_{12}(\varepsilon)L_i(\varepsilon)V_iL_i^T(\varepsilon)\}$$
$$-\mathrm{Tr}\{-(\mathscr{H}_i^1)_{21}(\varepsilon)L_i(\varepsilon)V_iL_i^T(\varepsilon) + (\mathscr{H}_i^1)_{22}(\varepsilon)(G_i(\varepsilon)W_iG_i^T(\varepsilon) + L_i(\varepsilon)V_iL_i^T(\varepsilon))\}$$
$$-\mathrm{Tr}\{(\mathscr{H}_i^1)_{33}(\varepsilon)L_{ic}(\varepsilon)V_{ic}L_{ic}^T(\varepsilon) - (\mathscr{H}_i^1)_{34}(\varepsilon)L_{ic}(\varepsilon)V_{ic}L_{ic}^T(\varepsilon)\}$$
$$-\mathrm{Tr}\{(\mathscr{H}_i^1)_{44}(\varepsilon)(G_{ic}(\varepsilon)W_{ic}G_{ic}^T(\varepsilon) + L_{ic}(\varepsilon)V_{ic}L_{ic}^T(\varepsilon)) - (\mathscr{H}_i^1)_{43}(\varepsilon)L_{ic}(\varepsilon)V_{ic}L_{ic}^T(\varepsilon)\}$$
$$-2(\breve{\mathscr{D}}_i^1)_{11}^T(\varepsilon)\Big[-B_{ii}(\varepsilon)R_{ii}^{-1}(\varepsilon)B_{ii}^T(\varepsilon)\sum_{s=1}^{k_i}\frac{\mu_i^s}{\mu_i^1}(\breve{\mathscr{D}}_i^s)_{11}(\varepsilon)$$

$$-C_{ii}(\varepsilon)R_{zi}^{-1}(\varepsilon)C_{ii}^{T}(\varepsilon)\sum_{s=1}^{k_i}\frac{\mu_i^s}{\mu_i^1}(\breve{\mathscr{D}}_i^s)_{11}(\varepsilon)+\sum_{j=1}^{N_i}B_{ij}(\varepsilon)u_{ij}^{*}(\varepsilon)\Big]-2(\breve{\mathscr{D}}_i^1)_{13}^T(\varepsilon)B_{ic}(\varepsilon)u_c(\varepsilon)$$

$$-\zeta_i^T(\varepsilon)Q_i(\varepsilon)\zeta_i(\varepsilon)-\sum_{r=1}^{k_i}\frac{\mu_i^r}{\mu_i^1}(\breve{\mathscr{D}}_i^r)_{11}^T(\varepsilon)B_{ii}(\varepsilon)R_{ii}^{-1}(\varepsilon)B_{ii}^T(\varepsilon)\sum_{s=1}^{k_i}\frac{\mu_i^s}{\mu_i^1}(\breve{\mathscr{D}}_i^s)_{11}(\varepsilon)$$

$$-\sum_{r=1}^{k_i}\frac{\mu_i^r}{\mu_i^1}(\breve{\mathscr{D}}_i^r)_{11}^T(\varepsilon)C_{ii}(\varepsilon)R_{zi}^{-1}(\varepsilon)C_{ii}^T(\varepsilon)\sum_{s=1}^{k_i}\frac{\mu_i^s}{\mu_i^1}(\breve{\mathscr{D}}_i^s)_{11}(\varepsilon) \quad (72)$$

$$\frac{d}{d\varepsilon}\mathscr{D}_i^r(\varepsilon)=-\mathrm{Tr}\{(\mathscr{H}_i^r)_{11}(\varepsilon)L_i(\varepsilon)V_iL_i^T(\varepsilon)-(\mathscr{H}_i^r)_{12}(\varepsilon)L_i(\varepsilon)V_iL_i^T(\varepsilon)\}$$

$$-\mathrm{Tr}\{-(\mathscr{H}_i^r)_{21}(\varepsilon)L_i(\varepsilon)V_iL_i^T(\varepsilon)+(\mathscr{H}_i^r)_{22}(\varepsilon)(G_i(\varepsilon)W_iG_i^T(\varepsilon)+L_i(\varepsilon)V_iL_i^T(\varepsilon))\}$$

$$-\mathrm{Tr}\{(\mathscr{H}_i^r)_{33}(\varepsilon)L_{ic}(\varepsilon)V_{ic}L_{ic}^T(\varepsilon)-(\mathscr{H}_i^r)_{34}(\varepsilon)L_{ic}(\varepsilon)V_{ic}L_{ic}^T(\varepsilon)\}$$

$$-\mathrm{Tr}\{(\mathscr{H}_i^r)_{44}(\varepsilon)(G_{ic}(\varepsilon)W_{ic}G_{ic}^T(\varepsilon)+L_{ic}(\varepsilon)V_{ic}L_{ic}^T(\varepsilon))-(\mathscr{H}_i^r)_{43}(\varepsilon)L_{ic}(\varepsilon)V_{ic}L_{ic}^T(\varepsilon)\}$$

$$-2(\breve{\mathscr{D}}_i^r)_{13}^T(\varepsilon)B_{ic}(\varepsilon)u_c(\varepsilon)-2(\breve{\mathscr{D}}_i^r)_{11}^T(\varepsilon)\Big[-B_{ii}(\varepsilon)R_{ii}^{-1}(\varepsilon)B_{ii}^T(\varepsilon)\sum_{s=1}^{k_i}\frac{\mu_i^s}{\mu_i^1}(\breve{\mathscr{D}}_i^s)_{11}(\varepsilon)$$

$$-C_{ii}(\varepsilon)R_{zi}^{-1}(\varepsilon)C_{ii}^T(\varepsilon)\sum_{s=1}^{k_i}\frac{\mu_i^s}{\mu_i^1}(\breve{\mathscr{D}}_i^s)_{11}(\varepsilon)+\sum_{j=1}^{N_i}B_{ij}(\varepsilon)u_{ij}^*(\varepsilon)\Big],\quad 2\le r\le k_i \quad (73)$$

where the terminal-value conditions $(\mathscr{H}_i^1)_{11}(t_f)=Q_i^f$, $(\mathscr{H}_i^r)_{11}(t_f)=0$ for $2\le r\le k_i$; $(\breve{\mathscr{D}}_i^r)_{11}(t_f)=-Q_i^f\zeta_i(t_f)$, $(\breve{\mathscr{D}}_i^r)_{11}(t_f)=0$ for $2\le r\le k_i$; and $\mathscr{D}_i^r(t_f)=\zeta_i(t_f)Q_i^f\zeta_i(t_f)$, $\mathscr{D}_i^r(t_f)$ for $2\le r\le k_i$. Thus, whenever the coupled backward-in-time differential equations (68)–(73) admit the matrix-valued solutions $\{(\mathscr{H}_i^r)_{11}(\cdot)\}_{r=1}^{k_i}$, vector-valued solutions $\{(\breve{\mathscr{D}}_i^r)_{11}(\cdot)\}_{r=1}^{k_i}$, and scalar-valued solutions $\{\mathscr{D}_i^r(\cdot)\}_{r=1}^{k_i}$, then the existence of the matrix-valued solutions $\{\mathscr{E}_i^r(\cdot)\}_{r=1}^{k_i}$, vector-valued solutions $\{\breve{\mathscr{T}}_i^r(\cdot)\}_{r=1}^{k_i}$, and scalar-valued solutions $\{\mathscr{T}_i^r(\cdot)\}_{r=1}^{k_i}$ satisfying the coupled forward-in-time differential equations (62)–(67) are assured.

By comparing the time-forward differential equations (62)–(67) to those of time-backward differential equations (68)–(73), one may recognize that these sets of differential equations are related to one another by

$$\frac{d}{d\varepsilon}\mathscr{E}_i^r(\varepsilon)=-\frac{d}{d\varepsilon}(\mathscr{H}_i^r)_{11}(\varepsilon);\quad \frac{d}{d\varepsilon}\breve{\mathscr{T}}_i^r(\varepsilon)=-\frac{d}{d\varepsilon}(\breve{\mathscr{D}}_i^r)_{11}(\varepsilon)$$

$$\frac{d}{d\varepsilon}\mathscr{T}_i^r(\varepsilon)=-\frac{d}{d\varepsilon}\mathscr{D}_i^r(\varepsilon),\qquad \varepsilon\in[t_0,t_f].$$

Enforcing the initial-value conditions of $\mathscr{E}_i^r(t_0) = 0$, $\breve{\mathscr{T}}_i^r(t_0) = 0$ and $\mathscr{T}_i^r(t_0) = 0$ uniquely implies the following results

$$\mathscr{E}_i^r(\varepsilon) = (\mathscr{H}_i^r)_{11}(t_0) - (\mathscr{H}_i^r)_{11}(\varepsilon); \quad \breve{\mathscr{T}}_i^r(\varepsilon) = (\breve{\mathscr{D}}_i^r)_{11}(t_0) - (\breve{\mathscr{D}}_i^r)_{11}(\varepsilon)$$
$$\mathscr{T}_i^r(\varepsilon) = \mathscr{D}_i^r(t_0) - \mathscr{D}_i^r(\varepsilon)$$

for all $\varepsilon \in [t_0, t_f]$ and yields a value function

$$\mathscr{W}_i(\varepsilon, \mathscr{Y}_i, \breve{\mathscr{Z}}_i, \mathscr{Z}_i) = \sum_{r=1}^{k_i} \mu_i^r \left[(x_i^0)^T (\mathscr{H}_i^r)_{11}(t_0) x_i^0 + 2(x_i^0)^T (\breve{\mathscr{D}}_i^r)_{11}(t_0) + \mathscr{D}_i^r(t_0) \right]$$

for which the sufficient condition (53) of the verification theorem is satisfied. Therefore, the extremal person-by-person equilibrium policy (58)–(61) minimizing (47) become optimal

$$K_{x_i}^*(\varepsilon) = -R_{ii}^{-1}(\varepsilon) B_{ii}^T(\varepsilon) \sum_{r=1}^{k_i} \hat{\mu}_i{}^r \mathscr{H}_{i*}^r(\varepsilon), \tag{74}$$

$$K_{z_i}^*(\varepsilon) = -R_{zi}^{-1}(\varepsilon) C_{ii}^T(\varepsilon) \sum_{r=1}^{k_i} \hat{\mu}_i{}^r \mathscr{H}_{i*}^r(\varepsilon), \tag{75}$$

$$p_{x_i}^*(\varepsilon) = -R_{ii}^{-1}(\varepsilon) B_{ii}^T(\varepsilon) \sum_{r=1}^{k_i} \hat{\mu}_i{}^r \breve{\mathscr{D}}_{i*}^r(\varepsilon), \tag{76}$$

$$p_{z_i}^*(\varepsilon) = -R_{zi}^{-1}(\varepsilon) C_{ii}^T(\varepsilon) \sum_{r=1}^{k_i} \hat{\mu}_i{}^r \breve{\mathscr{D}}_{i*}^r(\varepsilon), \qquad \hat{\mu}_i^r = \frac{\mu_i^r}{\mu_i^1} \tag{77}$$

The goals in this research investigation have been methodological. A noncooperative game-theoretic methodology for coordination control of distributed stochastic systems is successfully sought for theory building in contexts in which signaling effects are issued by a coordinator and distributed person-by-person equilibrium strategies by autonomous agents i and $i \in \bar{I}$ are placed toward performance robustness. At this point, it makes sense to integrate all of the contending results into the following unified theorem.

Theorem 5 (Person-by-Person Equilibrium Strategies). *Consider a distributed stochastic system governed by (3)–(16) whose pairs (A_{ii}, B_{ii}) and (A_{ii}, C_{ii}) are uniformly stabilizable on $[t_0, t_f]$. An N-tuple $\{(u_1^*, z_1^*), \ldots, (u_N^*, z_N^*)\}$ of control policies constitutes a feedback Nash equilibrium for the class of distributed stochastic system considered here. Furthermore, 2-tuple (u_i^*, z_i^*) imposing a person-by-person equilibrium strategy for the corresponding agent i and $i \in \bar{I}$ is implemented forwardly in time by*

$$u_i^*(t) = K_{x_i}^*(t)x_i(t) + p_{x_i}^*(t) \tag{78}$$

$$z_i^*(t) = K_{z_i}^*(t)x_i(t) + p_{z_i}^*(t), \quad t = t_f + t_0 - \varepsilon, \quad \varepsilon \in [t_0, t_f], \tag{79}$$

which strives to optimize the risk-value aware performance index (47) composed by a preferential set of mathematical statistics of the chi-squared cost random variable (16). The construction of the person-by-person equilibrium for each agent i is determined backwardly in time; e.g.,

$$K_{x_i}^*(\varepsilon) = -R_{ii}^{-1}(\varepsilon)B_{ii}^T(\varepsilon)\sum_{r=1}^{k_i}\hat{\mu}_i{}^r\mathscr{H}_{i*}^r(\varepsilon), \tag{80}$$

$$K_{z_i}^*(\varepsilon) = -R_{zi}^{-1}(\varepsilon)C_{ii}^T(\varepsilon)\sum_{r=1}^{k_i}\hat{\mu}_i{}^r\mathscr{H}_{i*}^r(\varepsilon), \tag{81}$$

$$p_{x_i}^*(\varepsilon) = -R_{ii}^{-1}(\varepsilon)B_{ii}^T(\varepsilon)\sum_{r=1}^{k_i}\hat{\mu}_i{}^r\breve{\mathscr{D}}_{i*}^r(\varepsilon), \tag{82}$$

$$p_{z_i}^*(\varepsilon) = -R_{zi}^{-1}(\varepsilon)C_{ii}^T(\varepsilon)\sum_{r=1}^{k_i}\hat{\mu}_i{}^r\breve{\mathscr{D}}_{i*}^r(\varepsilon), \tag{83}$$

wherein the normalized preferences $\hat{\mu}_i^r \triangleq \mu_i^r/\mu_i^1$'s are mutually chosen by each incumbent agent i for risk-averse coordinations toward co-design of individual performance robustness. The optimal set of supporting solutions satisfies the time-backward, matrix, vector, and scalar-valued differential equations

$$\begin{aligned}\frac{d}{d\varepsilon}(\mathscr{H}_{i*}^1)_{11}(\varepsilon) &= -[A_{ii}(\varepsilon) + B_{ii}(\varepsilon)K_{x_i}^*(\varepsilon) + C_{ii}(\varepsilon)K_{z_i}^*(\varepsilon)]^T(\mathscr{H}_{i*}^1)_{11}(\varepsilon)\\ &\quad - (\mathscr{H}_{i*}^1)_{11}(\varepsilon)[A_{ii}(\varepsilon) + B_{ii}(\varepsilon)K_{x_i}^*(\varepsilon) + C_{ii}(\varepsilon)K_{z_i}^*(\varepsilon)] - Q_{ii}(\varepsilon) - Q_i(\varepsilon)\\ &\quad - K_{x_i}^{*T}(\varepsilon)R_{ii}(\varepsilon)K_{x_i}^*(\varepsilon) - K_{z_i}^{*T}(\varepsilon)R_{zi}(\varepsilon)K_{z_i}^*(\varepsilon), \quad (\mathscr{H}_{i*}^1)_{11}(t_f) = Q_i^f\end{aligned} \tag{84}$$

$$\begin{aligned}\frac{d}{d\varepsilon}(\mathscr{H}_{i*}^r)_{11}(\varepsilon) &= -[A_{ii}(\varepsilon) + B_{ii}(\varepsilon)K_{x_i}^*(\varepsilon) + C_{ii}(\varepsilon)K_{z_i}^*(\varepsilon)]^T(\mathscr{H}_{i*}^r)_{11}(\varepsilon)\\ &\quad - (\mathscr{H}_{i*}^r)_{11}(\varepsilon)[A_{ii}(\varepsilon) + B_{ii}(\varepsilon)K_{x_i}^*(\varepsilon) + C_{ii}(\varepsilon)K_{z_i}^*(\varepsilon)]\\ &\quad - \sum_{v=1}^{r-1}\frac{2r!}{v!(r-v)!}\Big\{(\mathscr{H}_{i*}^v)_{11}(\varepsilon)L_i(\varepsilon)V_iL_i^T(\varepsilon)(\mathscr{H}_{i*}^{r-v})_{11}(\varepsilon)\\ &- (\mathscr{H}_{i*}^v)_{12}(\varepsilon)L_i(\varepsilon)V_iL_i^T(\varepsilon)(\mathscr{H}_{i*}^{r-v})_{11}(\varepsilon) - (\mathscr{H}_{i*}^v)_{11}(\varepsilon)L_i(\varepsilon)V_iL_i^T(\varepsilon)(\mathscr{H}_{i*}^{r-v})_{21}(\varepsilon)\\ &+ (\mathscr{H}_{i*}^v)_{12}(\varepsilon)G_i(\varepsilon)W_iG_i^T(\varepsilon)\mathscr{H}_{i*}^{r-v})_{21}(\varepsilon) + (\mathscr{H}_{i*}^v)_{12}(\varepsilon)L_i(\varepsilon)V_iL_i^T(\varepsilon)(\mathscr{H}_{i*}^{r-v})_{21}(\varepsilon)\end{aligned}$$

$$+(\mathcal{H}_{i*}^{v})_{13}(\varepsilon)L_{ic}(\varepsilon)V_{ic}L_{ic}^{T}(\varepsilon)(\mathcal{H}_{i*}^{r-v})_{31}(\varepsilon)-(\mathcal{H}_{i*}^{v})_{14}(\varepsilon)L_{ic}(\varepsilon)V_{ic}L_{ic}^{T}(\varepsilon)(\mathcal{H}_{i*}^{r-v})_{31}(\varepsilon)$$
$$-(\mathcal{H}_{i*}^{v})_{13}(\varepsilon)L_{ic}(\varepsilon)V_{ic}L_{ic}^{T}(\varepsilon)(\mathcal{H}_{i*}^{r-v})_{41}(\varepsilon)+(\mathcal{H}_{i*}^{v})_{14}(\varepsilon)G_{ic}W_{ic}G_{ic}^{T}(\varepsilon)(\mathcal{H}_{i}^{r-v})_{41}(\varepsilon)$$
$$+(\mathcal{H}_{i*}^{v})_{14}(\varepsilon)L_{ic}(\varepsilon)L_{ic}L_{ic}^{T}(\varepsilon)(\mathcal{H}_{i*}^{r-v})_{41}(\varepsilon)\Big\}, \quad (\mathcal{H}_{i*}^{r})_{11}(t_f)=0;\ 2\le r\le k_i \quad (85)$$

$$\frac{d}{d\varepsilon}(\mathcal{H}_{i*}^{1})_{12}(\varepsilon)=-[A_{ii}(\varepsilon)+B_{ii}(\varepsilon)K_{x_i}^{*}(\varepsilon)+C_{ii}(\varepsilon)K_{z_i}^{*}(\varepsilon)]^{T}(\mathcal{H}_{i*}^{1})_{12}(\varepsilon)$$
$$-(\mathcal{H}_{i*}^{1})_{11}(\varepsilon)L_i(\varepsilon)C_i(\varepsilon)-(\mathcal{H}_{i*}^{1})_{12}(A_{ii}(\varepsilon)-L_i(\varepsilon)C_i(\varepsilon)), \quad (\mathcal{H}_{i*}^{1})_{12}(t_f)=0 \quad (86)$$

$$\frac{d}{d\varepsilon}(\mathcal{H}_{i*}^{r})_{12}(\varepsilon)=-[A_{ii}(\varepsilon)+B_{ii}(\varepsilon)K_{x_i}^{*}(\varepsilon)+C_{ii}(\varepsilon)K_{z_i}^{*}(\varepsilon)]^{T}(\mathcal{H}_{i*}^{r})_{12}(\varepsilon)$$
$$-(\mathcal{H}_{i*}^{r})_{11}(\varepsilon)L_i(\varepsilon)C_i(\varepsilon)-(\mathcal{H}_{i*}^{r})_{12}(A_{ii}(\varepsilon)-L_i(\varepsilon)C_i(\varepsilon))$$
$$-\sum_{v=1}^{r-1}\frac{2r!}{v!(r-v)!}\Big\{(\mathcal{H}_{i*}^{v})_{11}(\varepsilon)L_i(\varepsilon)V_iL_i^{T}(\varepsilon)(\mathcal{H}_{i*}^{r-v})_{12}(\varepsilon)$$
$$-(\mathcal{H}_{i*}^{v})_{12}(\varepsilon)L_i(\varepsilon)V_iL_i^{T}(\varepsilon)(\mathcal{H}_{i*}^{r-v})_{12}(\varepsilon)-(\mathcal{H}_{i*}^{v})_{11}(\varepsilon)L_i(\varepsilon)V_iL_i^{T}(\varepsilon)(\mathcal{H}_{i*}^{r-v})_{22}(\varepsilon)$$
$$+(\mathcal{H}_{i*}^{v})_{12}(\varepsilon)G_i(\varepsilon)W_iG_i^{T}(\varepsilon)(\mathcal{H}_{i*}^{r-s})_{22}(\varepsilon)+(\mathcal{H}_{i*}^{v})_{12}(\varepsilon)L_i(\varepsilon)W_iL_i^{T}(\varepsilon)(\mathcal{H}_{i*}^{r-s})_{22}(\varepsilon)$$
$$+(\mathcal{H}_{i*}^{v})_{13}(\varepsilon)L_{ic}(\varepsilon)V_{ic}L_{ic}^{T}(\varepsilon)(\mathcal{H}_{i*}^{r-v})_{32}(\varepsilon)-(\mathcal{H}_{i*}^{v})_{14}(\varepsilon)L_{ic}(\varepsilon)V_{ic}L_{ic}^{T}(\varepsilon)(\mathcal{H}_{i*}^{r-v})_{32}(\varepsilon)$$
$$-(\mathcal{H}_{i*}^{v})_{13}(\varepsilon)L_{ic}(\varepsilon)V_{ic}L_{ic}^{T}(\varepsilon)(\mathcal{H}_{i*}^{r-v})_{42}(\varepsilon)+(\mathcal{H}_{i*}^{v})_{14}(\varepsilon)G_{ic}W_{ic}G_{ic}^{T}(\varepsilon)(\mathcal{H}_{i*}^{r-v})_{42}(\varepsilon)$$
$$+(\mathcal{H}_{i*}^{v})_{14}(\varepsilon)L_{ic}(\varepsilon)V_{ic}L_{ic}^{T}(\varepsilon)(\mathcal{H}_{i*}^{r-v})_{42}(\varepsilon)\Big\}, \quad (\mathcal{H}_{i*}^{r})_{12}(t_f)=0;\ 2\le r\le k_i \quad (87)$$

$$\frac{d}{d\varepsilon}(\mathcal{H}_{i*}^{1})_{13}(\varepsilon)=-[A_{ii}(\varepsilon)+B_{ii}(\varepsilon)K_{x_i}^{*}(\varepsilon)+C_{ii}(\varepsilon)K_{z_i}^{*}(\varepsilon)]^{T}(\mathcal{H}_{i*}^{1})_{13}(\varepsilon)$$
$$-(\mathcal{H}_{i*}^{1})_{13}(\varepsilon)A_{ic}(\varepsilon)+2K_{z_i}^{*T}(\varepsilon)R_{zi}(\varepsilon), \quad (\mathcal{H}_{i*}^{1})_{13}(t_f)=0 \quad (88)$$

$$\frac{d}{d\varepsilon}(\mathcal{H}_{i*}^{r})_{13}(\varepsilon)=-[A_{ii}(\varepsilon)+B_{ii}(\varepsilon)K_{x_i}^{*}(\varepsilon)+C_{ii}(\varepsilon)K_{z_i}^{*}(\varepsilon)]^{T}(\mathcal{H}_{i*}^{r})_{13}(\varepsilon)$$
$$-(\mathcal{H}_{i*}^{r})_{13}(\varepsilon)A_{ic}(\varepsilon)-\sum_{v=1}^{r-1}\frac{2r!}{v!(r-v)!}\Big\{(\mathcal{H}_{i*}^{v})_{11}(\varepsilon)L_i(\varepsilon)V_iL_i^{T}(\varepsilon)(\mathcal{H}_{i*}^{r-v})_{13}(\varepsilon)$$
$$-(\mathcal{H}_{i*}^{v})_{12}(\varepsilon)L_i(\varepsilon)V_iL_i^{T}(\varepsilon)(\mathcal{H}_{i*}^{r-v})_{13}(\varepsilon)-(\mathcal{H}_{i*}^{v})_{11}(\varepsilon)L_i(\varepsilon)V_iL_i^{T}(\varepsilon)(\mathcal{H}_{i*}^{r-v})_{23}(\varepsilon)$$
$$+(\mathcal{H}_{i*}^{v})_{12}(\varepsilon)G_i(\varepsilon)W_iG_i^{T}(\varepsilon)(\mathcal{H}_{i*}^{r-v})_{23}(\varepsilon)+(\mathcal{H}_{i*}^{v})_{12}(\varepsilon)L_i(\varepsilon)V_iL_i^{T}(\varepsilon)(\mathcal{H}_{i*}^{r-v})_{23}(\varepsilon)$$

$$+(\mathscr{H}_{i*}^{v})_{13}(\varepsilon)L_{ic}(\varepsilon)V_{ic}L_{ic}^{T}(\varepsilon)(\mathscr{H}_{i*}^{r-v})_{33}(\varepsilon)-(\mathscr{H}_{i*}^{v})_{14}(\varepsilon)L_{ic}(\varepsilon)V_{ic}L_{ic}^{T}(\varepsilon)(\mathscr{H}_{i*}^{r-v})_{33}(\varepsilon)$$
$$-(\mathscr{H}_{i*}^{v})_{13}(\varepsilon)L_{ic}(\varepsilon)V_{ic}L_{ic}^{T}(\varepsilon)(\mathscr{H}_{i*}^{r-v})_{43}(\varepsilon)+(\mathscr{H}_{i*}^{v})_{14}(\varepsilon)G_{ic}W_{ic}G_{ic}^{T}(\varepsilon)(\mathscr{H}_{i*}^{r-v})_{43}(\varepsilon)$$
$$+(\mathscr{H}_{i*}^{v})_{14}(\varepsilon)L_{ic}(\varepsilon)V_{ic}L_{ic}^{T}(\varepsilon)(\mathscr{H}_{i*}^{r-v})_{43}(\varepsilon)\Big\}, \quad (\mathscr{H}_{i*}^{r})_{13}(t_f)=0;\ 2\le r\le k_i \quad (89)$$

$$\frac{d}{d\varepsilon}(\mathscr{H}_{i*}^{1})_{14}(\varepsilon)=-[A_{ii}(\varepsilon)+B_{ii}(\varepsilon)K_{x_i}^{*}(\varepsilon)+C_{ii}(\varepsilon)K_{z_i}^{*}(\varepsilon)]^{T}(\mathscr{H}_{i*}^{1})_{14}(\varepsilon)$$
$$-(\mathscr{H}_{i*}^{1})_{14}(\varepsilon)(A_{ic}(\varepsilon)-L_{ic}(\varepsilon)C_{ic}(\varepsilon))$$
$$-(\mathscr{H}_{i*}^{1})_{13}(\varepsilon)L_{i}(\varepsilon)C_{i}(\varepsilon), \quad (\mathscr{H}_{i*}^{1})_{14}(t_f)=0 \quad (90)$$

$$\frac{d}{d\varepsilon}(\mathscr{H}_{i*}^{r})_{14}(\varepsilon)=-[A_{ii}(\varepsilon)+B_{ii}(\varepsilon)K_{x_i}^{*}(\varepsilon)+C_{ii}(\varepsilon)K_{z_i}^{*}(\varepsilon)]^{T}(\mathscr{H}_{i*}^{r})_{14}(\varepsilon)$$
$$-(\mathscr{H}_{i*}^{r})_{13}(\varepsilon)L_{i}(\varepsilon)C_{i}(\varepsilon)-(\mathscr{H}_{i*}^{r})_{14}(\varepsilon)(A_{ic}(\varepsilon)-L_{ic}(\varepsilon)C_{ic}(\varepsilon))$$
$$-\sum_{v=1}^{r-1}\frac{2r!}{v!(r-v)!}\Big\{(\mathscr{H}_{i*}^{v})_{11}(\varepsilon)L_{i}(\varepsilon)V_{i}L_{i}^{T}(\varepsilon)(\mathscr{H}_{i*}^{r-v})_{14}(\varepsilon)$$
$$-(\mathscr{H}_{i*}^{v})_{12}(\varepsilon)L_{i}(\varepsilon)V_{i}L_{i}^{T}(\varepsilon)(\mathscr{H}_{i*}^{r-v})_{14}(\varepsilon)-(\mathscr{H}_{i*}^{v})_{11}(\varepsilon)L_{i}(\varepsilon)V_{i}L_{i}^{T}(\varepsilon)(\mathscr{H}_{i*}^{r-v})_{24}(\varepsilon)$$
$$+(\mathscr{H}_{i*}^{v})_{12}(\varepsilon)G_{i}(\varepsilon)W_{i}G_{i}^{T}(\varepsilon)(\mathscr{H}_{i*}^{r-v})_{24}(\varepsilon)+(\mathscr{H}_{i*}^{v})_{12}(\varepsilon)L_{i}(\varepsilon)V_{i}L_{i}^{T}(\varepsilon)(\mathscr{H}_{i*}^{r-v})_{24}(\varepsilon)$$
$$+(\mathscr{H}_{i*}^{v})_{13}(\varepsilon)L_{ic}(\varepsilon)V_{ic}L_{ic}^{T}(\varepsilon)(\mathscr{H}_{i*}^{r-v})_{34}(\varepsilon)-(\mathscr{H}_{i*}^{v})_{14}(\varepsilon)L_{ic}(\varepsilon)V_{ic}L_{ic}^{T}(\varepsilon)(\mathscr{H}_{i*}^{r-v})_{34}(\varepsilon)$$
$$-(\mathscr{H}_{i*}^{v})_{13}(\varepsilon)L_{ic}(\varepsilon)V_{ic}L_{ic}^{T}(\varepsilon)(\mathscr{H}_{i*}^{r-v})_{44}(\varepsilon)+(\mathscr{H}_{i*}^{v})_{14}(\varepsilon)G_{ic}W_{ic}G_{ic}^{T}(\varepsilon)(\mathscr{H}_{i*}^{r-v})_{44}(\varepsilon)$$
$$+(\mathscr{H}_{i*}^{v})_{14}(\varepsilon)L_{ic}(\varepsilon)V_{ic}L_{ic}^{T}(\varepsilon)(\mathscr{H}_{i*}^{r-v})_{44}(\varepsilon)\Big\}, \quad (\mathscr{H}_{i*}^{r})_{14}(t_f)=0;\ 2\le r\le k_i \quad (91)$$

$$\frac{d}{d\varepsilon}(\mathscr{H}_{i*}^{1})_{21}(\varepsilon)=-(\mathscr{H}_{i*}^{1})_{21}(\varepsilon)[A_{ii}(\varepsilon)+B_{ii}(\varepsilon)K_{x_i}^{*}(\varepsilon)+C_{ii}(\varepsilon)K_{z_i}^{*}(\varepsilon)]$$
$$-(A_{ii}(\varepsilon)-L_{i}(\varepsilon)C_{i}(\varepsilon))^{T}(\mathscr{H}_{i*}^{1})_{21}(\varepsilon)$$
$$-(L_{i}(\varepsilon)C_{i}(\varepsilon))^{T}(\mathscr{H}_{i*}^{1})_{11}(\varepsilon), \quad (\mathscr{H}_{i*}^{1})_{21}(t_f)=0 \quad (92)$$

$$\frac{d}{d\varepsilon}(\mathscr{H}_{i*}^{r})_{21}(\varepsilon) = -(\mathscr{H}_{i*}^{r})_{21}(\varepsilon)[A_{ii}(\varepsilon) + B_{ii}(\varepsilon)K_{x_i}^{*}(\varepsilon) + C_{ii}(\varepsilon)K_{z_i}^{*}(\varepsilon)]$$
$$- (L_i(\varepsilon)C_i(\varepsilon))^T(\mathscr{H}_{i*}^{r})_{11}(\varepsilon) - (A_{ii}(\varepsilon) - L_i(\varepsilon)C_i(\varepsilon))^T(\mathscr{H}_{i*}^{r})_{21}(\varepsilon)$$
$$- \sum_{v=1}^{r-1} \frac{2r!}{v!(r-v)!} \Big\{ (\mathscr{H}_{i*}^{v})_{21}(\varepsilon)L_i(\varepsilon)V_iL_i^T(\varepsilon)(\mathscr{H}_{i*}^{r-v})_{11}(\varepsilon)$$
$$- (\mathscr{H}_{i*}^{v})_{22}(\varepsilon)L_i(\varepsilon)V_iL_i^T(\varepsilon)(\mathscr{H}_{i*}^{r-v})_{11}(\varepsilon) - (\mathscr{H}_{i*}^{v})_{21}(\varepsilon)L_i(\varepsilon)V_iL_i^T(\varepsilon)(\mathscr{H}_{i*}^{r-v})_{21}(\varepsilon)$$
$$+ (\mathscr{H}_{i*}^{v})_{22}(\varepsilon)G_i(\varepsilon)W_iG_i^T(\varepsilon)(\mathscr{H}_{i*}^{r-v})_{21}(\varepsilon) + (\mathscr{H}_{i*}^{v})_{22}(\varepsilon)L_i(\varepsilon)V_iL_i^T(\varepsilon)(\mathscr{H}_{i*}^{r-v})_{21}(\varepsilon)$$
$$+ (\mathscr{H}_{i*}^{v})_{23}(\varepsilon)L_i(\varepsilon)V_iL_i^T(\varepsilon)(\mathscr{H}_{i*}^{r-v})_{31}(\varepsilon) - (\mathscr{H}_{i*}^{v})_{24}(\varepsilon)L_{ic}(\varepsilon)V_{ic}L_{ic}^T(\varepsilon)(\mathscr{H}_{i*}^{r-v})_{31}(\varepsilon)$$
$$- (\mathscr{H}_{i*}^{v})_{23}(\varepsilon)L_{ic}(\varepsilon)V_{ic}L_{ic}^T(\varepsilon)(\mathscr{H}_{i*}^{r-v})_{41}(\varepsilon) + (\mathscr{H}_{i*}^{v})_{24}(\varepsilon)G_{ic}W_{ic}G_{ic}^T(\varepsilon)(\mathscr{H}_{i*}^{r-v})_{41}(\varepsilon)$$
$$+ (\mathscr{H}_{i*}^{v})_{24}(\varepsilon)L_{ic}(\varepsilon)V_{ic}L_{ic}^T(\varepsilon)(\mathscr{H}_{i*}^{r-v})_{41}(\varepsilon) \Big\}, \quad (\mathscr{H}_{i*}^{r})_{21}(t_f) = 0;\ 2 \le r \le k_i \quad (93)$$

$$\frac{d}{d\varepsilon}(\mathscr{H}_{i*}^{1})_{22}(\varepsilon) = -(A_{ii}(\varepsilon) - L_i(\varepsilon)C_i(\varepsilon))^T(\mathscr{H}_{i*}^{1})_{22}(\varepsilon) - (L_i(\varepsilon)C_i(\varepsilon))^T(\mathscr{H}_{i*}^{1})_{12}(\varepsilon)$$
$$- (\mathscr{H}_{i*}^{1})_{22}(\varepsilon)(A_{ii}(\varepsilon) - L_i(\varepsilon)C_i(\varepsilon))$$
$$- (\mathscr{H}_{i*}^{1})_{21}(\varepsilon)L_i(\varepsilon)C_i(\varepsilon), \quad (\mathscr{H}_{i*}^{1})_{22}(t_f) = 0 \quad (94)$$

$$\frac{d}{d\varepsilon}(\mathscr{H}_{i*}^{r})_{22}(\varepsilon) = -(A_{ii}(\varepsilon) - L_i(\varepsilon)C_i(\varepsilon))^T(\mathscr{H}_{i*}^{r})_{22}(\varepsilon) - (L_i(\varepsilon)C_i(\varepsilon))^T(\mathscr{H}_{i*}^{r})_{12}(\varepsilon)$$
$$- (\mathscr{H}_{i*}^{r})_{22}(\varepsilon)(A_{ii}(\varepsilon) - L_i(\varepsilon)C_i(\varepsilon)) - (\mathscr{H}_{i*}^{r})_{21}(\varepsilon)L_i(\varepsilon)C_i(\varepsilon)$$
$$- \sum_{v=1}^{r-1} \frac{2r!}{v!(r-v)!} \Big\{ (\mathscr{H}_{i*}^{v})_{21}(\varepsilon)L_i(\varepsilon)V_iL_i^T(\varepsilon)(\mathscr{H}_{i*}^{r-v})_{12}(\varepsilon)$$
$$- (\mathscr{H}_{i*}^{v})_{22}(\varepsilon)L_i(\varepsilon)V_iL_i^T(\varepsilon)(\mathscr{H}_{i*}^{r-v})_{12}(\varepsilon) - (\mathscr{H}_{i*}^{v})_{21}(\varepsilon)L_i(\varepsilon)V_iL_i^T(\varepsilon)(\mathscr{H}_{i*}^{r-v})_{22}(\varepsilon)$$
$$+ (\mathscr{H}_{i*}^{v})_{22}(\varepsilon)G_i(\varepsilon)W_iG_i^T(\varepsilon)(\mathscr{H}_{i*}^{r-v})_{22}(\varepsilon) + (\mathscr{H}_{i*}^{v})_{22}(\varepsilon)L_i(\varepsilon)V_iL_i^T(\varepsilon)(\mathscr{H}_{i*}^{r-v})_{22}(\varepsilon)$$
$$+ (\mathscr{H}_{i*}^{v})_{23}(\varepsilon)L_{ic}(\varepsilon)V_{ic}L_{ic}^T(\varepsilon)(\mathscr{H}_{i*}^{r-v})_{32}(\varepsilon) - (\mathscr{H}_{i*}^{v})_{24}(\varepsilon)L_{ic}(\varepsilon)V_{ic}L_{ic}^T(\varepsilon)(\mathscr{H}_{i*}^{r-v})_{32}(\varepsilon)$$
$$- (\mathscr{H}_{i*}^{v})_{23}(\varepsilon)L_{ic}(\varepsilon)V_{ic}L_{ic}^T(\varepsilon)(\mathscr{H}_{i*}^{r-v})_{42}(\varepsilon) + (\mathscr{H}_{i*}^{v})_{24}(\varepsilon)G_{ic}W_{ic}G_{ic}^T(\varepsilon)(\mathscr{H}_{i*}^{r-v})_{42}(\varepsilon)$$
$$+ (\mathscr{H}_{i*}^{v})_{24}(\varepsilon)L_{ic}(\varepsilon)V_{ic}L_{ic}^T(\varepsilon)(\mathscr{H}_{i*}^{r-v})_{42}(\varepsilon) \Big\}, \quad (\mathscr{H}_{i*}^{r})_{22}(t_f) = 0;\ 2 \le r \le k_i \quad (95)$$

$$\frac{d}{d\varepsilon}(\mathscr{H}_{i*}^{1})_{23}(\varepsilon) = -(A_{ii}(\varepsilon) - L_i(\varepsilon)C_i(\varepsilon))^T(\mathscr{H}_{i*}^{1})_{23}(\varepsilon) - (\mathscr{H}_{i*}^{1})_{23}(\varepsilon)A_{ic}(\varepsilon)$$
$$- (L_i(\varepsilon)C_i(\varepsilon))^T(\mathscr{H}_{i*}^{1})_{13}(\varepsilon), \quad (\mathscr{H}_{i*}^{1})_{23}(t_f) = 0 \quad (96)$$

$$\frac{d}{d\varepsilon}(\mathscr{H}_{i*}^{r})_{23}(\varepsilon) = -(A_{ii}(\varepsilon) - L_i(\varepsilon)C_i(\varepsilon))^T (\mathscr{H}_{i*}^{r})_{23}(\varepsilon) - (\mathscr{H}_{i*}^{r})_{23}(\varepsilon)A_{ic}(\varepsilon)$$

$$- (L_i(\varepsilon)C_i(\varepsilon))^T (\mathscr{H}_{i*}^{r})_{13}(\varepsilon) - \sum_{v=1}^{r-1} \frac{2r!}{v!(r-v)!} \Big\{ (\mathscr{H}_{i*}^{v})_{21}(\varepsilon)L_i(\varepsilon)V_i L_i^T(\varepsilon)(\mathscr{H}_{i*}^{r-v})_{13}(\varepsilon)$$

$$- (\mathscr{H}_{i*}^{v})_{22}(\varepsilon)L_i(\varepsilon)V_i L_i^T(\varepsilon)(\mathscr{H}_{i*}^{r-v})_{13}(\varepsilon) - (\mathscr{H}_{i*}^{v})_{21}(\varepsilon)L_i(\varepsilon)V_i L_i^T(\varepsilon)(\mathscr{H}_{i*}^{r-v})_{23}(\varepsilon)$$

$$+ (\mathscr{H}_{i*}^{v})_{22}(\varepsilon)G_i(\varepsilon)W_i G_i^T(\varepsilon)(\mathscr{H}_{i*}^{r-v})_{23}(\varepsilon) + (\mathscr{H}_{i*}^{v})_{22}(\varepsilon)L_i(\varepsilon)V_i L_i^T(\varepsilon)(\mathscr{H}_{i*}^{r-v})_{23}(\varepsilon)$$

$$+ (\mathscr{H}_{i*}^{v})_{23}(\varepsilon)L_{ic}(\varepsilon)V_{ic}L_{ic}^T(\varepsilon)(\mathscr{H}_{i*}^{r-v})_{33}(\varepsilon) - (\mathscr{H}_{i*}^{v})_{24}(\varepsilon)L_{ic}(\varepsilon)V_{ic}L_{ic}^T(\varepsilon)(\mathscr{H}_{i*}^{r-v})_{33}(\varepsilon)$$

$$- (\mathscr{H}_{i*}^{v})_{23}(\varepsilon)L_{ic}(\varepsilon)V_{ic}L_{ic}^T(\varepsilon)(\mathscr{H}_{i*}^{r-v})_{43}(\varepsilon) + (\mathscr{H}_{i*}^{v})_{24}(\varepsilon)G_{ic}W_{ic}G_{ic}^T(\varepsilon)(\mathscr{H}_{i*}^{r-v})_{43}(\varepsilon)$$

$$+ (\mathscr{H}_{i*}^{v})_{24}(\varepsilon)L_{ic}(\varepsilon)V_{ic}L_{ic}^T(\varepsilon)(\mathscr{H}_{i*}^{r-v})_{43}(\varepsilon) \Big\}, \quad (\mathscr{H}_{i*}^{r})_{23}(t_f) = 0;\ 2 \le r \le k_i \quad (97)$$

$$\frac{d}{d\varepsilon}(\mathscr{H}_{i*}^{1})_{24}(\varepsilon) = -(A_{ii}(\varepsilon) - L_i(\varepsilon)C_i(\varepsilon))^T (\mathscr{H}_{i*}^{1})_{24}(\varepsilon) - (\mathscr{H}_{i*}^{1})_{23}(\varepsilon)L_{ic}(\varepsilon)C_{ic}(\varepsilon)$$

$$- (\mathscr{H}_{i*}^{1})_{24}(\varepsilon)(A_{ii}(\varepsilon) - L_i(\varepsilon)C_i(\varepsilon))$$

$$- (L_{ic}(\varepsilon)C_{ic}(\varepsilon))^T (\mathscr{H}_{i*}^{1})_{14}(\varepsilon), \quad (\mathscr{H}_{i*}^{1})_{24}(t_f) = 0 \quad (98)$$

$$\frac{d}{d\varepsilon}(\mathscr{H}_{i*}^{r})_{24}(\varepsilon) = -(A_{ii}(\varepsilon) - L_i(\varepsilon)C_i(\varepsilon))^T (\mathscr{H}_{i*}^{r})_{24}(\varepsilon) - (\mathscr{H}_{i*}^{r})_{23}(\varepsilon)L_{ic}(\varepsilon)C_{ic}(\varepsilon)$$

$$- (L_{ic}(\varepsilon)C_{ic}(\varepsilon))^T (\mathscr{H}_{i*}^{r})_{14}(\varepsilon) - (\mathscr{H}_{i*}^{r})_{24}(\varepsilon)(A_{ii}(\varepsilon) - L_i(\varepsilon)C_i(\varepsilon))$$

$$- \sum_{v=1}^{r-1} \frac{2r!}{v!(r-v)!} \Big\{ (\mathscr{H}_{i*}^{v})_{21}(\varepsilon)L_i(\varepsilon)V_i L_i^T(\varepsilon)(\mathscr{H}_{i*}^{r-v})_{14}(\varepsilon)$$

$$- (\mathscr{H}_{i*}^{v})_{22}(\varepsilon)L_i(\varepsilon)V_i L_i^T(\varepsilon)(\mathscr{H}_{i*}^{r-v})_{14}(\varepsilon) - (\mathscr{H}_{i*}^{v})_{21}(\varepsilon)L_i(\varepsilon)V_i L_i^T(\varepsilon)(\mathscr{H}_{i*}^{r-v})_{24}(\varepsilon)$$

$$+ (\mathscr{H}_{i*}^{v})_{22}(\varepsilon)G_i(\varepsilon)W_i G_i^T(\varepsilon)(\mathscr{H}_{i*}^{r-v})_{24}(\varepsilon) + (\mathscr{H}_{i*}^{v})_{22}(\varepsilon)L_i(\varepsilon)V_i L_i^T(\varepsilon)(\mathscr{H}_{i*}^{r-v})_{24}(\varepsilon)$$

$$+ (\mathscr{H}_{i*}^{v})_{23}(\varepsilon)L_{ic}(\varepsilon)V_{ic}L_{ic}^T(\varepsilon)(\mathscr{H}_{i*}^{r-v})_{34}(\varepsilon) - (\mathscr{H}_{i*}^{v})_{24}(\varepsilon)L_{ic}(\varepsilon)V_{ic}L_{ic}^T(\varepsilon)(\mathscr{H}_{i*}^{r-v})_{34}(\varepsilon)$$

$$- (\mathscr{H}_{i*}^{v})_{23}(\varepsilon)L_{ic}(\varepsilon)V_{ic}L_{ic}^T(\varepsilon)(\mathscr{H}_{i*}^{r-v})_{44}(\varepsilon) + (\mathscr{H}_{i*}^{v})_{24}(\varepsilon)G_{ic}W_{ic}G_{ic}^T(\varepsilon)(\mathscr{H}_{i*}^{r-v})_{44}(\varepsilon)$$

$$+ (\mathscr{H}_{i*}^{v})_{24}(\varepsilon)L_{ic}(\varepsilon)V_{ic}L_{ic}^T(\varepsilon)(\mathscr{H}_{i*}^{r-v})_{44}(\varepsilon) \Big\}, \quad (\mathscr{H}_{i*}^{r})_{24}(t_f) = 0;\ 2 \le r \le k_i \quad (99)$$

$$\frac{d}{d\varepsilon}(\mathscr{H}_{i*}^{1})_{31}(\varepsilon) = -A_{ic}^T(\varepsilon)(\mathscr{H}_{i*}^{1})_{31}(\varepsilon)$$

$$- (\mathscr{H}_{i*}^{1})_{31}(\varepsilon)[A_{ii}(\varepsilon) + B_{ii}(\varepsilon)K_{x_i}^*(\varepsilon) + C_{ii}(\varepsilon)K_{z_i}^*(\varepsilon)], \quad (\mathscr{H}_{i*}^{1})_{31}(t_f) = 0 \quad (100)$$

$$\frac{d}{d\varepsilon}(\mathscr{H}_{i*}^{r})_{31}(\varepsilon) = -(\mathscr{H}_{i*}^{r})_{31}(\varepsilon)[A_{ii}(\varepsilon)+B_{ii}(\varepsilon)K_{x_i}^{*}(\varepsilon)+C_{ii}(\varepsilon)K_{z_i}^{*}(\varepsilon)]$$
$$-A_{ic}^{T}(\varepsilon)(\mathscr{H}_{i*}^{r})_{31}(\varepsilon) - \sum_{v=1}^{r-1}\frac{2r!}{v!(r-v)!}\Big\{(\mathscr{H}_{i*}^{v})_{31}(\varepsilon)L_i(\varepsilon)V_iL_i^T(\varepsilon)(\mathscr{H}_{i*}^{r-v})_{11}(\varepsilon)$$
$$-(\mathscr{H}_{i*}^{v})_{32}(\varepsilon)L_i(\varepsilon)V_iL_i^T(\varepsilon)(\mathscr{H}_{i*}^{r-v})_{11}(\varepsilon)-(\mathscr{H}_{i*}^{v})_{31}(\varepsilon)L_i(\varepsilon)V_iL_i^T(\varepsilon)(\mathscr{H}_{i*}^{r-v})_{21}(\varepsilon)$$
$$+(\mathscr{H}_{i*}^{v})_{32}(\varepsilon)G_i(\varepsilon)W_iG_i^T(\varepsilon)(\mathscr{H}_{i*}^{r-v})_{21}(\varepsilon)+(\mathscr{H}_{i*}^{v})_{32}(\varepsilon)L_i(\varepsilon)V_iL_i^T(\varepsilon)(\mathscr{H}_{i*}^{r-v})_{21}(\varepsilon)$$
$$+(\mathscr{H}_{i*}^{v})_{33}(\varepsilon)L_{ic}(\varepsilon)V_{ic}L_{ic}^T(\varepsilon)(\mathscr{H}_{i*}^{r-v})_{31}(\varepsilon)-(\mathscr{H}_{i*}^{v})_{34}(\varepsilon)L_{ic}(\varepsilon)V_{ic}L_{ic}^T(\varepsilon)(\mathscr{H}_{i*}^{r-v})_{31}(\varepsilon)$$
$$-(\mathscr{H}_{i*}^{v})_{33}(\varepsilon)L_{ic}(\varepsilon)V_{ic}L_{ic}^T(\varepsilon)(\mathscr{H}_{i*}^{r-v})_{41}(\varepsilon)+(\mathscr{H}_{i*}^{v})_{34}(\varepsilon)G_{ic}W_{ic}G_{ic}^T(\varepsilon)(\mathscr{H}_{i*}^{r-v})_{41}(\varepsilon)$$
$$+(\mathscr{H}_{i*}^{v})_{34}(\varepsilon)L_{ic}V_{ic}L_{ic}^T(\varepsilon)(\mathscr{H}_{i*}^{r-v})_{41}(\varepsilon)\Big\}, \quad (\mathscr{H}_{i*}^{r})_{31}(t_f)=0;\ 2\le r\le k_i \quad (101)$$

$$\frac{d}{d\varepsilon}(\mathscr{H}_{i*}^{1})_{32}(\varepsilon) = -A_{ic}^{T}(\varepsilon)(\mathscr{H}_{i*}^{1})_{32}(\varepsilon)-(\mathscr{H}_{i*}^{1})_{31}(\varepsilon)L_i(\varepsilon)C_i(\varepsilon)$$
$$-(\mathscr{H}_{i*}^{1})_{32}(\varepsilon)(A_{ii}(\varepsilon)-L_i(\varepsilon)C_i(\varepsilon)), \quad (\mathscr{H}_{i*}^{1})_{32}(t_f)=0 \quad (102)$$

$$\frac{d}{d\varepsilon}(\mathscr{H}_{i*}^{r})_{32}(\varepsilon) = -A_{ic}^{T}(\varepsilon)(\mathscr{H}_{i*}^{r})_{32}(\varepsilon)-(\mathscr{H}_{i*}^{r})_{32}(\varepsilon)(A_{ii}(\varepsilon)-L_i(\varepsilon)C_i(\varepsilon))$$
$$-(\mathscr{H}_{i*}^{r})_{31}(\varepsilon)L_i(\varepsilon)C_i(\varepsilon)-\sum_{v=1}^{r-1}\frac{2r!}{v!(r-v)!}\Big\{(\mathscr{H}_{i*}^{v})_{31}(\varepsilon)L_i(\varepsilon)V_iL_i^T(\varepsilon)(\mathscr{H}_{i*}^{r-v})_{12}(\varepsilon)$$
$$-(\mathscr{H}_{i*}^{v})_{32}(\varepsilon)L_i(\varepsilon)V_iL_i^T(\varepsilon)(\mathscr{H}_{i*}^{r-v})_{12}(\varepsilon)-(\mathscr{H}_{i*}^{v})_{31}(\varepsilon)L_i(\varepsilon)V_iL_i^T(\varepsilon)(\mathscr{H}_{i*}^{r-v})_{22}(\varepsilon)$$
$$+(\mathscr{H}_{i*}^{v})_{32}(\varepsilon)G_i(\varepsilon)W_iG_i^T(\varepsilon)(\mathscr{H}_{i*}^{r-v})_{22}(\varepsilon)+(\mathscr{H}_{i*}^{v})_{32}(\varepsilon)L_i(\varepsilon)V_iL_i^T(\varepsilon)(\mathscr{H}_{i*}^{r-v})_{22}(\varepsilon)$$
$$+(\mathscr{H}_{i*}^{v})_{33}(\varepsilon)L_{ic}(\varepsilon)V_{ic}L_{ic}^T(\varepsilon)(\mathscr{H}_{i*}^{r-v})_{32}(\varepsilon)-(\mathscr{H}_{i*}^{v})_{34}(\varepsilon)L_{ic}(\varepsilon)V_{ic}L_{ic}^T(\varepsilon)(\mathscr{H}_{i*}^{r-v})_{32}(\varepsilon)$$
$$-(\mathscr{H}_{i*}^{v})_{33}(\varepsilon)L_{ic}(\varepsilon)V_{ic}L_{ic}^T(\varepsilon)(\mathscr{H}_{i*}^{r-v})_{42}(\varepsilon)+(\mathscr{H}_{i*}^{v})_{34}(\varepsilon)G_{ic}W_{ic}G_{ic}^T(\varepsilon)(\mathscr{H}_{i*}^{r-v})_{42}(\varepsilon)$$
$$+(\mathscr{H}_{i*}^{v})_{34}(\varepsilon)L_{ic}V_{ic}L_{ic}^T(\varepsilon)(\mathscr{H}_{i*}^{r-v})_{42}(\varepsilon)\Big\}, \quad (\mathscr{H}_{i*}^{r})_{32}(t_f)=0;\ 2\le r\le k_i \quad (103)$$

$$\frac{d}{d\varepsilon}(\mathscr{H}_{i*}^{1})_{33}(\varepsilon) = -A_{ic}^{T}(\varepsilon)(\mathscr{H}_{i*}^{1})_{33}(\varepsilon)-(\mathscr{H}_{i*}^{1})_{33}(\varepsilon)A_{ic}(\varepsilon)$$
$$-R_{zi}(\varepsilon), \quad (\mathscr{H}_{i*}^{1})_{33}(t_f)=0 \quad (104)$$

$$\frac{d}{d\varepsilon}(\mathscr{H}_{i*}^{r})_{33}(\varepsilon) = -A_{ic}^{T}(\varepsilon)(\mathscr{H}_{i*}^{r})_{33}(\varepsilon) - (\mathscr{H}_{i*}^{r})_{33}(\varepsilon)A_{ic}(\varepsilon)$$
$$- \sum_{v=1}^{r-1} \frac{2r!}{v!(r-v)!} \Big\{ (\mathscr{H}_{i*}^{v})_{31}(\varepsilon)L_i(\varepsilon)V_iL_i^T(\varepsilon)(\mathscr{H}_{i*}^{r-v})_{13}(\varepsilon)$$
$$- (\mathscr{H}_{i*}^{v})_{32}(\varepsilon)L_i(\varepsilon)V_iL_i^T(\varepsilon)(\mathscr{H}_{i*}^{r-v})_{13}(\varepsilon) - (\mathscr{H}_{i*}^{v})_{31}(\varepsilon)L_i(\varepsilon)V_iL_i^T(\varepsilon)(\mathscr{H}_{i*}^{r-v})_{23}(\varepsilon)$$
$$+ (\mathscr{H}_{i*}^{v})_{32}(\varepsilon)G_i(\varepsilon)W_iG_i^T(\varepsilon)(\mathscr{H}_{i*}^{r-v})_{23}(\varepsilon) + (\mathscr{H}_{i*}^{v})_{32}(\varepsilon)L_i(\varepsilon)V_iL_i^T(\varepsilon)(\mathscr{H}_{i*}^{r-v})_{23}(\varepsilon)$$
$$+ (\mathscr{H}_{i*}^{v})_{33}(\varepsilon)L_{ic}(\varepsilon)V_{ic}L_{ic}^T(\varepsilon)(\mathscr{H}_{i*}^{r-v})_{33}(\varepsilon) - (\mathscr{H}_{i*}^{v})_{34}(\varepsilon)L_{ic}(\varepsilon)V_{ic}L_{ic}^T(\varepsilon)(\mathscr{H}_{i*}^{r-v})_{33}(\varepsilon)$$
$$- (\mathscr{H}_{i*}^{v})_{33}(\varepsilon)L_{ic}(\varepsilon)V_{ic}L_{ic}^T(\varepsilon)(\mathscr{H}_{i*}^{r-v})_{43}(\varepsilon) + (\mathscr{H}_{i*}^{v})_{34}(\varepsilon)G_{ic}W_{ic}G_{ic}^T(\varepsilon)(\mathscr{H}_{i*}^{r-v})_{43}(\varepsilon)$$
$$+ (\mathscr{H}_{i*}^{v})_{34}(\varepsilon)L_{ic}V_{ic}L_{ic}^T(\varepsilon)(\mathscr{H}_{i*}^{r-v})_{43}(\varepsilon) \Big\}, \quad (\mathscr{H}_{i*}^{r})_{33}(t_f) = 0;\ 2 \le r \le k_i \quad (105)$$

$$\frac{d}{d\varepsilon}(\mathscr{H}_{i*}^{1})_{34}(\varepsilon) = -A_{ic}^{T}(\varepsilon)(\mathscr{H}_{i*}^{1})_{34}(\varepsilon) - (\mathscr{H}_{i*}^{1})_{34}(\varepsilon)(A_{ic}(\varepsilon) - L_{ic}(\varepsilon)C_{ic}(\varepsilon))$$
$$- (\mathscr{H}_{i*}^{1})_{33}(\varepsilon)L_{ic}(\varepsilon)C_{ic}(\varepsilon), \quad (\mathscr{H}_{i*}^{1})_{34}(t_f) = 0 \quad (106)$$

$$\frac{d}{d\varepsilon}(\mathscr{H}_{i*}^{r})_{34}(\varepsilon) = -A_{ic}^{T}(\varepsilon)(\mathscr{H}_{i*}^{r})_{34}(\varepsilon) - (\mathscr{H}_{i*}^{r})_{34}(\varepsilon)(A_{ic}(\varepsilon) - L_{ic}(\varepsilon)C_{ic}(\varepsilon))$$
$$- (\mathscr{H}_{i*}^{r})_{33}(\varepsilon)L_{ic}(\varepsilon)C_{ic}(\varepsilon) - \sum_{v=1}^{r-1} \frac{2r!}{v!(r-v)!} \Big\{ (\mathscr{H}_{i*}^{v})_{31}(\varepsilon)L_i(\varepsilon)V_iL_i^T(\varepsilon)(\mathscr{H}_{i*}^{r-v})_{14}(\varepsilon)$$
$$- (\mathscr{H}_{i*}^{v})_{32}(\varepsilon)L_i(\varepsilon)V_iL_i^T(\varepsilon)(\mathscr{H}_{i*}^{r-v})_{14}(\varepsilon) - (\mathscr{H}_{i*}^{v})_{31}(\varepsilon)L_i(\varepsilon)V_iL_i^T(\varepsilon)(\mathscr{H}_{i*}^{r-v})_{24}(\varepsilon)$$
$$+ (\mathscr{H}_{i*}^{v})_{32}(\varepsilon)G_i(\varepsilon)W_iG_i^T(\varepsilon)(\mathscr{H}_{i*}^{r-v})_{24}(\varepsilon) + (\mathscr{H}_{i*}^{v})_{32}(\varepsilon)L_i(\varepsilon)V_iL_i^T(\varepsilon)(\mathscr{H}_{i*}^{r-v})_{24}(\varepsilon)$$
$$+ (\mathscr{H}_{i*}^{v})_{33}(\varepsilon)L_{ic}(\varepsilon)V_{ic}L_{ic}^T(\varepsilon)(\mathscr{H}_{i*}^{r-v})_{34}(\varepsilon) - (\mathscr{H}_{i*}^{v})_{34}(\varepsilon)L_{ic}(\varepsilon)V_{ic}L_{ic}^T(\varepsilon)(\mathscr{H}_{i*}^{r-v})_{34}(\varepsilon)$$
$$- (\mathscr{H}_{i*}^{v})_{33}(\varepsilon)L_{ic}(\varepsilon)V_{ic}L_{ic}^T(\varepsilon)(\mathscr{H}_{i*}^{r-v})_{44}(\varepsilon) + (\mathscr{H}_{i*}^{v})_{34}(\varepsilon)G_{ic}W_{ic}G_{ic}^T(\varepsilon)(\mathscr{H}_{i*}^{r-v})_{44}(\varepsilon)$$
$$+ (\mathscr{H}_{i*}^{v})_{34}(\varepsilon)L_{ic}V_{ic}L_{ic}^T(\varepsilon)(\mathscr{H}_{i*}^{r-v})_{44}(\varepsilon) \Big\}, \quad (\mathscr{H}_{i*}^{r})_{34}(t_f) = 0;\ 2 \le r \le k_i \quad (107)$$

$$\frac{d}{d\varepsilon}(\mathscr{H}_{i*}^{1})_{41}(\varepsilon) = -(A_{ii}(\varepsilon) - L_i(\varepsilon)C_i(\varepsilon))^T(\mathscr{H}_{i*}^{1})_{41}(\varepsilon) - (L_i(\varepsilon)C_i(\varepsilon))^T(\mathscr{H}_{i*}^{1})_{31}(\varepsilon)$$
$$- (\mathscr{H}_{i*}^{1})_{41}(\varepsilon)[A_{ii}(\varepsilon) + B_{ii}(\varepsilon)K_{x_i}^{*}(\varepsilon) + C_{ii}(\varepsilon)K_{z_i}^{*}(\varepsilon)], \quad (\mathscr{H}_{i*}^{1})_{41}(t_f) = 0 \quad (108)$$

$$
\begin{aligned}
&\frac{d}{d\varepsilon}(\mathscr{H}_{i*}^{r})_{41}(\varepsilon) = -(A_{ic}(\varepsilon) - L_{ic}(\varepsilon)C_{ic}(\varepsilon))^{T}(\mathscr{H}_{i*}^{r})_{41}(\varepsilon) \\
&- (L_{ic}(\varepsilon)C_{ic}(\varepsilon))^{T}(\mathscr{H}_{i*}^{r})_{31}(\varepsilon) - (\mathscr{H}_{i*}^{r})_{41}(\varepsilon)[A_{ii}(\varepsilon) + B_{ii}(\varepsilon)K_{x_i}^{*}(\varepsilon) + C_{ii}(\varepsilon)K_{z_i}^{*}(\varepsilon)] \\
&- \sum_{v=1}^{r-1} \frac{2r!}{v!(r-v)!}\Big\{(\mathscr{H}_{i*}^{v})_{41}(\varepsilon)L_i(\varepsilon)V_iL_i^T(\varepsilon)(\mathscr{H}_{i*}^{r-v})_{11}(\varepsilon) \\
&- (\mathscr{H}_{i*}^{v})_{42}(\varepsilon)L_i(\varepsilon)V_iL_i^T(\varepsilon)(\mathscr{H}_{i*}^{r-v})_{11}(\varepsilon) - (\mathscr{H}_{i*}^{v})_{41}(\varepsilon)L_i(\varepsilon)V_iL_i^T(\varepsilon)(\mathscr{H}_{i*}^{r-v})_{21}(\varepsilon) \\
&+ (\mathscr{H}_{i*}^{v})_{42}(\varepsilon)G_i(\varepsilon)W_iG_i^T(\varepsilon)(\mathscr{H}_{i*}^{r-v})_{21}(\varepsilon) + (\mathscr{H}_{i*}^{v})_{42}(\varepsilon)L_i(\varepsilon)V_iL_i^T(\varepsilon)(\mathscr{H}_{i*}^{r-v})_{21}(\varepsilon) \\
&+ (\mathscr{H}_{i*}^{v})_{43}(\varepsilon)L_{ic}(\varepsilon)V_{ic}L_{ic}^T(\varepsilon)(\mathscr{H}_{i*}^{r-v})_{31}(\varepsilon) - (\mathscr{H}_{i*}^{v})_{44}(\varepsilon)L_{ic}(\varepsilon)V_{ic}L_{ic}^T(\varepsilon)(\mathscr{H}_{i*}^{r-v})_{31}(\varepsilon) \\
&- (\mathscr{H}_{i*}^{v})_{43}(\varepsilon)L_{ic}(\varepsilon)V_{ic}L_{ic}^T(\varepsilon)(\mathscr{H}_{i*}^{r-v})_{41}(\varepsilon) + (\mathscr{H}_{i*}^{v})_{44}(\varepsilon)G_{ic}W_{ic}G_{ic}^T(\varepsilon)(\mathscr{H}_{i*}^{r-v})_{41}(\varepsilon) \\
&+ (\mathscr{H}_{i*}^{v})_{44}(\varepsilon)L_{ic}V_{ic}L_{ic}^T(\varepsilon)(\mathscr{H}_{i*}^{r-v})_{41}(\varepsilon)\Big\}, \quad (\mathscr{H}_{i*}^{r})_{41}(t_f) = 0;\ 2 \le r \le k_i \qquad (109)
\end{aligned}
$$

$$
\begin{aligned}
\frac{d}{d\varepsilon}(\mathscr{H}_{i*}^{1})_{42}(\varepsilon) &= -(A_{ii}(\varepsilon) - L_i(\varepsilon)C_i(\varepsilon))^{T}(\mathscr{H}_{i*}^{1})_{42}(\varepsilon) - (L_i(\varepsilon)C_i(\varepsilon))^{T}(\mathscr{H}_{i*}^{1})_{32}(\varepsilon) \\
&- (\mathscr{H}_{i*}^{1})_{42}(\varepsilon)(A_{ii}(\varepsilon) - L_i(\varepsilon)C_i(\varepsilon)) \\
&- (\mathscr{H}_{i*}^{1})_{41}(\varepsilon)L_i(\varepsilon)C_i(\varepsilon), \quad (\mathscr{H}_{i*}^{1})_{42}(t_f) = 0 \qquad (110)
\end{aligned}
$$

$$
\begin{aligned}
&\frac{d}{d\varepsilon}(\mathscr{H}_{i*}^{r})_{42}(\varepsilon) = -(A_{ii}(\varepsilon) - L_i(\varepsilon)C_i(\varepsilon))^{T}(\mathscr{H}_{i*}^{r})_{42}(\varepsilon) - (L_i(\varepsilon)C_i(\varepsilon))^{T}(\mathscr{H}_{i*}^{r})_{32}(\varepsilon) \\
&- (\mathscr{H}_{i*}^{r})_{41}(\varepsilon)L_i(\varepsilon)C_i(\varepsilon) - (\mathscr{H}_{i*}^{r})_{42}(\varepsilon)(A_{ii}(\varepsilon) - L_i(\varepsilon)C_i(\varepsilon)) \\
&- \sum_{v=1}^{r-1} \frac{2r!}{v!(r-v)!}\Big\{(\mathscr{H}_{i*}^{v})_{41}(\varepsilon)L_i(\varepsilon)V_iL_i^T(\varepsilon)(\mathscr{H}_{i*}^{r-v})_{12}(\varepsilon) \\
&- (\mathscr{H}_{i*}^{v})_{42}(\varepsilon)L_i(\varepsilon)V_iL_i^T(\varepsilon)(\mathscr{H}_{i*}^{r-v})_{12}(\varepsilon) - (\mathscr{H}_{i*}^{v})_{41}(\varepsilon)L_i(\varepsilon)V_iL_i^T(\varepsilon)(\mathscr{H}_{i*}^{r-v})_{22}(\varepsilon) \\
&+ (\mathscr{H}_{i*}^{v})_{42}(\varepsilon)G_i(\varepsilon)W_iG_i^T(\varepsilon)(\mathscr{H}_{i*}^{r-v})_{22}(\varepsilon) + (\mathscr{H}_{i*}^{v})_{42}(\varepsilon)L_i(\varepsilon)V_iL_i^T(\varepsilon)(\mathscr{H}_{i*}^{r-v})_{22}(\varepsilon) \\
&+ (\mathscr{H}_{i*}^{v})_{43}(\varepsilon)L_{ic}(\varepsilon)V_{ic}L_{ic}^T(\varepsilon)(\mathscr{H}_{i*}^{r-v})_{32}(\varepsilon) - (\mathscr{H}_{i*}^{v})_{44}(\varepsilon)L_{ic}(\varepsilon)V_{ic}L_{ic}^T(\varepsilon)(\mathscr{H}_{i*}^{r-v})_{32}(\varepsilon) \\
&- (\mathscr{H}_{i*}^{v})_{43}(\varepsilon)L_{ic}(\varepsilon)V_{ic}L_{ic}^T(\varepsilon)(\mathscr{H}_{i*}^{r-v})_{42}(\varepsilon) + (\mathscr{H}_{i*}^{v})_{44}(\varepsilon)G_{ic}W_{ic}G_{ic}^T(\varepsilon)(\mathscr{H}_{i*}^{r-v})_{42}(\varepsilon) \\
&+ (\mathscr{H}_{i*}^{v})_{44}(\varepsilon)L_{ic}V_{ic}L_{ic}^T(\varepsilon)(\mathscr{H}_{i*}^{r-v})_{42}(\varepsilon)\Big\}, \quad (\mathscr{H}_{i*}^{r})_{42}(t_f) = 0;\ 2 \le r \le k_i \qquad (111)
\end{aligned}
$$

$$\frac{d}{d\varepsilon}(\mathscr{H}_{i*}^{1})_{43}(\varepsilon) = -(A_{ii}(\varepsilon) - L_i(\varepsilon)C_i(\varepsilon))^T(\mathscr{H}_{i*}^{1})_{43}(\varepsilon) - (L_i(\varepsilon)C_i(\varepsilon))^T(\mathscr{H}_{i*}^{1})_{33}(\varepsilon) - (\mathscr{H}_{i*}^{1})_{43}(\varepsilon)A_{ic}(\varepsilon)\,, \quad (\mathscr{H}_{i*}^{1})_{43}(t_f) = 0 \quad (112)$$

$$\begin{aligned}\frac{d}{d\varepsilon}(\mathscr{H}_{i*}^{r})_{43}(\varepsilon) &= -(A_{ii}(\varepsilon) - L_i(\varepsilon)C_i(\varepsilon))^T(\mathscr{H}_{i*}^{r})_{43}(\varepsilon) - (L_i(\varepsilon)C_i(\varepsilon))^T(\mathscr{H}_{i*}^{r})_{33}(\varepsilon)\\
&- (\mathscr{H}_{i*}^{r})_{43}(\varepsilon)A_{ic}(\varepsilon) - \sum_{v=1}^{r-1}\frac{2r!}{v!(r-v)!}\Big\{(\mathscr{H}_{i*}^{v})_{41}(\varepsilon)L_i(\varepsilon)V_iL_i^T(\varepsilon)(\mathscr{H}_{i*}^{r-v})_{13}(\varepsilon)\\
&- (\mathscr{H}_{i*}^{v})_{42}(\varepsilon)L_i(\varepsilon)V_iL_i^T(\varepsilon)(\mathscr{H}_{i*}^{r-v})_{13}(\varepsilon) - (\mathscr{H}_{i*}^{v})_{41}(\varepsilon)L_i(\varepsilon)V_iL_i^T(\varepsilon)(\mathscr{H}_{i*}^{r-v})_{23}(\varepsilon)\\
&+ (\mathscr{H}_{i*}^{v})_{42}(\varepsilon)G_i(\varepsilon)W_iG_i^T(\varepsilon)(\mathscr{H}_{i*}^{r-v})_{23}(\varepsilon) + (\mathscr{H}_{i*}^{v})_{42}(\varepsilon)L_i(\varepsilon)V_iL_i^T(\varepsilon)(\mathscr{H}_{i*}^{r-v})_{23}(\varepsilon)\\
&+ (\mathscr{H}_{i*}^{v})_{43}(\varepsilon)L_{ic}(\varepsilon)V_{ic}L_{ic}^T(\varepsilon)(\mathscr{H}_{i*}^{r-v})_{33}(\varepsilon) - (\mathscr{H}_{i*}^{v})_{44}(\varepsilon)L_{ic}(\varepsilon)V_{ic}L_{ic}^T(\varepsilon)(\mathscr{H}_{i*}^{r-v})_{33}(\varepsilon)\\
&- (\mathscr{H}_{i*}^{v})_{43}(\varepsilon)L_{ic}(\varepsilon)V_{ic}L_{ic}^T(\varepsilon)(\mathscr{H}_{i*}^{r-v})_{43}(\varepsilon) + (\mathscr{H}_{i*}^{v})_{44}(\varepsilon)G_{ic}W_{ic}G_{ic}^T(\varepsilon)(\mathscr{H}_{i*}^{r-v})_{43}(\varepsilon)\\
&+ (\mathscr{H}_{i*}^{v})_{44}(\varepsilon)L_{ic}V_{ic}L_{ic}^T(\varepsilon)(\mathscr{H}_{i*}^{r-v})_{43}(\varepsilon)\Big\}, \quad (\mathscr{H}_{i*}^{r})_{43}(t_f) = 0;\ 2 \le r \le k_i \quad (113)\end{aligned}$$

$$\begin{aligned}\frac{d}{d\varepsilon}(\mathscr{H}_{i*}^{1})_{44}(\varepsilon) &= -(A_{ii}(\varepsilon) - L_i(\varepsilon)C_i(\varepsilon))^T(\mathscr{H}_{i*}^{1})_{44}(\varepsilon) - (L_i(\varepsilon)C_i(\varepsilon))^T(\mathscr{H}_{i*}^{1})_{34}(\varepsilon)\\
&- (\mathscr{H}_{i*}^{1})_{44}(\varepsilon)(A_{ic}(\varepsilon) - L_{ic}(\varepsilon)C_{ic}(\varepsilon))\\
&- (\mathscr{H}_{i*}^{1})_{43}(\varepsilon)L_{ic}(\varepsilon)C_{ic}(\varepsilon)\,, \quad (\mathscr{H}_{i*}^{1})_{44}(t_f) = 0 \quad (114)\end{aligned}$$

$$\begin{aligned}\frac{d}{d\varepsilon}(\mathscr{H}_{i*}^{r})_{44}(\varepsilon) &= -(A_{ii}(\varepsilon) - L_i(\varepsilon)C_i(\varepsilon))^T(\mathscr{H}_{i*}^{r})_{44}(\varepsilon) - (L_i(\varepsilon)C_i(\varepsilon))^T(\mathscr{H}_{i*}^{r})_{34}(\varepsilon)\\
&- (\mathscr{H}_{i*}^{r})_{43}(\varepsilon)L_{ic}(\varepsilon)C_{ic}(\varepsilon) - (\mathscr{H}_{i*}^{r})_{44}(\varepsilon)(A_{ic}(\varepsilon) - L_{ic}(\varepsilon)C_{ic}(\varepsilon))\\
&- \sum_{v=1}^{r-1}\frac{2r!}{v!(r-v)!}\Big\{(\mathscr{H}_{i*}^{v})_{41}(\varepsilon)L_i(\varepsilon)V_iL_i^T(\varepsilon)(\mathscr{H}_{i*}^{r-v})_{14}(\varepsilon)\\
&- (\mathscr{H}_{i*}^{v})_{44}(\varepsilon)L_{ic}(\varepsilon)V_{ic}L_{ic}^T(\varepsilon)(\mathscr{H}_{i*}^{r-v})_{14}(\varepsilon) - (\mathscr{H}_{i*}^{v})_{41}(\varepsilon)L_i(\varepsilon)V_iL_i^T(\varepsilon)(\mathscr{H}_{i*}^{r-v})_{24}(\varepsilon)\\
&+ (\mathscr{H}_{i*}^{v})_{42}(\varepsilon)G_i(\varepsilon)W_iG_i^T(\varepsilon)(\mathscr{H}_{i*}^{r-v})_{24}(\varepsilon) + (\mathscr{H}_{i*}^{v})_{42}(\varepsilon)L_i(\varepsilon)V_iL_i^T(\varepsilon)(\mathscr{H}_{i*}^{r-v})_{24}(\varepsilon)\\
&+ (\mathscr{H}_{i*}^{v})_{43}(\varepsilon)L_{ic}(\varepsilon)V_{ic}L_{ic}^T(\varepsilon)(\mathscr{H}_{i*}^{r-v})_{34}(\varepsilon) - (\mathscr{H}_{i*}^{v})_{44}(\varepsilon)L_{ic}(\varepsilon)V_{ic}L_{ic}^T(\varepsilon)(\mathscr{H}_{i*}^{r-v})_{34}(\varepsilon)\\
&- (\mathscr{H}_{i*}^{v})_{43}(\varepsilon)L_{ic}(\varepsilon)V_{ic}L_{ic}^T(\varepsilon)(\mathscr{H}_{i*}^{r-v})_{44}(\varepsilon) + (\mathscr{H}_{i*}^{v})_{44}(\varepsilon)G_{ic}W_{ic}G_{ic}^T(\varepsilon)(\mathscr{H}_{i*}^{r-v})_{44}(\varepsilon)\\
&+ (\mathscr{H}_{i*}^{v})_{44}(\varepsilon)L_{ic}V_{ic}L_{ic}^T(\varepsilon)(\mathscr{H}_{i*}^{r-v})_{44}(\varepsilon)\Big\}, \quad (\mathscr{H}_{i*}^{r})_{44}(t_f) = 0;\ 2 \le r \le k_i \quad (115)\end{aligned}$$

$$\frac{d}{d\varepsilon}(\breve{\mathscr{D}}^1_{i*})_{11}(\varepsilon) = -(\mathscr{H}^1_{i*})_{13}(\varepsilon)B_{ic}(\varepsilon)u_c(\varepsilon) - K^{*T}_{x_i}(\varepsilon)R_{ii}(\varepsilon)p^*_{x_i}(\varepsilon)$$
$$- [A_{ii}(\varepsilon) + B_{ii}(\varepsilon)K^*_{x_i}(\varepsilon) + C_{ii}(\varepsilon)K^*_{z_i}(\varepsilon)]^T(\breve{\mathscr{D}}^1_{i*})_{11}(\varepsilon)$$
$$- (\mathscr{H}^1_{i*})_{11}(\varepsilon)[B_{ii}(\varepsilon)p^*_{x_i}(\varepsilon) + C_{ii}(\varepsilon)p^*_{z_i}(\varepsilon) + \sum_{j=1}^{N_i} B_{ij}(\varepsilon)u^*_{ij}(\varepsilon)]$$
$$- K^{*T}_{z_i}R_{zi}(\varepsilon)p^*_{z_i}(\varepsilon) + Q_i(\varepsilon)\zeta_i(\varepsilon), \quad (\breve{\mathscr{D}}^1_{i*})_{11}(t_f) = -Q^f_i\zeta_i(t_f) \quad (116)$$

$$\frac{d}{d\varepsilon}(\breve{\mathscr{D}}^r_{i*})_{11}(\varepsilon) = -[A_{ii}(\varepsilon) + B_{ii}(\varepsilon)K^*_{x_i}(\varepsilon) + C_{ii}(\varepsilon)K^*_{z_i}(\varepsilon)]^T(\breve{\mathscr{D}}^r_{i*})_{11}(\varepsilon)$$
$$- (\mathscr{H}^r_{i*})_{11}(\varepsilon)[B_{ii}(\varepsilon)p^*_{x_i}(\varepsilon) + C_{ii}(\varepsilon)p^*_{z_i}(\varepsilon) + \sum_{j=1}^{N_i} B_{ij}(\varepsilon)u^*_{ij}(\varepsilon)]$$
$$- (\mathscr{H}^r_{i*})_{13}(\varepsilon)B_{ic}(\varepsilon)u_c(\varepsilon), \quad (\breve{\mathscr{D}}^r_{i*})_{11}(t_f) = 0, \quad 2 \le r \le k_i \quad (117)$$

$$\frac{d}{d\varepsilon}(\breve{\mathscr{D}}^r_{i*})_{21}(\varepsilon) = -(A_{ii}(\varepsilon) - L_i(\varepsilon)C_i(\varepsilon))^T(\breve{\mathscr{D}}^r_{i*})_{21}(\varepsilon) - (L_i(\varepsilon)C_i(\varepsilon))^T(\breve{\mathscr{D}}^r_{i*})_{11}(\varepsilon)$$
$$- (\mathscr{H}^r_{i*})_{21}(\varepsilon)[B_{ii}(\varepsilon)p^*_{x_i}(\varepsilon) + C_{ii}(\varepsilon)p^*_{z_i}(\varepsilon) + \sum_{j=1}^{N_i} B_{ij}(\varepsilon)u^*_{ij}(\varepsilon)]$$
$$- (\mathscr{H}^r_{i*})_{23}(\varepsilon)B_{ic}(\varepsilon)u_c(\varepsilon), \quad (\breve{\mathscr{D}}^r_{i*})_{21}(t_f) = 0, \quad 1 \le r \le k_i \quad (118)$$

$$\frac{d}{d\varepsilon}(\breve{\mathscr{D}}^1_{i*})_{31}(\varepsilon) = -A^T_{ic}(\varepsilon)(\breve{\mathscr{D}}^1_{i*})_{31}(\varepsilon) + R_{zi}(\varepsilon)p_{z_i}(\varepsilon)$$
$$- (\mathscr{H}^1_{i*})_{31}(\varepsilon)[B_{ii}(\varepsilon)p^*_{x_i}(\varepsilon) + C_{ii}(\varepsilon)p^*_{z_i}(\varepsilon) + \sum_{j=1}^{N_i} B_{ij}(\varepsilon)u^*_{ij}(\varepsilon)]$$
$$- (\mathscr{H}^1_{i*})_{33}(\varepsilon)B_{ic}(\varepsilon)u_c(\varepsilon), \quad (\breve{\mathscr{D}}^1_{i*})_{31}(t_f) = 0 \quad (119)$$

$$\frac{d}{d\varepsilon}(\breve{\mathscr{D}}^r_{i*})_{31}(\varepsilon) = -A^T_{ic}(\varepsilon)(\breve{\mathscr{D}}^r_{i*})_{31}(\varepsilon)$$
$$- (\mathscr{H}^r_{i*})_{31}(\varepsilon)[B_{ii}(\varepsilon)p^*_{x_i}(\varepsilon) + C_{ii}(\varepsilon)p^*_{z_i}(\varepsilon) + \sum_{j=1}^{N_i} B_{ij}(\varepsilon)u^*_{ij}(\varepsilon)]$$
$$- (\mathscr{H}^r_{i*})_{33}(\varepsilon)B_{ic}(\varepsilon)u_c(\varepsilon), \quad (\breve{\mathscr{D}}^r_{i*})_{31}(t_f) = 0, \quad 2 \le r \le k_i \quad (120)$$

$$\frac{d}{d\varepsilon}(\breve{\mathscr{D}}^r_{i*})_{41}(\varepsilon) = -(L_{ic}(\varepsilon)C_{ic}(\varepsilon))^T(\varepsilon)(\breve{\mathscr{D}}^r_{i*})_{31}(\varepsilon) - (\mathscr{H}^r_{i*})_{43}(\varepsilon)B_{ic}(\varepsilon)u_c(\varepsilon)$$
$$- (\mathscr{H}^r_{i*})_{41}(\varepsilon)[B_{ii}(\varepsilon)p^*_{x_i}(\varepsilon) + C_{ii}(\varepsilon)p^*_{z_i}(\varepsilon) + \sum_{j=1}^{N_i} B_{ij}(\varepsilon)u^*_{ij}(\varepsilon)]$$
$$- (A_{ic}(\varepsilon) - L_{ic}(\varepsilon)C_{ic}(\varepsilon))^T(\breve{\mathscr{D}}^r_{i*})_{41}(\varepsilon), \quad (\breve{\mathscr{D}}^r_{i*})_{41}(t_f) = 0, \quad 1 \le r \le k_i \quad (121)$$

$$\frac{d}{d\varepsilon}\mathscr{D}^1_{i*}(\varepsilon) = -\mathrm{Tr}\{(\mathscr{H}^1_{i*})_{11}(\varepsilon)L_i(\varepsilon)V_iL_i^T(\varepsilon) - (\mathscr{H}^1_{i*})_{12}(\varepsilon)L_i(\varepsilon)V_iL_i^T(\varepsilon)\}$$
$$- \mathrm{Tr}\{-(\mathscr{H}^1_{i*})_{21}(\varepsilon)L_i(\varepsilon)V_iL_i^T(\varepsilon) + (\mathscr{H}^1_{i*})_{22}(\varepsilon)(G_i(\varepsilon)W_iG_i^T(\varepsilon) + L_i(\varepsilon)V_iL_i^T(\varepsilon))\}$$
$$- \mathrm{Tr}\{(\mathscr{H}^1_{i*})_{33}(\varepsilon)L_{ic}(\varepsilon)V_{ic}L_{ic}^T(\varepsilon) - (\mathscr{H}^1_{i*})_{34}(\varepsilon)L_{ic}(\varepsilon)V_{ic}L_{ic}^T(\varepsilon)\} - \zeta_i^T(\varepsilon)Q_i(\varepsilon)\zeta_i(\varepsilon)$$
$$- \mathrm{Tr}\{-(\mathscr{H}^1_{i*})_{43}(\varepsilon)L_{ic}(\varepsilon)V_{ic}L_{ic}^T(\varepsilon) + (\mathscr{H}^1_{i*})_{44}(\varepsilon)(G_{ic}W_{ic}G_{ic}^T(\varepsilon) + L_{ic}(\varepsilon)V_{ic}L_{ic}^T(\varepsilon))\}$$
$$- 2(\breve{\mathscr{D}}^1_{i*})^T_{11}(\varepsilon)[B_{ii}(\varepsilon)p_{x_i}(\varepsilon) + C_{ii}(\varepsilon)p_{z_i}(\varepsilon) + \sum_{j=1}^{N_i} B_{ij}(\varepsilon)u^*_{ij}(\varepsilon)] - p^{*T}_{x_i}(\varepsilon)R_{ii}(\varepsilon)p^*_{x_i}(\varepsilon)$$
$$- 2(\breve{\mathscr{D}}^1_{i*})^T_{31}(\varepsilon)B_{ic}(\varepsilon)u_c(\varepsilon) - p^{*T}_{z_i}(\varepsilon)R_{zi}(\varepsilon)p^*_{z_i}(\varepsilon), \quad \mathscr{D}^1_{i*}(t_f) = 0 \quad (122)$$

$$\frac{d}{d\varepsilon}\mathscr{D}^r_{i*}(\varepsilon) - \mathrm{Tr}\{(\mathscr{H}^r_{i*})_{11}(\varepsilon)L_i(\varepsilon)V_iL_i^T(\varepsilon) - (\mathscr{H}^r_{i*})_{12}(\varepsilon)L_i(\varepsilon)V_iL_i^T(\varepsilon)\}$$
$$- \mathrm{Tr}\{-(\mathscr{H}^r_{i*})_{21}(\varepsilon)L_i(\varepsilon)V_iL_i^T(\varepsilon) + (\mathscr{H}^r_{i*})_{22}(\varepsilon)(G_i(\varepsilon)W_iG_i^T(\varepsilon) + L_i(\varepsilon)V_iL_i^T(\varepsilon))\}$$
$$- \mathrm{Tr}\{(\mathscr{H}^r_{i*})_{33}(\varepsilon)L_{ic}(\varepsilon)V_{ic}L_{ic}^T(\varepsilon) - (\mathscr{H}^r_{i*})_{34}(\varepsilon)L_{ic}(\varepsilon)V_{ic}L_{ic}^T(\varepsilon)\}$$
$$- \mathrm{Tr}\{-(\mathscr{H}^r_{i*})_{43}(\varepsilon)L_{ic}(\varepsilon)V_{ic}L_{ic}^T(\varepsilon) + (\mathscr{H}^r_{i*})_{44}(\varepsilon)(G_{ic}W_{ic}G_{ic}^T(\varepsilon) + L_{ic}(\varepsilon)V_{ic}L_{ic}^T(\varepsilon))\}$$
$$- 2(\breve{\mathscr{D}}^r_{i*})^T_{11}(\varepsilon)[B_{ii}(\varepsilon)p^*_{x_i}(\varepsilon) + C_{ii}(\varepsilon)p^*_{z_i}(\varepsilon) + \sum_{j=1}^{N_i} B_{ij}(\varepsilon)u^*_{ij}(\varepsilon)]$$
$$- 2(\breve{\mathscr{D}}^r_{i*})^T_{31}(\varepsilon)B_{ic}(\varepsilon)u_c(\varepsilon), \quad \mathscr{D}^r_{i*}(t_f) = 0, \quad 2 \le r \le k_i. \quad (123)$$

Notice that as for comparison with other state-of-the-art research, the principal distinguishing feature of the research investigation herein is the pervasive use of noncooperative game theory and person-by-person equilibrium strategies (78) and (79) across the hierarchy for coordinated control of distributed systems. The emphasis is the recognition of the presence of a coordinator and incumbent systems and thus, addressing an important challenge in performance analysis supported by (84)–(123) for intra- and inter-interactions considered at the outset to achieve the attributes of "desired effects" and "tailored performance."

5 Conclusions

The present research investigation results in significant contributions to coordination control science's existing portfolio of methodologies. This portfolio contains a coordinator which directs two or more interconnected stochastic systems. Thinking about risk-averse attitudes toward performance uncertainty suggests new ideas for extending existing theories of distributed control and multiperson decision analysis. In this sense, the present research article suggested that making decisions using the proposed method protects decision makers and/or controller designers from overly optimistic design decisions that may not be the best under uncertainty. To account for mutual influence from immediate neighbors that give rise to interaction complexity such as potential noncooperation, each incumbent system or self-directed agent autonomously focuses on the search for a person-by-person equilibrium which is in turn remotely supported by local observers. Further discussions showed that the person-by-person equilibrium is equivalent to the concept of feedback Nash strategy. Another research issue discussed includes adjusting risk-averse attitudes via risk-value aware performance indices. The process of adjustment for performance risk aversion imposes some computational requirements as needed by the construction of the states of the person-by-person equilibrium.

Future work will focus on distributed multiscale modeling and control with explicit communications and partial information patterns, wherein research issues are: (a) how the feedback of incumbent systems would affect macroscales and macrostates of dominant coordinators? (b) how fast, small-scale behavior of incumbent systems could potentially trigger conformation changes of dominant coordinators? and (c) reliable and effective pathways for transferring information and knowledge from dominant players to fringe players and vice versa?

References

1. Friedman JW (1990) Game theory with applications to economics. 2nd edn. New York: Oxford University Press
2. Fudenberg D and Levine DK (1998) The theory of learning in games. Cambridge, MA: MIT Press
3. Huang M (2010) Large-population LQG games involving a major player: the Nash certainty equivalence principle. SIAM Journal of Control Optimization, 48(5):3318–3353
4. Pham KD (2008) Non-cooperative outcomes for stochastic nash games: decision strategies towards multi-attribute performance robustness. The 17th World Congress International Federation on Automatic Control, pp. 11750–11756
5. Pham KD (2008) New results in stochastic cooperative games: strategic coordination for multi-resolution performance robustness. In: Hirsch MJ, Pardalos PM, Murphey R, Grundel D (eds.) Optimization and Cooperative Control Strategies. Series Lecture Notes in Control and Information Sciences, Vol. 381, pp. 257–285, Springer Berlin, Heidelberg
6. Pham KD, Liberty SR and Jin G (2008) Multi-cumulant and Pareto solutions for tactics change prediction and performance analysis in stochastic multi-team non-cooperative games. In: Won C-H, Schrader C and Michel AN (eds.) Advances in statistical control, algebraic

systems theory and dynamic systems characteristics. Systems & Control: Foundations and Applications, pp. 65–97. Birkhauser, Boston
7. Pham KD (2011) Performance-reliability aided decision making in multiperson quadratic decision games against jamming and estimation confrontations. Journal of Optimization Theory and Applications, 149(3):559–629
8. Brockett RW (1970) Finite dimensional linear systems. Wiley, New York
9. Jacobson DH (1973) Optimal stochastic linear systems with exponential performance criteria and their relation to deterministic games. IEEE Transactions on Automatic Control, 18: 124–131
10. Whittle P (1990) Risk sensitive optimal control. John Wiley & Sons, New York
11. Fleming WH and Rishel RW (1975) Deterministic and stochastic optimal control. New York: Springer-Verlag

Modulating Communication to Improve Multi-agent Learning Convergence

Paul Scerri

Abstract The problem of getting interacting agents to learn simultaneously to improve their joint performance has received significant attention in the literature. One of the key challenges is to manage the system-wide effects that occur due to learning in a non-stationary environment. In this paper, we look at the impact on the system-wide dynamics and the learning convergence due to communication between the agents. Specifically, we look at the problem of learning routes between locations in a graph in the case where agents using the same edge at the same time slow each other down. We implemented and empirically examined a model where the agents simply try to model each edge in the graph as being either slow, medium, or fast due to the other agents using that edge. Communication on a fixed social network occurs only when an agent changes the speed category it has for a particular link, e.g., when it changes from believing a link is slow to believing it is medium. We find that the system dynamics are very sensitive to the ratio between the influence of direct observations on local beliefs to the influence of communicated beliefs. For some values of this ratio, convergence to good behavior can occur very quickly, but for others a brief period of good performance is followed by wild oscillations.

Keywords Multi-agent learning • Information sharing • Congestion games • Social networks

---

P. Scerri (✉)
Robotics Institute, Carnegie Mellon University, Pittsburgh, PA 15213, USA
e-mail: pscerri@cs.cmu.edu

A. Sorokin and P.M. Pardalos (eds.), *Dynamics of Information Systems: Algorithmic Approaches*, Springer Proceedings in Mathematics & Statistics 51,
DOI 10.1007/978-1-4614-7582-8_7,

1 Introduction

Many interesting domains require that robots or agents simultaneously learn and interact with one another for their mutual benefit over time. When the actions of one agent impact the outcomes of another agent, individual learning often leads to complex system dynamics. A canonical example of this problem is cooperative path planning [1, 7], where agents using the same routes negatively interfere with one other, but many other domains have been studied including soccer [17] and markets [25].

One of the core problems in multi-agent learning is that when one agent changes its behavior, it impacts the outcome for another agent. When all the agents simultaneously learn, the collective behavior and individual rewards can vary wildly and unpredictably. Controlling these system dynamics in a way that efficiently or even eventually has good behavior has received much attention. Typically, the approach is to somehow change local learning so that undesirable system effects are damped [6,22]. For example, some agents can hold their behavior fixed while others learn or agents can be made to learn slowly to provide a more stable environment for the other agents to learn.

In this chapter, we look at how modulating communication between agents can be used to control the system dynamics. Specifically, we hypothesize that if the agents share less information and have more localized models of the overall performance, they might naturally slow their learning rate and naturally allow the weakest individuals to replan against a more stable environment. However, it will clearly also be the case that less shared information means less ability to work out the right thing to do. We empirically investigate this hypothesis with a simple model of a graph and agents needing to repeatedly get between two locations on the graph, and agents using the same edges negatively interfere with one another. We find that communication does change the overall learning dynamics and that those dynamics are very sensitive to how the agents use communicated information to update their local models.

The simplest model for the agents is a history of times taken on particular edges, decayed over time to account for change. Whenever they traverse an edge, they communicate the time taken with some other members of the system. The agents used this model to plan the fastest route to their goal. An alternative model, which turns out to require much less communication to keep up to date, is for agents to simply model edges as *fast*, *medium*, or *slow*. The agents only communicate when they change from believing an edge is in one category to believing it is in another. An agent can change belief, either due to a local observation or a communicated message. Due to the fact that beliefs can change based on communicated information, a single observation may lead to a cascade of belief changes through the network. Intuitively, the ternary model focuses the agents on only communicating coarse information and slows some of the system dynamics because one slow or fast traversal of a link, by an agent will not necessarily be enough to cause it to change belief categories; however, if an edge is consistently slow or fast, that information will get propagated widely. We show empirically that the ternary model can get agents to good collective behavior more quickly than the averaging model, although over many iterations the solutions are not as good, because the agents have less fine-grained information to work with.

While the ternary model can get to good solutions faster, it can also lead to wild system dynamics. It turns out that the behavior of the ternary model is very sensitive to the weighting agents give to messages communicated from others, relative to the weighting they give to their own observations. If they weight information from communication low, the dynamics are stable but it takes longer to find good solutions, while higher weightings lead to quicker reactions but bad oscillations. The best system behavior is observed when this weighting is decayed over time, allowing the agents to initially share a lot of information but then stabilize the system dynamics once a reasonable solution is found.

2 Model

Our model consists of agents A, places P and edges G over some number of iterations. Each agent $a \in A$ has some place, $p_{\text{home}} \in P$ where it starts each iteration and some place $p_{\text{work}} \in P$ where it must get to each iteration. To get to p_{work} it must use edges connecting places. Individual edges $g \in G$ connect exactly two places. The agents task is to get from p_{home} to p_{work} most quickly each iteration.

The time that it will take an agent to traverse an edge depends purely on the number of agents already on the edge when it gets to the edge. Specifically, the time taken by an agent is $10 + n_{\text{already}}^3$, where n_{already} is the number of agents on the edge when the agent reaches it. The simulation randomizes the order the agents execute so that in one iteration an agent might be the first on the edge and have a very short travel time and another iteration it might be tenth onto the edge and have a very long travel time, even if none of the agents change their routes.

This model has two important features. First, the agents will get very different perspectives on speed of a edge, based on exactly when they get onto the edge. Hence, either many iterations or cooperation is needed to create an accurate model. Second, busy edges heavily penalize the agents, just a few extra agents on a edge will dramatically slow the last few agents down.

For experimental purposes, in all but one of the experimental cases below, all agents have the same p_{home} and p_{work}. This makes for more interesting traffic congestion problems and requires more coordination among the agents, but as Fig. 7 shows, changing this does not change the overall dynamics.

In every iteration, each agent uses a model of the graph to plan a path from p_{home} to p_{work}. The agents use a simple A* algorithm [24] to do the planning based on their current model of edge traversal times Agents are risk neutral, trying to minimize expected travel time. They then execute their plan without adapting to observed conditions. At the end of an iteration, the agents can communicate about what they observed. The model the agent plans with and the information it communicates are described below.

It is assumed that each agent plans selfishly but communicates truthfully and cooperatively. We are interested in two primary metrics. First, the average time it takes for an agent to get from p_{home} to p_{work}. Second, the volume of communication. As the agents build their models and adapt their plans to the changing models, the average transit time will change. As a secondary measure, we are interested in the change in average transit time over time.

Communication Network

The agents are organized into a social network where they can only communicate directly with a small subset of the rest of the agents. Information is propagated through the network in a peer-to-peer manner. Empirically we evaluate different network structures to understand the impact of how the information moves. Unless otherwise noted below, we use a random network with degree 5 to connect the agents.

2.1 Agent Reasoning

The agents have to choose a route that will most quickly get them to their destination, based on experiences so far and from experiences communicated from other agents. The optimal thing to do would be game theoretic reasoning that considers likely plans by others and the changes they will make, given their previous experiences. However, this is typically infeasible. Cooperative agents with low cost communication might coordinate in advance to balance the routes, but for the purposes of this work we assume that to be infeasible also. Moreover, if communication has any non-negligible cost, any agent will only have partial information about traffic on edges over time.

Below we describe two models for reasoning about the road network, the first uses a simple moving average of expected times for each edge, the second having the agents only characterize a road as *slow*, *medium*, or *fast*. Using either of these models the agents estimate the time taken to use a particular road and use A* to compute their expected fastest route, excluding any reasoning about how other agents might change their behavior. Notice that the agents are generally moving to a Nash Equilibrium, where, at least according to their local models, they have no incentive to change behavior. However, as has been noted before, even if the agents do reach a Nash Equilibrium, it may be the case that the outcome is far from the social optimal outcome [13, 18].

2.1.1 Averaging Model

The simplest model an agent can have of the graph is to store the average time taken by agents traversing that edge. Since the utilization of an edge will change over time, a moving average is used to keep the model updated with respect to the current situation.

Communication using the averaging model is constrained by the problems of double counting. If agents simply share their current estimates with each other, where those estimates use both their own observations and communication from other agents there is a possibility that individual observations can end up being taken into account multiple times and skewing the averages. Various consensus algorithms that essentially ignore this effect have been developed and shown to work reasonably well [21,28]. In this work, we take the conservative approach and require that agents share actual observations, which is expensive in terms of communication, but leads to principled, accurate averages. Hence, every time an agent traverses an edge, it communicates the time it took to traverse that edge to its direct neighbors in the social network.

The agents estimate for an edge is simply $e_i' = \alpha e_i + (1-\alpha)\text{obs}$, where e_i is the current estimate for the edge and obs is the new observation for the edge, whether communicated or observed locally. In this paper, we use $\alpha = 0.95$.

2.1.2 Ternary Model

In the ternary model, agents only track whether they believe a edge is *slow*, *medium*, or *fast*. The agents keep a normalized frequency distribution of the observations for each of the edges, decayed over time. Specifically, for each edge e, the agent has model $M_e = \{p_{\text{slow}}, p_{\text{medium}}, p_{\text{fast}}\}, p_{\text{slow}} + p_{\text{medium}} + p_{\text{fast}} = 1$. When an agent gets an observation of a particular category it adds β_{local} for a local observation and β for a communicated observation to the relevant p and then normalizes. For example, initially $M_e = p_{\text{slow}} = 0.33, p_{\text{medium}} = 0.33, p_{\text{fast}} = 0.33\}$, $\beta_{\text{local}} = 0.1$ and the agent observes an edge to be fast, $M' = \{p_{\text{slow}} = 0.302, p_{\text{medium}} = 0.302, p_{\text{fast}} = 0.395\}$.

The agents take the most probable category, $\max M$, and plan as if that was the case. In the experiments below, an edge in a particular category is assumed to take time 300, 156, and 12 for $p_{\text{slow}}, p_{\text{medium}}$, and p_{fast}, respectively, corresponding to the average time when approximately 3, 7, and 11 agents also use the edge. These were chosen to have a useful distribution for the default parameters. It might be more accurate to compute a value weighted by the different probabilities, but this is left for future work to allow for cleaner isolation of major effects. When $\max M$ changes for an edge, i.e., when the agent's belief about an edge changes categories, it communicates the new category to its direct neighbors in the social network.

This model was designed simply to make communication easier. The agents communicate whenever their model changes from believing the edge falls into one category to believing the edge falls into another speed category. Agents receiving communications about category changes need to decide how to integrate the measurement into their model. A communication will occur based on a number of observations building up belief in some category, so it could be weighted more heavily than a local observation, which is a single data point. However, as the experiments show, this leads to some undesirable system wide effects. The critical parameter is the ratio of the weighting of local observations to communication observations when integrating with the filter, β. Notice that the double counting effect can occur with this communication model, since an agent might receive messages from different network neighbors about the same category change, but those changes may have been caused by the same single observation (potentially taken by a completely other agent). As the experiments show, this effect is both real and important in impacting the system dynamics. Notice that when an agent receives a communication that another agent now believes an edge is in a certain category, it does not know anything about the observations that led the agent to reach that conclusion and therefore it is difficult to use it in a completely principled way. It may be that it receives multiple messages about the same edge from two different neighbors that reached their conclusion based on the observations of a third neighbor. Thus, the weighting the agent uses is a heuristic that balances the value in using communication to build their own model and the risk of being very sensitive to few observations.

3 Empirical Investigation

In this section, we present a detailed experimental investigation of the model and communication protocol presented above. The investigation shows that the ternary model can lead the agents to much more quickly find good solutions, but the higher the β the wilder and worse the oscillations that occur over time. Unless otherwise stated, the following parameters were used across all the experiments (Table 1).

The graph network was a small world network created by randomly placing locations on a unit square, connecting all locations within 0.2 of each other and then creating ten random edges. The time taken to traverse an edge is independent of the length of the edge. For each of the graphs, the x-axis shows the iteration number and the y-axis shows the average travel time for all agents averaged over all runs, therefore lower is better performance.

In the following, we present a series of experiments designed to understand the model. First, Fig. 1 compares the averaging model with two ternary models, using $\beta = 0.05$ and $\beta = 0.15$. Notice that when $\beta = 0.15$, i.e., when agents place more weight on communicated information the early performance is better than the averaging algorithm, but after about 100 iterations wild dynamics begin to occur and average performance falls off very badly. This phenomena is qualitatively the same as a phenomena known as *bursting* in adaptive control [2, 15] and is likely caused by similar dynamics, though the distributed system state and high uncertainty makes this difficult to confirm.

When communication from neighbors is weighted less heavily, i.e., $\beta = 0.05$, the system gets to good solutions much more slowly, but remains stable over time. Even at this lower β the averaging model does better, because it eventually has more detailed information, i.e., it has average times for each edge, not just one of three categories. The averaging model has far more consistent behavior, at first improving greatly and then slipping back before finally getting good performance. We can conclude from this that the ternary model can work very quickly, though the β value needs to be chosen carefully.

Table 1 Default parameter values

Parameter	Value
No. of iterations	300
No. of agents	100
Graph network	Small world
No. of places	100
No. of random links	10
β	0.15
β_{local}	0.2
Social network	Random
Runs	200

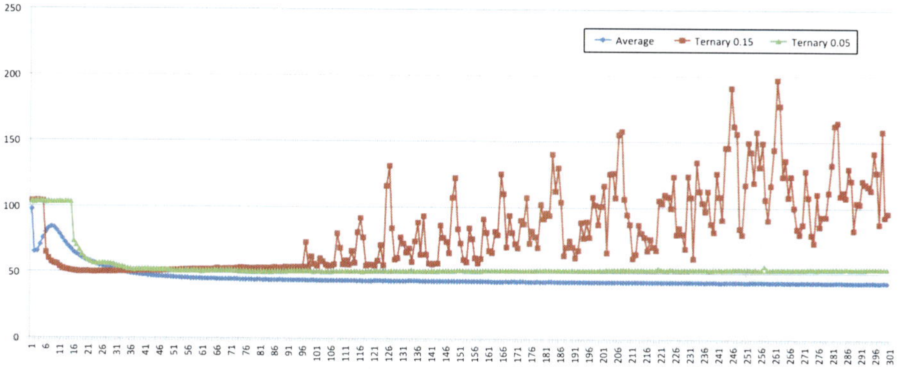

Fig. 1 Comparison of consensus model with ternary model with two different neighbor weightings. Notice that the ternary model with the higher β finds good solutions quickly but then becomes unstable. With the lower β, the ternary model does not outperform the averaging model

Next we compared ten different values for β, ranging from very low, i.e., 0.01, representing almost no influence by neighbors to high enough so that most of the agent's information comes from shared information. Figure 2 shows the results with Fig. 3 showing the first few iterations of the same data. There is a clear pattern as β goes from low to high. The higher β, the earlier the system finds good solutions, but also the earlier it becomes unstable. The very lowest β values, 0.01 and 0.06, do not become unstable for thousands of iterations (we cannot say for certain that it remains stable forever). There appears to be a clear trade-off between the speed at which good solutions are found and how stable the system is over time.

The social network connecting the agents restricts how information moves across the system. Therefore it is reasonable to expect that changing the structure of the network impacts the system dynamics. Figure 4 shows the dynamics with three different network structures. The *Ring* network singly connects the agents into a ring, resulting in the least connectivity. The *Small Worlds* network doubly connects the agents into a ring and includes a small number of random links. The *Random* network is as described above. Notice that the *Small Worlds* and *Ring* networks perform stably and identically, while the *Random* network exhibits the oscillations. We can conclude that the network structure can also damp the flow of information,

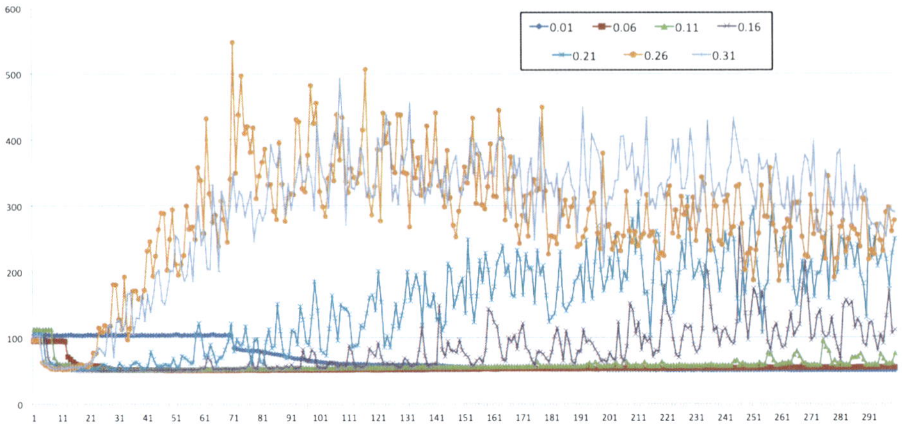

Fig. 2 Comparison of different neighbor weighting rates, β. Notice the higher rates have the agents finding good solutions more quickly but later the system oscillates chaotically, while lower rates get to good solutions more slowly, but are more stable over time

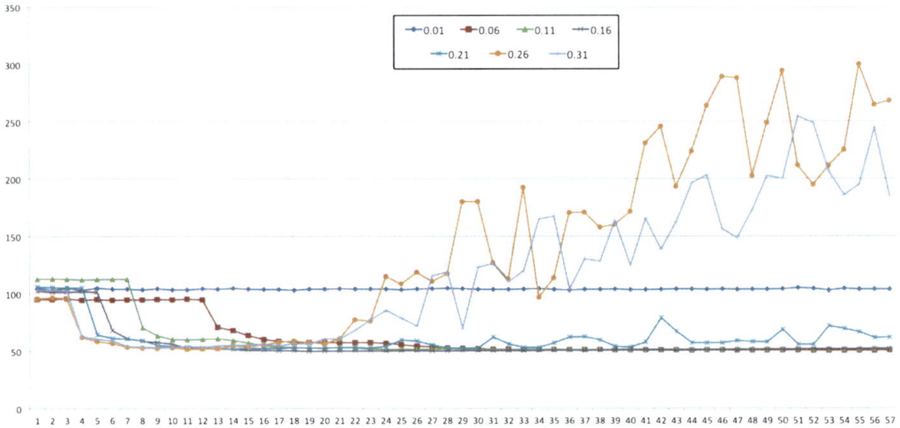

Fig. 3 Comparison of different neighbor rates over the first few iterations, highlighting the convergence

in the same way as changing β and therefore stabilizes the system. In Figs. 5 and 6 we compare the details of the social network experiments versus the averaging case. In Fig. 5 the number of agents that change routes from the iteration before is shown. These are agents that have received enough new information to have found a different best path. The averaging case exhibits qualitatively different behavior to the ternary cases, with the number of agents changing routes slowly decreasing over time. In the ternary cases, no agents change behavior until enough evidence has built up to cause a small number of agents to change. Curiously, the wild oscillations observed in the *Random* network case are caused by relatively few agents changing routes. Figure 6 shows the amount of communication used by each of the algorithms, with a log-scale on the y-axis. The averaging case sends a message

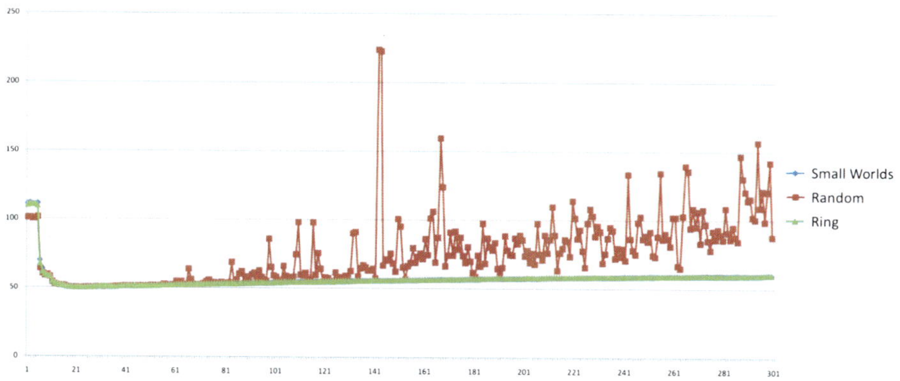

Fig. 4 Comparison of three different social networks

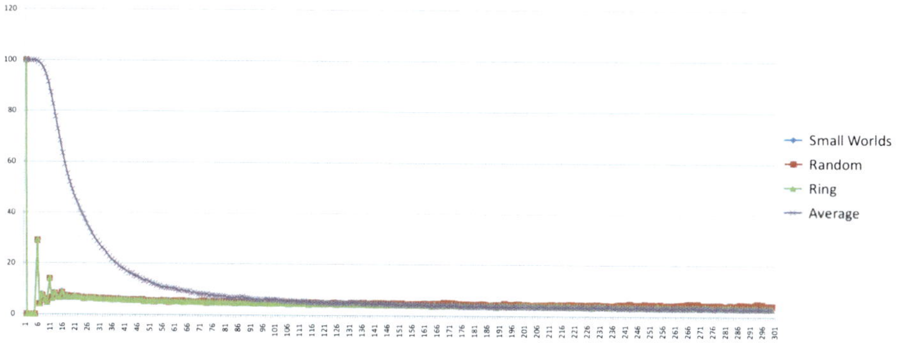

Fig. 5 The number of agents changing their routes from the iteration before, for averaging case and ternary case with three different social networks

for every edge traversed, which may not be necessary if a smarter consensus algorithm was used. However, the ternary models use an order of magnitude fewer messages. The different social networks also have qualitatively different behavior: in the *Random* network case, the communication drifts up over time, as the system oscillates causing changes in belief, while for the *Ring* and *Small Worlds* case, the communication level drifts down as the system stabilizes.

The previous experiments have all the agents using the same start and end locations. This means the agents all want to use the same edges. In the next experiment, we relaxed this, allowing more than one start and end location. Figure 7 shows the dynamics with between one and ten start and end locations. It appears that this makes no difference to the behavior, even though the congestion varies. It may be that the random links in the small worlds network are the key links regardless of the start and end locations, this is a question for future work.

Given the apparent trade-off and stability in the choice of β, we hypothesized that decreasing it over time might give the best of both worlds, good early performance followed by stability later. Figure 8 shows the result, starting with $\beta = 0.3$ and

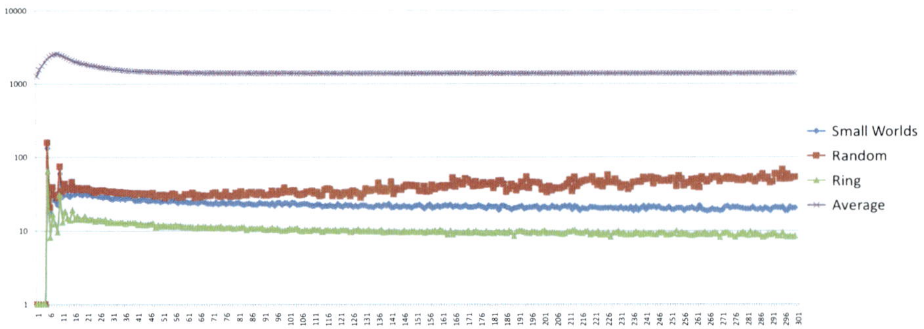

Fig. 6 The volume of communication per iteration, for averaging case and ternary case with three different social networks

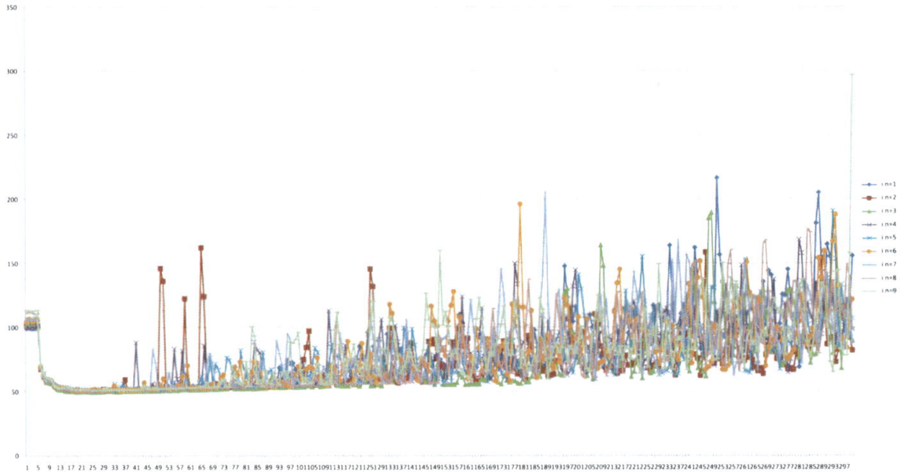

Fig. 7 Comparison of performance as the number of different start and end locations is changed, changing the rate of overlap between agents

decreasing by a fixed decay rate each iteration. A high enough decay rate can in fact lead to stable performance over time. However, even picking the empirically identified ideal decay rate, 0.007, does not allow the ternary model to outperform the averaging model in the long term, because the model is coarser providing less information for the agents to plan with. With the decay, however, over approximately 30 iterations, the ternary model is far superior (Fig. 9).

In the experiments above, the system dynamics were looked at over 300 iterations. In the next set of experiments, we looked in detail at what happens over a much longer period of time, when the system is given a chance to settle out of its oscillating behavior. Figures 10–12 show various metrics for the case of $\beta \in 0.1, 0.2, 0.3$, i.e., weightings of communicated information that cause stable through unstable behavior. In each case, each point along the x-axis is the average over 10 iterations, so the 200 points represent 2,000 iterations.

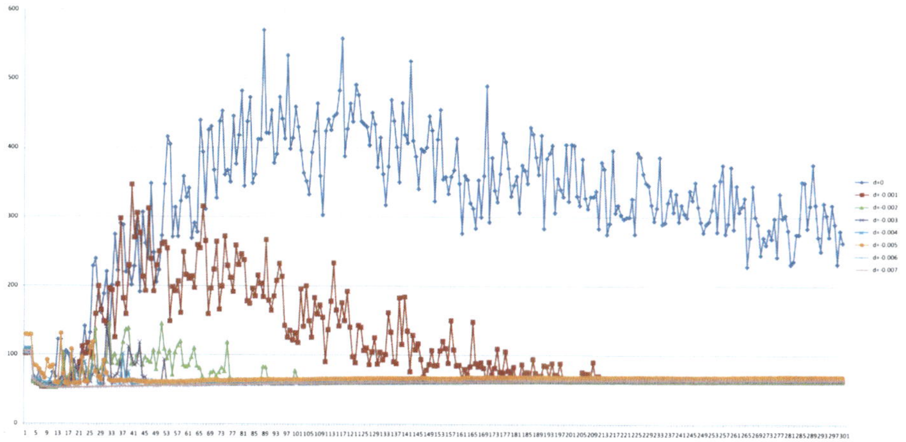

Fig. 8 Decaying the neighbor weighting rate over time. An appropriate rate of decay can balance between quick convergence and stability

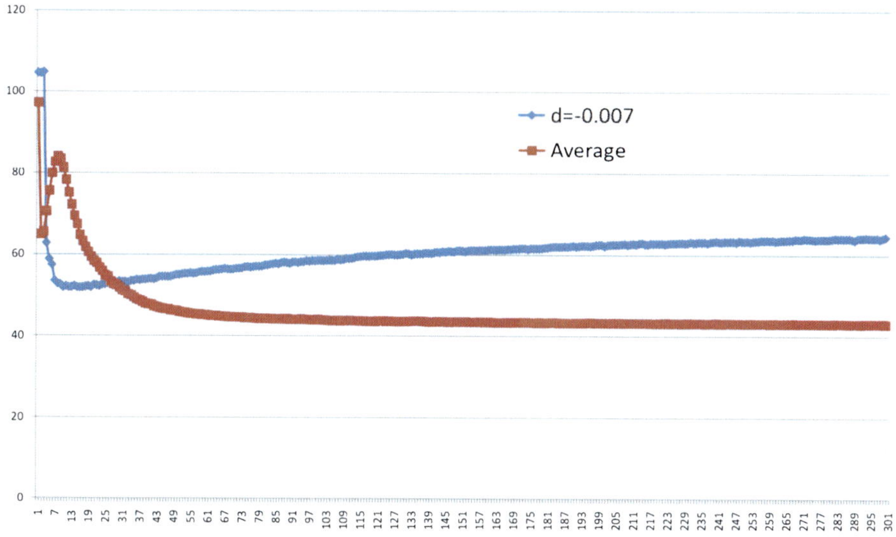

Fig. 9 Comparison of ternary model with a decay factor 0.007 on neighbor weights, β, with the averaging model

The figures show the qualitative differences in behavior, despite the long-run behavior being approximately the same. Figure 10 shows the average transit times. For $\beta = 0.3$, the oscillations occur but soon settle down so that average times are approximately the same as for the more stable cases. However, the next four figures show that the dynamics of getting to this relatively stable behavior and the behavior itself are qualitatively different for the three cases. Figure 11 shows the number of agents that choose a different path from the iteration before. This is a

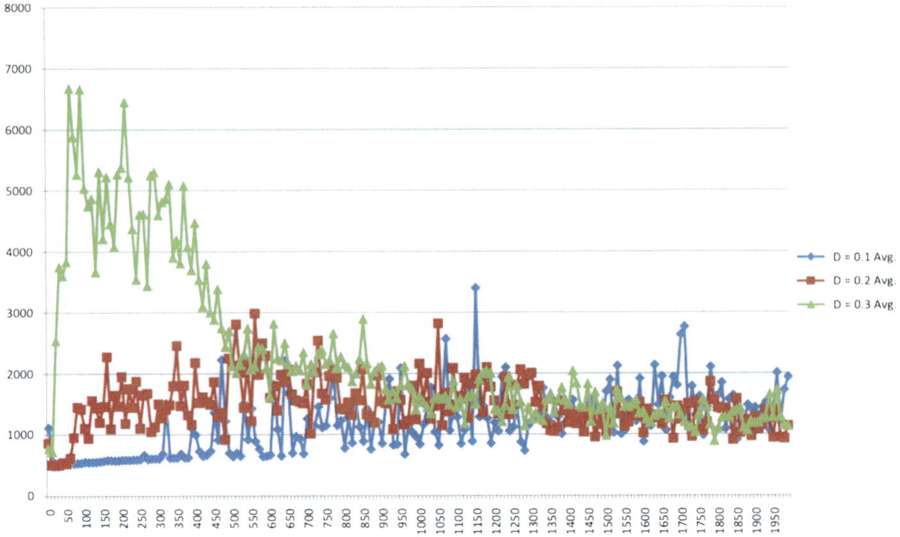

Fig. 10 Average transit times over 2,000 iterations with $\beta = D \in 0.1, 0.2, 0.3$

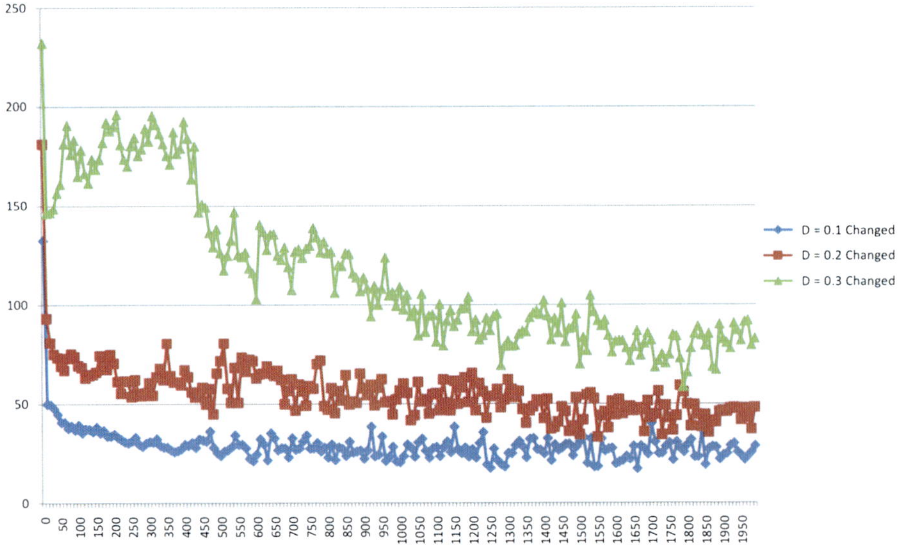

Fig. 11 The number of agents changing routes from the previous iteration over 2,000 iterations with $\beta = D \in 0.1, 0.2, 0.3$

measure of how much the information sharing is impacting the agents, since they will change routes whenever they believe a different route will be faster. Notice that the more communication is valued, higher β, the more agents that typically change. This is to be expected, since more information should lead to more changes.

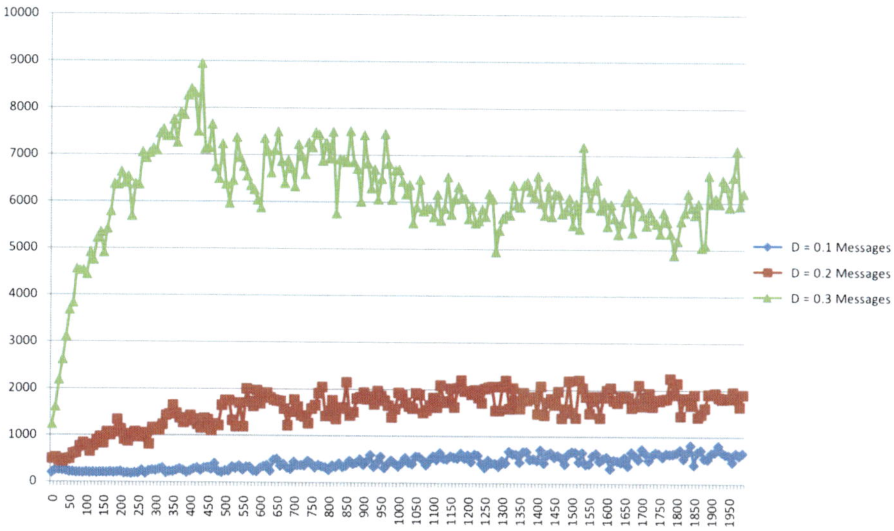

Fig. 12 The average number of messages at an iteration over 2,000 iterations with $\beta = D \in 0.1, 0.2, 0.3$

However, curiously, even after more than 1,000 iterations, when the average transit time is approximately equal for each of the cases, the number of agents changing at each step is still significantly different and apparently stably so. The different message weightings lead to different amounts of churn. This is supported in Fig. 12 which shows the number of messages at each iteration. Although they use the same communication algorithm, the different weightings lead to different volumes of communication, which stabilize over time, as a function of the noise in the system due to the randomization of which agent gets to the edge first. The next two graphs hint that while the average times are stable and approximately the same, there are still qualitatively different dynamics going on. Figure 13 shows the time of the last agent to complete its route, the outlier of the system and Fig. 14 shows the time of the median agent. The time of the last agent actually goes up over time when the communication weight is low, while it eventually consistently comes down when communicated information is weighted more highly. This suggests that the communication is good at getting rid of some of the outliers, although it caused more early on, during the unstable phase. The median times show a significant and qualitative change over time. In the cases where communication is weighted higher, the median fall consistently, while when there is less weight on communication, median behavior is at first better but then drifts higher. When communication is weighted higher, there is more coherence across the system, bringing the median down.

In the final experiment, we looked at whether it was the use of three categories that made the difference or simply the use of categories versus averaging. The results are shown in Figs. 15–20. Figures 15–17 show the average transit times for

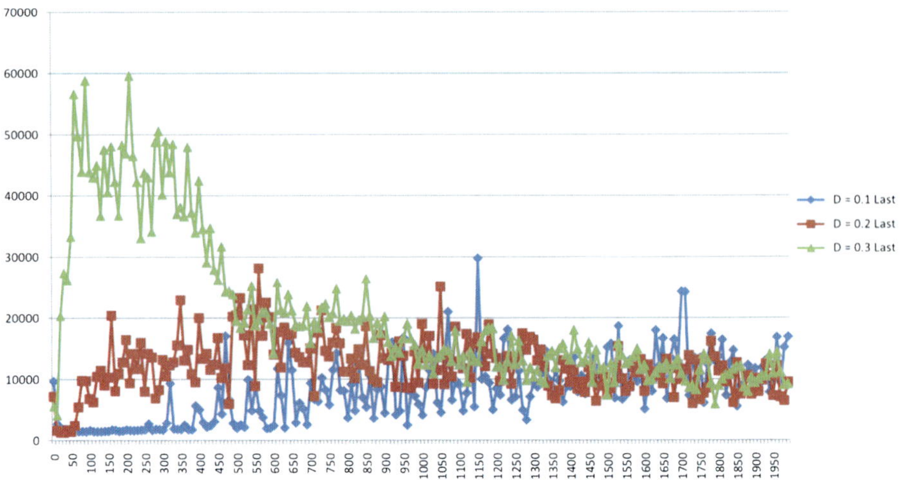

Fig. 13 The time of the last agent completing its traversal over 2,000 iterations with $\beta = D \in 0.1, 0.2, 0.3$

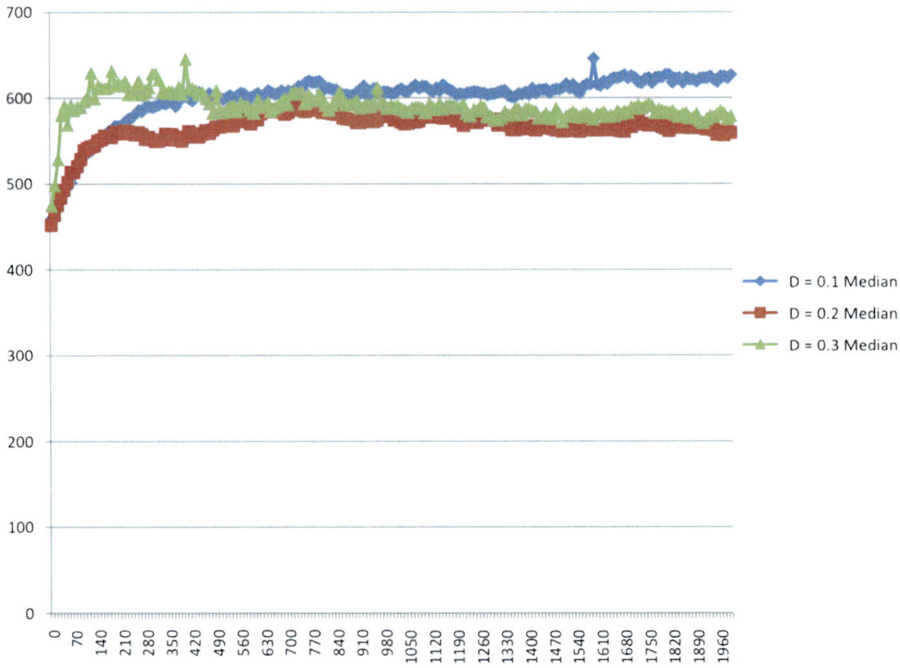

Fig. 14 The time of the median traversal over 2,000 iterations with $\beta = D \in 0.1, 0.2, 0.3$

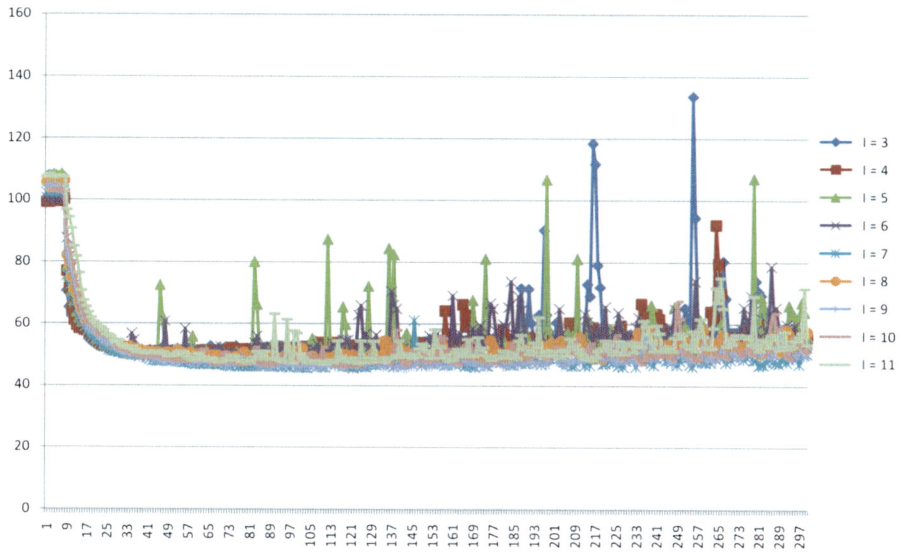

Fig. 15 Average transit times using different numbers of categories, I for edge traversal times, with $\beta = 0.1$

between three and eleven categories for $\beta = 0.1$, $\beta = 0.2$, and $\beta = 0.3$, respectively. For the lowest communication weighting, the extra categories do not make any difference; however, when communication is weighted more highly, there is an interesting bifurcation. With up to six categories, the unstable behavior discussed above is observed, but with more than six categories the system is more stable, acting more like the averaging model. While in the limit, with an infinite number of categories, it is clear that the behavior should approach that of the averaging model, it is less clear why the bifurcation should occur. Figures 18–20 shed some light on this. They show the average number of messages exchanged in each iteration. The bifurcation is quite dramatic at six categories. What appears to happen is that with enough categories, each agent gets enough messages and is sensitive enough to them to communicate further. This leads to a lot of information being exchanged and the overall system acting in a fundamentally different way. Thus, the effect is emergently the same as more communication or a denser social network in having the overall system be more coherent, though it is achieved here through a different mechanism.

4 Related Work

Using agents to manage congestion in road networks has been addressed from a variety of perspectives [4]. Learning of traffic light patterns has been of particular interest [10, 19]. Bazzan has previously found that sharing information between

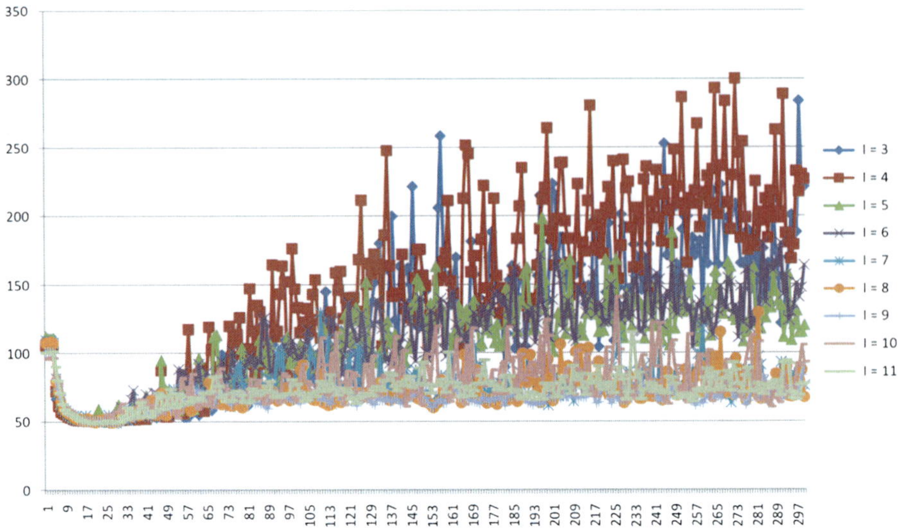

Fig. 16 Average transit times using different numbers of categories, I for edge traversal times, with $\beta = 0.2$

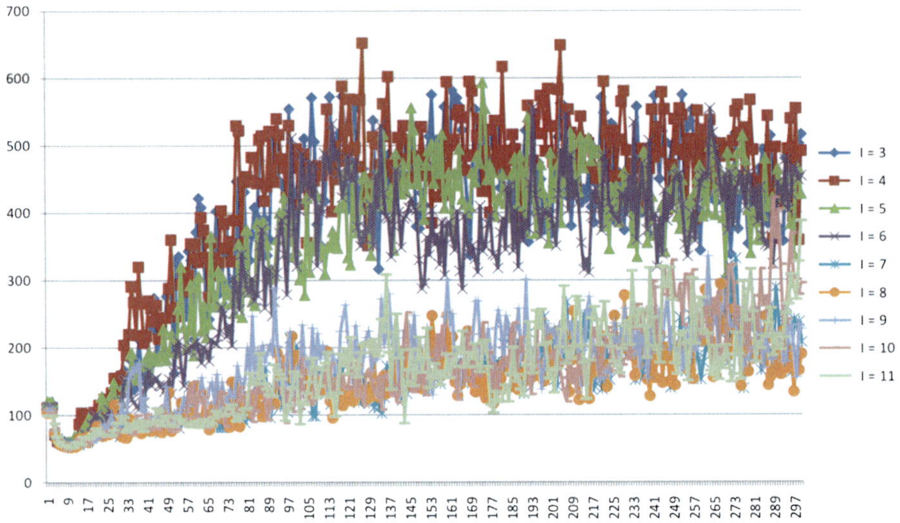

Fig. 17 Average transit times using different numbers of categories, I for edge traversal times, with $\beta = 0.3$

traffic lights does not necessarily help performance [9]. More general management of congestion has also been looked at extensively [3, 5]. Multi-agent learning is an extensively studied problem [8, 30]. Most work focuses on how individual agents

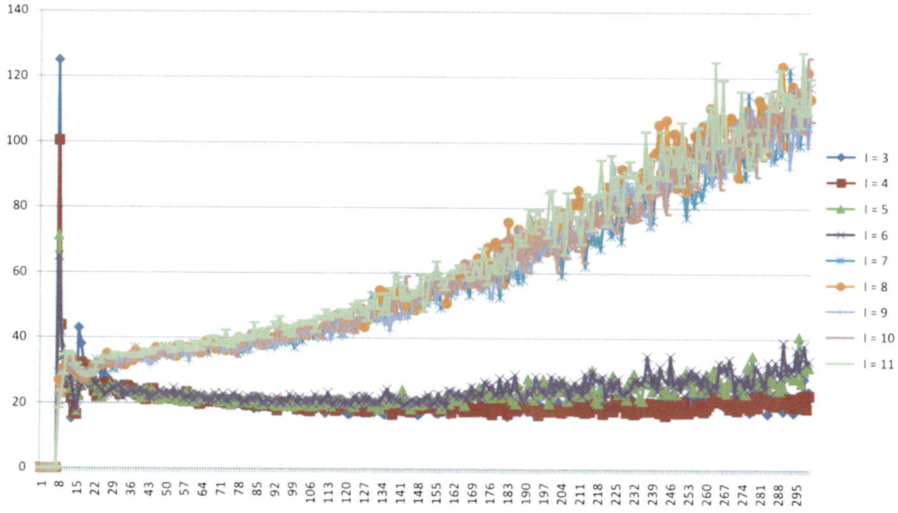

Fig. 18 Number of messages per iteration using different numbers of categories, I for edge traversal times, with $\beta = 0.1$

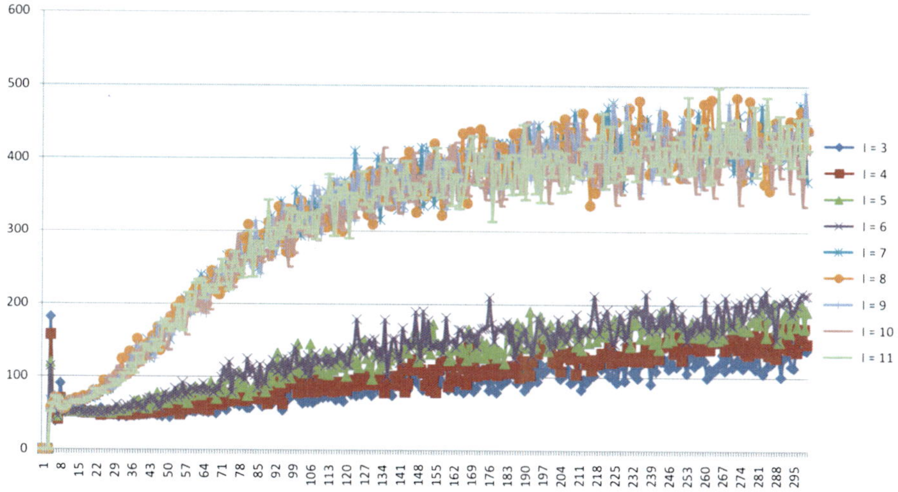

Fig. 19 Number of messages per iteration using different numbers of categories, I for edge traversal times, with $\beta = 0.2$

should learn in the context of the team, e.g., [16, 29]. Multi-agent versions of reinforcement learning has been a particularly popular approach [8, 22, 26].

Watts [27] empirically studies the global cascades under random networks with different interpersonal influences. Hirshleifer [14] studies the information sharing and propagation in social networks. Nekovee et al. [20] builds a stochastic model for spreading of rumors. They use mean-field analysis equation to describe the

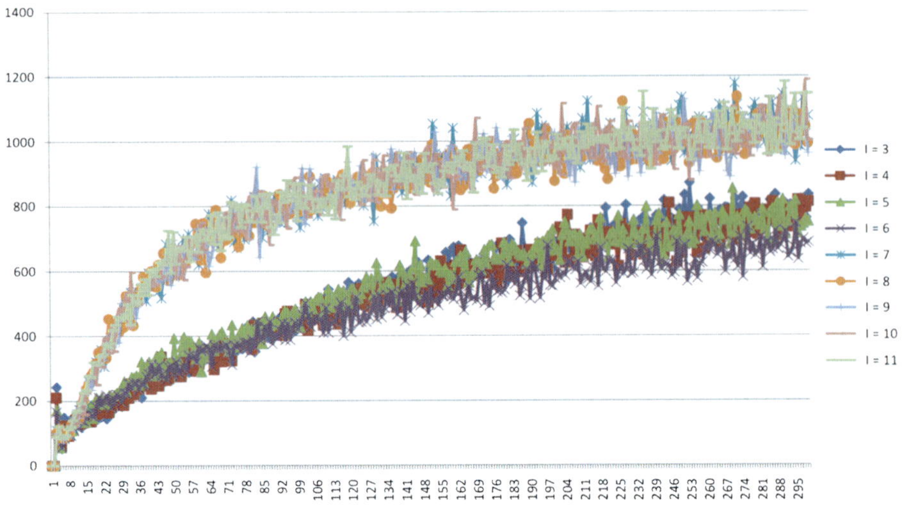

Fig. 20 Number of messages per iteration using different numbers of categories, I for edge traversal times, with $\beta = 0.3$

dynamics of the model for complex social network and show that large complex networks has critical threshold in the rumor spreading rate. Glinton et al. [11, 12] have studied the emergent behavior of the dynamics of large-scale networks. Their model is based on a network of small number of sensors and a team of agents who share information about a single fact. They have shown that system behaves optimally in a very small range of system parameters. When the parameters deviate from that range, the system performance degrades dramatically. They have shown that this behavior is caused by cascades of belief changes due to a single sensor reading. All these works show the emergent behavior of information cascading, but they have not studied the multidimensional facts or the relation of the dynamic behavior and the correlation of different facts. Reece et al. [23] have provided a multi-dimensional trust model to allow agents to share correlated multi-dimensional contracts. They have developed an approach based on Kalman filter to fuse relevant information from other agents. They have shown that their approach improves significantly over the simple approach based on single dimension of trust.

5 Conclusions and Future Work

In this paper, we showed the potential for a simple ternary model and communication scheme to help a simultaneously learning agents find good joint solutions faster. However, we also showed that a model where agents keep and share more detailed information eventually leads to better solutions. A planned direction for future work is look at combining the models, using the ternary model for quick early solutions and then using the detailed averaging model to find the best solutions. The ternary model uses dramatically less communication and propagates coarse information relatively slowly. Theoretically, we would like to understand better exactly how much communication is required to get the best multi-agent learning performance and why less information is apparently helpful. Finally, we intend to combine this technique for managing learning dynamics with more traditional multi-agent learning approaches to see if even better performance can be achieved.

Acknowledgment This research has been funded in part by the AFOSR MURI grant FA9550-08-1-0356.

References

1. A. Ahmed, P. Varakantham, and S.F. Cheng. Uncertain congestion games with assorted human agent populations. 2012.
2. B. Anderson. Adaptive systems, lack of persistency of excitation and bursting phenomena. *Automatica*, 21(3):247–258, 1985.
3. A.L.C. Bazzan. *Multi-agent systems for traffic and transportation engineering*. Information Science Publishing, 2009.
4. A.L.C. Bazzan. Opportunities for multiagent systems and multiagent reinforcement learning in traffic control. *Autonomous Agents and Multi-Agent Systems*, 18(3):342–375, 2009.
5. N. Bhouri, S. Haciane, and F. Balbo. A multi-agent system to regulate urban traffic: Private vehicles and public transport. In *Intelligent Transportation Systems (ITSC), 2010 13th International IEEE Conference on*, pages 1575–1581. IEEE, 2010.
6. M. Bowling and M. Veloso. Multiagent learning using a variable learning rate. *Artificial Intelligence*, 136(2):215–250, 2002.
7. W. Burgard, M. Moors, D. Fox, R. Simmons, and S. Thrun. Collaborative multi-robot exploration. In *Robotics and Automation, 2000. Proceedings. ICRA'00. IEEE International Conference on*, volume 1, pages 476–481. IEEE, 2000.
8. L. Busoniu, R. Babuska, and B. De Schutter. A comprehensive survey of multiagent reinforcement learning. *Systems, Man, and Cybernetics, Part C: Applications and Reviews, IEEE Transactions on*, 38(2):156–172, 2008.
9. D. De Oliveira and A.L.C. Bazzan. Multiagent learning on traffic lights control: effects of using shared information. *Multi-agent systems for traffic and transportation engineering*, 2009.
10. S. El-Tantawy and B. Abdulhai. An agent-based learning towards decentralized and coordinated traffic signal control. In *Intelligent Transportation Systems (ITSC), 2010 13th International IEEE Conference on*, pages 665–670. IEEE, 2010.
11. R.T. Glinton, P. Scerri, and K. Sycara. Towards the understanding of information dynamics in large scale networked systems. In *Information Fusion, 2009. FUSION'09. 12th International Conference on*, pages 794–801. IEEE, 2009.
12. R. Glinton, P. Scerri, and K. Sycara. Exploiting scale invariant dynamics for efficient information propagation in large teams. In *Proceedings of the 9th International Conference on Autonomous Agents and Multiagent Systems: volume 1-Volume 1*, pages 21–30. International Foundation for Autonomous Agents and Multiagent Systems, 2010.
13. J.N. Hagstrom and R.A. Abrams. Characterizing braess's paradox for traffic networks. In *Intelligent Transportation Systems, 2001. Proceedings. 2001 IEEE*, pages 836–841. IEEE, 2001.
14. D. Hirshleifer. The Blind Leading the Blind: Social Influence, Fads, and Informational Cascades. *University of California at Los Angeles, Anderson Graduate School of Management*, 1993.

15. BA Huberman and E. Lumer. Dynamics of adaptive systems. *Circuits and Systems, IEEE Transactions on*, 37(4):547–550, 1990.
16. M. Kaisers and K. Tuyls. Frequency adjusted multi-agent q-learning. In *Proceedings of the 9th International Conference on Autonomous Agents and Multiagent Systems: volume 1-Volume 1*, pages 309–316. International Foundation for Autonomous Agents and Multiagent Systems, 2010.
17. S. Kalyanakrishnan, Y. Liu, and P. Stone. Half field offense in robocup soccer: A multiagent reinforcement learning case study. *RoboCup 2006: Robot Soccer World Cup X*, pages 72–85, 2007.
18. Y.A. Korilis, A.A. Lazar, and A. Orda. Avoiding the braess paradox in non-cooperative networks. *Journal of Applied Probability*, 36(1):211–222, 1999.
19. S. Lämmer and D. Helbing. Self-stabilizing decentralized signal control of realistic, saturated network traffic. Santa Fe Institute, 2010.
20. M. Nekovee, Y. Moreno, G. Bianconi, and M. Marsili. Theory of rumour spreading in complex social networks. *Physica A: Statistical Mechanics and its Applications*, 374(1):457–470, 2007.
21. R. Olfati-Saber, J.A. Fax, and R.M. Murray. Consensus and cooperation in networked multi-agent systems. *Proceedings of the IEEE*, 95(1):215–233, 2007.
22. L. Panait and S. Luke. Cooperative multi-agent learning: The state of the art. *Autonomous Agents and Multi-Agent Systems*, 11(3):387–434, 2005.
23. S. Reece, S. Roberts, A. Rogers, and N.R. Jennings. A multi-dimensional trust model for heterogeneous contract observations. In *PROCEEDINGS OF THE NATIONAL CONFERENCE ON ARTIFICIAL INTELLIGENCE*, volume 22, page 128. Menlo Park, CA; Cambridge, MA; London; AAAI Press; MIT Press; 1999, 2007.
24. S. Russell, P. Norvig, and A. Artificial Intelligence. A modern approach. *Artificial Intelligence. Prentice-Hall, Egnlewood Cliffs*, 1995.
25. L. Tesfatsion and K.L. Judd. *Handbook of computational economics: agent-based computational economics*, volume 2. North Holland, 2006.
26. M. Vasirani and S. Ossowski. A computational market for distributed control of urban road traffic systems. *Intelligent Transportation Systems, IEEE Transactions on*, (99):1–9, 2011.
27. D.J. Watts. A simple model of global cascades on random networks. *Proceedings of the National Academy of Sciences of the United States of America*, 99(9):5766, 2002.
28. F. Xiao and L. Wang. Asynchronous consensus in continuous-time multi-agent systems with switching topology and time-varying delays. *Automatic Control, IEEE Transactions on*, 53(8):1804–1816, 2008.
29. C. Zhang and V. Lesser. Multi-agent learning with policy prediction. In *Proceedings of the 24th National Conference on Artificial Intelligence (AAAI10)*, 2010.
30. C. Zhang, V. Lesser, and P. Shenoy. A multi-agent learning approach to online distributed resource allocation. In *IJCAI 2009, Proceedings of the Twenty-first International Joint Conference on Artificial Intelligence*, pages 361–366, 2009.

Minimum-Risk Maximum Clique Problem

Maciej Rysz, Pavlo A. Krokhmal, and Eduardo L. Pasiliao

Abstract In this work, we consider the minimum-risk maximum clique problem on stochastic graphs. Namely, assuming that each vertex of the graph is associated with a random variable describing a cost or a loss, such that the joint distribution of all variables on the graph is known, the goal is to determine a clique in the graph that has the lowest risk, given a specific risk measure. It is shown that in the developed problem formulation, minimization of risk is facilitated through inclusion of additional vertices in the partial solution, whereby an optimal solution represents a maximal clique in the graph. In particular, two instances of risk-averse maximum clique problems are considered, where risk exposures of a graph's vertices are "isolated" (i.e., not dependent on risk profiles of other vertices) and "neighbor-dependent," or dependent on the risk profiles of adjacent vertices. Numerical experiments on randomly generated Erdos–Renyi demonstrating properties of optimal risk-averse maximum cliques are conducted.

Keywords Maximum clique problem • Stochastic graphs • Coherent risk measures • Conditional value-at-risk

M. Rysz • P.A. Krokhmal (✉)
Department of Mechanical and Industrial Engineering, The University of Iowa, 3131 Seamans Center, Iowa City, IA 52242, USA
e-mail: maciej-rysz@uiowa.edu; krokhmal@engineering.uiowa.edu

E.L. Pasiliao
Munitions Directorate, Air Force Research Lab, 101 West Eglin Blvd, Eglin AFB, FL 32542, USA
e-mail: eduardo.pasiliao@eglin.af.mil

A. Sorokin and P.M. Pardalos (eds.), *Dynamics of Information Systems: Algorithmic Approaches*, Springer Proceedings in Mathematics & Statistics 51,
DOI 10.1007/978-1-4614-7582-8_8,

1 Introduction and Motivation

Network problems in the presence of uncertainties have been studied extensively in various areas of operations research, industrial engineering, computer science, and other fields. Stochastic factors arising in network optimization problems can manifest in various forms and may drastically impact the overall network topology, flow distribution, as well as incur unforeseeable costs/losses. In this work, we consider a setting in which uncertainties are induced by stochastic factors associated with network nodes, which differs from many traditional stochastic network models existing in the literature that attribute uncertainties to networks' arcs.

We first present a descriptive formulation of the class of *risk-averse graph theoretic problems* that we are interested in, which includes the minimum-risk maximum clique problem that is the subject of this study. Let $G = (V, E)$ be a graph where each node $i \in V$ has an associated random value X_i representing cost or loss (in other words, smaller realizations of X_i's are preferable), and assume that the joint distribution of X_i's is known. Then, given a risk measure ρ (see Sect. 2 for a formal definition of risk measures), consider the problem of finding the minimum-risk subgraph of G that has a prescribed property $\mathscr{Q}$:

$$\begin{aligned} \min_{S\subseteq V(G),\, w} \quad & \rho\Big(\sum_{i\in S} w_i X_i\Big) \\ \text{s. t.} \quad & \sum_{i\in S} w_i = 1 \\ & w_i \geq 0, \;\; i \in V \\ & S[G] \in \mathscr{Q}_G, \end{aligned} \tag{1}$$

where $S[G]$ is the subgraph induced by a subset S of nodes $V(G)$, and $\mathscr{Q}_G$ is the set of all subgraphs of G with the desired property $\mathscr{Q}$. In the present work, $\mathscr{Q}_G$ represents the set of all *complete subgraphs*, or *cliques*, in G

$$\mathscr{Q}_G = \{S \subseteq V(G) \mid \forall i, j \in S : (i, j) \in E(G)\}. \tag{2}$$

The variables w_i in (1) represent the weights with which vertices of the minimum-risk induced subgraph of G are selected. From a mathematical perspective, non-unity weights in (1) ensure that the problem is well posed, or nontrivial, in the sense that an optimal subgraph would not reduce to a single node. A formal justification of the well-posedness of the minimum-risk formulation (1) is furnished in Sect. 3, but an intuitive illustration can be given as follows. For instance, in the case when $\mathscr{Q}_G$ is defined as in (2), X_i for $i \in V$ are iid with a finite second moment, risk measure ρ is chosen as variance, $\rho(X) = \sigma^2(X)$, and the weights are restricted to be uniform, i.e., $w_i = 1/|S|$ for $i \in S$ and $w_i = 0$ otherwise, then formulation (1) reduces to the (deterministic) maximum clique problem owing to the well-known fact that

$$\sigma^2\left(\frac{1}{n}\sum_{i=1}^{n} X_i\right) = \frac{\sigma^2(X_i)}{n} \to 0, \quad n \to \infty.$$

It is worth noting that in the graph-theoretical literature the vertices' weights are traditionally considered as input parameters of the problem and are fixed.

From a modeling perspective, the presence of weights in (1) can be motivated as follows. Consider a network of sensors that can generate information of uncertain quality, and assume that quality of a sensor's output depends on the length of time that the sensor is "live." The links between sensors allow them to share information and potentially improve the quality of their output. Then, given a certain budget, it is of interest to distribute a resource among the sensors (e.g., energy supplies) in such a way that the risk of receiving poor quality information from the sensor network is minimized. This setting reduces to the above risk-averse maximum clique problem, where one has to select a set of sensors that form a complete subgraph, and to assign weights to the selected sensors that correspond to the proportions of the total energy budget, so as to minimize the risk of information loss.

While network uncertainties are more often associated with network arcs in the literature, a number of studies considered uncertainties relative to nodes demands, etc. Here we briefly discuss several cases focusing on various stochastic effects on networks that are more closely related to the present work. Ukkusuri and Mathew [13] confirmed that long-term demand uncertainties in a network design problems significantly affect network properties compared to their equivalent deterministic counterparts. In another study, Atamturk and Zhang [3] discussed management of uncertain node demands by solving a two-stage stochastic optimization problem, thereby deferring network flow decisions until after realizations of demand materialized. Similarly, Glockner and Nemhauser [7] represented a dynamic network flow problem with arc capacity randomness as a multistage stochastic linear program. They propose other applications focusing on cases where flow through the network is affected by uncertainties attributed to arcs. Several studies examined the effects of stochastic arc failures on networks. Aneja et al. [1] analyzed flow patterns that maximize residual flow under probabilistic arc failure. Verweij et al. [14] used a sample average approximation method to solve several two-stage stochastic routing problems subject to arc failures and unexpected delays. Boginski et al. [4] and Sorokin et al. [12] proposed a mathematical programming approach minimizing flow losses through a network by capturing the impact of probabilistic arc failures relative to conditional expectation of worst-case outcomes. To the authors' knowledge they were the first to introduce Conditional Value-at-Risk (CVaR) [10, 11] into a classical network flow optimization problem as a means of managing risk and collateral losses under arc failures.

In this study we introduce the *risk averse,* or *minimum-risk maximum clique problem*, i.e., the problem of finding a clique with the lowest risk in a given graph, whose vertices represent random variables with a known joint distribution, and discuss some properties of minimum-risk cliques. To this end, Sect. 2 reviews the relevant definitions of risk measures. Section 3 presents a general formulation of the risk averse maximum clique problem and introduces mathematical programming formulations of two special cases, when the risk exposure of a node in the network does or does not depend on the risk exposures of its neighbors (minimum-risk maximum clique problems with "neighborhood-dependent" and "isolated" risk exposures, respectively). Finally, Sect. 4 presents results of numerical experiments conducted on randomly generated graphs.

2 Coherent Risk Measures in Stochastic Programming

Formally, *risk measure* $\rho(X)$ associated with some random outcome X from probability space $(\Omega, \mathscr{F}, \mathsf{P})$ can be defined as a mapping $\rho : \mathscr{X} \to \mathbb{R}$, where $\mathscr{X}$ is a space of bounded $\mathscr{F}$-measurable functions $X : \Omega \mapsto \mathbb{R}$. In what follows, it is assumed that X represents a cost or a loss, whereby its larger realizations are considered "riskier." To be of practical use, however, the above definition of risk measure must be complemented with additional properties that would make utilization of such a risk measure meaningful in a specific application.

Historically, developments of methods for measuring "risk" in decision-making problems under uncertainty was mostly application-driven, or ad hoc. While such an approach allows for tailoring of the risk preferences as induced by a particular choice of risk measure to the application-specific requirements, it may lead to situations when the constructed risk measure lacks certain properties that are commonly considered as mandatory in the risk management community. A notorious example of this kind is served by a risk measure that is well known in financial literature under the name of Value-at-Risk (VaR), which is widely considered as a de facto standard for measuring risk in the banking industry. Mathematically, VaR with confidence level $\alpha \in (0,1)$ is defined as the α-quantile of the loss distribution:

$$\mathrm{VaR}_{\alpha}(X) = \inf\{\eta \mid \mathsf{P}[X \leq \eta] \geq \alpha\}, \tag{3}$$

and is therefore generally *non-convex* in X, thereby not allowing for proper *risk reduction via diversification*, which constitutes a fundamental principle in risk management practice.

Owing to a large degree to the failings of VaR, recent advances in risk theory pioneered by Artzner et al. [2] spawned an axiomatic approach to the construction of risk measures by postulating desirable properties that a "good" risk measures should possess. Namely, in [2, 5] the authors have identified four properties, or axioms, and termed the functionals conforming to all four properties as *coherent risk measures*:

(A1) *Monotonicity:* $X \leq 0 \Rightarrow \rho(X) \leq 0$ for all $X \in \mathscr{X}$
(A2) *Sub-additivity:* $\rho(X+Y) \leq \rho(X) + \rho(Y)$ for all $X, Y \in \mathscr{X}$
(A3) *Positive homogeneity:* $\rho(\lambda X) = \lambda \rho(X)$ for all $X \in \mathscr{X}$ and $\lambda > 0$
(A4) *Transitional invariance:* $\rho(X+a) = \rho(X) + a$ for all $X \in \mathscr{X}$ and $a \in \mathbb{R}$

Note that assuming (A3) holds, (A2) may be replaced by the *convexity* axiom

(A2′) *Convexity:* $\rho(\lambda X + (1-\lambda)Y) \le \lambda\rho(X) + (1-\lambda)\rho(Y)$ for all $X, Y \in \mathscr{X}$, $\lambda \in [0,1]$

Axiom (A1) ensures that lower losses bear lower risk. The subadditivity property (A2) and its derivative property of convexity (A2′) are of fundamental importance from both methodological and mathematical viewpoints: in the risk management context, they express the "risk reduction through diversification" principle, while from mathematical perspective, convexity opens the door for use of efficient optimization methods for control and optimization of risk using coherent risk measures. Axiom (A3) implies that scaling of X by a positive factor scales the risk $\rho(X)$ correspondingly. Finally, axiom (A4) guarantees that constant changes to X impact its risk by the same amount. The solid methodological foundation has made the concept of coherent risk measures very popular in both theory and practice of modern risk management; in general, the axiomatic approach to construction of risk measures became the dominant framework in the field during the last decade [9].

Observe, however, that the above axiomatic definition of the class of coherent risk measures is non-constructive, in the sense it does not provide a functional form of coherent risk measures. Moreover, the ability to employ coherent measures of risk in optimization problems depends on the availability of a functional representation, typically in the formalism of convex analysis, that is conducive to implementation in mathematical programming formulations.

To this end, Krokhmal [8] has proposed a representation for coherent measures of risk in the form of convolution of some function $\phi : \mathscr{X} \mapsto \mathbb{R}$ that satisfies axioms (A1)–(A3), and is a lower semicontinuous function with an additional property of $\phi(\eta) > \eta$ for all real $\eta \neq 0$. Then, the optimal value of the following (convex) stochastic programming problem exists and is a proper coherent measure of risk:

$$\rho(X) = \min_{\eta \in \Re} \eta + \phi(X - \eta). \tag{4}$$

In [8] it was shown that coherent risk measures which admit representation (4) can be efficiently incorporated in objectives and constraints of mathematical programming models. A well-known instance of (4) that has been used widely in stochastic optimization in recent years is the CVaR [10, 11]

$$\mathrm{CVaR}_\alpha(X) = \min_{\eta \in \Re} \eta + (1-\alpha)^{-1}\mathbb{E}(X-\eta)^+, \quad \alpha \in [0,1], \tag{5}$$

where $X^+ = \max\{0, X\}$. Intuitively, CVaR with confidence level α, $\mathrm{CVaR}_\alpha(X)$, corresponds to the expected loss that exceeds the $\mathrm{VaR}_\alpha(X)$ level:

$$\mathrm{CVaR}_\alpha(X) = \mathbb{E}[X \mid X \ge \mathrm{VaR}_\alpha(X)]. \tag{6}$$

It is important to emphasize that relation (6) is not exact and holds only in certain special cases, for instance, when the loss function X is continuously distributed. For the definition of CVaR in the case of general loss distributions, see [11]; alternatively, expression (5) can be considered as the definition of CVaR. In the context of stochastic programming, when the distribution of loss function X is given by scenario set $\mathcal{N}$, such that scenario probabilities $\mathsf{P}\{X = X_s\} = p_s$, $s \in \mathcal{N}$, the optimization problem in (5) reduces to stochastic programming problems of the form

$$\begin{aligned} \min \quad & \eta + (1-\alpha)^{-1} \sum_{s \in \mathcal{N}} p_s t_s \\ \text{s. t.} \quad & t_s \geq X_s - \eta, \quad s \in \mathcal{N} \\ & t_s \geq 0, \quad s \in \mathcal{N}, \end{aligned} \tag{7}$$

where t_s are auxiliary variables associated with scenarios $s \in \mathcal{N}$. In this study, we will use CVaR as a risk measure ρ in risk-averse maximum clique problem, but the general approach is applicable to a broad class of coherent risk measures that admit representation (4).

3 Risk-Averse Maximum Clique Problems

In this section we first formalize the descriptive definition of minimum-risk maximum clique problem (1) and then present two special cases of the general formulation.

Consistent with the above discussion of risk measures, let us define the risk $\mathscr{R}(S)$ of selecting subgraph $S[G]$ of the given graph G as

$$\mathscr{R}(S) = \min\left\{ \rho\left(X_G(S;w)\right) \;\middle|\; \sum_{i \in S} w_i = 1;\ w_i \geq 0\ \forall\, i \in S \right\}, \tag{8}$$

where $\rho(X)$ is a (coherent) measure of risk, and $X = X_G(S;w)$ is the cost/loss function associated with the subset S of nodes in G that also depends on the weights w of nodes in S. For instance, in (1) the implicitly defined loss function has the form $X_G(S;w) = \sum_{i \in S} X_i w_i$. Then, the problem of determining a minimum-risk subgraph of G with the prescribed property $\mathscr{Q}$ can be presented in the form

$$\min\left\{ \mathscr{R}(S) \;\middle|\; S[G] \in \mathscr{Q}_G \right\}. \tag{9}$$

In general, $\mathscr{Q}_G$ may denote the set of subgraphs of G with any desired property, e.g., the set of all complete subgraphs (2), or the set of all *independent sets* in G:

$$\mathscr{Q}_G = \{ S \subseteq V(G) \mid \forall i, j \in S : (i,j) \notin E(G) \},$$

or the set of subgraphs spanning a *path* from s to t in G:

$$\mathscr{Q}_G = \{S \subseteq V(G) \mid S = \{s \equiv i_0, i_1, \ldots, i_{n-1}, i_n \equiv t\};\ (i_{k-1}, i_k) \in E(G),\ 1 \le k \le n\},$$

and so on.

Obviously, (9) can be equivalently written as

$$\begin{aligned} \min_{S \subseteq V(G),\, w} \quad & \rho\big(X_G(S;w)\big) \\ \text{s. t.} \quad & \sum_{i \in S} w_i = 1 \\ & w_i \ge 0,\ \ i \in V \\ & S[G] \in \mathscr{Q}_G. \end{aligned} \tag{10}$$

In the case when the property $\mathscr{Q}$ in denotes completeness of subgraph $S[G]$, (10) represents the general formulation of the risk-averse maximum clique problem.

In this work, we consider two cases of loss function $X_G(S;w)$ in (10) that describe "propagation" of uncertainties and risks within the network. In the first case, risk exposures of the network nodes are "isolated" in the sense that the loss (risk) profile at node i is determined only by the random factor X_i at that node. In other words, risk profiles of individual nodes are independent of stochastic factors at other nodes.

In the second case, it is assumed that the risk exposure of node i depends on its own loss profile X_i as well as losses of the adjacent nodes, thus the overall risk of selected subset S depends not only on the stochastic factors X_i at individual nodes but also on their exposure to neighboring nodes within S. This assumption reflects risk interdependencies observed in many applications, for example, in the financial context, where inter-bank lending heavily exposes counterparties.

The next two sections present mixed integer programming formulations of risk averse maximum clique problem with "isolated" and "neighbor-dependent" stochastic effects. For concreteness, we consider risk measure ρ in (10) to be chosen as the CVaR, $\rho(X) = \mathrm{CVaR}_\alpha(X)$. We assume that losses X_i associated with vertices $i \in V$ have a discrete joint distribution that can be represented by scenario set $\mathscr{N}$, such that X_{is} is the realization of a stochastic factor X_i under scenario $s \in \mathscr{N}$.

3.1 Risk-Averse Maximum Clique Problem with Isolated Risk Exposures

According to the discussion above, the loss function $X_G(S;w)$ corresponding to isolated risk exposures is defined simply as a weighted sum of losses X_i among selected nodes $i \in S$:

$$X_G(S;w) = \sum_{i \in S} w_i X_i. \tag{11}$$

By introducing binary decision variables x_i, $i \in V$, such that

$$x_i = \begin{cases} 1, & i \in S, \\ 0, & i \notin S, \end{cases}$$

the *risk-averse maximum clique problem with isolated risk exposures* can be formulated as a mixed integer programming problem of form

$$\min \quad \rho\Big(\sum_{i \in V} w_i x_i X_i\Big) \tag{12a}$$

$$\text{s. t.} \quad \sum_{i \in V} w_i = 1 \tag{12b}$$

$$w_i \leq x_i, \quad \forall i \in V \tag{12c}$$

$$x_i + x_j \leq 1, \quad \forall (i,j) \in \overline{E} \tag{12d}$$

$$x_i \in \{0,1\}, \; w_i \geq 0, \quad \forall i \in V. \tag{12e}$$

Constraint (12c) ensures that weights w_i can be nonzero only for the vertices i that are included in the solution S, while constraint (12d) maintains that the set of selected nodes forms a complete subgraph, or a clique. Observe that due to the presence of constraint (12c) the nonlinearity in the objective function (12a) attributed to the products $w_i x_i$ can be eliminated by replacing $w_i x_i$ with just w_i, so that the objective of (12) takes the form

$$\rho\Big(\sum_{i \in V} w_i X_i\Big).$$

When the joint distribution of stochastic factors X_i, $i \in V$, is given by scenario set $\{X_{is}\}_{s \in \mathscr{N}}$, and risk measure ρ is chosen as CVaR_α, the risk-averse maximum clique problem (12) reduces to the following 0–1 mixed integer stochastic programming problem

$$\min \quad \eta + \frac{1}{1-\alpha} \sum_{s \in \mathscr{N}} p_s t_s \tag{13a}$$

$$\text{s. t.} \quad \sum_{i \in V} w_i = 1 \tag{13b}$$

$$w_i \leq x_i, \quad \forall\, i \in V \tag{13c}$$

$$x_i + x_j \leq 1, \quad \forall\, (i,j) \in \overline{E} \tag{13d}$$

$$t_s \geq \sum_{i \in V} w_i X_{is} - \eta, \quad \forall\, s \in \mathscr{N} \tag{13e}$$

$$x_i \in \{0,1\}, \; w_i \geq 0, \; \forall\, i \in V; \; t_s \geq 0 \; \forall s \in \mathscr{N}, \tag{13f}$$

where p_s is the probability of scenario $s \in \mathcal{N}$, i.e.

$$\mathsf{P}\Big\{\bigcap_{i \in V} X_i = X_{is}\Big\} = p_s, \quad s \in \mathcal{N},$$

and, naturally, one has $\sum_{s \in \mathcal{N}} p_s = 1$.

3.2 *Risk-Averse Maximum Clique Problem with Neighbor-Dependent Risk Exposures*

Loss function (11) only considers isolated stochastic effects, meaning that changes affecting one vertex do not impact its neighbors and vice versa. However, as discussed above, many physical network structures frequently do exhibit shared risk exposure. We next consider a form of loss function X_G that reflects this observation, allowing the risk exposure of a selected node i to depend on its own loss profile X_i in addition to the loss profiles of the adjacent nodes included in the selected subset S:

$$X_G(S; w) = \sum_{i \in S} \left(w_i X_i + \sum_{j \in S \setminus i} \theta_{ij} w_j X_j \right), \tag{14}$$

where the parameters θ_{ij} denote the degree of exposure of vertex i to vertex j. It is natural to assume that exposure θ_{ij} is nonzero only if an edge exists between i and j: $(i, j) \in E$. Although the meaning of θ_{ij} ultimately depends on the model application, for simplicity we assume that each vertex in V has uniform exposure to its neighbors:

$$\theta_{ij} = \begin{cases} 1/|V(G)|, & \text{if } (i, j) \in E(G) \\ 0, & \text{otherwise.} \end{cases}$$

Observe that (14) can equivalently be expressed in the form

$$X_G(S; w) = \sum_{i \in S} w_i X_i \left(1 + \sum_{j \in S \setminus i} \theta_{ji} \right), \tag{15}$$

which is similar to the form of loss function (11) with isolated exposures if one considers the stochastic factor at node i to be defined as $\tilde{X}_i = X_i\big(1 + \sum_{j \in S \setminus i} \theta_{ji}\big)$. Note, however, that defined in such a way risk profile $\tilde{X}_i$ is *dependent on the selected subset of nodes* S, and, consequently, on the risk profiles of all neighbors of i, since S is a clique.

Introducing binary variables x_i as before, the *risk-averse maximum clique problem with neighbor-dependent risk exposures* can be formulated as

$$\min \quad \rho\left(\sum_{i\in V}\left(w_i x_i X_i + \sum_{j:(i,j)\in E} \theta_{ij} x_i x_j w_j X_j\right)\right) \tag{16a}$$

$$\text{s. t.} \quad \sum_{i\in V} w_i = 1 \tag{16b}$$

$$w_i \le x_i, \quad \forall\, i \in V \tag{16c}$$

$$x_i + x_j \le 1, \quad \forall\,(i,j) \in \overline{E} \tag{16d}$$

$$x_i \in \{0,1\},\; w_i \ge 0,\; \forall\, i \in V. \tag{16e}$$

Once again, constraint (16c) allows for replacing products $w_i x_i$ in the objective with just w_i. Also selecting risk measure ρ in the objective as CVaR_α, problem (16) reduces to a nonlinear 0–1 mixed integer stochastic optimization problem of the form

$$\min \quad \eta + (1-\alpha)^{-1} \sum_{s\in\mathcal{N}} p_s t_s \tag{17a}$$

$$\text{s. t.} \quad t_s \ge \sum_{i\in V}\left(X_{is} w_i + \sum_{j:(i,j)\in E} \theta_{ij} X_{js} w_j x_i\right) - \eta, \quad \forall\, s \in \mathcal{N} \tag{17b}$$

$$\sum_{i\in V} w_i = 1 \tag{17c}$$

$$w_i \le x_i, \quad \forall\, i \in V \tag{17d}$$

$$x_i + x_j \le 1, \quad \forall\,(i,j) \in \overline{E} \tag{17e}$$

$$x_i \in \{0,1\},\; w_i \ge 0,\; \forall\, i \in V;\; t_s \ge 0, \quad \forall\, s \in \mathcal{N}. \tag{17f}$$

The remaining nonlinear terms $w_j x_i$ in the constraint (17b) can be linearized by introduction of auxiliary variables γ_{ij} as follows:

$$\gamma_{ij} \le w_j, \quad \forall\, i,j \in V$$
$$\gamma_{ij} \le x_i, \quad \forall\, i,j \in V$$
$$\gamma_{ij} \ge w_j + x_i - 1, \quad \forall\, i,j \in V$$
$$\gamma_{ij} \ge 0, \quad \forall\, i,j \in V.$$

The following mixed-integer linear formulation for problem (16) is then obtained:

$$\min \quad \eta + (1-\alpha)^{-1} \sum_{s\in\mathcal{N}} p_s t_s \tag{18a}$$

$$\text{s. t.} \quad t_s \ge \sum_{i\in V}\left(X_{is} w_i + \sum_{j:(i,j)\in E} \theta_{ij} X_{js} \gamma_{ij}\right) - \eta, \quad \forall\, s \in \mathcal{N} \tag{18b}$$

$$\sum_{i \in V} w_i = 1 \tag{18c}$$

$$w_i \leq x_i, \quad \forall\, i \in V \tag{18d}$$

$$x_i + x_j \leq 1, \quad \forall\, (i,j) \in \overline{E} \tag{18e}$$

$$\gamma_{ij} \leq w_j, \quad \forall\, i,j \in V \tag{18f}$$

$$\gamma_{ij} \leq x_i, \quad \forall\, i,j \in V \tag{18g}$$

$$\gamma_{ij} \geq w_j + x_i - 1, \quad \forall\, i,j \in V \tag{18h}$$

$$x_i \in \{0,1\},\ w_i \geq 0,\ \gamma_{ij} \geq 0,\ \forall\, i,j \in V;\ t_s \geq 0,\ \forall\, s \in \mathscr{N}. \tag{18i}$$

Observe that the additional complexity of formulation (18) in comparison with (13) attributes to the fact that the risk exposure of a node in solution set S depends not only on its own risk profile but also on risk profiles of adjacent nodes included in the solution set.

Finally, we show that the adopted definition (8) of risk $\mathscr{R}(S)$ for subgraph S and the chosen loss functions $X_G(S;w)$ of form (11), (14) are consistent with the sub-additivity property of coherent risk measures. Namely, we demonstrate that the following is true.

Proposition 1. *Consider definition* (8) *of risk for subset S of vertices in graph $G = (V,E)$, where each vertex $i \in V$ is associated with a random element X_i. If risk measure ρ in* (8) *is coherent, and the loss function associated with selecting $S \subseteq V$ is given by* (11) *or* (14)*, then risk $\mathscr{R}$ satisfies*

$$\mathscr{R}(S') \leq \mathscr{R}(S) \quad \textit{for all} \quad S' \supseteq S. \tag{19}$$

Proof. Consider, without loss of generality, $S = \{1,\ldots,n\} \subset V$ and $S' = S \cup \{n+1\}$. Then, for loss function $X_G(S;w)$ in the form (11), $X_G(S;w) = \sum_{i \in S} w_i X_i$, denote

$$\begin{aligned}(w_1^*,\ldots,w_n^*) &\in \arg\min \left\{ \rho\left(X_G(S;w)\right) \,\middle|\, \sum_{i \in S} w_i = 1;\ w_i \geq 0\ \forall\, i \in S \right\} \\ &= \arg\min \left\{ \rho\Big(\sum_{i=1}^{n} w_i X_i\Big) \,\middle|\, \sum_{i=1}^{n} w_i = 1;\ w_i \geq 0,\ \ i = 1,\ldots,n \right\},\end{aligned}$$

and

$$(w_1^{**},\ldots,w_{n+1}^{**}) \in \arg\min \left\{ \rho\Big(\sum_{i=1}^{n+1} w_i X_i\Big) \,\middle|\, \sum_{i=1}^{n+1} w_i = 1;\ w_i \geq 0,\ \ i = 1,\ldots,n+1 \right\}.$$

Then, one immediately has

$$\mathscr{R}(S') = \rho\left(\sum_{i=1}^{n+1} w_i^{**} X_i\right) \leq \rho\left(\sum_{i=1}^{n} w_i^{*} X_i + 0 \cdot X_{n+1}\right) = \mathscr{R}(S).$$

The same arguments can be applied to loss function $X_G(S;w)$ given by (14). □

Corollary 1. *Proposition* (1) *implies that an optimal solution of risk-averse maximum clique problems* (12), (16) *represents a* maximal *clique of the underlying graph G (however, not necessarily its* maximum *clique).*

4 Numerical Experiments

In this section we conduct numerical experiments demonstrating solution properties of problems (13) and (18). Due to the fact that the deterministic maximum clique problem is NP-hard itself, its risk-averse versions (13) and (18) are NP-hard as well, thus requiring significant computational efforts to solve even for moderate-sized graph instances.

In particular, of specific interest in the presented case study were the sizes of the optimal risk-averse cliques produced by formulations (13) and (18) in comparison with the maximum clique size in the respective graphs obtained without considering stochastic effects, i.e., as a solution of problem

$$\begin{aligned} \max \quad & \sum_{i \in V} x_i \\ \text{s. t.} \quad & x_i + x_j \leq 1, \quad \forall\, (i,j) \in \overline{E} \\ & x_i \in \{0,1\}, \quad \forall\, i \in V, \end{aligned} \tag{20}$$

with variables x_i defined as before.

4.1 Implementation, Scenario Data, Graphs, and Numerical Results

The stochastic mixed integer programming problems (13) and (18) were coded in Python 2.6.6 and solved using CPLEX 12.2 on a quad-core 2.20 GHz PC with 16 GB RAM.

We use randomly generated Erdos–Renyi graphs [6] $G(V,p)$ where every edge is independently formed with prescribed a probability p. Random scenario data corresponding to each vertex $i \in V$ were generated according to a uniform distribution over an interval $[-0.5, 0.5]$.

Table 1 Average optimal clique sizes and computation times in seconds obtained by problems (13) and (20)

p	$\|V\|$	$\|\mathscr{N}\|$	Maximum clique size	Time (s)	Maximum risk-averse clique size	Time (s)
0.25	25	50	3.6	0.004	3.5	0.06
	50	50	4.9	0.06	3.9	0.38
	75	50	5.2	0.75	4.2	1.41
	100	50	5.4	2.74	4.5	7.69
	150	50	6.1	21.75	4.6	34.97
	200	50	6.6	196.95	4.7	114.25
0.5	25	50	5.7	0.003	5.1	0.05
	50	50	7.8	0.16	6.3	0.52
	75	50	8.4	0.90	6.4	3.80
	100	50	9.2	2.68	6.6	10.16
	150	50	10.1	32.36	7.2	60.59
	200	50	11	207.07	7.8	251.38
0.75	25	50	9.6	0.003	8	0.05
	50	50	12.7	0.18	9.7	0.77
	75	50	15.6	0.61	11.1	3.356
	100	50	16.7	3.52	12.5	10.36
	150	50	19.1	87.81	12.8	58.67
	200	50	20.9	1,524.75	14.4	263.01

A total of 20 instances of (13) or (18) have been solved for each combination of graph vertex/scenario set size. To demonstrate the effect of the degree of risk aversity, as given by the confidence level α of CVaR measure, on the size of risk-averse maximum clique in (13), we also solve 20 instances of (13) and report the average clique size for each value of α. We also compared the average sizes of risk-averse maximum cliques as given by (13) or (18) with the average size of deterministic maximum clique as given by (20) over the same randomly generated graph instances. Finally, we compare optimal clique sizes in problems (13) and (18) at a single graph size/scenario set size instance with varying levels of parameter α.

4.2 *Risk-Averse Maximum Clique Problem with Isolated Risk Exposures*

Table 1 demonstrates averaged optimal clique sizes with respective computation times for the discussed implementations of problems (13) and (20) for randomly generated graphs of sizes $|V| = 50, 75, 100, 150, 200$ and average densities $p = 0.25, 0.5, 0.75$. In all instances, the number of scenarios (i.e., realizations of the vector $(X_1, \ldots, X_{|V|})$) was fixed at 50, and the confidence level of the CVaR was

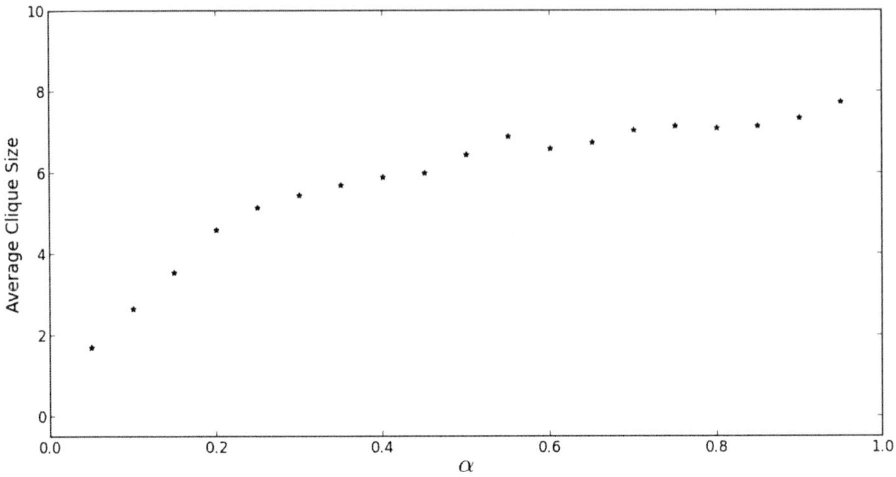

Fig. 1 Average optimal clique sizes at varying risk tolerances α using graphs with 150 vertices and 50 scenarios ($p = 0.50$)

chosen as $\alpha = 0.9$. One of the observations that can be drawn from Table 1 is that the average sizes of risk-averse maximum cliques are smaller than the corresponding sizes of risk-neutral (deterministic) maximum cliques.

We next examine optimal subgraph sizes in relation to varying risk tolerance levels (note that larger values of α correspond to more risk-averse preferences), with α taking values within the range $[0.05, 0.95]$ at increments of 0.05 in randomly generated graphs with 150 vertices and average density of $p = 0.50$ and 50 scenarios. Figure 1 establishes a strong relation between α and average optimal subgraphs. Noteworthy are low α levels (e.g., $\alpha = 0.05$) which occasionally reduce the optimal subgraph to a single vertex, confirming that lax risk requirements prevent appropriate, if any, diversification among multiple vertices. Furthermore, a transition from low risk tolerance inducing no diversification ($\alpha \approx 0.05$) towards effectual levels ($\alpha \geq 0.1$) expands optimal clique sizes at high rates over interval $\alpha \in [0.05, 0.25]$, while consecutive restrictions have a more moderate impact. This outcome is consistent with properties of coherent risk measures, such as CVaR, which allow for efficient diversification due to sub-additivity/convexity property.

The results shown in Fig. 1 can be illustrated through financial setting, where lax risk constraints are commonly associated with little or no asset diversification, whereas the tighter risk constraints, and, correspondingly, increased risk aversity lead to improved diversification. In a network setting, specifically (13), we can analogously express lacking diversification over vertices for insufficiently large α-levels corresponding to low degree of risk aversity. Initial incremental increases in α are reflected in steep clique size growth rates, with a dissipating effect as $\alpha \to 1$.

Table 2 Average optimal clique sizes and computation times in seconds obtained from (18) and (20)

p	$\|V\|$	$\|\mathscr{N}\|$	Maximum clique size	Time (s)	Maximum risk-averse clique size	Time (s)
0.25	25	50	3.6	0.004	3.1	1.69
	50	50	4.9	0.06	3.3	36.83
	75	50	5.2	0.75	3.9	289.21
0.5	25	50	5.7	0.003	4.2	6.22
	50	50	7.8	0.16	4.8	195.95
	75	50	8.4	0.90	6.1	2,395.69

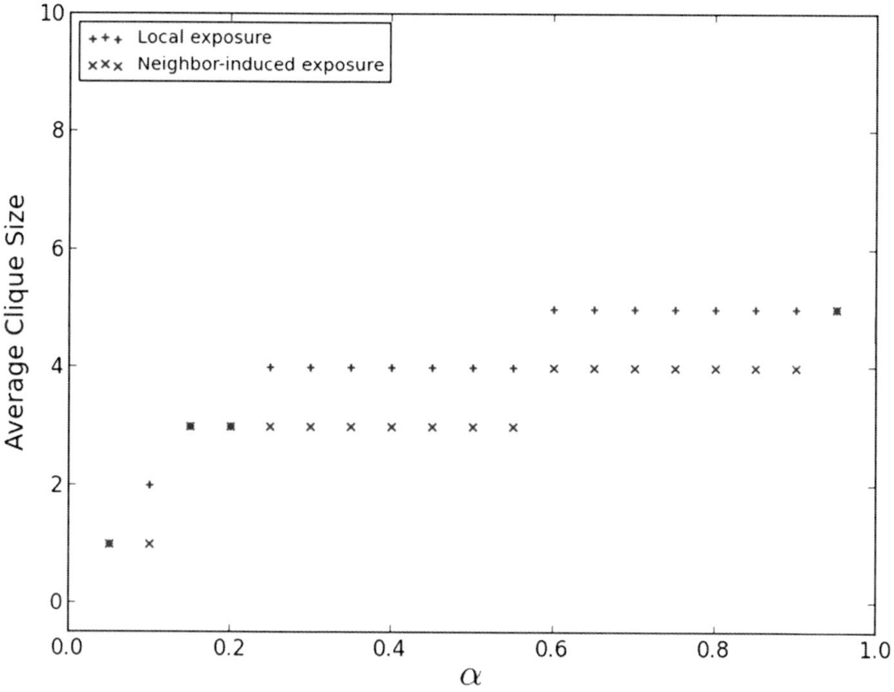

Fig. 2 Comparative optimal clique sizes at ranging risk tolerances α for problems (13) and (18) using a single graph with 50 vertices and 50 scenarios

4.3 *Risk-Averse Maximum Clique Problem with Neighbor-Dependent Exposure*

For construction of problem (17) we impose $\theta_{ij} = 1/|V|$ over all vertices $i \in V$, and conduct computational simulations for graphs of sizes $|V| = 25, 50, 75$ and average densities of $p = 0.25, 0.50$. In each problem instance, distribution of uncertainties was modeled using 50 scenarios, and the confidence level α of the CVaR measure was set at $\alpha = 0.9$. Table 2 reports the resulting average clique sizes and corresponding computation times when risk exposure of a node depends on the risk profiles of its neighbors. Observe that optimal risk-averse clique sizes are significantly smaller on average in comparison with the same instances in Table 1. Furthermore, in Fig. 2 we demonstrate that problem (18) consistently requires higher levels of α to attain similar optimal subgraph sizes.

5 Conclusions

In this work, we have introduced minimum-risk maximum clique problem, i.e. a risk-averse maximum clique problem on stochastic graphs. A distinguishing feature of the problem setting considered in this study is the assumption that stochastic factors in the underlying networks are associated with vertices, as opposed to the prevalent literature settings where uncertainties are attributed to network arcs. Two formulations of risk-averse maximum clique problems corresponding to "isolated" and "neighbor-dependent" risk exposures at the nodes are presented. It is shown that optimal solutions of risk-averse maximum clique problems are represented by maximal cliques. Numerical experiments on randomly generated Erdos–Renyi demonstrating properties of optimal risk-averse maximum cliques have been conducted.

References

1. Aneja YP, Chandrasekaran R, Nair KPK (2001) Maximizing residual flow under an arc destruction. Networks 38(4):194–198
2. Artzner P, Delbaen F, Eber F, and Heath (1999) Coherent measures of risk. Mathematical Finance 9(3): 203–228
3. Atamturk A, Zhang M (2007) Two-stage robust network flow and design under demand uncertainty. Operations Research 55(4): 662–673
4. Boginski V, Commander C, Turko T (2009) Polynomial-time identification of robust network flows under uncertain arc failures. Optimization Letters 3(3):461–473
5. Delbaen F (2002) Coherent risk measures on general probability spaces. In: Sandmann K, Schonbucher PJ (Eds.), Advances in Finance and Stochastics: Essay in Honour of Dieter Sondermann. Springer, pp. 1–37
6. Erdös P, Rényi A (1960) On the Evolution of Random Graphs. Publication of the Mathematical Institute of the Hungarian Academy of Sciences 5:17–61
7. Glockner GD, Nemhauser GL (2000) A dynamic network flow problem with uncertain arc capacities: formulation and problem structure. Operations Research 48(2): 233–242
8. Krokhmal P (2007) Higher moment risk measures. Quantitative Finance 7(4): 77–91
9. Krokhmal P, Zabarankin M, Uryasev S (2011) Modeling and optimization of risk. Surveys in Operations Research and Management Science 16(2): 49–66.
10. Rockafellar R, Uryasev S (2000) Optimization of conditional value-at-risk. Journal of Risk 2: 21–41
11. Rockafellar R, Uryasev S (2002) Conditional value-at-risk for general loss distributions. Journal of Banking and Finance 26(7): 1443–1471

12. Sorokin A, Boginski V, Nahapetyan A et al (2011) Computational risk management techniques for fixed charge network flow problems with uncertain arc failures. J Comb Optim, doi: 10.1007/s1087801194222
13. Ukkusuri S, Mathew T (2007) Robust transportation network design under demand uncertainty. Computer-Aided Civil and Infrastructure Engineering 22: 6–18
14. Verweij B, Ahmed S, Kleywegt AJ et al (2003) The sample average approximation method applied to stochastic routing problems: a computational study. Journal Computational Optimization and Applications 24(2–3): 289–333

Models for Assessing Vulnerability in Imperfect Sensor Networks

Sibel B. Sonuç and J. Cole Smith

Abstract We examine a directed network on which sensors exist at a subset of the nodes. At each node, a target (e.g., an intruder in a physical network, a fire in a building, or a virus in a computer network) may or may not exist. Sensors detect the presence of these targets by monitoring the nodes at which they are located and all nodes adjacent from their positions. However, in practical settings a limited-cardinality subset of sensors might fail. A failed sensor may report false positives or negatives and, as a result, the network owner may not be able to ascertain whether or not targets exist at some nodes. If it is not possible to deduce whether or not a target exists at a node with a given set of sensor readings, then the node is said to be ambiguous. We show that a network owner must solve a series of combinatorial optimization problems to determine which nodes are ambiguous. Furthermore, we determine the worst-case number of ambiguous nodes by optimizing over the set of all sensor readings that could possibly arise. We also present mathematical programming formulations for these problems under varying assumptions on how sensors fail, and on what assumptions a network owner makes on how sensors fail. Our computational results illustrate how these varying assumptions impact the number of ambiguous nodes.

Keywords Ambiguity assessment • Ambiguity order • Fault-tolerant sensor networks • Sensor failure options • Interdiction

S.B. Sonuç • J.C. Smith (✉)
Department of Industrial and Systems Engineering, University of Florida,
P.O. Box 116595, Gainesville, FL 32611-6595, USA
e-mail: sibel.bilge@ufl.edu; cole@ise.ufl.edu

A. Sorokin and P.M. Pardalos (eds.), *Dynamics of Information Systems: Algorithmic Approaches*, Springer Proceedings in Mathematics & Statistics 51,
DOI 10.1007/978-1-4614-7582-8_9,

1 Introduction

We consider sensor networks that attempt to monitor a collection of several critical locations in some environment in which various types of threats may exist. These threats may represent the physical presence of an entity in the environment (e.g., a fire or intruder), or some virtual entity such as a computer virus. These types of problems are often modeled by a network whose nodes represent locations that must be monitored by the network owner, where an arc exists between two nodes if a sensor colocated at one node is able to determine whether there is an intruder at the other node. In this paper, the *status* of a node indicates whether or not a target exists at that node. The information sent from the sensors is collected as a set of *readings* at an administrative center, and made available to the network owner to assess the statuses of the nodes. However, due to the technical limits of the equipment used in the sensor systems (e.g., equipment failure, power shortage, equipment-specific capabilities) and also to the environmental factors that may degrade sensor function (e.g., hills, clouds, thunderstorms) the information received from the sensors may not always be accurate. The study of *fault-tolerant* sensor systems focuses on sensor networks in which a subset of deployed sensors might fail, and a faulty sensor might give a false positive or negative reading. In this paper, we study seven different cases on how the sensors might fail, and how the network owner utilizes knowledge of sensor functionality to assess where targets must or must not exist in the network.

We say that a node is *ambiguous* if and only if it is not possible to verify the status of that node with the current sensor readings. It is useful to envision an *attacker* that has somehow gained the ability to control the readings of any sensor that has failed, in order to maximize the number of ambiguous nodes. Hence, in this paper, we say that the attacker "hijacks" a set of sensors. Our class of problems can be represented as a *Stackelberg game* [23] in which the attacker acts first by hijacking a limited number of the sensors on the network and manipulating their readings. The *defender* then ascertains, via the solution of a series of optimization problems, whether or not each node in the network is ambiguous. A critical consideration in determining node statuses is the maximum number of sensors that can simultaneously fail, which is a parameter, κ, that the defender utilizes to infer subsets of sensors that have failed. The role of interdiction here may actually represent an adversarial entity that seeks damage to the sensor network, but could alternatively be viewed as a (worst-case) set of simultaneous sensor failures that could occur.

We model the relationship between the sensors and the nodes being monitored by a directed network $G = (N,A)$ with node set N and arc set A. An arc (u,v) belongs to A if and only if a sensor at node u is capable of monitoring node v. We denote the index set of sensor locations by $S \subseteq N$, and of faulty sensors by $H \subseteq S$. In this paper, we assume that a sensor can monitor itself, i.e., $(u,u) \in A$ if $u \in S$. For the sake of simplicity, we will use "sensor u" and "sensor at node u" interchangeably. For $u \in N$, the *forward star* $\text{FS}(u) = \{v \in N : (u,v) \in A\}$ and *reverse star* $\text{RS}(u) = \{v \in N : (v,u) \in A\}$ give the set of nodes that would be monitored by a

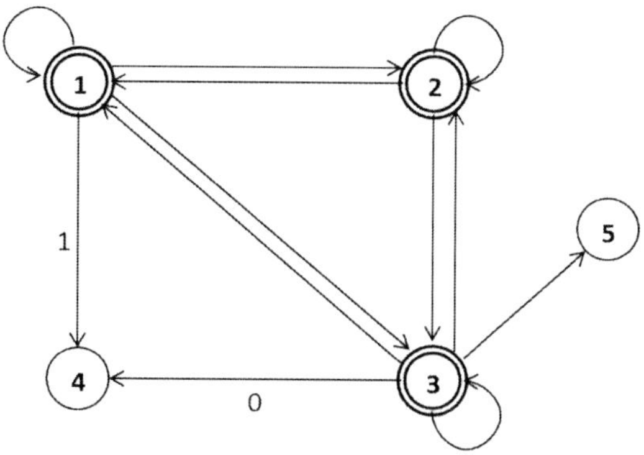

Fig. 1 A sensor network with five nodes and $S = \{1,2,3\}$ and $\kappa = 1$

sensor located at node u, and the set of potential sensor locations that could monitor node u, respectively.

Figure 1 shows a five-node network with $S = \{1,2,3\}$, where $\kappa = 1$. In our figures, we designate nodes having sensors with double lines and nodes without sensors with a single line. The direction of the arcs depicts the direction of monitoring. The numbers next to each arc show the values of the corresponding sensor readings, where a 1 indicates that the sensor detects a target and a 0 indicates that the sensor does not detect a target. (Some readings are omitted in the figures for clarity.) We define notation r_{ij}, which equals 1 if sensor $i \in S$ reports a target at node $j \in N$, and equals 0 otherwise. We say that sensor u and sensor v *conflict* if $r_{uj} \neq r_{vj}$ for some $j \in N$. Observe in Fig. 1 that sensors 1 and 3 conflict, because they disagree on the presence of a target at node 4. Even without looking at the other readings, the network owner can conclude that either sensor 1 or sensor 3 is faulty. Since $\kappa = 1$, we have that sensor 2 is *accurate* (i.e., non-faulty), and all nodes in $\text{FS}(2)$ $(= \{1,2,3\})$ are unambiguous. It may also be possible to determine the status of nodes 4 and 5 as well, depending on readings of sensors 1 and 3.

In addition to accurately obtaining sensor readings, another issue in sensor systems seeks to determine the exact location of a target on the network in the case that a sensor system can detect, but not locate, a target (e.g., a smoke detector signaling a fire nearby, but without knowledge of precisely where the fire is located). Slater [19] and Harary and Melter [7] discuss *single-fault-tolerant locating-dominating* sets (called *metric bases* in [7]) in which only one of the sensors fails and there is a single target in the network. In their work, they assume that a working sensor at node u can detect a target in its neighborhood $\text{FS}(u)$ and gives the exact location of the target only if the target is at node u itself. In an extension of this work, Slater [20] examines *single-fault-tolerant* sensor networks in which a working sensor u can determine the exact location of a target at any node in its neighborhood $\text{FS}(u)$. By contrast, our study assumes that there are no restrictions on the number of targets in the network. This setting yields a problem that is more complex than the special case in which the number of targets is known. Moreover, we consider multiple failures in the set of deployed sensors. Our paper is also based on the

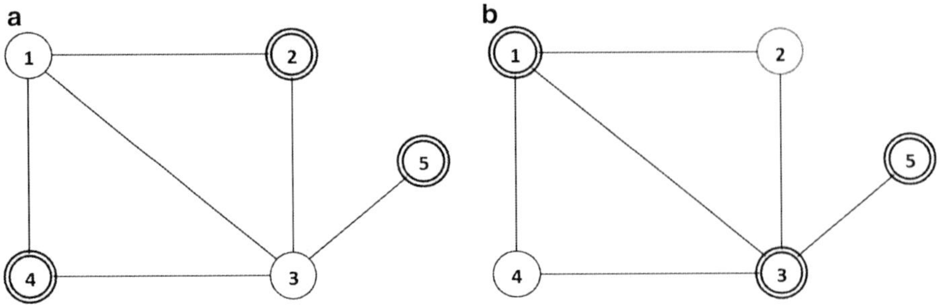

Fig. 2 Multiple domination on a five-node graph. **(a)** Two-domination with dominating set $D_a = \{2,4,5\}$. Only nodes 1 and 3 are dominated twice or more. **(b)** Two-tuple (double) domination with dominating set $D_b = \{1,3,5\}$. All nodes on the graph are dominated twice or more

assumption that a working sensor assesses whether or not a target exists specifically at each of its neighbors, as opposed to the case in which sensors report whether a target exists at some unspecified neighbor.

In sensor network problems, the network owner's general aim is to place sensors on a subset of the nodes such that every node is monitored by at least one sensor. Hence, the sensor allocation problem (with perfect sensor operations) can be modeled exactly as the *dominating set* problem. The dominating set problem is well studied in the literature [8, 9, 17]. The earliest graph domination problems look at finding a minimum cardinality dominating set [3, 14], which is known to be NP-complete [6]. For graphs with no isolated nodes, Ore [14] shows that there exists a partition of vertices into two disjoint dominating sets. Cockayne and Hedetniemi [4] define a graph's *domatic number* as the maximum number of disjoint dominating sets on the graph, and extend Ore's result by showing that the domatic number of any graph with no isolated nodes is at least two. One subject in domination research that is of particular relevance to our study is *multiple domination*, where each node is required to be dominated by more than one node in the dominating set. In *n-domination* problems [5], multiple domination (by at least n nodes) is required only for those nodes not in the dominating set, whereas in *n-tuple domination* it is required for all nodes in the network. Figure 2 illustrates these multiple domination concepts on a five-node graph, in which the double-lined nodes are in the dominating set.

In fault-tolerant sensor networks, the overall aim is to design the network so that the failures' effect on the performance of the sensor system is limited. In this case, a simple dominating set approach may not be appropriate because single-domination solutions are vulnerable to sensor failures. Slater [20] introduces the *liar's domination* problem, in which a dominating set is able to locate an intruder when one of the sensors fails. Although no multiple-domination condition is explicitly enforced in the problem, it is a natural result at optimality: every liar's dominating set is double dominating, and every triple-dominating set gives a liar's domination. Survivable networks and failures in telecommunication network systems are applications that have inspired network interdiction research. Soni et al. [22] provide a detailed survey of survivable network design and its applications. Rajan and Atamturk [15] introduce a mathematical formulation for designing flow networks, based on multiple domination and rerouting options for flows on disrupted paths, allowing the transfer of data flow to a different group of routers. Smith et al. [21] look at optimization of flow networks as a Stackelberg game where the arc failures happen as a result of an attacker's action. They study mathematical formulations based on different scenarios regarding an attacker's abilities and strategy.

Sensor networks and wireless communication networks are two additional areas that benefit from dominating set optimization research [1, 10, 12, 16]. A typical communication network consists of several source nodes, some of which also serve as router points to transfer information through the network. Lim and Kim [11] study optimal *flooding* on wireless communication networks where information from a source node is sent to all other nodes in the network. A typical application of flooding lies in broadcasting networks, where a node passes information to its neighbors in a single broadcast until the information is delivered to all nodes in the network. This problem searches for a flooding tree that minimizes the number of broadcasts, which is equivalent to solving the minimum connected dominating set problem. Shen and Smith [18] consider a broadcast domination problem in which each link has a construction cost and each router has a power indicating how far it can deliver the information. Their formulation for broadcast domination minimizes the total construction cost of the router system, which is given by the total link cost and power assignments to the selected routers. Cockayne and Hedetniemi [4] model communication networks as a multiple domination problem with domatic number d. Their work provides a model to determine minimum-cost links to connect a given set of nodes by d disjoint sets of routers, so that the failure of a set of routers does not disconnect the network. For a detailed survey of graph-theory applications in wireless telecommunication networks, we refer the reader to [2].

The remainder of the paper is organized as follows. Section 2 provides a detailed description of our problem, including the various assumptions that govern sensor failures. Section 3 gives the mathematical formulations for each of the cases introduced in Sect. 2, depending on how the sensors may actually fail, and on whether or not the defender is aware of how the sensors fail (or whether the defender assumes a worst-case failure behavior). Computational results are presented in Sect. 4 that study how efficiently a mixed-integer programming solver can optimize our prescribed formulations, and compare the number of ambiguous nodes on instances having varying assumptions on sensor failure behaviors. We conclude the paper in Sect. 5 by examining opportunities for future research.

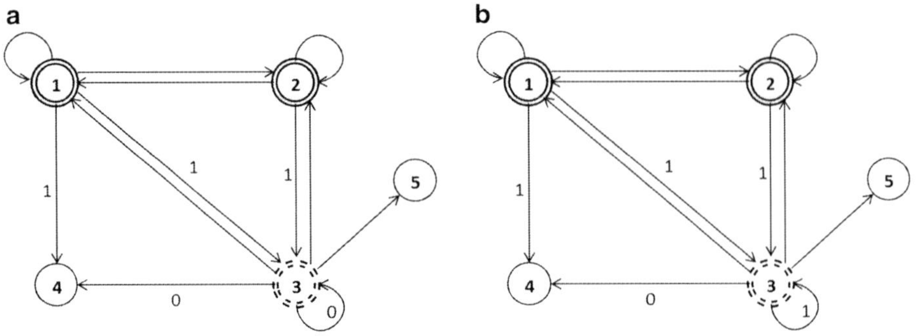

Fig. 3 $S = \{1,2,3\}$, $H = \{3\}$, and $\kappa = 1$. **(a)** If $r_{33} = 0$, then node 5 is ambiguous. **(b)** If $r_{33} = 1$, then nodes 4 and 5 are ambiguous

2 Problem Definition

We assume that the maximum number of sensors that can simultaneously fail (κ) is known, which allows the defender to make inferences regarding which sensors have failed, and where targets must, or must not, necessarily exist in the network. Furthermore, the cases we study in this paper encompass four different options for the ability of the attacker to control the readings of hijacked sensors.

Figure 3 illustrates how the node statuses are assessed when the defender assumes that the attacker can change all readings of each hijacked sensor (sensor 3 in this case) to any value. In Fig. 3a, sensor 3 reports no target at node 3, and sensors 1 and 2 report a target at node 3. Since $\kappa = 1$, sensors 1 and 2 cannot be faulty at the same time, and so the defender knows that sensor 3 must be faulty. In this example, there is no accurate sensor monitoring node 5, which is therefore an ambiguous node. All other nodes are monitored by at least one accurate sensor (sensors 1 and 2). Given that the attacker aims to maximize the number of ambiguous nodes, it may report several correct readings in order to conceal the identity of faulty sensors. Figure 3b represents this action for sensor 3, where r_{33} is set to 1 in order to agree with sensors 1 and 2 on the status of node 3. In this case, both nodes 4 and 5 are ambiguous. (Note that r_{35} is irrelevant in both cases, and hence its value is omitted in Fig. 3a, b.)

The foregoing examples show that the attacker does not wish to have hijacked sensors always report false readings (as it is optimal to have $r_{33} = 1$), or always report true readings (as $r_{34} = 0$ is optimal). Furthermore, we observe that even when the defender ascertains which nodes belong to H, it is still possible for ambiguous nodes to exist (as is the case in Fig. 3a). On the other hand, it is also possible to have no ambiguous nodes even if we cannot identify all faulty sensors (e.g., if G is a clique with $\kappa + 1$ nodes, sensors on all nodes, and $r_{ij} = 0$, $\forall i, j \in N$; however, this scenario corresponds to a suboptimal attacker action).

We consider seven model variations based on different capabilities that the attacker may have, and based on the assumptions that the network owner makes regarding the attacker's abilities. There are four different options regarding the attacker's capabilities, each of which assumes that the attacker can hijack at most κ sensors, and cannot affect readings from sensors that are not hijacked. The four options that govern the attacker's capabilities are summarized as follows:

Option A The attacker possesses the ability to change the readings of all hijacked sensors to any value on any monitored node.
Option B A hijacked sensor always reports false readings on all nodes that it monitors.
Option C Each hijacked sensor reports at most one false reading.
Option D There exists an upper bound $\tau \geq 0$ on the total number of false readings on the network.

The example given in Fig. 3 assumes that the attacker has option A capabilities. Option A has the least amount of restrictions for the attacker among the options we consider in this paper. Option B omits the flexibility for a hijacked sensor to report a true reading. (Recall that the capability to control the readings is a key feature of option A because a hijacked sensor might wish to report a true reading on some of their neighbors in order to conceal the identity of faulty sensors.) Option C imposes an additional constraint that limits the number of false readings on each hijacked sensor to be no more than one. Option D puts a limit on the total number of false readings over all of the hijacked sensors. Note that option A is a special case of option D, in which τ equals the number of nodes in the network. Option C is more restrictive than option D if the number of sensors deployed in the network is not more than τ. No such comparison between option C and D can be made when τ is less than the number of sensors, because option D allows an uneven distribution of false readings among the sensors while option C allows a larger number of false readings in the network in this case.

The models we study in this paper also examine assumptions that the network owner might make regarding the attacker's capabilities. The first group of models assumes that the network owner is aware of the limitations of the attacker, and thus knows which option governs the attacker's actions. (Hence, we study four such models.) On the other hand, in practice, the network owner may not have access to full information regarding the limitations of the attacker and hence, a worst-case scenario (i.e., option A) has to be taken as the default option for a risk-averse defender. This results in three more models where the attacker follows option B (or C, or D) but the defender assumes that the attacker has option-A capabilities. Observe that any attacker's action made under options B, C, or D is also feasible under option A. Suppose that the defender assumes an option-A attack, although the attacker is limited by option B, C, or D. In this case, the defender may conclude that a node is ambiguous, given a set of sensor readings, whereas with correct information as to the actual attacker option (B, C, or D), the defender could have ascertained whether or not a target exists at the node. In this paper, we analyze how much the defender's knowledge about the attacker's capabilities affects ambiguity in the sensor network.

In the remainder of this paper, each case will be denoted as $[X_a, X_d]$ where X_a corresponds to the actual option of the attacker and X_d gives the option assumed by the defender (e.g., [B, A] gives the case where the attacker follows option B but the defender assumes that the attacker has option A). Note that an opposite approach to these options (i.e., assuming that the attacker's option is more restrictive than it actually is) may underestimate the number of ambiguous nodes. Therefore, we restrict our attention to the seven following cases: [A, A], [B, A], [C, A], [D, A], [B, B], [C, C], and [D, D].

3 Mathematical Formulations

We now develop mathematical formulations for the seven variations of the attacker/defender problems presented in Sect. 2. We begin in Sect. 3.1 by exploring the *ambiguity assessment problem*, which determines whether or not each node is ambiguous, given a set of sensor readings (i.e., r_{ij}-values) and assumptions on the attacker's capabilities (corresponding to options A, B, C, and D). The ambiguity assessment model forms the basis for the attacker's model, which we explore in Sect. 3.2 for each of the seven cases in the previous section.

3.1 Ambiguity Assessment for Option A

To begin, recall that the defender is unaware of which sensors have been hijacked, and seeks to determine *plausible* candidates for H (i.e., those for which $|H| \leq \kappa$, and no pair of sensors in $S \setminus H$ conflict). A node $k \in N$ is ambiguous, given a set of r-values, if there exist two plausible sets H^1 and H^2: One in which $r_{ik} = 1$, $\forall i \in (S \setminus H^1) \cap \mathrm{RS}(k)$, and one in which $r_{ik} = 0$, $\forall i \in (S \setminus H^2) \cap \mathrm{RS}(k)$. To represent each choice of assignment for all nodes on the network, we model the ambiguity assessment model on a scenario-based approach. The scenario denoted by $\{k1\}$ assumes that a target exists at node k; similarly, scenario $\{k0\}$ assumes that no target exists at node k.

To determine if a scenario can exist, given a set of readings, the defender must hypothesize a set of failed sensors along with a set of target locations that could (simultaneously) exist in the network. Accordingly, for scenario $\{ka\}$, $\forall k \in N$, $a \in \{0,1\}$, define variables x_i^{ka} that equal 1 if sensor i is assumed to be faulty in scenario $\{ka\}$, and 0 otherwise. The binary variable t_j^{ka} indicates a feasible status reading of node j in scenario $\{ka\}$ with respect to given sensor readings $\{r_{ij}\}_{i\in S, j\in N}$ and sensor attacks $\{x_i^{ka}\}_{i\in S}$. That is, $r_{ij} = t_j^{ka}$, $\forall i \in S:\ x_i^{ka} = 0$, $\forall j \in \mathrm{FS}(i)$. The following set of linear inequalities imposes these conditions in scenario $\{ka\}$:

$$r_{ij} - x_i^{ka} \leq t_j^{ka} \leq r_{ij} + x_i^{ka} \qquad \forall i \in S,\ j \in \mathrm{FS}(i). \tag{1}$$

Inequalities (1) force $r_{ij} = t_j^{ka}$ when $x_i^{ka} = 0$, and places no restrictions on t_j^{ka} (beyond $0 \le t_j^{ka} \le 1$) when $x_i = 1$. For example, given the readings in Fig. 3b, the defender must consider scenarios $\{40\}$ and $\{41\}$ to assess the status of node 4. This analysis could give the following values on (some) scenario-variables: $x_1^{40} = 1$, $x_3^{40} = 0$, $t_4^{40} = 0$ for scenario $\{40\}$ and $x_1^{41} = 0$, $x_3^{41} = 1$, $t_4^{41} = 1$ for scenario $\{41\}$, implying that both scenarios $\{40\}$ and $\{41\}$ are feasible and node 4 is ambiguous. On the other hand, in Fig. 3a, sensor 3 is proven to be faulty and $x_3^{ka} = 1$ for all scenarios $\{ka\}$. This results in $t_4^{40} = 1$, indicating the infeasibility of scenario $\{40\}$ (while scenario $\{41\}$ remains feasible).

Let binary variable z_k equal 1 if and only if node $k \in N$ is ambiguous, and 0 otherwise. Because z_k should equal one only when it is possible for a target to exist at k in scenario $\{k0\}$, while no target exists at node k in scenario $\{k1\}$, we wish to set $z_k = \min\{1 - z_k^{k0}, z_k^{k1}\}$. This condition can be stated within a *linear* mixed-integer program by constraining z_k as:

$$t_k^{k1} \ge z_k, \tag{2a}$$

$$t_k^{k0} \le 1 - z_k, \tag{2b}$$

and then maximizing z_k in the objective to force z_k to take its largest value allowed by (2a) and (2b).

Formulation (3) determines the number of ambiguous nodes by utilizing this scenario-based approach, where $a(r)$ is said to be the *ambiguity order* of the graph with respect to r, assuming that the attacker is constrained by option A.

$$a(r) = \max \sum_{k \in N} z_k \tag{3a}$$

$$\text{s.t.} \sum_{i \in S} x_i^{ka} \le \kappa \qquad \forall k \in N,\ a \in \{0,1\} \tag{3b}$$

$$r_{ij} - x_i^{ka} \le t_j^{ka} \le r_{ij} + x_i^{ka} \qquad \forall i \in S,\ j \in \text{FS}(i),\ k \in N,\ a \in \{0,1\} \tag{3c}$$

$$t_k^{k1} \ge z_k, \qquad \forall k \in N \tag{3d}$$

$$t_k^{k0} \le 1 - z_k, \qquad \forall k \in N \tag{3e}$$

$$x_i^{ka} \in \{0,1\} \qquad \forall i \in S,\ k \in N,\ a \in \{0,1\} \tag{3f}$$

$$0 \le t_j^{ka} \le 1 \qquad \forall j \in N,\ k \in N,\ a \in \{0,1\}. \tag{3g}$$

The objective function (3a), along with (3d) and (3e), determines the number of ambiguous nodes, as justified above. Constraints (3b) ensure that the defender never assesses more than κ sensors to be hijacked under any scenario. Constraints (3c) relate the x-variables to the t-variables as described above, and Constraints (3f) and (3g) give logical constraints and bounds on the x- and t-variables, respectively.

Note that there exists an optimal solution in which $t_j^{ka} \in \{0,1\}$ for all $j \in N$, $k \in N$, $a \in \{0,1\}$ and $z_j \in \{0,1\}$. Given binary x-values, (3c) places simple integer bounds on the t-values. Any binary solution for which the t-variables belong to the range stipulated by these constraints is optimal, so long as t_k^{k1} takes its largest possible value, and t_k^{k0} takes its smallest possible value, for all $k \in N$. Optimality forces z_j to take the minimum of 1, t_k^{k1}, and $1 - t_k^{k0}$; the minimum of these values is binary.

Next, observe that the model given in (3) can be equivalently solved by considering each scenario $\{ka\}$ separately, where we would fix $t_k^{k0} = 0$ ($t_k^{k1} = 1$) in scenario $\{k0\}$ ($\{k1\}$). If a feasible solution exists to both problems, then node k is ambiguous (and otherwise it is not). However, as we will show in the next subsection, this model is no longer separable when r_{ij}-values not are given, and instead become variables in the attacker's problem.

Observe that (3) is a linear mixed-integer program, and so our suggested algorithm to assess whether or not any node is ambiguous has a worst-case exponential time complexity. Thus, a reasonable question would ask whether or not a polynomial-time algorithm exists for this problem. The following theorem shows that even the simpler problem of determining a single plausible set of sensor locations, given r, is difficult. First define the following decision problem.

> PLAUSIBLE SET: Given a directed network $G(N,A)$, sensor set S, and readings r, does there exist a plausible set of sensor locations $H \subseteq S$ such that $|H| \leq \kappa$, for some specified positive integer κ?

Theorem 1. PLAUSIBLE SET *is NP-complete.*

Proof. We prove this by reduction from the vertex cover problem, defined as follows [6].

> VERTEX COVER: Given an undirected graph $G(N,E)$, and a positive integer η, does there exist a subset $V \subseteq N$, $|V| \leq \eta$, such that for every $(i,j) \in E$, either $i \in N$ or $j \in N$?

To show that PLAUSIBLE SET belongs to NP, consider a candidate solution H, with $|H| \leq \kappa$. One can verify that H is plausible by showing that no pair of accurate sensors conflicts at any node in N by examining $\mathrm{RS}(j)$ for all $j \in N$, and checking whether all r_{ij}-values match, for all $i \in \mathrm{RS}(j) \cap (S \setminus H)$. This verification can be done in $O(|N|^2)$ time, noting that $|\mathrm{RS}(j) \cap (S \setminus H)|$ is bounded by $|N|$ for all $j \in N$.

Given a VERTEX COVER instance with undirected graph $G' = (N', E')$, and integer η, we construct a PLAUSIBLE SET instance having a directed sensor network $G'' = (N'', A'')$ with sensor set S, readings r, and integer κ, such that G'' has a feasible selection of κ faulty sensors if and only if G' has a vertex cover of size η. First, let $\kappa = \eta$, and let N'' consist of $|N'| + |E'|$ nodes, one node j for every $j \in N'$, and one node w_{ij} for every edge $(i,j) \in E'$, $i < j$. Sensor set S consists of all nodes in N'' that correspond to nodes in N'. Also, for each edge $(u,v) \in E'$, create arcs (u, w_{uv}) and $(v, w_{uv}) \in A''$, and set $r_{uw_{uv}} = 0$ and $r_{vw_{uv}} = 1$. That is, each edge on graph G' corresponds to a conflicting pair of sensors on G''. Figure 4

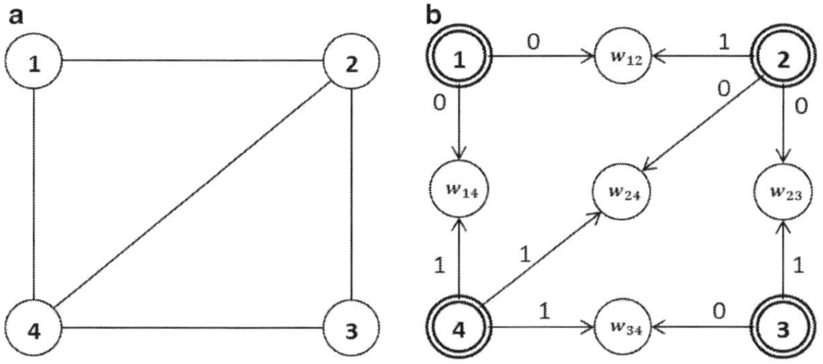

Fig. 4 Reduction of a VERTEX COVER instance to PLAUSIBLE SET instance. An edge (u,v) on G' transfers to $u,v \in S$, $w_{uv} \in N'' \setminus S$, $(u,w_{uv}),(v,w_{uv}) \in A''$ on G''. **(a)** Graph G' with four nodes. **(b)** Sensor network G'' with $S = \{1,2,3,4\}$

illustrates the transformation on a four-node graph. This transformation can be done in polynomial time, because $|N''| = |N'| + |E'|$ and $|A''| = 2|E'|$.

Suppose that G' has a vertex cover $V \subseteq N'$ of size κ. This selection of nodes on G' corresponds to a selection of sensors on G'' where at least one of each conflicting pair of sensors belongs to V. To show that setting $H = \{$nodes in N'' corresponding to $V\}$ is a feasible PLAUSIBLE SET solution, first note that $|H| = \eta = \kappa$. Next, note that a node u hosting a sensor is only monitored by its own sensor on G'', and hence no pair of sensor nodes in $S \setminus H$ could give conflicting reports at u. Now consider $w_{ij} \in N''$ for which $r_{iw_{ij}} \neq r_{jw_{ij}}$. Because V is a feasible vertex cover, either i or j belongs to H. Therefore, at most one sensor in $S \setminus H$ monitors w_{ij}, and no conflict among sensors could exist at w_{ij}. Thus, V must correspond to a PLAUSIBLE SET solution.

Now, suppose that PLAUSIBLE SET has a solution, H. We show that setting $V = H$ is a feasible VERTEX COVER solution as well. First, $|V| = \kappa = \eta$, satisfying the cardinality limit. Second, note that each sensor $u \in S$ is in conflict with sensor $v \in S$ regarding the status of node w_{uv} if there exists an edge $(u,v) \in E'$. Therefore, at least one of u and v is faulty (i.e., belongs to H), and H gives a vertex cover for G'. Hence, the VERTEX COVER instance is equivalent to the transformed PLAUSIBLE SET instance. Moreover, since all numerical data used in this transformation is bounded by $|N'|$, we have that PLAUSIBLE SET is NP-complete in the strong sense. □

3.2 *Formulation of the Attacker Problems*

We first analyze in Sect. 3.2.1 the attack formulation in the worst-case scenario for the defender, corresponding to case [A, A]. We then examine the cases in which the attacker's actions are limited, although the defender is unaware of these restrictions and assumes that the attacker is restricted as in option A. The three cases [X_a, A], for $X_a = $ B, C, D, in which the defender does not have the knowledge on the attacker's limitations, are discussed in Sects. 3.2.2, 3.2.3, and 3.2.4, respectively. Finally for the three cases [X_a, $X_d = X_a$], $X_a = $ B, C, D, in which the defender has full information on which option constrains the attacker's actions, we develop formulations in Sects. 3.2.5, 3.2.6, and 3.2.7, respectively.

3.2.1 Case [A, A]

Formulation (3) gives the ambiguity assessment model in which r-values are input parameters of the problem. In order to prescribe a set of r-values that maximize ambiguity, the attacker must simultaneously determine a potential set of target locations on the nodes, along with the readings that emanate from κ sensors that the attacker chooses to hijack. Let binary variables x_i equal 1 if sensor $i \in S$ is hijacked by the attacker, and 0 otherwise. Binary variable t_i, $\forall i \in N$, equals 1 if the worst-case scenario considered by the attacker involves a target that actually exists at node i, and equals 0 otherwise. Also, the r-values, which were treated as fixed values in the ambiguity assessment problem, are now released to be binary attacker variables. We then obtain the following mathematical programming formulation.

$$\max \ a(r) \tag{4a}$$

$$\text{s.t.} \ \sum_{i \in S} x_i \leq \kappa \tag{4b}$$

$$r_{ij} - x_i \leq t_j \leq r_{ij} + x_i \qquad \forall i \in S, \ j \in \text{FS}(i) \tag{4c}$$

$$x_i \in \{0,1\} \qquad \forall i \in S \tag{4d}$$

$$r_{ij} \in \{0,1\} \qquad \forall i \in S, \ j \in \text{FS}(i) \tag{4e}$$

$$t_i \in \{0,1\} \qquad \forall i \in N. \tag{4f}$$

The objective function (4a) maximizes the network owner's ambiguity given in (3). Constraints (4b) limit the number of sensors that the attacker can hijack. Constraints (4c) force the reading variables r_{ij} to be accurate when $x_i = 0$ (sensor i is accurate), and allow them to be 0 or 1 when $x_i = 1$. Constraints (4d)–(4f) give the binary restrictions on the variables. Observe that because (3) is a maximization problem, (4) can be stated as a mixed-integer program by optimizing (3a), subject to all constraints in (3) and (4), where r_{ij} is treated as a binary variable, $\forall i \in S$, $j \in \text{FS}(i)$.

Note that intuitively, the attacker's choice of target deployments on the graph should have no effect on the defender's ambiguity assessment. The defender essentially seeks to determine which sensor readings could contradict the actual presence or absence of a target; the case in which a target is present is symmetrical to the case in which a target is not present. This feature of the problem implies that

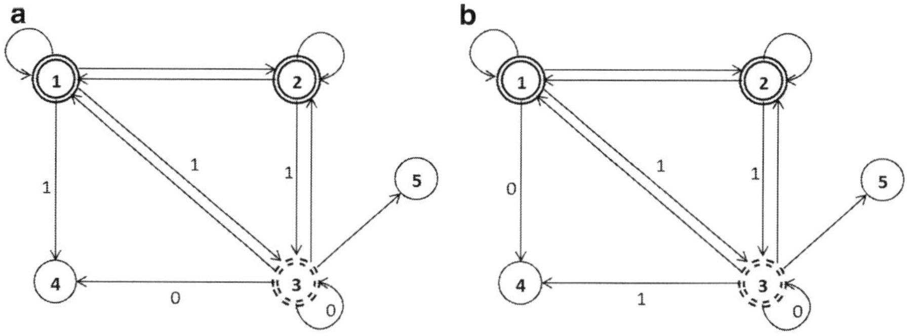

Fig. 5 Two equivalent r-solutions with respect to the status of node 4. Sensor 3 is faulty and reports an inaccurate reading, and sensor 1 reports an accurate reading, in both instances

we can remove the t-variables from the problem fixing $t_i = 0,\ \forall i \in N$, without loss of optimality, as proven below.

Theorem 2. *There exists an optimal solution to* (4) *in which* $t_i = 0,\ \forall i \in N$.

Proof. Consider any optimal solution (x^*, r^*, t^*) to (4) in which $t_i^* = 1$, for some $i \in N$, and examine an alternative solution $(\hat{x}, \hat{r}, \hat{t})$ in which $\hat{x} = x^*$; $\hat{t}_i = 0,\ \forall i \in N$; and $\hat{r}_{ij} = r_{ij}^* + (1 - 2r_{ij}^*)t_j^*,\ \forall i \in S,\ j \in \mathrm{FS}(i)$. (That is, all readings at node j are reversed in the new solution if $t_j^* = 1$.)

Let $\bar{x}^{ka}$ denote sensor failures assessed by the defender in scenario $\{ka\}$ in response to solution x^*. We show that if some node $k \in N$ was ambiguous given readings r^*, then it must still be ambiguous given readings $\hat{r}$. Because k was ambiguous, the defender found variable values $\bar{x}_i^{k1},\ \forall i \in S$, such that $\sum_{i \in S} \bar{x}_i^{k1} \le \kappa$ and $\bar{x}_i^{k1} = 1$ whenever $r_{ik}^* = 0,\ \forall i \in \mathrm{RS}(k)$, due to (3b), (3c), and (3d). Similarly, the defender found $\bar{x}^{k0}$-values such that $\sum_{i \in S} \bar{x}_i^{k0} \le \kappa$ and $\bar{x}_i^{k0} = 1$ whenever $r_{ik}^* = 1,\ \forall i \in \mathrm{RS}(k)$, due to (3b), (3c), and (3e). Given $\hat{r}$, the defender would still assess k as ambiguous, by choosing $\hat{x}_i^{k1} = \bar{x}_i^{k0}$, and $\hat{x}_i^{k0} = \bar{x}_i^{k1},\ \forall i \in S$.

Because (x^*, r^*, t^*) is optimal, and $(\hat{x}, \hat{r}, \hat{t})$ is a feasible solution that yields the same objective as (x^*, r^*, t^*), we have that $(\hat{x}, \hat{r}, \hat{t})$ must also be optimal. Furthermore, noting that $\hat{t}_j = 0,\ \forall j \in N$, this completes the proof. □

Figure 5 illustrates the transformation employed by Theorem 2 on node 4. A similar transformation can be done to place a target on an idle node, hence, there exist at least $2^{|N|}$ alternative optimal solutions to (4) due to the symmetric solutions corresponding to various t-values. Therefore, we fix all t_j-values to zero. In doing so, formulation (4) simplifies to:

$$\max \quad a(r) \tag{5a}$$

$$\text{s.t.} \quad \text{Constraints (4b), (4d), and (4e)} \tag{5b}$$

$$r_{ij} \le x_i \qquad \forall i \in S,\ j \in \mathrm{FS}(i). \tag{5c}$$

Note that the second inequality of (4c) becomes redundant with all t-values fixed to 0, and the first inequality of (4c) reduces to (5c). Because Theorem 2 can be equivalently applied to all attacker options, we hence fix the attacker's t-variables to zero in all subsequent models.

3.2.2 Case [B, A]

We now examine case [B, A], which restricts the attacker to report a false reading in all hijacked sensors. All other assumptions regarding the attacker's capabilities remain the same as case [A, A]. Case [B, A] affects constraints (5c) because the readings r_{ij} are forced to obey the condition $r_{ij} = |x_i - t_j| = x_i$. Hence, constraints (5c) are replaced with the following:

$$r_{ij} = x_i \qquad \forall i \in S,\ j \in \mathrm{FS}(i). \tag{6}$$

3.2.3 Case [C, A]

Unlike option B, options C and D allow faulty sensors to report any reading value, although they impose bounds on the number of false readings. Option C limits a faulty sensor $i \in H$ to give a false reading for at most one of the nodes it monitors. Recall that a reading of $r_{ij} = 1$, for some $i \in S,\ j \in \mathrm{FS}(i)$, implies that sensor i gives a false reading for node j (by Theorem 2). Hence, we can state the following inequality for option C:

$$\sum_{j \in \mathrm{FS}(i)} r_{ij} \le x_i \qquad \forall i \in S, \tag{7}$$

where the left-hand side of (7) gives the number of positive (false) readings by sensor i, and the right-hand side limits this number to 1 if sensor i is faulty and 0 otherwise.

3.2.4 Case [D, A]

Option D restricts the attacker to issue at most τ false readings in the network, instead of setting a limit per sensor. Hence, we aggregate the left-hand side of (7) over all $i \in S$ to constrain the total number of positive readings, and ensure that false readings emanate only from faulty sensors. Thus, we add the following constraints to (5):

$$\sum_{i \in S} \sum_{j \in \mathrm{FS}(i)} r_{ij} \le \tau, \tag{8a}$$

$$\sum_{j \in \mathrm{FS}(i)} r_{ij} \le \min\{\tau, |\mathrm{FS}(i)|\} x_i \qquad \forall i \in S. \tag{8b}$$

Note that constraints (8b) are the aggregation of the inequalities $r_{ij} \leq x_i$, which ensure that false (i.e., positive) readings emanate only from faulty sensors. The disaggregated inequalities require a larger constraint set, but generally induce tighter linear programming relaxations [13]. We will test the efficacy of solving [D, A] models using (8b) versus disaggregated inequalities $r_{ij} \leq x_i$ in our computational experiments.

3.2.5 Case [B, B]

In the last three cases, we re-analyze the cases given in Sects. 3.2.2–3.2.4 by allowing the network owner to have full knowledge of the attacker's capabilities. The defender's knowledge of the attacker's option changes the way the nodes' statuses are evaluated, and hence the way ambiguity is assessed. The first case we consider is [B, B], where a faulty sensor always gives false readings and the network owner is aware of this faulty behavior. Hence, whenever the defender decides that a sensor may be faulty, all readings from this sensor are assumed to be incorrect in the ambiguity assessment model.

In this case, note that $t_j^{ka} = |x_i^{ka} - r_{ij}|$ in the ambiguity assessment problem, under the assumption that a failed sensor always provides inaccurate readings. Therefore, all t-variables in (3) are bounded as follows, in addition to the constraints (3c):

$$t_j^{ka} \geq x_i^{ka} - r_{ij} \qquad \forall i \in S,\ j \in \text{FS}(i),\ k \in N,\ a \in \{0,1\} \tag{9a}$$

$$t_j^{ka} \leq 2 - x_i^{ka} - r_{ij} \qquad \forall i \in S,\ j \in \text{FS}(i),\ k \in N,\ a \in \{0,1\}. \tag{9b}$$

Observe that if $x_i^{ka} = 0$, then (9a) and (9b) are redundant, while (3c) forces $t_j^{ka} = r_{ij}$. Likewise, if $x_i^{ka} = 1$, then (3c) is redundant, and (9a) and (9b) force $t_j^{ka} = (1 - r_{ij})$.

3.2.6 Case [C, C]

In this case, the network owner is aware that the attacker can issue only one false report from each faulty sensor. In this case, the defender must determine which sensors could be faulty, and from those sensors assumed to be faulty, which reading is inaccurate. Accordingly, we define binary decision variables w_{ij}^{ka}, which equal 1 if sensor $i \in S$ is assumed to give a false reading at node $j \in N$ under scenario $\{ka\}$, for all $k \in N$ and $a \in \{0,1\}$. For each scenario $\{ka\}$, the defender assumes at most one false report per failed sensor, and hence we add the following constraints to the assessment problem:

$$\sum_{j \in \text{FS}(i)} w_{ij}^{ka} \leq x_i^{ka} \qquad \forall i \in S,\ k \in N,\ a \in \{0,1\}. \tag{10}$$

The target assessment conditions enforced by (3c) need to be revised, because now the defender knows that a faulty sensor cannot report a false reading on all monitored nodes. Therefore, we replace constraints (3c) with

$$r_{ij} - w_{ij}^{ka} \leq t_j^{ka} \leq r_{ij} + w_{ij}^{ka}$$
$$\forall i \in S,\ j \in \text{FS}(i),\ k \in N,\ a \in \{0,1\}. \tag{11}$$

3.2.7 Case [D, D]

This case allows the defender to be aware that the attacker is able to report at most τ false readings over all failed sensors. We define variables w_{ij}^{ka} as in [C, C]. We once again replace (3c) with (11), and add the following inequalities to the ambiguity assessment model (3):

$$\sum_{i\in S}\sum_{j\in \text{FS}(i)} w_{ij}^{ka} \leq \tau \qquad \forall k \in N,\ a \in \{0,1\}, \tag{12a}$$

$$\sum_{j\in \text{FS}(i)} w_{ij}^{ka} \leq \min\{\tau, |\text{FS}(i)|\} x_i^{ka}$$
$$\forall i \in S,\ k \in N,\ a \in \{0,1\}. \tag{12b}$$

Similar to (8b), an inequality in (12b) is aggregation of inequalities $w_{ij}^{ka} \leq x_i^{ka}$ over $j \in \text{FS}(i)$, which ensure that sensor i reports a false reading only if it has been hijacked.

4 Computational Experiments

We tested our seven models on a set of randomly generated instances having various sizes and graph densities. The models are implemented using CPLEX 12.4, using a C++ implementation via CPLEX Concert Technology. The computations are performed on an HP Pavilion with an AMD Athlon II Neo K325 Dual Core 1.30 GHz processor and 4.0 GB memory on a 64-bit Windows platform.

The results of our computational experiments are given in Table 1. The first column gives the instance parameters, where n, d, and i refer to number of nodes, edge density, and index label of the corresponding graph (referred as Gn-d-i), respectively. For each instance, the number of sensors equal one half of the number of nodes. In Table 1 the number of faulty sensors is bounded by 20 % of number sensors (i.e., $\kappa = \lfloor n/10 \rfloor$). For option D, the limit on the number of false readings is set to equal 2κ. Table 1 provides the optimal ambiguity order (z) and the CPU time (in seconds) required to solve each instance to optimality. We determined that for case [D, A], the aggregated constraints (8b) tend to yield a model that is easiest to

Table 1 Test instances and results of seven case models

Instance			[A, A]		[B, A]		[B, B]		[C, A]		[C, C]		[D, A]		[D, D]	
n	*d*	*i*	*z*	CPU	*z*	CPU	*z*	CPU	*z*	CPU	*z*	CPU	*z*	CPU	*z*	CPU
16	30	1	9	5	9	4	0	2	9	1	9	1	9	2	9	2
		2	6	8	6	3	0	4	6	3	6	1	6	4	6	2
		3	11	3	11	2	0	3	11	3	11	1	11	1	11	2
		4	7	10	7	11	0	3	7	3	7	1	7	3	7	3
		5	2	308	2	44	0	3	2	17	2	3	2	6	2	7
	50	1	2	20	0	48	0	4	1	2	1	1	2	3	2	10
		2	0	3,050	0	195	0	7	0	102	0	124	0	190	0	395
		3	1	271	0	60	0	5	1	3	1	3	1	21	1	18
		4	3	499	3	34	1	4	3	5	3	20	3	60	3	11
		5	2	78	1	56	0	3	2	1	1	7	2	5	2	8
18	30	1	10	18	10	8	0	3	10	6	10	2	10	5	10	3 %
		2	7	151	7	11	0	3	7	4	7	3	7	9	7	9
		3	2	258	0	519	0	7	1	1	1	2	2	5	2	12
		4	4	28	2	127	0	4	3	1	2	4	4	2	3	7
		5	6	174	6	31	0	4	6	7	6	1	6	11	6	5
	50	1	1	635	0	480	0	5	1	2	1	3	1	8	1	15
		2	1	1,149	0	1,189	0	5	1	56	1	2	1	45	1	4
		3	1	752	0	643	0	6	1	134	1	29	1	18	1	5
		4	1	1,230	0	1,631	0	8	1	20	1	2	1	1,002	1	16
		5	0	15,248	0	1,693	0	5	OOM		0	15	0	490	0	34
20	30	1	N/A		10	216	2	11	10	2,585	10	9,810	10	11,541	10	6,190
		2			11	174	0	14	11	1,856	11	1,394	11	2,940	11	2,509

N/A not available, *OOM* out of memory

solve by CPLEX, whereas for case [D, D], the disaggregated version of constraints (12b) is preferable on a majority of instances. Hence, the results we obtain in Table 1 correspond to the aggregated constraint model for [D, A] and to the disaggregated form for [D, D].

However, we note that the aggregated formulation was not uniformly more effective than the disaggregated formulation on all [D, A] instances. For instance, on G20-30-1, the disaggregated formulation solved in 8,017 CPU seconds, as opposed to 11,541 s for the aggregated formulation. On the other hand, in solving the ambiguity assessment model, the disaggregated constraints exhibited slightly better performance than the aggregated constraints in almost all instances. Hence, the results given in column [D, D] correspond to formulation with the aggregated constraints for the attacker, and disaggregated constraints for the ambiguity assessment problem.

Our results indicate that case [A, A] requires the most computational time among all cases. The large feasible region is one of the major factors causing the high computational time for this model as compared to the other cases. For all cases, the optimal ambiguity order is inversely correlated with the elapsed CPU time.

Intuitively, this behavior seems to stem from the fact that when the ambiguity order is high, it is not difficult to force many nodes to become ambiguous, and hence the identification of an optimal attacker solution is not difficult. An intriguing result presented in this table is the effect of the knowledge of the defender on the attacker's option. This effect is the most significant on option B, because the identification of a faulty sensor in case [B, B] allows the defender to ascertain whether or not targets exist at each node monitored by the faulty sensor. Note that the ambiguity order drops to zero in case [B, B] for all instances but two: G16-50-4 and G20-30-1. By contrast, the extent of the defender's knowledge of the attacker's option has no effect on the attacker's objective for options C and D, with the only exception being, for instance, G18-30-4. On the other hand, the cases where the defender has full knowledge on the attacker's capabilities tend to require significantly less computational time to solve on denser networks (i.e., $d = 50$) compared to their counterparts. This reduction in computational time may be due to the reduction of the search space in these cases, even though such models require an increase in the number of variables (with variables w_{ij}^{ka}) for cases [C, C] and [D, D].

Table 1 also shows that the restrictions imposed by options C and D on the attacker, while the defender assumes that the attacker has option A, do not have a significant impact on the attacker's objective in most instances. On the other hand, cases [C, A] and [D, A] reduce the computation time compared to case [A, A]. This observation implies that as long as the defender is not aware of the attacker's restrictions and the attacker is able to at least partially manipulate the readings of the hijacked sensors (as opposed to option B), investing additional effort to evaluate all possible options does not return much additional value to the attacker.

For all cases, the CPU time required increases rapidly with the network size (both in number of nodes and edges). On the other hand, network size is not the sole indicator of the difficulty of an instance: the high difference between the minimum and maximum CPU times required among equally sized instances (e.g., 635 s for G18-50-1 and 15,248 s for G18-50-5 for case [A, A]) suggest that the graph topology is also an important factor on the difficulty of the corresponding instance. Different structural characteristics such as existence of large cliques or leaf nodes (and their relative distribution on the network) might cause additional challenges due to the nature of the problem. In addition, the ratio of κ to the total number of sensors is another indicator for the difficulty of an instance. Note that our 18-node instances hosts 9 sensors, of which only one is faulty, and we are able to solve majority of 18-node instances (except G18-50-5 in case [C, A]). We present two instances with 20-nodes having reasonable solving times (note that a 20-node graph has 2 faulty sensors out of 10 sensors due to our 20 %-rule), while several other instances of the same size could not be solved within several hours. This rapid increment in CPU times between 18-node instances and 20-node instances suggest that the ratio of faulty sensors on the network is an important factor determining the difficulty of an instance. For this reason, and due to computational restrictions, the results for case [A, A] for the two 20-node instances here are omitted.

Table 2 gives computational results for the first two instances of G18-30 in which κ is now set to 2, as opposed to $\kappa = 1$ as in Table 1. The number next to the case

Table 2 Evaluation of τ-values for $\kappa = 2$ on G18-30 instances

Instance	z	CPU	z	CPU	z	CPU	z	CPU
	[D, A]-2		[D, A]-1.5		[D, A]-1		[D, A]-0.5	
1	14	38	14	24	14	57	14	10
2	13	20	13	17	13	11	13	5
	[D, D]-2		[D, D]-1.5		[D, D]-1		[D, D]-0.5	
1	14	45	14	42	14	24	10	1
2	13	54	13	24	13	19	7	3

label ("[D, A]" or "[D, D]") indicates the ratio of τ to κ (e.g., [D, A]-1.5 corresponds to case [D, A] where $\tau = 1.5\kappa$). This table presents some analysis on the sensitivity of our models with respect to different values of τ for option D. The results show that option D becomes restrictive to the attacker only when both τ is less than κ, and when the defender has full information on the attacker's option. Note, for instance, that even for the [D, A]-0.5 case, the defender is not able to reduce the number of ambiguous nodes beyond its ambiguity assessment when $\tau = 2\kappa$. On the other hand, [D, D]-0.5 shows that a relatively low τ-value does allow the defender to assess many fewer nodes as ambiguous, because the defender knows the attacker's option. In addition, the CPU times provided in Table 2 also show that a more restrictive τ-value reduces computational time while keeping the same objective function value. This lends further evidence to our observation in Table 2 that the attacker does not need to manipulate a large percentage of sensor readings in order to achieve maximum damage.

5 Conclusion

In this paper, we study sensor networks, in which each node might host a target. We propose four different failure options for deployed sensors, and assess the impact of sensor failures via the number of ambiguous nodes. We present mathematical formulations for determining a worst-case failure scenario for each option, where each failure option corresponds to two cases: either the defender has full knowledge on the failure option, or is unaware of the failure option and assumes a worst-case scenario. Our study also examines the impact of the defender's knowledge of these failure options. We show that for the defender, even identifying a plausible set of faulty sensors is NP-complete, using a reduction from vertex cover problem.

Our computational results indicate that the ambiguity order is often unaffected by the defender's ability to anticipate the restrictions of the attacker. An exception to this observation arises in case [B, B], where the attacker is forced to set the readings emanating from the faulty sensors to be false, and where the defender is able to anticipate these restrictions on the attacker. Moreover, for option D, in which τ total sensor readings can be false, the attacker's ability to create ambiguous nodes is usually unaffected by the limit τ, unless τ is small enough and the defender is aware of the attacker's restrictions. From a practical perspective, one important implication is that the use of the restricted-attack models [C, A], [C, C], [D, A], and [D, D] tend to report the same ambiguity number as [A, A], but in significantly less computation time. The implications of this objective are that the defender needs only to modify a small number of sensor readings to induce the maximum ambiguity order, and that the restricted-attack models may effectively be used in lieu of [A, A] in a heuristic scheme.

In addition, we see that an instance with relatively low optimal ambiguity order indicates a more challenging problem for the attacker and thus requires more computational time. Our computational results also show high variance among the computation times required to solve instances having the same number of nodes and edges. This indicates that the maximum ambiguity order problem (for the options discussed in this paper) seems to be highly sensitive to the structure of the sensor-network. Future research on how graph structures affect the difficulty of solving these problems is intriguing and represents a promising area of future investigation.

References

1. A. A. Abbasi and M. Younis. A survey on clustering algorithms for wireless sensor networks. *Computer Communications*, 30:2826–2841, 2007.
2. B. Balasundaram and S. Butenko. Graph domination, coloring and cliques in telecommunications. In M. G. C. Resende and P. M. Pardalos, editors, *Handbook of Optimization in Telecommunications*, pages 865–890. Springer, 2006.
3. N. Biggs. Perfect codes in graphs. *Journal of Combinatorial Theory, Series B*, 15(3):289–296, 1973.
4. E. Cockayne and S. Hedetniemi. Optimal domination in graphs. *IEEE Transactions on Circuits and Systems*, 22(11):855–857, 1975.
5. J. F. Fink and M. S. Jacobson. n-domination in graphs. In Y. Alavi, G. Chartrand, L. Lesniak, D. R. Lick, and C. E. Wall, editors, *Graph Theory with Applications to Algorithms and Computer Science*, pages 283–300. John Wiley & Sons, Inc., New York, 1985.
6. M. R. Garey and D. S. Johnson. *Computers and Intractability: A Guide to the Theory of NP-completeness*. W. H. Freeman & Co., Princeton, NJ, 1979.
7. F. Harary and R. A. Melter. On the metric dimension of a graph. *Ars Combinatoria*, 2:191–195, 1976.
8. T. W. Haynes, S. T. Hedetniemi, and P. J. Slater. *Fundamentals of Domination in Graphs*. Pure and Applied Mathematics. Marcel Dekker, New York, NY, 1998.
9. S. T. Hedetniemi and R. C. Laskar. *Topics on Domination*, volume 48 of *Annals of Discrete Mathematics*. North Holland, Amsterdam, 1991.
10. J. Kennington, E. Olinick, and D. Rajan, editors. *Wireless Network Design: Optimization Models and Solution Procedures*, volume 158 of *International Series in Operations Research & Management Science*. Springer, New York, NY, 2011.
11. H. Lim and C. Kim. Flooding in wireless ad hoc networks. *Computer Communications*, 24(3):353–363, 2001.
12. C. L. Liu. *Introduction to Combinatorial Mathematics*, volume 181. McGraw-Hill, New York, NY, 1968.

13. G. L. Nemhauser and L. A. Wolsey. *Integer and Combinatorial Optimization*. Wiley-Interscience, New York, NY, 1999.
14. O. Ore. *Theory of Graphs*, volume 38. American Mathematical Society, Providence, RI, Third edition, 1967.
15. D. Rajan and A. Atamturk. Survivable network design: Routing of flows and slacks. In G. Anandalingam and S. Raghavan, editors, *Telecommunications Network Design and Management*, pages 65–81. Kluwer, Norwell, MA, 2002.
16. M. G. C. Resende and P. M. Pardalos, editors. *Handbook of Optimization in Telecommunications*, volume 10. Springer, New York, NY, 2006.
17. S. Shen. Domination problems. In J. J. Cochran, editor, *Wiley Encyclopedia of Operations Research and Management Science*, pages 1470–1488. Wiley, Hoboken, NJ, 2010.
18. S. Shen and J.C. Smith. A decomposition approach for solving a broadcast domination network design problem. *Annals of Operations Research*, 1–28, Springer US, 2011.
19. P. J. Slater. Fault-tolerant locating-dominating sets. *Discrete Mathematics*, 249(1):179–189, 2002.
20. P. J. Slater. Liar's domination. *Networks*, 54(2):70–74, 2009.
21. J. C. Smith, C. Lim, and F. Sudargho. Survivable network design under optimal and heuristic interdiction scenarios. *Journal of Global Optimization*, 38:181–199, 2007.
22. S. Soni, R. Gupta, and H. Pirkul. Survivable network design: The state of the art. *Information Systems Frontiers*, 1(3):303–315, 1999.
23. H. von Stackelberg. *The Theory of the Market Economy*. William Hodge and Co., London, U.K., 1952.

Minimum Connected Sensor Cover and Maximum-Lifetime Coverage in Wireless Sensor Networks

Lidong Wu, Weili Wu, Kai Xing, Panos M. Pardalos, Eugene Maslov, and Ding-Zhu Du

Abstract Energy efficiency is an important issue in the study of wireless sensor networks. The minimum connected sensor cover problem and the maximum lifetime coverage problem are very well known in the literature on energy efficiency. In recent years, there are important developments in the study of these two problems through studying the relationship between the connected sensor cover and the group Steiner tree and the relationship between coverage and weighted dominating set. In this article, we introduce those relationships and related new developments on the minimum connected sensor cover problem and the maximum lifetime coverage problem.

Keywords Minimum weight • Connected sensor cover • Maximum lifetime • Coverage of sensor network • Constant-approximation • Group Steiner tree

L. Wu • W. Wu • K. Xing • D.-Z. Du (✉)
Department of Computer Science, University of Texas at Dallas, Richardson, TX 75080, USA
e-mail: lidong.wu@utdallas.edu; weiliwu@utdallas.edu; kai.xing@utdallas.edu; dzdu@utdallas.edu

P.M. Pardalos
Department of Industrial and System Engineering and Center for Applied Optimization, University of Florida, Gainesville, FL, USA
e-mail: pardalos@ufl.edu

E. Maslov
Laboratory of Algorithms and Technologies for Network Analysis (LATNA), National Research University Higher School of Economics, 136 Rodionova St., Nizhny Novgorod 603093, Russia
e-mail: lyriccoder@gmail.com

A. Sorokin and P.M. Pardalos (eds.), *Dynamics of Information Systems: Algorithmic Approaches*, Springer Proceedings in Mathematics & Statistics 51,
DOI 10.1007/978-1-4614-7582-8_10,

1 Introduction

The energy efficiency is an important issue in the study of wireless sensor networks because sensors have their power supplied with batteries, which have limited energy, and they are usually deployed into hostile environments, such as battlefield, underwater, and inside glaciers, so that recharging batteries is impossible. There exist two well-known optimization problems in the literature about the energy efficiency, the minimum connected sensor cover problem and the maximum lifetime coverage problem, arisen from different service requests in wireless sensor networks.

Consider a large number of sensors distributed in a region with duty for collecting information. When a request is to monitor the region for a long time, the maximum lifetime coverage problem occurs as follows.

> MAX-LC: Given a set $\mathscr{S}$ of sensors and a set $\mathscr{T}$ of target, find a proper schedule for active/sleep modes of sensors to maximize the lifetime of coverage, that is, the length of time period during which every target is monitored by at least one active sensor.

However, when a requested duty can be finished within a shorter time, especially within the lifetime of every sensor, the minimization on total energy consumption may be required. If all active sensors have the same energy consumption rate, that is, every sensor consumes the same amount of energy during the same time period, then the minimization of total energy is equivalent to the minimization of the number of active sensors. In such a situation, the minimum connected sensor cover problem occurs.

> MIN-CSC: Given a set $\mathscr{S}$ of sensors and a set $\mathscr{T}$ of target points (or a target area Ω), find the minimum subset of sensors to cover all targets.

MAX-LC and MIN-CSC have gained a lot of attention in the literature. In recent years, they received important progress, especially in computational complexity of approximation algorithms. Those results are obtained through the relationship between MIN-CSC and the group Steiner tree [33] and the relationship between MAX-LC and the weighted dominating set in unit disk graphs [16,22]. In this article, we introduce those relationships and related developments.

2 Connected Sensor Cover and Group Steiner Tree

The minimum connected sensor cover problem was first proposed by Gupta et al. [26] with requested target area. They presented a greedy algorithm with performance ratio $O(r \ln n)$ where n is the number of sensors and r is the link radius of the sensor network, i.e., for any two sensors with a sensing point in common, their hop distance is at most r. Zhang and Hou [36] found that when the communication radius R_c is at least twice of the sensing radius R_s, the coverage of a connected region implied the connectivity of subgraph induced by those sensors. Actually, area coverage can be transformed to target (point) coverage. For example, consider an area Ω which is divided by sensing areas of sensors in a sensor set $\mathscr{S}$ into small areas. Choose an interior point from each small area to form a target set $\mathscr{T}(\Omega)$. Then we have the following.

Lemma 1. *An area Ω is covered by $\mathscr{S}$ if and only if every target in $\mathscr{T}(\Omega)$ is covered by a sensor in $\mathscr{S}$.*

For target coverage, it is easy to show the following.

Lemma 2. *Let $\mathscr{S}$ be a set of sensors and $\mathscr{T}$ a set of targets. Let G be the bipartite graph with vertex sets $\mathscr{S}$ and $\mathscr{T}$ such that there exists an edge between $s \in \mathscr{S}$ and $a \in \mathscr{T}$ if and only if a can be covered by s, i.e., a can be covered by the sensing area of s. Suppose $R_c \geq 2R_s$ and G is connected. Then for any subset $\mathscr{S}'$ of $\mathscr{S}$, if all targets can be covered by $\mathscr{S}'$, then communication links between sensors in $\mathscr{S}'$ induce a connected graph with vertex set $\mathscr{S}'$.*

For a connected area Ω, if Ω is covered by the sensor set $\mathscr{S}$, then the bipartite graph with vertex sets $\mathscr{S}$ and $\mathscr{T}(\Omega)$ is clearly connected. Therefore, we have

Theorem 1 (Zhang and Hou [36]). *Suppose Ω is a connected area covered by a sensor set $\mathscr{S}$. Assume $R_c \geq 2R_s$. If Ω is covered by a sensor subset $\mathscr{S}'$, then communication links between sensors in $\mathscr{S}'$ induce a connected graph with vertex set $\mathscr{S}'$.*

This property is generalized by Zhou et al. [37] to the m-connectivity as follows.

Theorem 2 (Zou et al. [37]). *Suppose Ω is a connected area covered by a sensor set $\mathscr{S}$. Assume $R_c \geq 2R_s$. If every point of Ω is covered by at least m sensors (called* degree*) in a sensor subset $\mathscr{S}'$, then those sensors in $\mathscr{S}'$ would induce an m-connected communication network.*

Xing et al. [34] presented a coverage configuration protocol which can give different degree of coverage requested by applications. Bai et al. [3] studied a sensor deployment problem regarding the coverage and connectivity. Alam and Haas [1] studied this problem in three-dimensional sensor networks.

Funke et al. [21] improved approximation algorithms for the minimum connected sensor cover problem by allowing sensors to vary their sensing radius. With variable sensing radius and communication radius, Zhou et al. [38] designed a polynomial-time approximation with performance ratio $O(\log n)$. Chosh and Das [24] designed a greedy approximation using maximal independent set and Voronoi diagram. They determined the size of connected sensor cover produced by their algorithm. However, no comparison with optimal solution, that is, no analysis on approximation performance ratio is given.

In fact, for homogenous sensor networks with fixed sensing radius and communication radius, no theoretical approximation performance ratio has been improved before this paper. The reader may find more related information from a nice survey [25] on the connected sensor cover.

MIN-CSC is a special case of the minimum connected set cover problem as follows.

MIN-CSETC: Given a collection $\mathscr{C}$ of subsets of a finite set X and a graph G with vertex set $\mathscr{C}$ and edge set E, find a subcollection $\mathscr{C}' \subseteq \mathscr{C}$ such that $\mathscr{C}'$ covers every element in X and the subgraph of G induced by $\mathscr{C}'$ is connected.

MIN-CSETC is closely related to the group Steiner minimum tree problem as follows.

GSMT: Given an edge-weighted graph $G = (V,E)$, a root vertex $r \in V$ and k nonempty subsets of vertices, $g_1, g_2, \ldots, g_k$, find the minimum total weight tree containing r and at least one vertex from each subset g_i.

GSMT has been well studied [15, 20]. The following facts have been proven in the literature.

Theorem 3 (Halperin and Krauthgamer [27]). *GSMT has no polynomial-time $O(\log^{2-\varepsilon} n)$-approximation for any $\varepsilon > 0$ unless NP has quasi polynomial Las-Vega algorithm.*

Theorem 4 (Garg et al. [23]). *GSMT has a polynomial-time random algorithm which, with probability $1-\varepsilon$ (for any $\varepsilon > 0$), produces a $O(\log^3 n)$-approximation where n is the number of nodes in input graph.*

Above theorems are obtained with an important technique on metric space approximation [4, 5].

Next, we establish the relationship between CONNECTED SENSOR-COVER and GROUP STEINER TREE so that the above two results can be extended to CONNECTED SENSOR-COVER.

Theorem 5 (Wu et al. [33]). CONNECTED SENSOR-COVER *has no polynomial-time $O(\log^{2-\varepsilon} n)$-approximation for any $\varepsilon > 0$ unless NP has quasi polynomial Las-Vega algorithm*

Proof. We construct a reduction from GROUP STEINER TREE to CONNECTED SENSOR-COVER as follows.

Consider an input of GROUP STEINER TREE, a graph $G = (V,E)$ with edge weight $w: E \to Z_+$, a root vertex $r \in V$ and k nonempty subsets of vertices, $g_1, g_2, \ldots, g_k$. Define $X = \{g_0, g_1, \ldots, g_k\}$ where $g_0 = \{r\}$. For each node $u \in V$, define $S_u = \{g_i \mid u \in g_i\}$. For each edge $(u,v) \in E$, construct a path connecting S_u and S_v with $k \cdot w(u,v)$ intermediate nodes. Denote by H the obtained graph on $\mathscr{V} = \{S_v \mid v \in V\}$ and intermediate nodes. Suppose there is a polynomial-time $O(\log^{2-\varepsilon} n)$-approximation for CONNECTED SENSOR-COVER. Let D be such an approximation solution D on instance $(X, \mathscr{V}, H)$ and $\mathrm{opt}_{\mathrm{CSC}}$ the number of nodes in an optimal solution, i.e., the objective function value of CONNECTED SET-COVER. Then we have

$$|D| \le O(\log^{2-\varepsilon} n)\mathrm{opt}_{\mathrm{CSC}}.$$

Clearly, D is the node set of a tree T in H, which induces a tree T' in G. Denote by $w(T')$ the total edge weight of T'. Then

$$w(T') \le \frac{|D|}{k}.$$

Now, let T^* be a minimum group Steiner tree, which induces a tree T^{**}. Then the number of nodes in T^{**} is at most $w(T^*)(k+2)$ and is at least $\mathrm{opt}_{\mathrm{CSC}}$. Therefore,

$$w(T') \le O(\log^{2-\varepsilon} n) \cdot \frac{k+2}{k} \cdot w(T^*) = O(\log^{2-\varepsilon} n) \cdot w(T^*),$$

that is, T' is a polynomial-time $O(\log^{2-\varepsilon} n)$-approximation for GROUP STEINER TREE. By Theorem 3, *NP* has quasi polynomial Las-Vega algorithm.

In the above, we treat $(X, \mathscr{V}, H)$ as an instance of CONNECTED SENSOR-COVER. The reader may suspect this treatment because

(a) It is unclear how to represent each intermediate node as a subset of X and
(b) It is possible that $S_u = S_v$, but in the definition of CONNECTED SET-COVER, $\mathscr{U}$ is not a multiple subset collection.

We remark that (a) and (b) can be fixed easily. In fact, for each intermediate node x, we can introduce a new element e_x and let $S_x = \{e_x\}$ represent node x. For each $S_v, v \in V$, we can also introduce a new element e_v and add e_v into S_v. Finally, put all new elements into S_{r} and X. Note that g_0 is contained only in S_{r}. Therefore, any feasible solution of CONNECTED SET-COVER must contain S_{r}. This means that those subsets representing intermediate nodes are useless for covering elements and they are useful only in establishment of connectivity. Moreover, we can easily see that this modification does not affect the size of any solution of CONNECTED SENSOR-COVER. □

Theorem 6 (Wu et al. [33]). *There exists a polynomial-time $O(\log^3 n)$-approximation for* CONNECTED SENSOR-COVER *where n is the number of subsets in $\mathscr{U}$.*

Proof. Consider an instance of CONNECTED SENSOR-COVER, a finite set X, a collection $\mathscr{U}$ of subsets of X, and a graph $G = (\mathscr{U}, \mathscr{E})$. For each $x \in X$, define $g_x = \{S \in \mathscr{U} \mid x \in S\}$. Fixed a $z \in X$. We compute an approximation solution of CONNECTED SENSOR-COVER as follows:

Step 1. For each $R \in g_z$, we compute a $O(\log^3 n)$-approximation T_R for GROUP STEINER TREE on input consisting of graph G with edge weight one for every edge, the root R and groups g_x for $x \in X - \{z\}$.
Step 2. Among all T_R for $R \in g_z$, choose one T_{R^*} with the smallest number of nodes. Output the node set $\mathscr{A}$ of T_{R^*}.

Next, we show that $\mathscr{A}$ is a $O(\log^3 n)$-approximation solution of CONNECTED SENSOR-COVER. Suppose $\mathscr{A}^*$ is an optimal solution of CONNECTED SET-COVER and denote by T^* the tree interconnecting nodes in $\mathscr{A}^*$. Then the total edge weight of tree T^* is $|\mathscr{A}^*| - 1$. Let $R \in g_z \cap \mathscr{A}$. Let $\mathscr{A}_R$ be the node set of T_R. Then

$$|\mathscr{A}| - 1 \le |\mathscr{A}_R| - 1 \le O(\log^3 n) \cdot (|\mathscr{A}^*| - 1).$$

Thus,

$$|\mathscr{A}| \leq O(\log^3 n) \cdot |\mathscr{A}^*|. \qquad \square$$

3 Maximum Lifetime and Minimum Weight

When a monitoring task is requested frequently or for a long period, one may expect the sensor network to be able to do such a monitoring as long as possible. This expectation motivates many optimization problems with maximization of lifetime. The following is a typical one.

> MAX-LIFETIME COVERAGE: Given a set of (point) targets and a set of sensors, find a way to schedule sensors' working/sleeping time to maximize the lifetime of the sensor network where the sensor network is said to be alive if every target is covered by at least one working sensor. Such lifetime is also called the lifetime of coverage.

Initially, the lifetime of coverage is maximized through partitioning sensors into the maximum number of disjoint sensor covers [8, 10, 11] since in this way, the sensor can have a simple control device; after waking up, the sensor will keep working until its energy is exhausted. Cardei et al. [11] found that allowing exchanges between working/sleeping modes may make the lifetime of coverage longer.

For example, consider three sensors s_1, s_2, s_3 and three targets e_1, e_2, e_3. s_1 can cover e_2 and e_3, but not e_1. s_2 can cover e_1 and e_3, but not e_2. s_3 can cover e_1 and e_2, but not e_3. Clearly, a sensor set is a sensor cover if and only if it contains at least two sensors. Therefore, if all the three sensors are partitioned into disjoint sensor covers, then this partition contains only one sensor cover, which yields the lifetime one for coverage when we assume that every sensor has lifetime one. However, consider three time periods of length 0.5. In the first period, sensors s_1 and s_2 work and s_3 sleeps; in the second period, sensors s_2 and s_3 work and s_1 sleeps; in the third period, sensors s_1 and s_3 work and s_2 sleeps. Then, the lifetime of coverage would be 1.5.

Let S be the set of all sensors. Let $p_1, p_2, \ldots, p_k$ be all possible sensor covers. Let t_i denote the time of using sensor cover p_i. Define

$$a_{s,p} = \begin{cases} 1 & \text{if } s \in p, \\ 0 & \text{otherwise.} \end{cases}$$

Then MAX-LIFETIME COVERAGE can be represented by the following linear program.

$$\begin{aligned} \max\ & t_1 + \cdots + t_k \\ \text{s.t. } & a_{s,p_1} t_1 + \cdots + a_{s,p_k} t_k \leq 1 \quad \text{for } s \in S, \\ & t_i \geq 0 \quad \text{for } i = 1, \ldots, k. \end{aligned}$$

It is well known that the linear programming is polynomial-time solvable. Does this mean that MAX-LIFETIME COVERAGE is polynomial-time solvable? The answer is NO because the number k of possible sensor covers may be exponentially large with respect to the number of sensors, $|S|$. In fact, MAX-LIFETIME COVERAGE is NP-hard. There are many efforts [7–9, 11–14, 16, 29–32, 35, 36] made for design of approximations and heuristics for MAX-LIFETIME COVERAGE. Among them, some [12, 30–32, 36] actually study the area coverage. By area coverage, one means that the request is to monitor a finite area. However, the area coverage can be reduced into the target coverage. By target coverage, one usually means that the request is to monitor a finite set of targets points.

Cardei et al. [10] considered not only the coverage but also the connectivity. Du et al. [18] study the connected coverage with two active phase sensors. In this model, each sensor can be in active mode or sleep mode and each sensor in active mode can be in full active phase or semi-active phase. A full active sensor can monitor target and make connection between sensors while a semi-active sensor can only make connection between sensors. Suppose in a unit time, a full active sensor consumes energy u and a semi-active sensor consume energy v. Usually, $u \geq v$.

> MAX-LIFETIME CC: Given a set of targets and a set of sensors each with two active phases, full active phase and semi-active phase as described above, find a sleep/work schedule to maximize the lifetime of coverage and connectivity as follows:
>
> (a) Every target is sensed by at least one full active sensor.
> (b) The subgraph induced by active sensors is connected.

For this problem, a connected sensor cover p is a pair of a full-active sensor set p_1 and a semi-active sensor set p_2 such that every target is covered by a sensor in p_1 and the subgraph induced by $p_1 \cup p_2$ is connected. Let S be the set of all sensors and C the set of all connected sensor covers. Define

$$a_{s,p} = \begin{cases} u & \text{if } s \in p_1, \\ v & \text{if } s \in p_2, \\ 0 & \text{otherwise.} \end{cases}$$

Let x_p be the active time of connected sensor cover p. Then MAX-LIFETIME CC can be formulated as the following linear programming.

$$\begin{aligned} \text{(PLP)} \ \max \ & \sum_{p \in C} x_p \\ \text{s.t.} \ & \sum_{p \in C} a_{s,p} x_p \leq 1 \quad \text{for all } s \in S, \\ & x_p \geq 0 \quad \text{for all } p \in C. \end{aligned}$$

It is similar to MAX-LIFETIME COVERAGE that MAX-LIFETIME CC is also NP-hard. However, LP-representations of MAX-LIFETIME COVERAGE and MAX-LIFETIME CC give them the best-known approximation algorithms in the literature.

Those algorithms are of the primal–dual type; the design idea was initiated by Garg and Könemann [22]. Since MAX-LIFETIME COVERAGE can be considered as a special case of MAX-LIFETIME CC (when $v = 0$), MAX-LIFETIME CC will be used as an example to explain the main idea behind the design of such primal–dual algorithms.

First, we note that the dual linear programming of (PLP) is as follows:

$$\begin{aligned}
\text{(DLP)}\ \min \ & \sum_{s\in S} y_s \\
\text{s.t.}\ & \sum_{s\in S} a_{s,p} y_s \geq 1 \quad \text{for } p \in C, \\
& y_s \geq 0 \quad \text{for } s \in S.
\end{aligned}$$

By the duality theory, the maximum objective function value of the primal linear programming is equal to the minimum objective function value of the dual linear programming. Therefore, the objective function value of any primal feasible solution is a lower bound for the common optimal value and the objective function value of any dual feasible solution is an upper bound for the common optimal value.

The difference of two objective functions can be represented as follows:

$$\begin{aligned}
& \sum_{s\in S} y_s - \sum_{p\in C} x_p \\
& = \sum_{s\in S} y_s \left(1 - \sum_{p\in C} a_{s,p} x_p\right) + \sum_{p\in C} x_p \left(\sum_{s\in S} a_{s,p} y_s - 1\right).
\end{aligned}$$

This difference equal to 0, which implies

$$\begin{aligned}
\sum_{s\in S} y_s \left(1 - \sum_{p\in C} a_{s,p} x_p\right) &= 0 \\
\sum_{p\in C} x_p \left(\sum_{s\in S} a_{s,p} y_s - 1\right) &= 0,
\end{aligned}$$

is called the *complementary-slackness condition*. If $(x_p, p \in C)$ is a primal feasible solution, $(y_s, s \in S)$ is a dual feasible solution and the complementary-slackness condition holds, then $(x_p, p \in C)$ and $(y_s, s \in S)$ are optimal solutions for the primal linear programming and the dual linear programming, respectively.

Note that it is easy to obtain an initial primal feasible solution, *e.g.*, $x_p = 0$ for $p \in C$ form a trivial primal feasible solution. The classical primal–dual method for above linear programming may start with a primal feasible solution $(x_p, p \in C)$ and a dual infeasible solution $(y_s, s \in S)$; but they satisfy the complementary-slackness condition, *e.g.*, $x_p = 0$ for $p \in C$ and $y_s = 0$ for $s \in S$. In each iteration, the dual feasibility of $(y_s, s \in S)$ is improved while keeping the primal feasibility of $(x_p, p \in$

C) and the complementary-slackness condition holding until $(y_s, s \in S)$ becomes dual feasible.

The primal–dual method given by Garg and Könemann [22] uses a similar idea. However, since the classical primal–dual method computes the optimal solution and the method of Garg and Könemann is used for computing an approximation solution, there is a fundamental difference on the reserved relationship between $(x_p, p \in C)$ and $(y_s, s \in S)$. They are not required to satisfy the complementary-slackness condition. Instead, they are required to have smaller

$$\sum_{s \in S} y_s \left(1 - \sum_{p \in C} a_{s,p} x_p\right) + \sum_{p \in C} x_p \left(\sum_{s \in S} a_{s,p} y_s - 1\right). \quad (1)$$

Actually, when $(x_p, p \in C)$ is primal-feasible and $(y_s, s \in S)$ becomes dual-feasible, (1) gives an upper bound for the difference of the primal objective function value from the optimal and hence it gives an evaluation of approximation performance.

To increase the dual-feasibility of $(y_s, s \in S)$, initially set $y_s = \delta > 0$ for $s \in S$ instead of setting $y_s = 0$. In each iteration, in order to increase the dual-feasibility of $(y_s, s \in S)$, we need to increase the value of y_s for some $s \in S$. To keep $y_s(1 - \sum_{p \in C} a_{s,p} x_p)$ smaller, we may choose only those y_s to increase its value where $(1 - \sum_{p \in C} a_{s,p} x_p)$ decreases. To decrease $(1 - \sum_{p \in C} a_{s,p} x_p)$, we need to increase the value of x_p for some $p \in C$. However, increasing the value of x_p would cost increasing value of $x_p(\sum_{s \in S} a_{s,p} y_s - 1)$. This consideration results in a principle for choice of x_p, that is, choose x_p such that p gives the optimal solution of the following problem:

$$\min_{p \in C} \sum_{s \in S} a_{s,p} y_s.$$

This is exactly the weighted version of MIN-CSC as follows.

MINW-CSC: Given a set $\mathscr{S}$ of sensors with nonnegative weight $\mathscr{S} \to R^+$ and a set $\mathscr{T}$ of target points, find the minimum-weight subset of sensors to cover all targets.

This problem is NP-hard. Therefore, we may find an approximation solution instead of an optimal solution.

Let p^* be a connected sensor cover which is a ρ-approximation of MINW-CSC. We intend to increase x_{p^*}. Initially, $x_{p^*} = 0$. To keep $\sum_{p \in C} a_{s,p} x_p \leq 1$ for all $s \in S$, we need to compute s^* such that

$$a_{s^*,p^*} = \max_{s \in S} a_{s,p^*}$$

and set

$$x_{p^*} \leftarrow x_{p^*} + \frac{1}{a_{s^*,p^*}}. \quad (2)$$

When $x_p > 0$ for some $p \in C$, (2) may cost violation of the primal-feasibility of $(x_p, p \in C)$. However, it does not bring any issues. Note that the primal linear programming is equivalent to

$$\max_{x_p \geq 0, \forall p \in C} \frac{\sum_{p \in C} x_p}{\max_{s \in S} \sum_{p \in C} a_{s,p} x_p}.$$

Even if $(x_p, p \in C)$ is not primal-feasible,

$$\frac{x_p}{\max_{s \in S} \sum_{p \in C} a_{s,p} x_p}$$

is primal-feasible.

Now, for $s \in p^*$, $1 - \sum_{p \in C} a_{s,p} x_p$ is getting smaller and hence, we are able to increase y_s while keep $y_s(1 - \sum_{p \in C} a_{s,p} x_p)$ not increasing. Clearly, the increment of y_s should be proportional to the decrease in $(1 - \sum_{p \in C} a_{s,p} x_p)$. Hence,

$$y_s \leftarrow y_s + \frac{a_{s,p^*}}{a_{s^*,p^*}} \tag{3}$$

for $s \in p^*$. (Actually, (3) can hold for any $s \in S$ since $a_{s,p^*} = 0$ for $s \notin p^*$.)

From the above consideration, we may obtain the following primal–dual algorithm:

Primal–Dual Approximation for MAX-LC
Initially, set $x_p = 0$ for $p \in C$ and $y_s = \delta > 0$ for $s \in S$,
where δ is a positive constant.
repeat
(1) Compute ρ-approximation p^* of MINW-CSC
$\min_{p \in C} \sum_{s \in S} a_{s,p} y_s$;
(2) Compute $s^* \in p^*$ such that
$a_{s^*,p^*} = \max_{s \in p^*} a_{s,p^*}$;
(3) For $p \neq p^*$, unchange x_p, but update
$x_{p^*} \leftarrow x_{p^*} + \frac{1}{a_{s^*,p^*}}$;
(4) For every $s \in S$, update
$y_s \leftarrow y_s + \frac{a_{s,p^*}}{a_{s^*,p^*}}$;
until $(y_s, s \in S)$ is dual-feasible.
output $\frac{x_p}{\max_{s \in S} \sum_{p \in C} a_{s,p} x_p}$.

Du et al. [18] showed the following.

Theorem 7. *For any $\varepsilon > 0$, above primal–dual algorithm produces a polynomial-time $\rho(1+\varepsilon)$-approximation for* MAX-LC *when δ is chosen properly.*

By this theorem the approximation design for MAX-LC is reduced to the approximation design for MINW-CSC.

In a spatial case that every sensor has sensing radius R_s and communication radius $R_c \geq 2R_s$, Du et al. [18] showed that MINW-SCS has a polynomial-time $(7.475 + \varepsilon)$-approximation. This result is reached though efforts [2, 19, 28, 40] on partition techniques and [6, 39] on Steiner trees. The reader may also refer [17] for a quick understanding of partitions. Therefore, MAX-LC has a polynomial-time $(7.475 + \varepsilon)$-approximation, either.

4 Conclusion

This article is a survey on recent studies for two important optimization problems in wireless sensor networks: the minimum connected sensor cover problem and the maximum lifetime coverage problem. Both of them arise from consideration of energy efficiency and it was found recently that both of them have a polynomial-time constant-approximation. Two interesting relationships have been discovered: one is between connected set cover (a generalization of connected sensor cover) and a group of Steiner tree problems. Another is between the minimum weight sensor cover and the maximum lifetime coverage. The first relationship brings some results on complexity and approximation from group Steiner tree problem to connected set cover problem. The second relationship brings a constant-approximation from the minimum weight sensor cover to the maximum lifetime coverage problem. Those relationships may play important roles in further research on the related problems.

Acknowledgments This work was supported in part by National Science Foundation of USA under grants CNS0831579, CNS101630 and CCF0829993.

We wish to thank Dr. Alexey Sorokin for his insightful corrections and suggestions, especially modification on conclusion section.

References

1. S.M.N. Alam and Z.J. Haas, Coverage and connectivity in three-dimensional networks, *MobiCom'06*, 2006.
2. C. Ambühl, T. Erlebach, M. Mihalák and M. Nunkesser, Constant-approximation for minimum-weight (connected) dominating sets in unit disk graphs, *Proceedings of the 9th International Workshop on Approximation Algorithms for Combinatorial Optimization (APPROX 2006)*, LNCS 4110, Springer, 2006, pp. 3–14.
3. X. Bai, S. Kumar, D. Xuan, Z. Yun, and T.H. Lai, DEploying wireless sensors to achieve both coverage and connectivity, *MobiHoc'06*, 2006, pp.131–142.
4. Y. Bartal, Probability approximation of metric spaces and its algorithmic applications, *Proc of 37th FOCS*, 1996, pp. 184–193.
5. Y. Bartal, On approximating arbitrary metrics by tree metrics, *Proc of 30th STOC*, 1998, pp. 161–168.
6. Jaroslaw Byrka, Fabrizio Grandoni, Thomas Rothvoss and Laura Sanita, An improved LP-based approximation for Steiner tree, *STOC'10*, June 5–8, 2010, pp. 583–592.

7. M. Cardei, Coverage problems in sensor networks, *Handbook of Combinatorial Optimization (2nd Edition)*, P. M. Pardalos, D. Z. Du, and R. Graham (eds.), Springer, to appear.
8. M. Cardei and D.-Z. Du, Improving wireless sensor network lifetime through power aware organization, *ACM Wireless Networks*, Vol. 11, No. 3, pp. 333–340, May 2005.
9. M. Cardei and J. Wu, Coverage in wireless sensor networks, in *Handbook of Sensor Networks*, M.Ilyas and I. Mahgoub (eds.), CRC Press, 2004.
10. Mihaela Cardei, Dave MacCallum, Xiuzhen Cheng, Manki Min, Xiaohua Jia, Deying Li and Ding-Zhu Du, Wireless sensor networks with energy efficient organization, *Journal of Interconnection Networks* vol 3 no 3–4 (2002) 213–229.
11. Mihaela Cardei, My Thai, Yingshu Li, and Weili Wu, Energy-efficient target coverage in wireless sensor networks, *IEEE INFOCOM* (2005).
12. J. Carle and D. Simplot, Energy efficient area monitoring by sensor networks, *IEEE Computer* Vol 37, No 2 (2004) 40–46.
13. Maggie X. Cheng, Lu Ruan and Weili Wu, Achieving minimum coverage breach under bandwidth constraints in wireless sensor networks, *INFOCOM 2005*, pp. 2638–2645.
14. Maggie Xiaoyan Cheng, Lu Ruan, Weili Wu, Coverage breach problems in bandwidth-constrained sensor networks, TOSN 3(2): 12 (2007).
15. C. Chekuri, G. Even and G. Kortsarz, A greedy approximation algorithm for the group Steiner problem, *Discrete Applied Mathematics*, 154 (2006) 15–34.
16. Ling Ding, Weili Wu, James K. Willson, Lidong Wu, Zaixin Lu and Wonjun Lee, Constant-approximation for target coverage problem in wireless sensor networks, *Proc. of The 31st Annual Joint Conf. of IEEE Communication and Computer Society (INFOCOM)*, 2012.
17. D.-Z. Du, Ker-I Ko, and X. Hu, *Design and Analysis of Approximation Algorithms*, Springer, 2012, pp. 142–157.
18. Hongwei Du, Panos M. Pardalos, Weili Wu, Lidong Wu, Maximum Lifetime Connected Coverage with Two Active-Phase Sensors, *Journal of Global Optimization*, online in 2012.
19. T. Erlebach and M. Mihalk, A $(4+\varepsilon)$-approximation for the minimum-weight dominating set problem in unit disk graphs, *WAOA 2009*, pp. 135–146.
20. G. Even, G. Kortsarz, An approximation algorithm for the group Steiner problem, *Proceedings of SODA*, 2002, pp. 49–58.
21. S. Funke, A. Kesselman, F. Kuhn, Z. Lotker, and M. Segal, Improved approximation algorithms for connected sensor cover, *Wireless Networks* 13 (2007) 153–164.
22. N. Garg and J. Könemann, Faster and simpler algorithms for multicommodity flows and other fractional packing problems, *Proc. 39th Annual Symposium on the Foundations of Computer Science*, 1998, pp 300–309.
23. N. Garg, G. Konjevod and R. Ravi, A polylogarithmic approximation algorithm for the group Steiner tree problem, *SODA*, 2000.
24. A. Ghosh and S.K. Das, A distributed greedy algorithm for connected sensor cover in dense sensor networks, *LNCS* 3560 (2005) 340–353.
25. A. Ghosh and S.K. Das, Coverage and connectivity issues in wireless sensor networks, in *Movile, Wireless, and Sensor Networks: Technology, Applications, and Future Directions*, by R. Shorey, A.L. Ananda, M.C. Chan, and W.T. Ooi (eds.), John Wiley & Sons, Inc., 2006, pp. 221–256.
26. H. Gupta, S.R. Das and Q. Gu, Connected sensor cover: self-organization of sensor networks for efficient query execution, *MobiHoc'03*, 2003, pp. 189–200.
27. Eran Halperin, Robert Krauthgamer: Polylogarithmic inapproximability. STOC 2003: 585–594.
28. Y. Huang, X. Gao, Z. Zhang and W. Wu, A better constant-factor approximation for weighted dominating set in unit disk graph, *Journal of Combinatorial Optimization* 18 (2009) 174–194.
29. K. Kar and S. Banerjee, Node placement for connected coverage in sensor networks, *Proc. of WiOpt 2003: Modeling and Optimization in Mobile, Ad Hoc and Wireless Networks*, 2003.
30. S. Slijepcevic and M. Potkonjak, Power efficient organization of wireless sensor networks, *IEEE International Conference on Communications*, Jun. 2001, pp. 472–476.

31. D. Tian and N. D. Georganas, A coverage-preserving node se scheduling scheme for large wireless sensor networks, *Proc. of 1st ACM Workshop on Wireless Sensor Networks and Applications*, 2002.
32. X. Wang, G. Xing, Y. Zhang, C. Lu, R. Pless, C. Gill, Integrated coverage and connectivity configuration in wireless sensor networks, *Proceedings of the 1st International Conference on Embedded Networked Sensor Systems, Sensys.03*, Los Angeles, CA, November 2003, pp. 28–39.
33. L. Wu, H. Du, W. Wu, D. Li, J. Lv, and W. Lee, Approximations for Minimum Connected Sensor Cover, *IEEE Conference on Computer Communications*, 2013.
34. G. Xing, X. Wang, Y. Zhang, C. Liu, R. Pless, and C. Gill, Integrated coverage and connectivity configuration for energy conservation in sensor networks, *ACM Transactions on Sensor Networks*, vol. 1 no. 1 (2005) 36–72.
35. S. Yang, F. Dai, M. Cardei, J. Wu and F. Patterson, On connected multiple point coverage in wireless sensor networks, *Journal of Wireless Information Networks* 13 (2006), pp. 289–301
36. H. Zhang and J. C. Hou, Maintaining sensing coverage and connectivity in large sensor networks, *Ad Hoc & Sensor Wireless Networks* 1 (2005) 89–124.
37. Z. Zhou, S. Das, H. Gupta, Connected k-coverage problem in sensor networks, *Proceedings of the 13th International Conference on Computer Communications and Networks*, 2004, pp. 373–378.
38. Z. Zhou, S.R. Das, H. Gupta, Variable radii connected sensor cover in sensor networks, *ACM Transactions on Sensor Networks*, vol. 5 no. 1 (2009).
39. F. Zou, X. Li, S. Gao, and W. Wu, Node-weighted Steiner tree approximation in unit disk graphs, *J. Comb. Optim.* 18 (2009) 342–349.
40. Feng Zou, Yuexuan Wang, Xiaohua Xu, Hongwei Du, Xianyue Li, Pengjun Wan and Weili Wu, New approximations for weighted dominating sets and connected dominating sets in unit disk graphs, *Theoretical Computer Science* vol 412 no 3 (2011) 198–208.

Influence Diffusion, Community Detection, and Link Prediction in Social Network Analysis

Lidan Fan, Weili Wu, Zaixin Lu, Wen Xu, and Ding-Zhu Du

Abstract Social networks have received extensive attention among researchers across a wide range of disciplines such as computer science, physics, and sociology. This paper mainly overviews a variety of approaches for three problems in real-world life scenarios. The first problem is about influence diffusion, in which influence represents news, ideas, information, and so forth; the second one concerns with partitioning social networks into communities efficiently; and the third one is to predict the hidden or possible new links between individuals in the future based on the existing or observed information.

Keywords Influence diffusion • Community detection • Link prediction • Social network analysis • Social networks

1 Introduction

With the recent surge of online networks, people and organizations nowadays can interact and collaborate with each other more, providing the platform for the emergence of social networks in virtual environments. In a social network, nodes represent individuals or other entities embedded in a social context, and edges denote relationship between individuals. Since this kind of network is generally complex and highly dynamic, it is important to understand its behavior over time and particular interests can been seen in [70, 71]. Social network analysis (SNA) is a broad field of research that tries to exploit the social network structure and the

L. Fan • W. Wu • Z. Lu • W. Xu • D.-Z. Du (✉)
Department of Computer Science, University of Texas at Dallas, Richardson, TX 75080, USA
e-mail: ldfan28@gmail.com; weiliwu@utdallas.edu; zaixinlu@utdallas.edu; wen.xu@utdallas.edu; dzdu@utdallas.edu

A. Sorokin and P.M. Pardalos (eds.), *Dynamics of Information Systems: Algorithmic Approaches*, Springer Proceedings in Mathematics & Statistics 51,
DOI 10.1007/978-1-4614-7582-8_11,

dynamic actions within it. From the practical point of view, there are three lines of tasks that have received increasing interests from researchers recently, namely the influence diffusion problem, the community detection problem, and the link prediction problem.

As for the influence propagation problem, we mainly introduce several existing techniques from two aspects: influence maximization and multiple competitive influence dissemination. Examples of their applications include products promotion, which aims to select the most influential customers to attract more people to buy the product; political candidates election, in which both sides of the candidates try to "win" the most influential people to encourage more people to vote for them; rumor blocking or misinformation limitation, in which protectors are found to reduce or limit the spread of rumors as much as possible. The seminal work in influence diffusion is [40], and it is analyzed in detail in our paper.

With regard to the community detection problem, which tries to identify the community structure in a social network accurately, extensive attempts have been made. A measurement called modularity [52], which has a far-reaching impact on subsequential researches, was proposed to test the quality of a community partition. Later, most of the efforts were devoted to develop efficient and scalable algorithms. Moreover, several works addressed communities formation from the dynamic point of view. Since people in the same community have many common characteristics, knowing the community structure can help us understand the overall structural and functional properties of a large network. For instance, the communities in the blogspace often correspond to topics of interests. Thus, monitoring the aggregate trends and opinions exposed by these communities provides worthy insight to a number of business applications, such as marketing intelligence and competitive intelligence.

The link prediction problem aims at constructing a reliable link prediction model for uncovering missing links or estimating the formation of new links in the future. Examples of the link prediction problem include Collaborative Filtering recommender systems, which can be viewed as services predicting links between users and items within a user-item bipartite graph representing preferences or purchases; and protein/genetic interaction modeling, which can be viewed as predicting underlying protein–genetic interactions based on interaction observable in the network. Liben-Nowell et al. in [56,57] gave a nice survey about the structure-based measurements for link prediction, which was widely applied by later works. Besides network structure, node and edge attributes or the combination of both of them have been considered when design efficient algorithms. Furthermore, time was integrated to the problem as a parameter such that the problem better approximated to practical networks, which is essentially dynamic.

The remainder of the paper is organized as follows: in Sect. 2, we mainly introduce the influence diffusion problem. Section 3 describes the state-of-the-art research on community detection. Section 4 provides the approaches developed for link prediction problem. We conclude the paper and propose several possible directions for future research in Sect. 5 .

2 Influence Diffusion

The role of social networks as a massive medium becomes increasingly attractive with the rapid advance of online social networks such as Twitter, Facebook, and Blog. The influence such as opinions, ideas, and news can propagate in it. With respect to the influence diffusion, a well-known problem named Influence Maximization (IM), has been widely studied. IM is formally defined as: Given a graph $G = (V, E)$ as a social network, an influence spread model and an integer k, select the top-k nodes as seeds to maximize the expected influence spread (EIS). IM was first studied by Richardson and Domingos [18, 61] as an algorithmic problem. Later, it was investigated as a combinatorial optimization problem by Kempe et al. in [40, 41] under two most basic and widely studied diffusion models, namely the Independent Cascade (IC) model [25,26] and the Linear Threshold (LT) model [31, 64]. They achieved a series of theoretical results, especially, the EIS function was proved to be submodular and a $(1 - 1/e)$-approximation ratio for the greedy algorithm was obtained. This is a milestone work for IM in social networks; therefore, we would like to introduce the main contributions in this paper.

Firstly, we introduce the two basic diffusion models. In both IC and LT models, there are three common assumptions: (1) a node only has two states: active (accept an innovation) and inactive; (2) at the beginning, the nodes in the seed set are active and others are inactive; (3) when a node becomes active, it will no longer turn to inactive. As for the IC model, the cascade starts with an initial set of active nodes S_0, and the process proceeds in discrete steps according to the following randomized rule. When node v first turns to active in step t, it obtains a single chance to activate each currently inactive neighbor u and it succeeds with probability $p_{u,v}$, which is a parameter independent of history actions. When u has many active neighbors, the actions on u from them are arranged in an arbitrary order. If v succeeds, then u will become active in step $t+1$; if it does not succeed, it will have no chance to activate u in subsequent rounds. The process goes on until no more activations are possible. In the LT model, a node u is influenced by each neighbor v according to the probability $w_{u,v}$ such that $\sum_{v \in N(u)} w_{u,v} \leq 1$, where $N(u)$ is the neighbor set of u. Each node u chooses a threshold θ_u uniformly at random from the range $[0, 1]$. At any step t, if the total weight from the active neighbors of an inactive node u is no less than θ_u, then u becomes active at step $t+1$. Actually, θ_u reflects the likelihood of node u to accept an innovation.

Having introduced the two models, then, we will present the important conclusions obtained in [40]. Kempe et al. defined the influence of a set of nodes A to be the expected number of active nodes in the end of the activation process, denoted as $\sigma(A)$, where A is the initially active seed set. They proved that under both IC and LT models, $\sigma(A)$ is submodular, satisfying a natural "diminishing returns" property, that is, the marginal gain from adding an element to set A is at least as high as the marginal gain from adding the same element to a superset of A.

2.1 Submodularity for Independent Cascade [40]

To deal with the difficulty of computing the quantities of $\sigma(A)$, Kempe et al. formulated an equivalent view of the influence diffusion process. Consider the case that node v has just become active, and it attempts to activate its neighbor u, succeeding with probability $p_{u,v}$. The outcome of this random event can be viewed as being determined by flipping a coin of bias $p_{u,v}$. Since it does not matter whether the coin is flipped at the beginning of the whole process or at the moment that v is activated, thus, the authors assumed that the coin is flipped in advance. The edges that the coin flip indicates that an activation will be successful are declared to be *live*; the remaining edges are declared to be *blocked*.

Claim 1. *A node u ends up active if and only if there is a path from some node in A to u consisting entirely of live edges.*

Theorem 1. *For an arbitrary instance of the Independent Cascade model, the resulting influence function $\sigma(A)$ is submodular.*

Proof. Let X denote one sample point in the probability space which contains all the possible sets of outcomes for all the coin flips on the edges. Define $\sigma_X(A)$ to be the total number of nodes activated by the process under X provided that A is the seed set. Once X is fixed, then $\sigma_X(A)$ is a deterministic quantity. By **Claim 1**, $\sigma_X(A)$ is actually the number of nodes that can be reached through live-edge paths from any node in A. For $S \subseteq T \subseteq V$, consider $\sigma_X(S \cup \{v\}) - \sigma_X(S)$, which is the number of nodes that can be reached by v while cannot be reached from the nodes in S, it is at least as large as the quantity of $\sigma_X(T \cup \{v\}) - \sigma_X(T)$ since T is bigger. Thus, $\sigma_X(A)$ is submodular. Since a nonnegative linear combination of submodular functions is still submodular, therefore, $\sigma(A) = \Sigma_{\text{outcomes } X} \text{Prob}[X] \cdot \sigma_X(A)$ is submodular. □

2.2 Submodularity for Linear Threshold [40]

Different from the IC model, Kempe et al. constructed another equivalent diffusion process for influence diffusion in the LT model.

Claim 2. *For a given targeted set A, the following two distributions over sets of nodes are the same:* (1) *The distribution over active sets obtained by running the Linear Threshold process to completion starting from A;* (2) *The distribution over sets reachable from A via live-edge paths, under the random selection of live edges defined above.*

Proof. Case1. Graph G is directed and acyclic. Fix a topology order of nodes u_1, ..., u_n and set the distribution of active sets according to this order. For each node u_i, the distribution on active subsets of its neighbors has been determined, thus under LT diffusion model, the probability that a node u_i becomes active when its neighbor set S_i are active, is $\sum_{v \in S_i} w_{u_i,v}$. This is exactly the probability that the live incoming edge selected by u_i lies in S_i.

Case2. G is not acyclic. On the one hand, consider the Linear Threshold process. Define S_t to be the set of active nodes at the end of step t, $t = 0, 1, 2, \ldots$. If node u has not become active by the end of step t, then the probability that it becomes active in step $t+1$ is $\frac{\sum_{v \in S_t \setminus S_{t-1}} w_{u,v}}{1-\sum_{v \in S_{t-1}} w_{u,v}}$. On the other hand, the live-edge process which aims at identifying the live edges runs as follows. Start with a seed set S. For each node u with at least one edge from S, u is said to be reachable if u's live edge comes from S. At the end of the first stage, a set of reachable nodes S_1 is obtained, then the procedure is continued to exploit further reachable nodes on edges from S_1, and sets S_2, S_3, ... are generated. If node u has not been found to be reachable by the end of step t, then the probability that it is determined to be reachable in step $t+1$ is $\frac{\sum_{v \in S_t \setminus S_{t-1}} w_{u,v}}{1-\sum_{v \in S_{t-1}} w_{u,v}}$, which is the same as the distribution obtained by Linear Threshold process. □

Theorem 2. *For an arbitrary instance of the Linear Threshold model, the resulting influence function $\sigma(A)$ is submodular.*

Proof. From the conclusion in **Fact 2**, the submodularity can be proved similarly as in the proof of Theorem 1. □

Greedy Algorithm [40]

1. INPUT: A graph $G = (V, E)$, an integer $k > 0$;
2. OUTPUT: A seed set S.
3. Initialize $S = \phi$;
4. **While** $|S| < k$ **do**
5. Select $v = arg\max_{u \in V \setminus S}(\sigma(S \cup \{u\}) - \sigma(S))$;
6. $S = S \cup \{v\}$;
7. **EndWhile**
8. **Return** S.

EIS in [40] was obtained through Monte-Carlo simulation, and the exact computation of EIS was left as an open problem and was addressed subsequently by [10–12, 29, 30, 46].

For the IC model, Leskovec et al. [46] developed a cost effective lazy forward (CELF) algorithm, which demonstrated to be up to 700 times faster than standard greedy algorithm. However, CELF cannot be scaled to large social networks. To further reduce the running time, Chen et al. in [10] proposed two algorithms called NewGreedy and MixedGreedy, where NewGreedy aims at reducing the running time by deleting edges that have no contribution to the influence spread (similar idea was also proposed in [43]), and MixedGreedy combines NewGreedy and CELF, applying NewGreedy in the first stage and CELF for the remaining steps, and MixedGreedy was proved to be faster than both NewGreedy and CELF. In addition, they proposed a new diffusion structure, namely maximum influence arborescence (MIA), to reduce the running time on calculating EIS, and the efficiency of MIA was demonstrated in [11]. Goyal et al. [29] presented CELF++ which is an extension to the CELF, showing 0.35–0.55 faster than CELF. Besides selecting the influential nodes greedily, Wang et al. [74] developed a community-based algorithm to mine the top-k influential nodes, and Jiang et al. in [39] presented a heuristic algorithm based on Simulated Annealing.

In terms of the LT model, Chen et al. in [12] proved that the EIS under the LT model can be computed in linear time in a directed acyclic graph, and they proposed an algorithm based on local directed acyclic graph (LDAG). Given a general graph, they converted the original graph into small acyclic graphs, and then computed the marginal gain through only considering the EIS of a node within its local graph. In [50], Narayanam and Narahari developed an algorithm for the LT model that selected the nodes based on the Shapley Value. In [30], Goyal et al. proposed SIMPATH, which estimated the EIS by searching for the simple paths starting from seeds. Since it is computationally expensive to find all the simple paths, they adopted a parameter η to prune them. Furthermore, they applied the vertex cover optimization to cut down the number of iterations.

While IM is maximizing the influence of one cascade source, another problem, which includes two or more kinds of cascades resources, focuses on maximizing the influence of each cascade as much as possible, and it has been studied extensively in [4,5,8,45,72]. Bharathi et al. [4] studied competitive influence diffusion under the extension of the IC model. They proposed a $(1 - 1/e)$-factor algorithm for computing the best response to an opponent's strategy and gave an FPTAS for the problem of maximizing the influence of a single player provided the underlying graph is a tree. Kostka et al. [45] considered the rumors diffusion as a game theoretic problem under a much more restricted model compared with IC and LT. They showed that the first player did not always obtain benefit although he/she started earlier. Trpevski et al. [72] proposed a competitive rumors spreading model based on susceptible-infected-susceptible (SIS) model in epidemic domain, but they did not address the issue of influence maximization or rumor blocking. Borodin et al. in [5] studied competitive influence diffusion in several different models extended from LT. Chen et al. [13] addressed positive influence maximization under an extension of the IC model with negative opinions about the product or service quality.

Unlike the multi-cascades mentioned above, a problem concerning two opposite cascades, namely positive information (protectors) and negative information (rumors), aims at using the positive information to reduce the influence of negative information. Kimura et al. in [44] dealt with it through blocking a certain number of links in a network. The most recent works regarded with this problem include [7, 37, 55]. In [7], Budak et al. studied the eventual influence limitation (EIL) problem under the extension of IC model. They focused on the greedy algorithm and several simple heuristics, while did not search efficient and scalable methods that maintain good precision meanwhile. He et al. in [37] proposed a competitive linear threshold (CLT) model to address the influence blocking maximization (IBM) problem. They proved that this problem was submodular and theoretically obtained a $(1 - 1/e)$-approximation ratio from greedy algorithm. To reduce the computation time, He et al. further proposed the CLDAG algorithm that is similar to the LDAG algorithm in [12]. In [55], Nguyen et al. came up with the β_T^I-Node Protector diffusion problems, which are actually the extensions of the IBM problem, under LT and IC. The goal is to find the smallest set of highly influential nodes that can limit the viral spread of misinformation originated from set I to a desired rate $(1 - \beta)$ $(\beta \in [0, 1])$ in T time steps. They proposed a greedy viral stopper (GVS) algorithm that greedily adds nodes with the best influence gain for β-Node Protectors to the current solution. They also applied GVS to the network restricted to T-hop neighbors of the initial set I and reached a slightly better bound for β_T^I-Node Protector problems. Moreover, they proposed a community-based algorithm which outputted a good selection of nodes to control the rumors in a timely manner.

3 Community Detection

Social networks have shown interesting and attractive properties, such as the small-world property, power-law degree distributions, and local structures, also known as communities. In this section, we will focus on this community property, which is that the internal links within the same community are highly dense while the links between communities are comparatively sparse as shown in Fig. 1. The attributes of individuals in the same community are more similar than that of people who come from different communities, which means that they might share more on information, interests, experiences, and other useful resources. So discovering the underlying community structures will have direct impacts on optimizing and managing activities in a social network.

Many approaches mainly focusing on topological structures based on various criteria including betweenness [24, 52], modularity [52], normalized cut [65], structural density [77], and partition density [2] have been studied. In addition, please refer to [21] for more information.

3.1 Betweenness [24, 52]

Hierarchical clustering is the most widely used method among traditional community detection methods. The core of the hierarchical clustering method is the definition of a similarity measure between vertices. Once such a measure is determined, one can compute the similarity values for all pairs of nodes in the given network, no matter whether they are connected or not. Hierarchical clustering approaches aim to identify the groups with high similarity, and it can be classified into two categories:

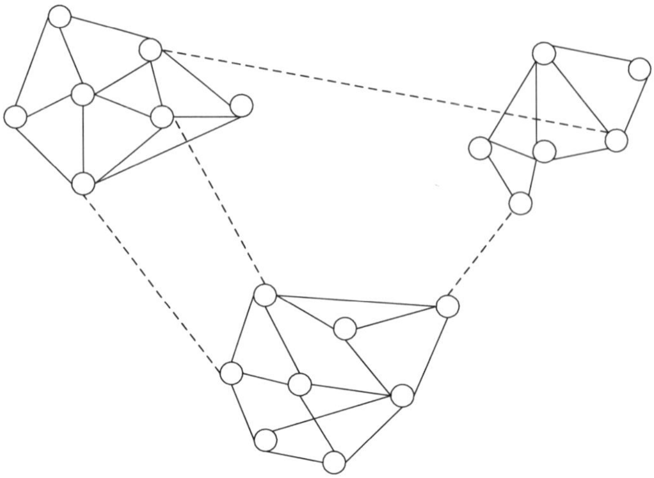

Fig. 1 Community structure

- Agglomerative algorithms: which start from the node set of the original graph. Edges are iteratively added to the corresponding node pairs according to the decreasing order of similarity values.
- Divisive algorithms: which start from the original graph. Edges are removed based on the increasing order of the similarity values.

The two algorithms operate in opposite directions, that is, agglomerative algorithms are bottom-up while divisive algorithms are top-down. Both of them can be well illustrated by a tree (dendrogram) as indicated in Fig. 2 for the original graph, in which the leaves represent all the nodes in the original graph and a non-leaf node denotes the community resulted from merging two smaller communities. The root would be the original graph. Horizontal cuts through this dendrogram at different levels provide divisions of the network into larger or smaller numbers of communities. Since the hierarchical clustering method does not provide the termination rule as well as the size and number of communities, the stopping conditions should be imposed under application scenarios.

The approaches proposed by Girvan and Newman [24, 52] are the most popular and seminal work in the community identification search track and gave new perspectives for the evolution of divisive algorithms. They constructed the communities by progressively removing edges from the original graph. Therefore, edge betweenness is employed for the removal operation, and the betweenness of an edge e is defined as the number of shortest paths between pairs of other vertices that run through e. If a network contains communities or groups that are only loosely connected by a few intergroup edges, then all shortest paths between different communities must go along one of these few edges. Thus, the edges connecting communities will have high edge betweenness. By removing these

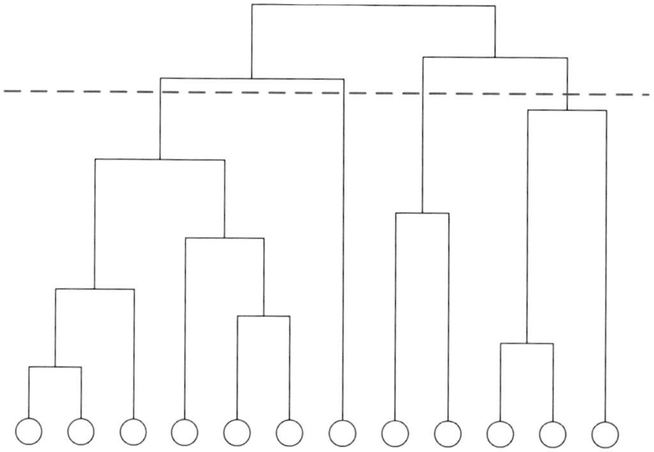

Fig. 2 Dendrogram

edges, they separate into groups from one another and so reveal the underlying community structure of the graph. The concrete algorithm is as follows:

1. For each edge, compute its edge betweenness and sort all these edges according to decreasing edge betweenness order.
2. Remove the edge with the largest edge betweenness, in case of ties with other edges, one of them is picked at random.
3. Recalculate the betweenness for all edges (to save computation time, it is advisable to only consider those edges whose betweenness values change after the removal.) Perform the sorting operation.
4. Repeat steps 2 and 3 until there is no edge left.

In [24], the authors had to deal with the whole hierarchy of partitions for no knowledge about how to choose the best partition. In a successive refinement [52], Newman et al. selected the partition with the largest value of modularity which has been frequently used ever since. Furthermore, [52] considered three alternative measures: geodesic edge betweenness, random-walk edge betweenness, and current-flow edge betweenness, all of which base on the same idea.

3.2 Modularity [52]

Newman in [52] proposed the concept of *modularity* which is the first measurement for partitions quality. The formal definition is described as follows.

For a given network $G = (V, E)$, consider a special division of G into n_c communities. Define an $n_c \times n_c$ symmetric matrix A whose element A_{ij} represents the fraction of all edges in the network that connect vertices in community i

to vertices in community j. Here we consider all the edges as in the original network. The trace $\mathrm{Tr}(A) = \sum A_{ii}$ of A provides the fraction of edges in the network that link vertices in the same community, and the higher value of this trace, the better the network division is. However, when all the vertices are placed in a single community, the trace value is maximized, that is $\mathrm{Tr}(A) = 1$, indicating no community structure at all. Then Newman et al. defined the row (or column) sums $R_i = \sum_j A_{ij}$ to denote the fraction of edges that are attached to vertices in community i, they showed that $A_{ij} = R_i R_j$ if the edges fall between vertices in a network without considering the communities that they reside. Then the modularity measure is defined as

$$Q = \sum_i (A_{ii} - R_i^2) = \mathrm{Tr}(A) - \|A^2\| \tag{1}$$

where $\|A\|$ represents the sum of the elements of matrix A. Intuitively, the modularity is calculating the number of edges within the same community minus the expected number of such edges if we randomly build the same number of connections between the vertices. If a particular division gives no more within-community edges than would be expected by random chance, this modularity is 0. Values other than 0 imply deviations from randomness, and values approaching the maximum $Q = 1$ indicate significant community structure. In practice, values usually fall in the range from 0.3 to 0.7, and higher values are rare to be obtained.

In general, maximizing Q is an NP-hard problem [6]. Hence, many heuristic approaches, which try to maximize the modularity value, were proposed [28]. Such approaches include greedy agglomeration [52,73], mathematical programming [1], spectral methods [66], simulated annealing [34], sampling techniques [63], etc.

The first method called greedy strategy, which falls in the category of agglomerative algorithms, was proposed by Newman et al. [51]. Initially, each node represents a single community with no edges presented, and edges are added one by one during the procedure to merge pairs of communities such that the modularity increases. At each step, the edge is chosen based on the principle that the partition gives the maximum increase (minimum decrease) of modularity with respect to the previous configuration. If the addition of an edge does not change the partition, i.e. the edge is internal to one of the communities previously formed, then modularity stays the same. The procedure continues until all the nodes are merged in one community and a dendrogram is generated at the same time, showing the order of the merging processes. In [14], Clauset et al. pointed out that since the adjacency matrix is sparse, an enormous number of operations implemented to update e_{ij} in Newman's algorithm are useless, that is, merging the communities having no between edges makes no contribution to the modularity variation ΔQ_{ij}, they focused on communities that are linked by some between edges. Additionally, they employed the data structures for sparse matrices and made the update much quicker than in Newman's technique. The greedy optimization of Clauset et al. was proved to be one of the few algorithms that can be used to estimate the modularity maximum on very large graphs. However, this algorithm tends to yield poor values of the modularity maxima since large communities are formed by sacrificing small ones. Danon et al. in [17] provided a modification of Newman's modularity-based algorithm to accommodate for communities of varying sizes without slowing down the computation. This strategy leads to better modularity optima, especially when communities are very different in size. Furthermore, it is observed that the choice of reasonable initial communities can significantly improve the accuracy of the greedy optimization [19, 76, 78].

Simulated Annealing (SA) is a stochastic procedure for global optimization with avoiding the risk that the system gets trapped in local optima. It searches all the possible states to find the global optimum for a function f. The transition between states happens with a probability according to the change of f. If f increases at the time the transition occurs, then the probability is 1, otherwise, the probability is $\exp(\beta\Delta f)$, where Δf is the decrease of f and β is an index of stochastic noise, a sort of inverse temperature which increases after each stage. Guimerà et al. [35] were the first to employ SA for submodularity optimization and later in [34] SA was adopted by coupling two types of "moves": local moves, in which a single vertex is randomly shifted from one community to another; global moves, which contain mergers and splits of communities.

However, modularity is not a scale-invariant measurement, meaning that it cannot detect small communities [22].

3.3 *Works Based on Dynamic Views*

Besides analyzing communities in static social networks, recently, finding communities and their evolutions in dynamic networks has gained an enormous amount of attention. Chakrabarti et al. [9] proposed a framework called *temporal smoothness* to capture the evolution of clusters. This framework assumed that the structure of clusters significantly changes in a very short time is less desirable, thus, it tried to smooth out each community at every step during clustering from two kinds of qualities: the snapshot quality and the history quality. Especially, through formulating the problem of analyzing community evolutions in terms of nonnegative matrix factorization, Lin et al. proposed the FacetNet algorithm [48] and developed an iterative algorithm that is guaranteed to converge to an optimal solution. Actually, the detection of community structure with temporal smoothness can be formulated as a multi-objective optimization problem. One is to maximize the cluster accuracy at the current time. Another is to minimize the clustering drift at the consecutive time steps. Folino et al. [20] proposed a dynamic multi-objective genetic algorithm to detect communities in dynamic networks by employing genetic algorithm. The two objectives to be optimized are formulated as Community Score and normalized mutual information (NMI), respectively. Another study is by Kim [42], who proposed adaptive integration of multi-objective evolutionary algorithms based on NSGA-II for networks, particularly for online social network clustering.

Palla et al. [59] analyzed a co-authorship network and a mobile phone network through the clique percolation method (CPM). A model was proposed by Tantipathananandh [68] et al., where the problem for finding the dynamic community was formulated as a graph coloring problem. And a heuristic technique that involves greedily matching pairs of nodes between states according to the descending order of groups similarity was employed. To identify and trace community structure of dynamic online social networks, Nguyen et al. in [53] proposed an adaptive modularity-based method called quick community adaptation (QCA). Taking advantage of the structure identified from previous network snapshots, they updated the network communities quickly and efficiently. In [54], the authors presented adaptive finding overlapping community structure (AFOCS), a two-phase adaptive framework for detecting, updating, and tracing the evolution of overlapping communities in dynamic mobile networks. It firstly identified all possible basic network communities with FOCS and then applied AFOCS to adaptively update these structures along with the evolution of the network. Rosvall et al. [62] developed a framework for identifying changes in dynamic networks. At each time, the original network observed is clustered. Subsequently, the network is perturbed through a bootstrap resampling process and re-clustered, and this is repeated for a large number of runs to quantify the significance of clusters generated at each time step. Finally, "alluvial" diagrams, which connect the associated clusters from different time steps, are employed to illustrate the progression of the clusters over time. In [32], the authors described a model for tracking communities, in which each community is characterized by a series of evolutionary events. Based on this model, they presented a scalable community-tracking strategy to identify dynamic communities efficiently. Gong et al. in [27] presented a novel multi-objective immune algorithm to identify communities in dynamic networks. It optimized the modularity and NMI through nondominated neighbor immune algorithm. Furthermore, the problem-specific knowledge was used by the genetic operators and local search to improve the effectiveness and efficiency of this algorithm.

4 Link Prediction

It is well known that social networks are highly dynamic and their structures change along with the creation of new links between individuals. Understanding the mechanics by which they evolve is a fundamental task in SNA. Thus, the problem, namely link prediction, has obtained intensive attention among researchers these years. Liben-Nowell et al. in [56, 57] introduced link prediction problem as: given a snapshot of a social network $G = (V, E)$ at time t and a future time t', the problem aims at predicting the new links that are likely to appear in the network within the time interval $[t, t']$. They introduced several methods adapted from techniques used in graph theory and SNA as shown in Table 1. The common feature is that a connection weight score(u, v) for pairs of nodes (u, v) is computed based on the

Table 1 Link prediction metrics

Metric	Definition of score
Common neighbor	$\lvert N(u) \cap N(v) \rvert$ [a]
Jaccard's coefficient Index	$\frac{\lvert N(u) \cap N(v) \rvert}{\lvert N(u) \cup N(v) \rvert}$
Preferential attachment Index	$\lvert N(u) \rvert \cdot \lvert N(v) \rvert$
Graph distance	(Negated) length of shortest path between u and v
Katz$_\beta$ Index	$\sum_{k=1}^{\infty} \beta^k \cdot \lvert \text{paths}_{u,v}^{<k>} \rvert$ [b]
Adamic/Adar Index	$\sum_{w \in N(u) \cap N(v)} \frac{1}{\log \lvert N(w) \rvert}$
Hitting time	$-T_{u,v}$
Stationary-normed	$-T_{u,v} \cdot \pi_v^{\text{c}}$
Commute time	$-(T_{u,v} + T_{v,u})$
Stationary-normed	$-(T_{u,v} \cdot \pi_v + T_{v,u} \cdot \pi_u)$
Rooted PageRank$_\beta$	Stationary distribution weight of y under the following random walk: (1) with probability β, jump to x; (2) with probability $1-\beta$, go to random neighbor of current node
SimRank$_\alpha$ [d]	(1) 1, if $x = y$; (2) $\alpha \cdot \frac{\sum_{a \in N(u)} \sum_{b \in N(v)} \text{score}(u,v)}{\lvert N(u) \rvert \cdot \lvert N(y) \rvert}$, otherwise

[a] $N(x)$:={all the neighbors of node x}
[b] $\text{paths}_{x,y}^{<k>}$:={paths of length exactly k from u to v}. weighted: $\text{paths}_{x,y}$:= number of collaborations between x and y; unweighted: $\text{paths}_{x,y}$:= 1 iff x and y collaborate
[c] $T_{x,y}$:= expected time for random walk from x to y; π_y := stationary distribution weight of y, proportion of time the random walk is at node y
[d] α is between 0 and 1

input graph, and then a ranked list in decreasing order of score(u,v) is produced. In other words, generating a measure of proximity or "similarity" between nodes u and v with respect to the network topology is the vital component. They evaluated these measures over the co-authorship network.

Since [56, 57] mainly focus on explaining the measures in Table 1 through experiments, here we briefly describe the experiment setup and its relative results.

4.1 Experimental Setup [56, 57]

Given a social network $G = (V,E)$, and edge $e = (u,v)$ means that u and v have relation at time $t(e)$. Multiple relations between u and v are regarded as parallel edges and different time-stamps are assigned correspondingly. $G[t,t']$ represents the subgraph of G containing all the edges that appear in $[t,t']$, where $t < t'$ and two time intervals with four times $t_0 < t_0' < t_1 < t_1'$ are defined: training interval $[t_0,t_0']$ and test interval $[t_1,t_1']$. Then an algorithm is applied to the network $G[t_0,t_0']$ to predict the edges that may present in the network $G[t_1,t_1']$ and do not appear in $G[t_0,t_0']$. To guarantee the efficiency of this experiment, a nodes set **Core** is defined based on two parameters κ_{training} and κ_{test}, all of which are set to 3. All the nodes in **Core** satisfy: (1) incident to at least κ_{training} edges in $G[t_0,t_0']$; (2)

incident to at least κ_{test} edges in $G[t_1, t_1']$. Denote $G = (V_1, E_{\text{old}})$ as the subgraph of a co-authorship network, and E_{new} is adopted to denote the edges set with endpoint in V_1 and co-authoring a paper during the test interval but not the training interval. Then for each predictor p, it must output a ranked node pairs list R_p in $V_1 \times V_1 - E_{\text{old}}$, indicating the new predicted collaborations in decreasing order of appearance probability. To address the meaningful component, they focused on the set **Core**, and defined $E^*_{\text{new}} := E_{\text{new}} \cap (\mathbf{Core} \times \mathbf{Core})$ with $k := |E^*_{\text{new}}|$. Then, they determined the performance measure for predictor p via taking the first k pairs in **Core** $\times$ **Core** which are obtained from the ranked list R_p,

4.2 Three Meta-Approaches [56, 57]

The following three meta-approaches can combine with any of the above methods included in Table 1.

- *Low-Rank Approximation.* All of the above link prediction methods have an equivalent formulation in terms of the adjacent matrix M of a graph. For instance, in the common neighbor method, each node u is mapped to the row $r(u)$ of M, and then $\text{score}(u, v)$ is defined as the inner product of the rows $r(u)$ and $r(v)$. For a large matrix M, a general technique used to analyze its structure is to choose a relatively small number m and compute the rank-m matrix M_k that best approximates M with regard to a number of standard matrix norms. Intuitively, considering M_k rather than M can be viewed as a type of "noise-reduction" technique that produces most of the structure in the matrix while uses a greatly simplified representation. In their experiments, they explored three applications of low-rank approximation: (1) rank by Katz measure, and M_k is used in the underlying formula; (2) rank by common neighbors, the score is obtained by inner products of rows in M_k; (3) (u, v) entry in M_k is defined as $\text{score}(u, v)$.
- *Unseen Bigrams.* Observing that link prediction is similar to the problem of estimating frequencies for unseen bigrams in language modeling-pairs of words that appear together in a test corpus, but not in the corresponding training corpus, the authors estimated $\text{score}(u, v)$ through the values of $\text{score}(w, v)$, in which nodes w are akin to v. The detail is as follows: Suppose the values of $\text{score}(u, v)$ computed by one of the measures above have been obtained. For $k \in Z^+$, S_u^k denotes the k nodes most related to u under $\text{score}(u, \cdot)$. The improved scores are as follows:

$$\text{score}^*_{\text{unweighted}}(u, v) = |w : w \in N(v) \cap S_u^k| \tag{2}$$

$$\text{score}^*_{\text{weighted}}(u, v) = \sum_{w : w \in N(v)} \text{score}(u, w). \tag{3}$$

- *Clustering*. Clustering procedure is firstly adopted to delete the "noisy" edges to enhance the quality of a predictor and then the predictor is implemented on these vital edges. Consider a method computing values of score(u,v): compute score(x,y) for all edges in E_{old} and delete β fraction of these edges with lowest scores. In the remaining subgraph, recompute score(u,v) for all pairs (u,v).

The experiments on large co-authorship demonstrated that information about future connections could be obtained from network topology alone and the measurements used to detect node proximity outperform more direct measures, specially, they found that the Adamic–Adar measure of node similarity performs best.

While the work of [56, 57] only focused on the network structure, Taskar et al. [69] relied on machine learning techniques and used personal information (attributes) of users (music, books, hobbies, etc.) to increase the accuracy of predictions. Later, O'Madadhain et al. [49] took the geographic location as an attribute to predict events (interactions) between entities. Hasan et al. [36] abstracted several nodal and topological attributes for link prediction problem and applied a variety of classifiers such as support vector machines and decision tree to predict interactions in bibliographic databases.

Backstrom et al. [3] proposed a method based on Supervised Random Walks, which combines the information of the network structure with node and edge attributes to predict the links efficiently. The node and edge features are used to learn edge strengths (i.e., random walk transition probabilities) such that the random walk on a network is more likely to visit "positive" than "negative" nodes. Here positive nodes are the ones that new edges will be created to in the future, and the negative are the remaining nodes. They summarized a supervised learning task provided a source node s and training examples about which nodes s will establish links to in the future. Then they formulated the problem that studies a function that assigns a strength (i.e., random walk transition probability) to each edge such that the random walk scores in the network nodes to which s creates new links have higher scores than nodes to which s does not create links.

4.3 Supervised Random Walks

- *Classification Aspect.* Exploit a classifier that predicts pairs of nodes that s is going to create links and denote them as positive training examples, the other nodes are denoted as negative training examples. It needs to consider how to extract the nodes and edges features such as age, gender, and creation time.
- *Nodes Ranking Aspect.* Design an approach that will assign higher scores to nodes which s create links to than to those that s does not link to. This method mainly takes advantage of the structure of the network.

4.4 Optimization Problem Formulation

Given a directed graph $G = (V, E)$, a source node s, a destination nodes set $P = \{p_1, \ldots, p_l\}$ to which s will create edges in the future and the no-link nodes set $N = \{n_1, \ldots, n_k\}$ to which s does not create edges. Denote the candidate nodes as $C = c_i = P \cup N$. View nodes in P as positive and nodes in N as negative training examples. Each node and each edge in G is associated with a set of characteristics. For each edge (u, v), define its corresponding feature vector $\varphi_{u,v}$, which contains the attributes of nodes u and v (e.g., age, gender, hometown) and the interaction features (e.g., when the edge is created or how many messages u and v communicated). Then compute the strength $s_{u,v} = f_m(\varphi_{u,v})$ which models the random walk transition probability. f_w is employed to compute the first edge strengths of all edges. Then a random walk with restarts is run from s. Nodes are ordered by a probability p_u obtained from the stationary distribution p of the random walk and top ranked nodes are then viewed as destinations of future links of s. It is obvious that nodes connected to s through paths of strong edges (edges with large edge strength) will be visited by the random walk with high possibility and thus rank higher. The key task now is to study the parameter m with regard to function f_m. Then the optimization problem to find the optimal set of parameters m of function $f_m(\varphi_{u,v})$ is as follows:

$$\begin{aligned} &\min_m F(m) = \parallel m \parallel^2 \\ &s.t. \forall p_i \in P, n_j \in N : q_{n_j} < q_{p_i} \\ &i = 1, \cdots, l \ \ and \ \ j = 1, \cdots, k, \end{aligned} \tag{4}$$

where q_i is the vector of PageRank scores. Since it is hard to find a solution that satisfies all the constraints above, thus the authors made the constraints "soft" by a loss function g which penalizes violated constraints, then the optimization problem becomes:

$$\min_w F(m) = \parallel m \parallel^2 + \alpha \sum_{p_i \in P, n_j \in N} g(q_{n_j} - q_{p_i}), \tag{5}$$

where α is a parameter that balances the complexity (norm of m) and the fit of the model (how much the constraints can be violated). Moreover, if $q_{n_j} - q_{p_i} < 0$, then $g(\cdot) = 0$; and for $q_{n_j} - q_{p_i} > 0$, then $g(\cdot) > 0$.

In addition to the approaches mentioned above, probabilistic methods were explored to build a model that can represent a network much accurately. This kind of approaches examine the elements of the network through relational data models that are able to include relevant information from nodes, relationships, and the network as a whole. The main idea is to establish a probabilistic model based on a set of parameters α that is obtained according to the observed network. Then the existence of a link between a given pair of nodes u and v is determined by the conditional probability $P(e^{(u,v)}|\alpha)$ [75]. Corresponding to these methods, examples of networks are Relational Markov Networks, Relational Bayesian Networks, and Relational Dependency Networks.

Besides what have been introduced, various kinds of approaches have been developed to address the link prediction problem. Clauset et al. [15] proposed maximum likelihood-based methods that represent the clusters in the network as a hierarchy, which in turn are represented as a dendrogram. LinkBoost [16] explored the community structure by a novel degree-dependent cost function and showed that minimization of the associated risk can lead to more links predicted within communities than between communities. Huang and Lin [38], Potgieter et al. [60], and Soares and Prudêncio [67] tried to use time-aware techniques for link prediction problem, which took the dynamic perspectives of networks into consideration. Through following the experimental framework articulated by Guha et al. in their study of trust and distrust on Epinions [33], Leskovec et al. [47] extended their approach in a number of directions, one of which is to infer an individual's attitudes towards the relationship with others. To learn more about the link prediction problem, please refer to [23, 58].

5 Conclusion and Discussion

Social networks, which contain intricate structures and enormous information about its members, play a variety of roles for practical applications. For examples, it can be regarded as a platform for individual communications; it can be used to investigate structural pattern of certain groups; it can be exploited to predict interactions between individuals that will appear in the future, and so forth. Studies on these aspects lead promising steps for researchers across different fields. Therefore, in this survey, we have summarized several works with regard to three problems: influence diffusion, community detection, and link prediction. These represent some of the common threads emerging from a variety of domains like sociology, computer science, and physics.

Although a large amount of techniques or algorithms have been proposed for those problems respectively, there is still much room for further efforts. We propose several research directions related to the three problems as follows:

- Influence Diffusion Problem: Although several works have paid attention to the diffusion probability computation, most of them only applied the network structure. To obtain more convincing data, it is desirable to compute the influence propagation probabilities by considering the attributes of individuals, such as gender, age, interest, and location. Also, until now, a mass of works explored the rumor blocking problem over the IC and the LT models, studying it under models like SIS and susceptible-infected-recovered (SIR) is an interesting direction. In addition, since the network develops with time, it is more realistic to integrate the time variable to the influence dissemination process.

- Community Detection Problem: With the appearance of enormous number of online social networks and their wide application, it will do benefit for the individuals to identify communities in those networks like facebook, mobile social network, and weighted social networks. In previous works, the authors addressed the community detection problem with assumption that the complete connection information within the entire network is available. However, in many practical networks, such as enemy networks, the complete connection are very difficult or even impossible to obtain, therefore, it is of great challenge to distinguish communities provided that the network information is incomplete.
- Link Prediction Problem: Soares et al. [67] tested their time-aware information-based algorithm merely over the co-authorship networks, extending their algorithm to networks in other domains is a future line of research. In addition, many works on link prediction problem just concerned how to predict the links accurately that the relations between individuals are different, like positive or negative as in [47], is ignored, thus, it is promising to explore methods that efficiently predict those two kinds of links. Since community detection has close relation with link prediction, then with regard to the community robustness issue, it is worth trying to strengthen the existing connections between individuals via creating new relations using the methods for link prediction.

References

1. G. Agarwal and D. Kempe: Modularity-maximizing graph communities via mathematical programming. European Physical Journal B, 66:409–418, 2008.
2. Y. Ahn, J. Bagrow and S. Lehmann: Link communities reveal multiscale complexity in networks. Nature, 466:761–764, 2010.
3. L. Backstrom and J. Leskovec: Supervised random walks: predicting and recommending links in social networks. In WSDM '11, 2011.
4. S. Bharathi, D. Kempe and M. Salek: Competitive influence maximization in social networks. In WINE, pp. 306–311, 2007.
5. A. Borodin, Y. Filmus and J. Oren: Threshold models for competitive influence in social networks. In WINE, pp. 539–550, 2010.
6. U. Brandes, D. Delling, M. Gaertler, R. Gorke, M. Hoefer, Z. Nikoloski, and D. Wagner. On modularity clustering. IEEE Transactions on Knowledge and Data Engineering, 20(2):172188, 2008.
7. C. Budak, D. Agrawal and A. E. Abbadi: Limiting the spread of misinformation in social networks. In WWW, pp. 665–674, 2011.
8. T. Carnes, C. Nagarajan, S. M. Wild and A. van Zuylen: Maximizing influence in a competitive social network: a followers perspective. In ICEC, pp. 351–360, 2007.
9. D. Chakrabarti, R. Kumar and A. Tomkins: Evolutionary clustering. In Proc. In KDD, pp. 554–560, 2006.
10. W. Chen, Y. Wang, and S. Yang: Efficient Influence Maximization in Social Networks. the 2009 ACM SIGKDD Conference on Knowledge Discovery and Data Mining, pp. 199–208, 2009.
11. W. Chen, C. Wang and Y. Wang: Scalable Influence Maximization for Prevalent Viral Marketing in Large-scale Social Networks. the 2010 ACM SIGKDD Conference on Knowledge Discovery and Data Mining, pp. 1029–1038, 2010.

12. W. Chen, Y. Yuan and L. Zhang: Scalable Influence Maximization in Social Networks under the Linear Threshold Model. the 2010 International Conference on Data Mining, pp. 88–97, 2010.
13. W. Chen, A. Collins, R. Cummings, T. Ke, Z. Liu, D. Rincn, X. Sun, Y. Wang, W. Wei and Y. Yuan: Influence maximization in social networks when negative opinions may emerge and propagate. In SDM, pp. 379–390, 2011.
14. A. Clauset, M. E. J. Newman and C. Moore: Finding community structure in very large networks, Phys. Rev. E 70, 066111, 2004.
15. A. Clauset, C. Moore and M. E. J. Newman: Hierarchical structure and the prediction of missing links in networks. Nature, 453(7191):98-101, 2008.
16. P. M. Comar, P.N. Tan and A. K. Jain: LinkBoost: A novel cost-sensitive boosting framework for community-level network link prediction. In ICDM, pages 131–140, 2011.
17. L. Danon, A. Díaz-Guilera and A. Arenas: Effect of size heterogeneity on community identification in complex networks. J. Stat. Mech. P11010, 2006.
18. P. Domingos and M. Richardson: Mining the network value of customers. In KDD, pp. 57–66, 2001.
19. H. Du, M. W. Feldman, S. Li and X. Jin: An algorithm for detecting community structure of social networks based on prior knowledge and modularity. Complexity 12(3), pp. 53–60, 2007.
20. F. Folino and C. Pizzuti: A multi-objective and evolutionary clustering method for dynamic networks. In Proc. Int. Conf. Advances in Social Networks Analysis and Mining, pp. 256–263, August 2010.
21. S. Fortunato: Community detection in graphs. Physics Reports, 486(3–5), 2010.
22. S. Fortunato and M. Barthelemy: Resolution limit in community detection. Proceedings of The National Academy of Sciences, 104(1):36–41, 2007.
23. L. Getoor and C. P. Diehl: Link mining: a survey. SIGKDD Explor. Newsl., 7(2):3–12, 2005.
24. M. Girvan and M. E. J. Newman: Community structure in social and biological networks, PNAS, vol.99, no. 12, pp. 7821–7826, 2002.
25. J. Goldenberg, B. Libai and E. Muller: Using Complex Systems Analysis to Advance Marketing Theory Development. Academy of Marketing Science Review, 2001.
26. J. Goldenberg, B. Libai and E. Muller: Talk of the Network: A Complex Systems Look at the Underlying Process of Wordof- Mouth. Marketing Letters, 12(3): pp. 211–223, 2001.
27. M. G. Gong, L. J. Zhang and J. J. Ma, et al.: Community detection in dynamic social networks based on multiobjective immune algorithm, Journal of Computer Science and Technology 27, pp. 455–467, 2012.
28. B. H. Good, Y. A. de Montjoye and A. Clauset: The performance of modularity maximization in practical contexts. Physical Review E, 81:046106, 2010.
29. A. Goyal, W. Lu, and L. V. S. Lakshmanan: CELF++: Optimizing the Greedy Algorithm for Influence Maximization in Social Networks. the 2011 International World Wide Web Conference, pp. 47–48, 2011.
30. A. Goyal, W. Lu and L. V. S. Lakshmanan: SIMPATH: An Efficient Algorithm for Influence Maximization under the Linear Threshold Model. the 2011 IEEE International Conference on Data Mining, pp. 211–220, 2011.
31. M. Granovetter: Threshold Models of Collective Behavior. American Journal of Sociology, 83(6): pp. 1420–1443, 1978.
32. D. Greene, D. Doyle and P. Cunningham: Tracking the evolution of communities in dynamic social networks. In ASONAM, pp. 176–183, 2010.
33. R. V. Guha, R. Kumar, P. Raghavan and A. Tomkins: Propagation of trust and distrust. In Proc. 13th WWW, 2004.
34. R. Guimerà and L. N. Amaral: Functional cartography of complex metabolic networks. Nature, 433(7028):895–900, 2005.
35. R. Guimerà, M. SALES-PARDO and L. A. N. AND AMARAL: Modularity from fluctuations in random graphs and complex networks. Phys. Rev. E 70 art. no. 025101, 2004.
36. M. A. Hasan, V. Chaoji, S. Salem and M. Zaki: Link prediction using supervised learning. In Proc. of SDM 06 workshop on Link Analysis, Counterterrorism and Security, 2006.

37. X. He, G. Song, W. Chen and Q. Jiang: Influence blocking maximization in social networks under the competitive linear threshold model. SDM to appear, 2012.
38. Z. Huang and D. K. J. Lin: The time-series link prediction problem with applications in communication surveillance. INFORMS J. on Computing, 21:286-303, April 2009.
39. Q. Jiang, G. Song, G. Cong, Y. Wang, W. Si and K. Xie: Simulated Annealing Based in Influence Maximization in Social Networks. the 2011 AAAI Conference on Artificial Intelligence, 2011.
40. D. Kempe, J. Kleinberg and É. Tardos: Maximizing The Spread of Influence Through a Social Network. the 2003 ACM SIGKDD Conference on Knowledge Discovery and Data Mining, pp. 137–146, 2003.
41. D. Kempe, J. M. Kleinberg and É. Tardos: Influential nodes in a diffusion model for social networks. In ICALP, pp. 1127–1138, 2005.
42. K. Kim, R. McKay and B. R. Moon: Multi-objective evolutionary algorithms for dynamic social network clustering. In Proc. the 12th Conf. Genetic and Evolutionary Computation, pp. 1179–1186, July 2010.
43. M. Kimura, K. Saito and R. Nakano: Extracting Influential Nodes for Information Diffusion on Social Network. the 2007 AAAI Conference on Artificial Intelligence, pp. 1371–1376, 2007.
44. M. Kimura, k. Saito and H. Motoda: Minimizing the spread of contamination by blocking links in a network. In: Proceedings of the 23rd AAAI Conference on Artificial Intelligence, 2008.
45. J. Kostka, Y. A. Oswald and R. Wattenhofer: Word of mouth: Rumor dissemination in social networks. In SIROCCO, pp. 185–196, 2008.
46. J. Leskovec, A. Krause, C. Guestrin, C. Faloutsos, J. Van- Briesen and N. S. Glance: Cost-Effective Outbreak Detection in Networks. the 2007 ACM SIGKDD Conference on Knowledge Discovery and Data Mining, pp. 420–429, 2007.
47. J. Leskovec, D. Huttenlocher and J. Kleinberg: Predicting Positive and Negative Links in Online Social Networks, In Proceedings of WWW'2010, ACM Press, New York, 2010.
48. Y. R. Lin, Y. Chi, S. H. Zhu, H. Sundaram and B. L. Tseng: Facetnet: A framework for analyzing communities and their evolutions in dynamic networks. In Proc. the 17th Int. Conf. World Wide Web, pp. 685–694, April 2008.
49. J. O'Madadhain, J. Hutchins and P. Smyth: Prediction and ranking algorithms for event-based network data. ACM SIGKDD Exploration Newsletter, 7(2):23-30, 2005.
50. R. Narayanam and Y. Narahari: A Shapley Value Based Approach to Discover Influenctial Nodes in Social Networks. IEEE Transactions on Automation Science and Engineering, 8(1): pp. 130–147, 2011.
51. M. E. J. Newman: Fast algorithm for detecting community structure in networks. Phys. Rev. E 69, 066133 (2004).
52. M. E. Newman and M. Girvan: Finding and evaluating community structure in networks, Phys. Rev. E 69 (2), 026113, 2004.
53. N. P. Nguyen, T. N. Dinh, Y. Xuan and M. T. Thai: Adaptive algorithms for detecting community structure in dynamic social networks. INFOCOM, 2011.
54. N. P. Nguyen, T. N. Dinh, S. Tokala and M. T. Thai: Overlapping communities in dynamic networks: Their detection and mobile applications. MOBICOM, 2011.
55. N. P. Nguyen, G. Yan, M. T. Thai and S. Eidenbenz: Containment of Misinformation Spread in Online Social Networks. WebSci, 2012.
56. D. Liben-Nowell and J. M. Kleinberg: The link prediction problem for social networks. In CIKM, pp. 556–559, 2003.
57. D. Liben-Nowell and J. M. Kleinberg: The link-prediction problem for social networks. JASIST, 58(7), pp. 1019–1031, 2007.
58. L. Lü and T. Zhou: Link prediction in complex networks: A survey. Physica A, 390:1150–1170, 2011.
59. G. Palla, A. L. Barabasi and T. Vicsek: Quantifying social group evolution. Nature, 446(7136): 664–667, 2007.
60. A. Potgieter, K. A. April, R. J. E. Cooke and I. O. Osunmakinde: Temporality in link prediction: Understanding social complexity, 2007.

61. M. Richardson and P. Domingos: Mining knowledge-sharing sites for viral marketing. In KDD, pp. 61–70, 2002.
62. M. Rosvall and C. T. Bergstrom: Mapping change in large networks. PLoS ONE, 5, e8694, 2010.
63. M. Sales Pardo, R. Guimer, A. Moreirà and L. Amaral: Extracting the hierarchical organization of complex systems. Proceedings of the National Academy of Sciences, 104(39):15224–15229, 2007.
64. T. Schelling: Micromotives and Macrobehavior. Norton, 1978.
65. J. Shi and J. Malik: Normalized cuts and image segmentation. IEEE Transactions on Pattern Analysis and Machine Intelligence, 22(8): pp. 888–905, 2000.
66. M. Shiga, I. Takigawa and H. Mamitsuka: A spectral clustering approach to optimally combining numerical vectors with a modular network. In KDD, pp. 647–656, 2007.
67. P. R. D. S. Soares and R. B. C. Prudêncio: Time Series Based Link Prediction. Neural Networks (IJCNN), The 2012 International Joint Conference on 10–15 June 2012.
68. C. Tantipathananandh, T. Berger-Wolf and D. Kempe: A framework for community identification in dynamic social networks. In Proc. 13th ACM SIGKDD International conference on Knowledge Discovery and Data mining, pp. 717–726, ACM, 2007.
69. B. Taskar, M.F. Wong, P. Abbeel and D. Koller: Link prediction in relational data. In Neural Information Processing Systems, volume 15, 2003.
70. M. Thai and P. M. Pardalos: Handbook of Optimization in Complex Networks. Communication and Social Networks. Springer. Series: Springer Optimization and Its Applications, 2012. Vol. 58. ISBN 978-1-4614-0856-7.
71. M. Thai and P. M. Pardalos: Handbook of Optimization in Complex Networks. Theory and Applications. Springer. Series: Springer Optimization and Its Applications, 2012. Vol. 57. ISBN 978-1-4614-0753-9.
72. D. Trpevski, W. K. S. Tang and L. Kocarev: Model for rumor spreading over networks. Physics Review E, 2010.
73. K. Wakita and T. Tsurumi: Finding community structure in mega-scale social networks. In WWW, pp. 1275–1276, 2007.
74. Y. Wang, G. Cong, G. Song and K. Xie: Community-Based Greedy Algorithm for Mining Top-k Influential Nodes in Mobile Social Networks. the 2010 ACM SIGKDD Conference on Knowledge Discovery and Data Mining, pp. 1039–1048, 2010.
75. E. W. Xiang: A survey on link prediction models for social network data. Science And Technology, 2008.
76. B. Xiang, E. H. Chen and T. Zhou: Finding community structure based on subgraph similarity. Complex Networks, vol.207, pp. 73–81, 2009.
77. X. Xu, N. Yuruk, Z. Feng and T. A. J. Schweiger: Scan: A structural clustering algorithm for networks. In KDD, pp. 824–833, 2007.
78. Z. Ye, S. Hu and J. Yu: Adaptive clustering algorithm for community detection in complex networks. Physical Review E, 78:046115, 2008.

Comparative Analysis of Local Search Strategies for Transmission Network Expansion Planning

Alla Kammerdiner, Alex Fout, and Russell Bent

Abstract The demands for electricity and the electrical power generation in various areas may change significantly with time. These changes require additional transmission corridors to be installed in the existing electrical power network. The problem of electrical grid expansion can be formulated as a nonlinear mixed integer programming problem. The local search algorithms, which employ constructive heuristics for defining a neighborhood, could be used iteratively to find approximate solutions for this problem. In this study, we compare a number of local search strategies using statistical analysis techniques.

Keywords Transmission expansion planning • Nonlinear mixed integer programming • Local search • Heuristics • Explorative statistical analysis

1 Introduction

The demand for energy and the electrical power generation have been growing globally with increase in the world population, industrial development, access to transportation, and communication. Environmental and socioeconomic concerns related to these trends are important factors that must be taken into account. As a result, development of proper technology and infrastructure to reliably satisfy the world's growing demands for clean, sustainable, and economical energy has become a major global challenge of this century.

A. Kammerdiner • A. Fout
New Mexico State University, P.O. Box 30001, MSC 4230, Las Cruces, NM 88003-8001, USA
e-mail: alla@nmsu.edu

R. Bent
Los Alamos National Laboratory, PO Box 1663, Los Alamos, NM 87545

A. Sorokin and P.M. Pardalos (eds.), *Dynamics of Information Systems: Algorithmic Approaches*, Springer Proceedings in Mathematics & Statistics 51,
DOI 10.1007/978-1-4614-7582-8_12,

To be able to respond to the changing energy needs, the hierarchical, centrally controlled structure of the current power grid will undergo a radical transformation. The improved grid will utilize modern sensors, communication connections, and computational advances to enhance its own stability, flexibility, and efficiency. This future grid is often referred to as "smart grid."

Along with grid control and grid reliability, efficient grid design is vitally important in developing technology for the future smart grid. Smart grid design determines how to best upgrade and expand the electric power transmission network to meet growing energy demand utilizing sources of renewable, sustainable, and cheap energy. These sources may be located in non-contiguous, transmission-deficient areas. Solving these problems is encompassed in transmission network expansion planning (TNEP).

The goal of TNEP is to determine the optimal plan for power grid expansion. The plan must specify the number of new power lines to be installed in each transmission corridor and the number of new control components added at each bus. TNEP has been described with varying degrees of fidelity and has been studied from a wide scope of perspectives [8]. A recent review of many of the approaches, models, and algorithms for TNEP can be found in [10]. The problem of long-term transmission system planning based on the so-called direct current (DC) model is considered.

Due to constraints imposed by physical laws of the electrical power flows, the resulting optimization problem is a mixed-integer nonlinear programming (MINLP) problem characterized by high complexity, especially for large-scale and real-world problems. Whereas linear integer programming has developed into a mature discipline of mathematical optimization over the last 50 years, nonlinear mixed-integer programming still is generally considered a very young field [6]. In fact, majority of the problems and methods for MINLP are not as comprehensive or well developed as in the case of linear mixed-integer programs. Due to the lack of its own methodology, traditionally, MINLP problems were solved using global optimization. Hence, the focus was on numerical methods for finding solutions of nonlinear continuous optimization problems and the integrality constraints were only an afterthought and were typically handled using branch-and-bound on the integer decision variables. Only more recently, the interest of researchers in integer programming has shifted more to developing its own methodology for MINLP. Furthermore, even in the pure continuous case, nonlinear optimization is known to be NP-hard [6]. Therefore, for the sake of simplicity, we consider a relatively small power network with only six buses (i.e., a six-node network).

TNEP problem belongs to a larger class of MINLP problems related to network design. Network design problems arise in a wide range of applications, including telecommunications, logistics and transportation, and supply chain management [4]. For large scale systems, solving such problems is computationally difficult [1, 10]. In particular, finding exact solutions can be very time consuming. As a result, development of heuristic algorithms for such problems has attracted considerable amount of research attention [4]. A review of the literature on the heuristic search algorithms for network design indicates that "almost every case procedures that achieve a high level of performance take advantage of problem-specific structures" [4]. Hence, understanding the problem structure may aid in development of efficient heuristic search algorithms for solving these problems.

Application of a local search algorithm imposes a neighborhood structure on the problem. This structure describes how many search moves it takes to get from one feasible solution to another [9], and, together with the objective function, it constitutes a fitness landscape of the problem [11]. The problem structure of several computationally difficult discrete optimization problems, such as the flow-shop scheduling problem [3], the traveling salesman problem [5], and the quadratic assignment problem (QAP) [2] has been investigated empirically. Yet, to the best of our knowledge, the solution space structure of the network design related problems, specifically, that of TNEP, has received little attention. This chapter presents an empirical study of the TNEP solution space structure imposed by a local search.

Constructive heuristics and metaheuristics have been previously applied to study TNEP [8]. Local search procedures are often utilized to some extent by many modern metaheuristics either in their pure or hybridized forms. Obviously, the local search algorithms, which employ constructive heuristics for defining a neighborhood, may be used iteratively either by itself or in conjunction with some metaheuristic to find approximate solutions for this problem. Given multiple constructive heuristics for moving to a new solution from some current solution of the TNEP problem, the question arises:

Which of these heuristics (if any) would be more advantageous to use in the local search procedure, if, for instance, we want to reach a better quality solution?

In this chapter, we analyze the performance of multiple alternative versions of local search algorithm on the TNEP problem, which is based on a benchmark instance known as a Graver's six-bus system. Using a constructive heuristic for TNEP, a set of solutions, known as a neighborhood, can be produced from a given solution. Alternative heuristics may generate different neighborhoods, hence resulting in different versions of local search algorithm. Here, these alternative versions of local search are also called (search) strategies.

The performance of a given search strategy can be described using a number of characteristics, including the value of a local optimum reached in a specific iteration or the number of steps it took to reach a local optimum during an iteration. For any considered search strategy, an iteration is completely determined by its starting solution. Consequently, to understand differences in the performance of different search strategies, we collect the values of multiple characteristics for different starting solutions and compare the data sets obtained using alternative strategies. As it turns out, even for a small six-node network, the number of starting solutions for the TNEP problem is very large. Therefore, the statistical comparison is performed based on a sample of starting solutions and not the entire population. The sample statistics, diagnostic plots, and correlation analysis are among the tools that help us gain some insight into observable differences among the considered search strategies.

The chapter is organized as follows. In the following section, TNEP is given a nonlinear mixed integer programming formulation, and the solution space for the TNEP problem is described. In Sect. 3 the application of local search algorithms, which utilize constructive heuristics to define a neighborhood of a given solution, is briefly explained. Furthermore, the statistical techniques, which are used to analyze the characteristics of solutions produced by the considered local search strategies, are summarized. The results from the data analysis and discussion of the performed graphical diagnostics and statistical inferences about various characteristics of the solutions and neighborhoods produced by different local search strategies are presented in Sect. 4. Finally, Sect. 5 concludes the chapter.

2 Problem Formulation

Existing electrical power grid can be represented by a network with a set V of nodes and a set E of arcs or edges. In the context of TNEP, the nodes symbolize buses on a power grid, whereas the arcs denote transmission corridors connecting two buses. In the existing electrical power network, each bus has a current number B_i of components (e.g., B_i shunt capacitors for regulating AC power), whereas each transmission corridor (i, j) from bus i to bus j has a current number A_{ij} of electrical circuits. Suppose that at most N_{ij} additional circuits can be installed in the transmission corridor (i, j) in the excess of currently present A_{ij} lines between i and j, and the cost of installing each additional circuit in the corridor (i, j) is denoted by κ_{ij}, $(i, j) \in E$. In general, up to a given maximum, say M_i, number of components can be added at bus i, and the installation cost of each additional control component on the bus i is κ_i for any $i \in V$.

Our formulation for TNEP uses the DC model and does not incorporate any possible addition of control components on the network buses. We further modify the formulation in [1], where the TNEP problem is stated via multi-objective optimization with lexicographic cost function, by including both the total overload and the total cost of installing additional lines in the transmission corridors.

After introducing the decision variables:

y_{ij} (Nonnegative real-valued) overload in the corridor (i, j),
x_{ij} (Nonnegative integer) number of installed circuits in the corridor (i, j),
f_{ij} (Nonnegative real-valued) electrical flow in the corridor (i, j),
θ_i (Real-valued) voltage angle on the bus i,

TNEP is formulated via NLMIP as follows:

$$\min \sum_{(i,j)\in E} (y_{ij} + \kappa_{ij} x_{ij}) \tag{1}$$

subject to

$$f_{ij} - x_{ij} r_{ij} \le y_{ij}, \quad \forall (i, j) \in E, \tag{2}$$

$$f_{ij} = -f_{ji}, \quad \forall (i, j) \in E, \tag{3}$$

$$\sum_{j \in V} f_{ij} \leq g_i - l_i, \quad \forall i \in V, \tag{4}$$

$$f_{ij} - \gamma_{ij} x_{ij} (\theta_i - \theta_j) = 0, \quad \forall (i,j) \in E, \tag{5}$$

$$f_{ij} \in \mathbb{R}, \quad \forall (i,j) \in E, \tag{6}$$

$$\theta_i \in \mathbb{R}, \quad \forall i \in E, \tag{7}$$

$$x_{ij} \in \{A_{ij}, A_{ij}+1, \ldots, A_{ij}+N_{ij}\}, \quad \forall (i,j) \in E, \tag{8}$$

$$y_{ij} \geq 0, \quad \forall (i,j) \in E. \tag{9}$$

In addition, to the extra line installation cost κ_{ij}, $(i,j) \in E$, and the maximum number of extra circuits N_{ij}, $(i,j) \in E$, the above formulation includes other problem parameters:

r_{ij} The capacity of a single circuit in the corridor (i,j),
γ_{ij} The susceptance of a single circuit in the corridor (i,j),
g_i The generation (i.e., produced electricity) on the bus i,
l_i The load (i.e., demand for power) on the bus i.

The relationship between the flow through a transmission corridor, the total capacity for all lines in the corridor, and the respective overflow through this corridor is described by the constraint (2). The requirement

$$f_{ij} \leq x_{ij} r_{ij}, \quad \forall (i,j) \in E$$

that the total flow of electricity from one node to another does not exceed the total capacity between those two nodes was relaxed, allowing for overflow, y_{ij}. On the other hand, inclusion of the total overflows in all corridors into the objective function (1) ensures that together y_{ij}'s remain as small as possible. The constraint (4) says that the outflow from every node i cannot exceed the generation minus load at the node. It is a relaxed version of

$$g_i - l_i + \sum_{j \in V} f_{ij} = 0, \quad \forall (i,j) \in E,$$

which ensures the conservation of flow at each bus i according to *Kirchoff's law*. Whereas (3) ensures antisymmetry of flow in each corridor (i,j), and (5) represents the relationship between the phase angle and DC power according to *Ohm's law*.

Obviously, the TNEP formulation (1)–(9) is an NLMIP problem, because the x_{ij} decision variables are integer and the constraint (5) contains the product of decision variables x_{ij} and θ_i. As mentioned earlier, one could apply standard global optimization approaches, but the integrality of x_{ij}'s adds additional challenge to solving the TNEP problem. In fact, the constraint (8) implies that there are $\sum_{i,j \in V} N_{ij} + |E|$ different sets of integer variables (where $|E|$ is the cardinality of E, i.e., the number of corridors in the power grid). Let us denote $|V| = n$, then there are a total of $\binom{n}{2} = \frac{n(n-1)}{2}$ possible transmission corridors between a pair of buses, i.e., $\frac{n(n-1)}{2}$ integer decision variables in the solution space of the TNEP. When $N_{ij} = N$, then this simplifies to having a total of $(N+1)^{\frac{n(n-1)}{2}}$ different ways to set the values of all integer decision variables. For instance, in the case of Graver's six-bus system and assuming the ability to add at most one extra circuit to any transmission corridor in the grid, we have $n = 6$ and $N = 1$. Even these relatively small problem parameters already result in the TNEP solution space having $2^{15} \approx 32,000$ possible combinations of 15 integer decision variables.

3 Approach

As shown in Sect. 2, the size of the TNEP solution space part, which corresponds to integer decision variables in the problem, grows exponentially with increase in the number of buses in an initial power grid. Consequently, when solving TNEP problem instances for large realistic power networks, an exploration of all possible electrical line additions via exhaustive search would not be practical. In fact, current optimization approaches for TNEP (see, e.g., simulation optimization based method in [1]) typically do not attempt to find the exact solution of the problem and instead search for a good quality approximate solution.

The need for solving large realistic instances of hard optimization problems has led to increased interest in metaheuristics, which often proves to be faster than some of the more traditional, exact approaches for both global and discrete optimization. For instance, many metaheuristic and hybrid algorithms were applied to the QAP, a well-known hard problem in nonlinear optimization with discrete decision variables [7]. Most metaheuristics either already incorporate some type of local search procedure or allow themselves to be hybridized with a local search algorithm to improve their performance. Metaheuristic procedures typically involve two alternating stages: an exploration phase (which is designed to quickly move to new unexplored areas in the solution space) and an exploitation phase (which combs through the local areas in search of improved solution). The exploitation stage typically ends in a local optimum, then the algorithm switches back into the (global) exploration mode.

Taking into account the local search usefulness and the challenge of dealing with integer variables in addition to nonlinearity of the problem, the following method is proposed in the chapter. As an alternative to first using global optimization to solve the relaxed version of the TNEP problem, where integer decision variables are temporarily allowed to take on the real values, and then taking care of integrality constraint (8), we propose to first explore that portion of solution space, which is described by the integer decision variables x_{ij}, via a local search-based algorithm and then (given the x_{ij} values) solve a remaining problem. To improve the solution quality, local search would be used iteratively either on its own by using some type of restart procedure or in conjunction with a metaheuristic algorithm (typically, in the exploitation phase).

Local search is a general technique in discrete and global optimization for improving a local solution. Local search starts at some initially constructed solution as its current solution and works by systematically exploring the solutions that are similar to the current solution until it either finds a higher quality solution or determines that the current solution has the best objective value when compared to all other solutions that are sufficiently close to itself. The similarity or distance is imposed on the solution space by defining a neighborhood using some rule. For instance, when the solution space of an optimization problem is represented as a sequence of zeros and ones, a neighborhood may be defined as all solutions whose Hamming distance to the current solution is one (i.e., any 0–1 sequence of the appropriate length that is different from the 0–1 sequence representing the current solution in exactly one position).

Various versions of local search procedure may be obtained based on what type of neighborhood rule is chosen. Obviously, we would like to define a neighborhood in such a way that the resulting local search algorithm is likely to exhibit a good performance and, hopefully, outperform alternative versions of local search that are constructed using other neighborhood rules. We propose several local search versions based on alternative constructive heuristics for TNEP. The remainder of this chapter focuses on investigation of different properties of the considered search algorithms using sample statistics, diagnostic plots, and correlation analysis aiming to gain better insight into the alternative algorithms behavior.

3.1 Constructive Heuristics as Local Search Strategies

The versions of local search algorithm, studied here, are built based on fourteen alternative constructive heuristics. These heuristics are applied to some given TNEP solution so that, each time we modify this solution, a new solution is obtained. Each constructive heuristic specifies a rule that can be used in the local search algorithm to create a neighborhood of a current solution.

Definition 1. Given a solution $\mathscr{S}$ for an instance of optimization problem $\mathscr{P}$ and a distance $d(\cdot,\cdot)$ on the solution space of $\mathscr{P}$, a *neighborhood* $\mathscr{N}_{\mathscr{S}}$ of $\mathscr{S}$ is the set of all such solutions $\mathscr{S}_1$ that are exactly distance one away from $\mathscr{S}$, i.e., $d(\mathscr{S},\mathscr{S}_1) = 1$.

If a rule, which specifies a transformation from the solution space into itself, imposes a distance metric on the solution space, then the rule also defines a neighborhood relationship on the solution space. Moreover, a neighborhood of the solution $\mathscr{S}$ is simply a set of those solutions that are precisely one move away from $\mathscr{S}$ according to the rule. Based on the specified rule, a local search algorithm evaluates either all or a subset of solutions in the neighborhood of the current solution and then moves to a solution with an improved objective function (as compared to its values in the current solution and any of the current solution's immediate neighbors, whose objective were computed). If no such (improved) solution is found (among all of the solutions) in the neighborhood, then the current solution is a local optimal solution, and the algorithm should restart (by generating a new initial solution) in another area of the solution space.

We construct the following fourteen rules that can be used in the local search to move from one solution to another:

Strategy 1. The neighborhood consists of all solutions that add an arc that is adjacent to *anyone of the two nodes of the last arc* (i.e., the most recently added link in the network graph).

Strategy 2. Add an arc to the *highest degree* node of the most recently changed arc (i.e., the node that is adjacent to the most recently added arc and has more emanating arcs than the other node of that arc).

Strategy 3. Add an arc to the *lowest degree* node of the most recently changed arc (i.e., the node that is adjacent to the most recent arc and has less emanating arcs than the other node of that arc).

Strategy 4. Add an arc to the *highest weighted* degree node of the most recently changed arc, where the weight of a node j is computed as the sum of $g_i - l_i$ for all nodes i adjacent to j, i.e.,

$$w_j = \sum_{(i,j)\in E} (g_i - l_i). \tag{10}$$

Strategy 5. Add an arc to the *lowest weighted* degree node of the most recently changed arc.

Strategy 6. Add an arc that is adjacent to anyone of the two nodes of the most recent arc *as long as it creates a cycle.*

Strategy 7. Add an arc that is adjacent to one of the two nodes of the most recent arc *only if it avoids creating a cycle.*

Strategy 8. Delete an arc that is adjacent to *anyone of the two nodes of the most recent arc.*

Strategy 9. Delete an arc to the *highest degree* node of the most recently changed arc.

Strategy 10. Delete an arc to the *lowest degree* node of the most recently changed arc.

Strategy 11. Delete an arc to the *highest weighted* degree node of the most recently changed arc.

Strategy 12. Delete an arc to the *lowest weighted* degree node of the most recently changed arc.

Strategy 13. Delete an arc that is adjacent to anyone of the two nodes of the most recent arc *as long as it breaks a cycle.*

Strategy 14. Delete an arc that is adjacent to one of the two nodes of the most recent arc *only if it avoids avoids breaking a cycle.*

To illustrate the differences between some of the above strategies, let us consider an example of the network with a graph $G = (V, E)$ and the current solution $\mathscr{S}_0$ shown in Fig. 1. The vertex set of G is $V = \{1,2,3,4,5\}$, and the edge set is $E = \{(1,2),(1,3),(1,4),(2,3),(2,5),(3,4)\}$. Suppose that arc $(1,2)$ was added last to obtain the current solution $\mathscr{S}_0$. Also assume that

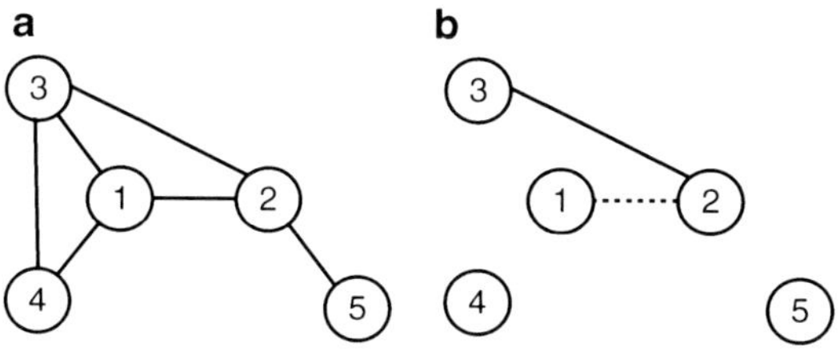

Fig. 1 An example of a graph G and the current solution $\mathscr{S}_0$ on G. The dotted line depicts the arc $(1,2)$ that was added last. (**a**) Graph G. (**b**) Current solution $\mathscr{S}_0$

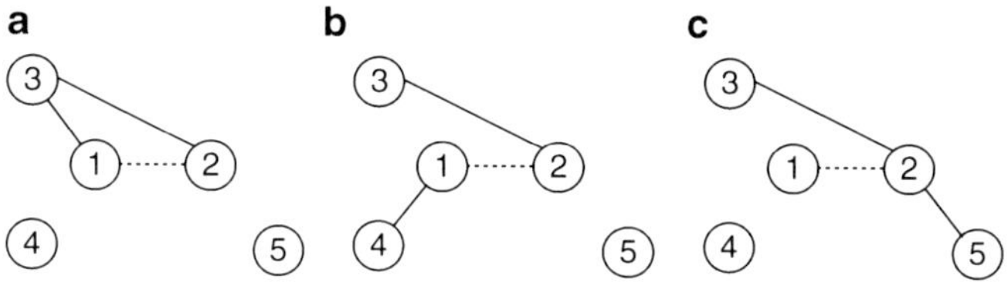

Fig. 2 Examples of new solutions, which can be obtained from the current solution (left subplot in Fig. 1) by applying some of the described strategies. (**a**) Arc $(1,3)$ added. (**b**) Arc $(1,4)$ added. (**c**) Arc $(2,5)$ added

$$g_1 - l_1 = 1, g_2 - l_2 = 3, g_3 - l_3 = 1, g_4 - l_4 = 2, g_5 - l_5 = 1.$$

Then the current solution's node weights are

$$w_1 = 3, w_2 = 2, w_3 = 3, w_4 = 0, w_5 = 0.$$

Hence, the neighborhoods produced by Strategies 1–7 include some of the solutions displayed in Fig. 2. In particular, the use of Strategy 1 on the current solution $\mathscr{S}_0$ results in the neighborhood, which consists of solutions depicted in Fig. 2a–c. Notice that node 1 has the degree of 1, while its weighted degree is 3. On the other hand, node 2 has the degree of 2, but its weighted degree is 2. In other words, node 1 is the lowest degree node but the highest weighted degree for the last added arc $(1,2)$. At the same time, node 2 is the highest degree node but the lowest weighted degree node for $(1,2)$. Solution in Fig. 2c is the only one in the neighborhood that is produced by Strategy 2. Strategy 3 gives the neighborhood that includes only the solutions in Fig. 2a, b. Strategy 4 produces the solutions in Fig. 2a, b. Strategy 5 gives only the solution in Fig. 2c. Strategy 6 results in the neighborhood that consists only of the solution in Fig. 2a, and Strategy 7 gives the neighborhood that includes the solutions in Fig. 2b, c.

It is easy to observe several relationships among the proposed strategies. In particular, Strategies 1–7 transform the electrical grid by adding a power line into a transmission corridor. In contrast, Strategies 8–14 work by deleting a single circuit between a pair of buses from the grid. As a matter of fact, Strategies 8–14 are the direct opposites of Strategies 1–7, respectively, since the former remove arcs/lines where the latter create them. For a given current solution, any of the neighborhoods obtained using Strategies 2–7 are proper subsets of the neighborhood resulting from using Strategy 1. Similarly, for a given current solution, its neighborhoods created using Strategies 9–14 are all proper subsets of the neighborhood obtained using Strategy 8. Of course, even with slight differences in the rules defining neighborhoods, each iteration offers the possibility for divergence among differing versions of the search algorithm. It is interesting to see what effect some differences in terms of the way these strategies are defined have on the algorithm performance. Observe that a node has a high weighted degree according to (10) when its adjacent nodes have a large surplus of electricity. Hence, we would also like to understand the effect this surplus of generated power at the adjacent buses has on the differences in the behavior and performance of local search implementations based on Strategies 4, 5, 11, and 12.

3.2 Characteristics of the Algorithms Behavior

We want to further explore how the differences in strategies impact the behavior of the search algorithms that utilize these strategies. To accomplish this, we consider several characteristics of the search process, including:

1. The average number ns of solutions contained in the neighborhood, where the average was taken over all iterations (i.e., until the local optimum was reached) of a given sample.
2. The average number f of feasible solutions contained in the neighborhood (over all iterations of a given sample).
3. The average proportion f/ns of feasible solutions to the total number of solutions in the neighborhood (over all iterations of a given sample).
4. The average number b of improving, feasible solutions contained in the neighborhood (over all iterations of a given sample).
5. The average proportion b/ns of improving feasible solutions to the total number of solutions in the neighborhood (over all iterations of a given sample).
6. The local optimum OV.
7. The average best-local-improvement bld in the objective value from one iteration to the next.
8. The average relative best-local-improvement bld/b. Where each iteration's relative best-local-improvement value is computed as a ratio of the best-local-improvement value to the number of improving feasible solutions in the neighborhood at each step (iteration), and the average was again taken over all iterations.
9. The total improvement d in the objective value for a given strategy after all iterations.
10. The maximum cost $Cost$ of installing additional lines after all iterations.

11. The number l of iterations, or steps, that the algorithm performed by before reaching a local optimum.

Observe that the above characteristics, in varying ways, describe the behavior and performance of a local search algorithm. For instance, the last characteristic l values represent the speed with which the algorithm is able to find a local optimum, whereas *OV* obviously gives us the quality of the solution found by the application of a local search algorithm.

3.3 Diagnostic and Explorative Statistical Analysis

Let us first explain how the data set for explorative statistical analysis is produced. A set I of randomly generated initial solutions of a given TNEP problem instance is used to initialize a local search algorithm. For each strategy σ_i, $i = 1, \dots, 14$, in Sect. 3.1, the respective search algorithm version is run on different initial solutions $\mathcal{S} \in I$ and the (eleven) characteristics in Sect. 3.2 are computed from the data collected during the algorithm's execution. This produces a vector $a_{\mathcal{S}} = (a_{\mathcal{S}1}, \dots, a_{\mathcal{S}11})$ of the characteristics' values for each initial solution $\mathcal{S}$, i.e., an algorithm run initialized from a given solution constitute a trial in this experiment. Hence, combining these vectors (i.e., different trials) into a matrix $A_\sigma = (a_{\mathcal{S}})_{\mathcal{S} \in I}$ for all initial solutions in the set I gives us a random sample for each considered strategy $\sigma \in \{\sigma_1, \dots, \sigma_{14}\}$.

Notice that each characteristic is treated as a random variable whose value changes depending on a choice of initial solution and search strategy. Furthermore, for a given strategy and a specified initial solution, together all the considered characteristics form an (11-dimensional) random vector. As noted in Sect. 3.2, the characteristics depict the search behavior and performance during the execution of an algorithm. Therefore, by considering strategy σ an independent variable taking values $\sigma_1, \dots, \sigma_{14}$, we can apply multivariate statistical techniques to see what effect a different choice of strategy has on the algorithms' performance and behavior.

The descriptive statistics (such as a sample mean vector, sample covariance, and correlation matrices.) help summarize the underlying random distribution of the search characteristics. In particular, we compute the sample mean vector as follows:

$$\mu = \frac{1}{|I|} \sum_{\mathcal{S} \in I} a_{\mathcal{S}}, \tag{11}$$

where $|I|$ denotes the cardinality of set I, μ is an 11-dimensional real vector

$$\mu = \begin{pmatrix} \mu_1 \\ \vdots \\ \mu_{11} \end{pmatrix}$$

and

$$\mu_j = \frac{1}{|I|} \sum_{\mathscr{S} \in I} a_{\mathscr{S} j}, \quad j = 1, \ldots, 11.$$

In addition, the sample variance–covariance matrix is calculated according to

$$\Phi = \frac{1}{|I| - 1} \sum_{\mathscr{S} \in I} (a_{\mathscr{S}} - \mu)(a_{\mathscr{S}} - \mu)^{\mathrm{T}}, \tag{12}$$

where ^{T} symbolizes the transposition operation (i.e., $(b_{ij})^{\mathrm{T}} = (b_{ji})$). Then the sample correlation matrix

$$\Psi = \left(V^{1/2}\right)^{-1} \Phi \left(V^{1/2}\right)^{-1}, \tag{13}$$

where

$$V = \mathrm{diag}(\Phi) = (\Phi_{jj})_{j=1,\ldots,11} = \begin{pmatrix} \Phi_{1,1} \\ \vdots \\ \Phi_{11,11} \end{pmatrix}$$

denotes the (11-dimensional) vector composed of the elements on the main diagonal of matrix Φ.

To visually detect patterns, it is convenient to represent a sample correlation matrix graphically by means of a temperature map. A temperature map depicts each element of the correlation matrix by a colored square, so that the higher the element's value the warmer is the corresponding square's color. For instance, the correlation of 1 is a red square, whereas an element that displays -1 is shown on a map by a blue square. In application to our analysis, the rows and columns of such a map symbolize the characteristics and the bottom-left to top-right diagonal would show the highest possible temperature of 1, since the squares of this diagonal simply denote the correlation of the respective characteristic with itself.

The univariate distributions of the considered characteristics are visualized via the correspondent box-and-whisker plots. The box plots display the distribution quartiles, with median as the center point and the first and third quartiles q_1, q_3 giving the bottom and top edges of the box, respectively. The data that are not considered outliers are shown by whiskers on a plot. The outliers are defined as any value outside of the $[q_1 \pm \omega(q_3 - q_1)]$ range, with $\omega = 1.5$.

4 Results and Discussion

This section presents the results of the numerical experiments that were conducted on a benchmark TNEP instance known as Graver's six-bus system. The system represents a smaller size power network and is well studied in the context of TNEP. This example allows us to gain some initial insight into common and diverging patterns in the behavior and performance of local search versions using alternative constructive heuristics. To accomplish this, we first apply the approaches outlined in Sect. 3 on the 6-bus system and then present the results of statistical analysis. The statistical methods in Sect. 3.3 are used as the means for explorative analysis of the impact different choice of strategy (constructive heuristic) has on the characteristics describing search behavior on the solution space of the TNEP instance. Summarizing and visualizing these results allows us to observe some patterns in the data. Possible explanations and interpretations for these observations are also given in this section.

For each of the fourteen alternative strategies described in Sect. 3.1, a local search algorithm utilizing the respective constructive heuristic was implemented in MATLAB 7.11.0 (http://www.mathworks.com/).

Recall that a local search algorithm starts at some initial solution. When an initial solution is infeasible, it provides no means of comparison for the first iteration of the algorithm. Hence, infeasible solutions were excluded from any further consideration. Out of 200 randomly generated initial solutions, six solutions were infeasible, and so, they were not included into set I. The other 194 randomly generated solutions formed set I of the initial solutions, which were used by a local search algorithm to generate the data sets for statistical analysis. Moreover, when using Strategies 6 and 13, none of the selected 194 initial solutions produced a feasible neighborhood (i.e., a neighborhood containing at least a single feasible solution). Consequently, these two strategies (6 and 13) were completely excluded from further analysis.

To obtain the characteristics described in Sect. 3.2, the search algorithm implementations based on Strategies 1–5, 7, 8–12, and 14 were run on the benchmark six-bus system instance of the TNEP problem (1)–(9), and, for each strategy (except excluded Strategies 6 and 13), the following data set was collected during a search process:

- The values for each of the decision variables x_{ij}, y_{ij}, f_{ij}, and θ_i.
- The number $ns_{\mathcal{N}}$ of solutions contained in the neighborhood $\mathcal{N}$.
- The number $f_{\mathcal{N}}$ of feasible solutions contained in the neighborhood $\mathcal{N}$.
- The number $b_{\mathcal{N}}$ of improving, feasible solutions contained in the neighborhood $\mathcal{N}$.
- The objective value $OV_{\mathcal{N}}$ of the best solution in the neighborhood $\mathcal{N}$.
- The best-local-improvement bld in the objective value from the previous step.
- The overall improvement d of the objective value from the initial solution to the iteration when the local optimum was reached.

Table 1 Sample means of the eleven characteristics for Strategies 1–5, 7, 8–12, and 14

Str	*ns*	Cost	l	*OV*	d	*b*/*ns*	*f*/*ns*	*bld*/*b*	bld	*f*	*b*
1	10	960.93	12.21	1,509.79	2,402.46	0.39	0.87	40.22	185.12	8.69	3.94
2	5.13	731.44	5.93	2,877.39	1,034.85	0.36	0.83	73.91	157.77	4.29	1.87
3	5.89	898.2	10.29	1,717.7	2,194.55	0.48	0.92	63.01	202.34	5.42	2.78
4	5.04	708.01	5.56	2,823.25	1,088.99	0.33	0.86	86.41	170.92	4.36	1.66
5	5.01	853.27	8.46	2,071.26	1,840.99	0.49	0.83	70.78	200.36	4.16	2.45
7	10	960.93	12.21	1,509.79	2,402.46	0.39	0.87	40.22	185.12	8.69	3.94
8	10	453.62	1.79	3,791.14	121.11	0.1	0.42	27.34	39.3	4.16	0.95
9	5.8	453.22	1.46	3,817.8	94.44	0.13	0.48	26.84	32.76	2.73	0.71
10	5.58	466.07	0.64	3,808.21	51.12	0.08	0.43	20.93	25.06	2.4	0.46
11	5.14	454.74	1.02	3,835.59	75.66	0.12	0.46	25.75	31.14	2.38	0.61
12	5.15	455.59	0.8	3,798.3	57.34	0.1	0.46	22.48	25.94	2.38	0.52
14	10	453.62	1.79	3,791.14	121.11	0.1	0.42	27.34	39.3	4.16	0.95

- The cost $Cost_{\mathscr{S}} = \sum_{(i,j)\in E} \kappa_{ij} x_{ij}$ of a given solution $\mathscr{S}$. *Cost* represents the part of objective function (1) that corresponds to the cost of installing additional lines in the power grid to satisfy changed load and generation parameters.

The characteristics in Sect. 3.2 were computed from the above data. The calculated values formed respective random samples for every considered strategy (i.e., 1–5,7,8–12,14) as explained in detail in Sect. 3.3. This allowed us to compute the corresponding sample mean values and the correlation matrix for the characteristics, as well as to draw box-and-whisker plot representations of a characteristic's univariate distribution for various strategies and all characteristics. We also used the calculated sample correlation matrix to create a temperature map representing the linear relationships between the pairs of characteristics.

Table 1 summarizes the sample means of characteristics *ns*, *Cost*, *l*, *OV*, *d*, *b*/*ns*, *f*/*ns*, *bld*/*b*, *bld*, *f*, and *b* for Strategies 1–5, 7, 8–12, and 14.

Examination of the sample mean vectors of different strategies shows a clear difference between Strategies 1–5,7, which install additional lines into transmission corridors, and Strategies 8–12,14, which remove power lines. In particular, sample means for the latter strategies appear to be more similar in values to each other, whereas the sample means for the former strategies tend to vary more in comparison. For instance, from Table 1, we can see that the means for Strategies 1–5, and 7 seem to differ dramatically from Strategies 8–12,14 in terms of the variables *OV*, *d*, and *bld*.

We use graphical representations of univariate distributions via box-and-whisker plots to visually detect similarities and differences among the fourteen strategies in terms of the distribution of values of each considered characteristic. By placing all the box-and-whiskers plots of a common characteristic for all strategies together on one figure, we can easily see which strategies produce similarly distributed values of that characteristic. Figure 3 shows eleven subfigures, each of which combines box-and-whisker plots of the respective characteristic for Strategies 1–5, 7, 8–12, and 14.

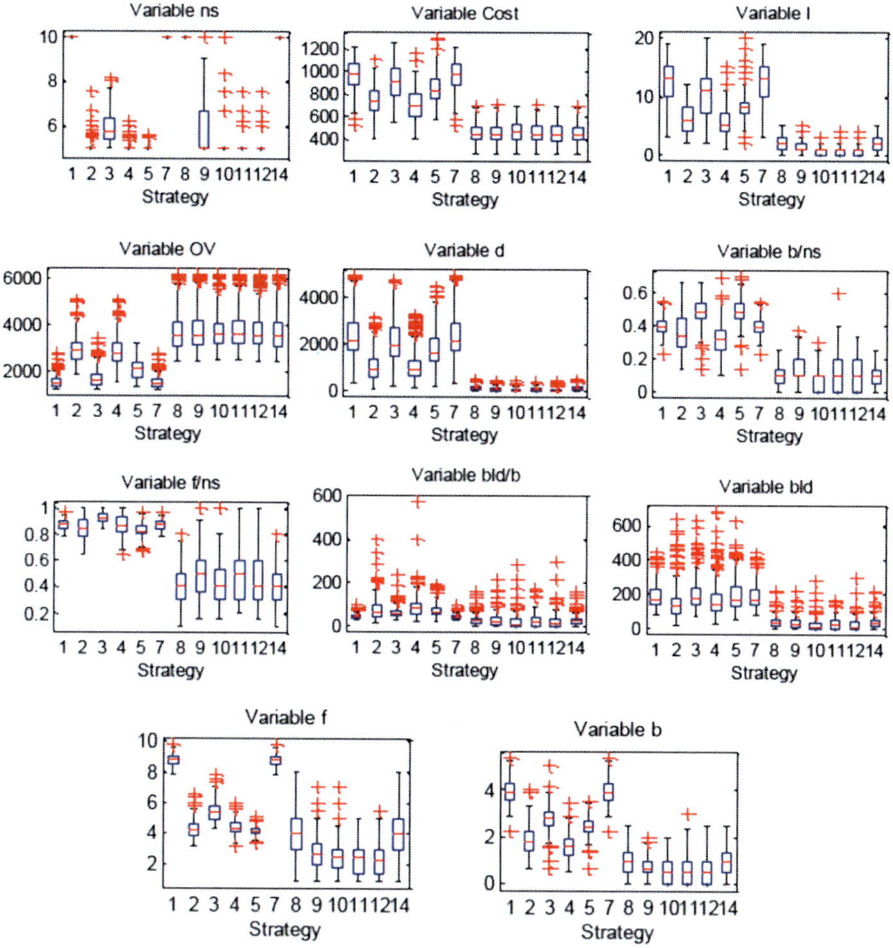

Fig. 3 Box-and-whisker plots of characteristics ns, $Cost$, l, OV, d, b/ns, f/ns, bld/b, bld, f, and b for Strategies 1–5, 7, 8–12, and 14.

The combined box-and-whisker plots in the eleven subfigures of Fig. 3 once again indicate a noticeable difference between those strategies (1–5,7) that add power lines in the corridors and those (8–12,14) that remove circuits across essentially all variables. This trend may reflect the fact that removing lines from the transmission corridors in the grid is qualitatively different as compared to adding. In fact, the TNEP problem involves expanding a network to meet demand, while removing lines clearly does the opposite. This explains why Strategies 8–14 perform so poorly in the minimization of the objective function (1).

Observe that Strategies 1 and 7 appear identical in both the box-and-whisker plots in Fig. 3 and with respect to their mean values in Table 1. As it turned out,

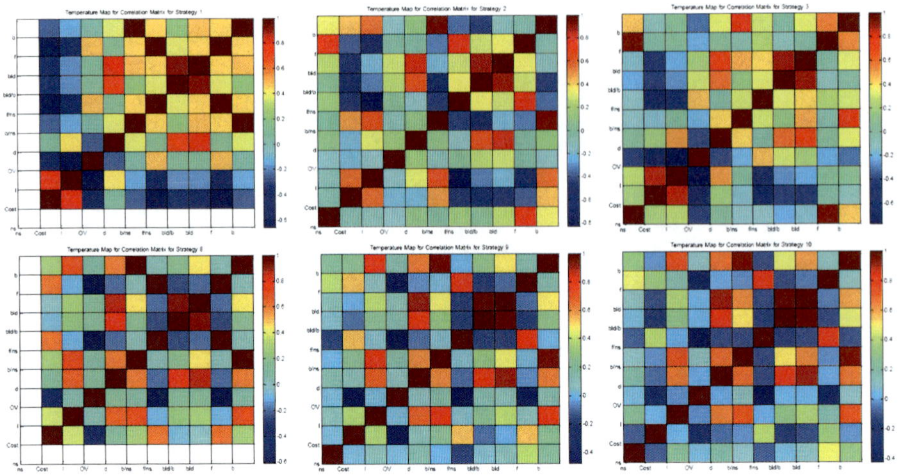

Fig. 4 Temperature map for correlation matrices for strategies 1, 2, 3 (*top*) and 8, 9, 10 (*bottom*), respectively. The correlation matrices were constructed using the data that describe characteristics ns, $Cost$, l, OV, d, b/ns, f/ns, bld/b, bld, f, and b

neither the 194 initial solutions nor the solutions created during an algorithm run created a cycle in the network. As a result, Strategy 7 was simply reduced to Strategy 1. The same is true for Strategy 14 with respect to Strategy 8. These results also explain why Strategies 6 and 13 (which were excluded) produced no successful trials, since, by definition, these strategies only accept solutions which create a cycle. Considerable differences appear to exist among Strategies 1–5,7 for most of the eleven characteristics. Although, for all of them, the variability in the characteristics' values is reduced in comparison with Strategies 8–12,14. It is noteworthy that, regardless of strategy, the average relative best-local-improvement bld/b appears to be roughly the same.

The calculated correlation matrices for pairs of characteristics are visualized in Figs. 4 and 5. The former figure contains six temperature maps for Strategies 1, 2, and 3 (top) and Strategies 8, 9, and 10 (bottom). The latter figure contains six temperature maps for Strategies 4, 5, and 7 (top) and Strategies 11, 12, and 14 (bottom). The warmer colors correspond to positive correlations and the cooler colors denote negative correlations. The twelve plots allow us to visually detect the patterns in these pairwise relationships.

The correlation matrices provide insight into the strength of the linear relationship between pairs of variables. Consequently, certain high correlation values are expected in the sample correlation matrix, such as the values on the reverse diagonal and those representing correlations between bld and bld/b, f and f/ns, b and b/ns. At the same time, other linear relationships can be seen in Figs. 4 and 5, which are unanticipated and therefore, far more interesting. For instance, there are strong negative correlations between f/ns and OV across all strategies, and strong positive

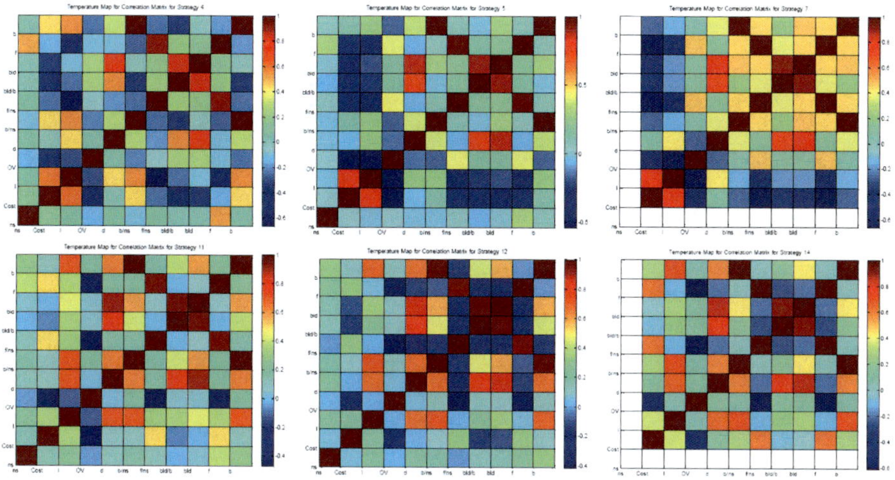

Fig. 5 Temperature map for correlation matrices for strategies 4, 5, 7 (*top*) and 11, 12, 14 (*bottom*), respectively. The correlation matrices were constructed using the data that describe characteristics ns, $Cost$, l, OV, d, b/ns, f/ns, bld/b, bld, f, and b

correlations between bld and d. In other words, a larger proportion of feasible solutions seems to allow for a lower overall objective value, and a higher local improvement in the objective value indicates a higher overall improvement through all iterations. Strategies 1, 7, 8, and 14, all have white rows for the characteristic ns, the average number of solutions per iteration. This is because for those strategies, there is no variability in ns. In all four cases $ns = 10$ for every single trial (i.e., initial solution from I). Because there is no variation, a correlation with that variable is undefined.

5 Conclusion

This chapter presented an approach aimed at understanding the behavior of a local search applied to the TNEP problem. Our approach utilized explorative statistical analysis and diagnostic plots to visually detect patterns in the data characterizing the algorithm performance. The interpretation of discovered differences and similarities helps gain initial insight into the solution space properties of the TNEP problem instance, which is based on a well-known benchmark power system. The small size of the considered network is one of the limitations of the study. A similar study on several instances based on larger, more realistic power networks would be necessary to confirm or disprove the observed properties.

References

1. R. Bent, A. Berscheid, and G.L. Toole. Transmission Network Expansion Planning with Simulation Optimization. *Proc. of the 24th AAAI Conference on Artificial Intelligence*, 21-26, 2010.
2. F. Chicano, G. Luque, E. Alba Elementary landscape decomposition of the quadratic assignment problem. *Proc. of the 12th annual conference GECCO on Genetic and evolutionary computation*, 1425–1432, 2010.
3. J. Czogalla, A. Fink Fitness landscape analysis for the no-wait flow-shop scheduling problem. *Journal of Heuristics* 18(1), 25–51, 2012.
4. I. Gamvros, B. Golden, S. Raghavan, and D. Stanojevic. Heuristic Search for Network Design. In H. Greenberg (ed.), *Operations Research and Technology: Tutorials from INFORMS 2004*, Kluwer Academic Press, 1–49, 2004.
5. D. Hains, L.D. Whitley, A.E. Howe. Revisiting the Big Valley Search Space Structure in the TSP, *Journal of Operations Research Society* 62, 305–312, 2010.
6. R. Hemmecke, M. Köppe, J. Lee and R. Weismantel Nonlinear Integer Programming. M. Jünger, T. Liebling, D. Naddef, G. Nemhauser, W. Pulleyblank, G. Reinelt, G. Rinaldi, and L. Wolsey (eds.), *50 Years of Integer Programming 1958-2008: The Early Years and State-of-the-Art Surveys*, Springer-Verlag, 2009, ISBN 3540682740.
7. A. Kammerdiner, T. Gevezes, E. Pasiliao, L. Pitsoulis, and P. Pardalos. Quadratic Assignment Problem. In S. Gass and M. Fu (eds.), *Encyclopedia of Operations Research and Management Science*, 3rd edition, Springer, 2013, to appear.
8. G. Latorre, R.D. Cruz, J.M. Areiza, and A. Villegas. Classification of Publications and Models on Transmission Expension Planning. *European Journal of Operations Research* 83, 1-20, 2003.
9. T. Schiavinotto and T. Stützle. A review of metrics on permutations for search landscape analysis. *Computers & Operations Research* 34, 3143–3153 2007.
10. A. Sorokin, J. Portella, P. M. Pardalos. Algorithms and Models for Transmission Expansion Planning. In A. Sorokin, S. Rebennack, P. M. Pardalos, N. Illiadis, M. Pereira (eds.), *Handbook of Networks in Power Systems*, 395-433, Springer, 2012.
11. P.F. Stadler. Fitness landscapes. In M. Lässig, A. Valleriani (eds.), *Biological evolution and statistical physics*. Springer, 187-207, 2002.

Printed by Publishers' Graphics LLC